An Introduction to Metabolic Pathways

An Introduction to Metabolic Pathways

S. Dagley
M.A. D.Sc. F.R.I.C.
Professor of Biochemistry
University of Minnesota

Donald E. Nicholson
B.Sc. Ph.D. F.I.Biol. F.R.I.C.
Senior Lecturer in Bacteriology
School of Medicine, University of Leeds

John Wiley & Sons Inc
New York

SBN 471 63706 8

Library of Congress Catalog

Card Number 76–99975

First published 1970

Printed in Great Britain

Contents

The Determination of Metabolic Pathways

Regulation of Metabolism

Introduction

The demand for a presentation of metabolic reactions in the form of charts may be evaluated from the fact that nearly three hundred thousand copies of 'Metabolic Pathways' (D. E. Nicholson 1969) have been distributed since the first edition appeared in 1960. As the demand for charts increased, so did the number of requests made to the author, from all parts of the world, to furnish references to publications relating to the various reactions formulated. In order to meet this demand it was decided to present, in book form, separate charts of pathways in which enzymes were named and references were cited. Further, to assist authors of research papers, the nomenclature approved by the Enzyme Commission has been given as far as this was possible, although as our work progressed it became increasingly clear that a large number of enzymes which have been thoroughly investigated have not yet been named by the Commission. Our attention has not been confined to the sequences of 'Metabolic Pathways'; many other metabolic relationships are also summarized in these pages.

This book has also been written to satisfy needs that arise when intermediary metabolism is taught to students at all levels. Although many good modern presentations of biochemistry are available, the enthusiastic teacher seldom wishes to follow a set text closely and he is usually convinced, and rightly so, that his own selection of material is the best for his own unique presentation of the subject. The use of the blackboard during the discussion of an extensive metabolic pathway can be fatiguing, and alternative visual aids are not without their drawbacks particularly with regard to note-taking by the students. It is very convenient for the student to have before him a chart of the particular metabolic relationships which the teacher has selected for discussion. Further, when revising for examinations, students find it valuable to have a concise summary of material.

When a metabolic sequence is shown on a chart, the need to economize in space raises problems of presentation. For this reason we have usually given the open chain structures of sugars and some of their phosphate esters, although this is obviously not possible for glucose-1-phosphate, for example, which must be shown in a ring structure; here, a linear presentation of the pyranose ring is often given.

A problem in stereochemistry arises as a result of the necessity to draw sugars as horizontal formulae instead of the more usual vertical ones. This is probably not serious except in the cases of compounds with a single asymmetric carbon atom, which cannot be rotated through 90° without altering the designation of the structure. Thus

$$\begin{array}{c} CHO \\ H.C.\ OH \\ CH_2\ OH \end{array}$$

is D-glyceraldehyde, but when turned through 90° and drawn as

$$\begin{array}{c} H \\ HOCH_2.C.CHO \\ OH \end{array}$$

it becomes, by convention, L-glyceraldehyde. Further rotation of 90° restores it to the original configuration. It is therefore important to realize that the *linear* structures of sugars and their related metabolites must be rotated through 90° before they become Fischer projections.

It is not good practice in teaching to give chemical equations that do not balance. It is our hope that in every reaction we have shown all the reactants and products, including water; and we have usually given overall balanced equations for complex sequences and cycles. Although the use abbreviations is desirable for economy in space, this often entails pitfalls for the student. Thus, it was only after considerable deliberation that we adopted the symbol Ⓟ to represent the phosphoryl group, PO_3H_2 or PO_3^{2-}. We are aware that this symbol has been a source of errors in texts: thus it is not uncommon to find the synthesis of ATP represented as:

Ⓟ + Ⓟ·Ⓟ—adenosine → Ⓟ·Ⓟ·Ⓟ—adenosine

This equation does not represent ATP correctly since the α and β phosphorus atoms are not combined as PO_3H_2 (or PO_3^{2-}); and it ignores a most fundamental biochemical fact, namely that when inorganic phosphate participates in the phosphorylation of ADP it is necessary for water to be abstracted. Even if inorganic phosphate were represented correctly, namely as HOⓅ, it would still be difficult for the beginner to understand how this water arises. Likewise, if Ⓟ is used in the abbreviation, pyrophosphate should be given as Ⓟ.OⓅ and not as ⓅⓅ. Such considerations as these led us to decide, with one exceptional category, to write the structures of polyphosphates in full, to restrict the use of Ⓟ to phosphate esters and to designate orthophosphate and pyrophosphate as P_i and PP^i respectively. The one exception, dictated by the need for economy of space, was the designation of compounds such as uridine diphosphate glucose by the symbols glucosyl-PPU, etc.

Although in the interest of brevity we might have shown $NADH_2$ as the product of reactions in which the coenzyme is reduced, we have preferred to write NADH + H^+ in accordance with our aim of showing all the products of reactions. The fact that protons arise when NAD^+ is reduced is sometimes important in practice. Thus a bacterial enzyme converts *meso*-tartrate into glycerate by oxidizing the substrate with NAD^+ to dihydroxyfumarate which is then converted by the reduced form of the same cofactor into glycerate and carbon dioxide. L. D. Kohn and W. B. Jakoby (*Biochem. biophys. Res. Commun.* (1966) **22,** 33) showed that amongst the conditions required to demonstrate the two enzymic activities, apparently catalysed by one and the same enzyme, it was necessary to observe the first reaction at pH 8.2, when the oxidation of the substrate was favoured, and the second at pH 6.5 which favoured the oxidation of the coenzyme.

A great deal of thought has been given to the presentation and design of the pathways cited in this book. Although the majority are fairly straightforward vertical sequences, several *cycles* have been included, and these have been carefully designed to give maximum information and understanding. However, cycles can often be sources of misconceptions. A survey of current textbooks of biochemistry demonstrates many inadequacies of presentation which produce confusion in the minds of students since many cycles apparently function to no purpose. This is particularly true with regard to the presentation of the pentose-phosphate cycle, which often appears as a complex but purposeless series of reactions in which nothing enters and nothing leaves the 'cycle'. Cycles are not the optimistic meanderings of imaginative biochemists in search of biochemical perpetual motion but are static symbols of dynamic processes in which reactants are metabolized and products are produced. In our presentations of cycles, care has been taken to demonstrate exactly what enters and what leaves each cycle, and to show that the purpose of a cycle is to regenerate one of the initial reactants upon which the perpetuation of the cycle depends. This is then summed up by a stoichiometrically-balanced equation.

The selection of references has been a major problem in the preparation of our book. Some pathways have a very scanty literature and present no problems of selection, but in other cases an enormous amount of literature is available. For the latter we have tried to select review articles where appropriate, since these are the bases from which more detailed literature searches may be launched. In addition we have included other references, usually relevant to individual steps of a pathway, which are of classical interest or of recent date of publication. The purpose in quoting recent rather than original references in many cases is that the former may be obtained from the latter and a more comprehensive survey obtained. In order to make the references of maximum value we have included titles as well as journal and volume references. We feel that this is a valuable aid to selection of references which may be helpful to the iess specialized student.

A rather unusual method has been adopted for the identification of references. In the chapters on the Determination and Regulation of Metabolism, they are arranged in page number and numerical sequence within each page. Thus the third reference on page 20 would be indexed as 20.3 unless it had been previously classified on an earlier page. In the charts, each compound has been given a reference letter by which it can be recognized and indexed when written alongside its relevant page number. Thus, for example, 2-acetolactate has the reference 119C, acetoin 119D and the enzyme which converts the one to the other is referred to as 119CD. Where general or review references occur they are indexed as such. The advantage of this method of indexing lies in its adaptability since additional entries can be made at almost any place without necessitating the re-numbering of previous or subsequent entries. It therefore makes possible the addition of very recent references to the index, and will greatly simplify the production of future editions of the book.

The Determination of Metabolic Pathways

The living cell: a unique chemical system

The ultimate goal of biochemistry is to arrive at a complete description of living organisms in chemical terms; and one aspect of this description is to formulate all the chemical reactions that living cells can accomplish. In this chapter we shall outline some of the general methods used to elucidate metabolic pathways and we shall illustrate these methods with specific examples. Such a selection, made from the vast range of biochemical literature, must be severely restricted: it will reflect our own personal interests and will inevitably exclude numerous studies of fundamental importance. Indeed, although the experiments we shall cite have all advanced our knowledge of metabolism, we shall choose them in the first place for what they teach us about the scope and limitations of the methods under discussion. It is also evident that a living cell is not adequately described by listing the chemical reactions it can utilize. This is true even when we understand the mechanisms by which the reactions in the list are regulated; for in addition to compounds that are readily formulated, the cell needs to synthesize the complex structures it contains; and some of these may serve as supports for ordered arrangements of the enzymes of metabolism. This feature introduces an uncertainty into enzymology, because a biochemist cannot be certain, when he removes an enzyme from the cell and submits it to extensive purification, that the protein he isolates has the same properties or even the same molecular structure as it possessed in its natural environment.

We realize, therefore, that a living cell cannot be regarded simply as a bag containing those enzymes, and many more besides, that catalyse the reactions listed in this book. Nevertheless it is useful to enquire what broad features are shown by these assemblies of biochemical reactions that set them apart from other chemical systems. The most striking feature of the chemistry of the living cell is the dynamic state of its constituent molecules. Cells contain large numbers of enzymes that catalyse the formation and breakdown of molecules that a chemist would find most difficult to synthesize or degrade if he used the conditions of temperature, pressure or acidity under which life flourishes. To take a simple example, if citric acid were isolated from bacteria it could be stored for an indefinite period of time at the temperature at which the bacteria had grown; but when it is inside the living cell the citrate molecule is broken into fragments within seconds and the carbon atoms are incorporated into molecules of other compounds whose existences may be just as transitory. Within minutes, bacteria that have been exposed to a trace of radioactive carbon dioxide will have incorporated these carbon atoms into a multitude of cell constituents. Even for higher animals Hamlet's wish that 'this too too solid flesh would melt' is, in a sense, already fulfilled: half the protein in the liver of a rat is broken down and replaced within about five days; and none of us is the same person, atom for atom, from one week to the next.

A glance at the metabolic map reveals an extensive network of reactions in which one reaction sequence is repeatedly linked to another through those chemical intermediates that are common to each one. A dozen sequences, for example, are linked through the common metabolite, pyruvate: some of them are degradative reactions that give rise to pyruvate whilst in other cases the compound initiates a pathway of biosynthesis. It is evident that the reaction network of the metabolic map provides a means by which pyruvate and many other compounds can be rapidly synthesized and degraded by enzymes. However, at any moment of time the concentrations of these metabolites within the cell remain, in general, nearly constant: not because the molecules are chemically stable but because a steady state is achieved with a balance maintained between the rates at which particular molecules are broken down and reformed. To maintain this steady state the cell requires a continual supply of energy and materials. This, as H.A. Krebs has pointed out [1] is the chemical background of excitability, one of the fundamental properties of living matter. A change in the environment of a cell may be described in the last resort in chemical terms and the very life of the cell may depend upon its response. This response is not likely to need some entirely new activity. It may draw upon one or more of the multitude of reactions already existing in the resting condition. The vast interrelated network of reactions maintained in the steady state confers upon the cell a chemical resilience upon which its survival in a changing environment may depend.

The necessity for a network of metabolic reactions

The necessity for linking one reaction sequence with another through shared metabolites can also be appreciated when we consider the energy requirements of the cell. In general, the events that occur spontaneously in Nature are those which entail a loss of ordered arrangements of atoms and an increase in disorder; but living cells continually reverse this trend. When certain bacteria are introduced into a solution that contains molecules of acetic acid and a few species of inorganic ions they rapidly assemble atoms of carbon, hydrogen, oxygen, nitrogen and phosphorus into those complex and ordered arrangements that constitute the macromolecules of their progeny. As the physical chemist G.N. Lewis observed [2], 'living organisms alone seem able to breast the great stream of apparently irreversible processes. These processes tear down, living things build up. While the rest of the world seems to move towards a dead level of uniformity, the living organism is evolving new substances and more and more intricate forms'. Earlier, Lord Kelvin had cautiously excluded the operations of 'animate agencies' when he formulated the Second Law of Thermodynamics. But living organisms do indeed conform to these laws, and many of their chemical transformations, like others found in inanimate nature, can accomplish decreases in entropy provided that energy is supplied to the system. The ultimate source of this energy is the food of the cell which is degraded by catabolic (energy-releasing) reactions that are linked with the anabolic processes of biosynthesis.

Transfer of energy in metabolism

A thorough discussion of energy transfers in living systems lies outside the scope of this chapter. The reader is referred to a lucid presentation of general principles by A.L. Lehninger [3] and to a detailed survey by H.A. Krebs and H.L. Kornberg [4]. Suffice it to say that adenosine triphosphate (ATP) is often one of the products formed by enzyme systems that release energy. This is a highly reactive, or 'energy rich' compound which is the source of free energy needed to drive to completion those biosynthetic reactions in which it participates. If we compare the tasks of the cell with those of the organic chemist, the latter is able to choose chemical reagents which he knows from experience, rather than from formal calculation are those able to supply the free energy required to accomplish a particular synthesis. The choice of the living cell, however, is restricted to those reagents, of which ATP is the most important, that are generated during the degradation of foodstuffs; but this limitation is compensated by the wide range of enzymes that can make the most of ATP and use it in a great many biosynthetic reactions.

The term 'energy rich' has been adopted by biochemists because the decrease in standard free energy that occurs when ATP is hydrolysed, to yield adenosine diphosphate (ADP) and inorganic phosphate, is greater than that which accompanies the hydrolysis of many other phosphate esters which are termed 'energy poor'. Changes in standard free energy, like changes in heat content, are additive. In consequence, if we consider the synthesis of sucrose according to the equation:

$$\text{glucose} + \text{fructose} \rightarrow \text{sucrose} + H_2O;\ \Delta G' = +5{,}500 \text{ cal} \quad \ldots\ldots\ (1)$$

we can predict that if ATP can take part in the synthesis as shown in equation (2), free energy will be liberated:

$$\text{ATP} + \text{glucose} + \text{fructose} \rightarrow \text{sucrose} + \text{ADP} + \text{phosphate};\ \Delta G' = -2{,}500 \text{ cal} \quad (2)$$

This prediction is made possible from a knowledge of the free energy of hydrolysis of ATP (equation 3); and equation (2) is obtained when equation (1) and equation (3) are added together.

$$\text{ATP} + H_2O \rightarrow \text{ADP} + \text{phosphate};\ \Delta G' = -8{,}000 \text{ cal} \quad \ldots\ldots\ (3)$$

We can therefore regard the free energy liberated when ATP is hydrolysed as being sufficient to compensate for that required to synthesize sucrose according to equation (1); and since a chemical change will not take place unless free energy is released, the participation of ATP renders synthesis possible, although the small overall decrease in free energy in reaction (2) suggests that the route would not be particularly effective.

It is legitimate to perform such a calculation in order to decide whether or not a proposed metabolic sequence is thermodynamically feasible, but it must not be inferred that equation (2) adequately describes the reactions involved in the synthesis of sucrose. Indeed it does not; for glucose first reacts with ATP to give glucose-6-phosphate and this compound must be converted into glucose-1-phosphate before reacting with fructose. Moreover, the standard free energy changes employed in the calculation refer only to specified conditions, namely to a chemical change at pH 7·0 for reactants at unit activity, giving products also at unit activity. Since these conditions are unlikely to be physiological, the values of $\Delta G'$ used in these calculations may bear little relation to the actual changes in free energy that occur in the cell.

It may also be mentioned that although, under favourable conditions, sucrose may be formed by microorganisms by these reactions, higher plants use a different route which involves the participation of uridine diphosphoglucose. This compound is formed from glucose-1-phosphate by a reaction that requires one molecule of uridine triphosphate, which, from the standpoint of energetics, is equivalent to utilizing an additional molecule of ATP to drive forward the synthesis of sucrose.

It is a misnomer to refer to the *bond* between the terminal phosphate group of ATP and the rest of the molecule as being 'energy rich', particularly when the phrase implies that 8 kcal of free energy are always made available when the bond breaks, regardless of environment. Nevertheless, if it is borne in mind that this represents the order of magnitude of the free energy change required for the efficient transfer of energy through the agency of ATP, we can appreciate why it is that metabolic pathways consist of so many individual reactions. If the transformation of one metabolite into another occurs with the liberation of considerably more free energy than 8 kcal, much will be lost as heat and little conserved for biosynthesis. For this reason the cell does not usually employ transformations that are thermodynamically irreversible and many of the reactions of a metabolic pathway are usually not far removed from chemical equilibrium. Thus, the conversion of glucose into lactate, which can be formulated simply as $C_6H_{12}O_6 \rightarrow 2\ C_3H_6O_3$, is accomplished in several stages by the multienzyme system of glycolysis. Biological oxidations often occur by the transfer of hydrogen atoms from a substrate to a pyridine dinucleotide such as nicotinamide adenine dinucleotide (NAD); and the oxidation of the reduced dinucleotide by molecular oxygen is accompanied by a fall in $\Delta G'$ of 52 kcal/mole. However in the respiratory chain the overall reaction is divided up into a series of coupled oxidations and reductions of the cytochromes, with the result that about 40 % of the energy liberated is harnessed in the form of three molecules of ATP per molecule of dinucleotide oxidized.

Reaction sequences may be broken by removing co-factors

The obligatory participation of ATP in many biochemical transformations, or the requirement that many enzymes have for coenzymes such as NAD, are factors that have assisted biochemists to isolate and study reaction sequences that would otherwise seem to be woven inextricably into the reaction network of metabolism. The classical studies that elucidated the glycolytic sequence may be taken as an example. The anaerobic fermentation of glucose by yeast can be represented by the equation:

$$\text{glucose} \rightarrow 2\ \text{ethanol} + 2\ CO_2 \qquad (4)$$

Reaction (4), together with side reactions, is catalysed by whole cells of yeast. The participation of ATP and phosphate esters is not revealed; for while there is a net synthesis of ATP in glycolysis there is no accumulation by the intact cell because the compound is used in other reactions. However, when yeast juice is used, the fermentation of glucose to ethanol is no longer coupled to such reactions and the overall equation then shows that inorganic phosphate (P_i) is needed:

$$\text{glucose} + 2\ \text{ADP} + 2\ \text{P}_i \rightarrow 2\ \text{ethanol} + 2\ \text{CO}_2 + 2\ \text{ATP} + 2\text{H}_2\text{O} \ldots \quad (5)$$

When Harden and Young first demonstrated this requirement for inorganic phosphate they did not, of course, add ADP to the reaction mixture since the compound was not then known. The extent of the reaction was therefore limited by the amount of ADP that happened to be present in the yeast juice: that is, part of the glucose was needed for the regeneration of ADP according to equation (6) and was not converted into ethanol by equation (5)

$$\text{glucose} + 2\ \text{ATP} \rightarrow \text{fructose-1,6-diphosphate} + 2\ \text{ADP} \ldots\ldots\ldots \quad (6)$$

Equation (7), which is the sum of equations (5) and (6), shows how the fructose-1,6-diphosphate which Harden and Young isolated was formed from glucose and inorganic phosphate:

$$2\ \text{glucose} + 2\ \text{P}_i \rightarrow 2\ \text{ethanol} + 2\ \text{CO}_2 + \text{fructose-1,6-diphosphate} + 2\text{H}_2\text{O} \quad (7)$$

In this example a reaction sequence was interrupted, and a metabolite accumulated in amounts sufficient for its identification, as a consequence of depleting the system of an essential reactant. The early studies of glycolysis were further assisted by the removal by dialysis of coenzymes such as NAD, magnesium ions and thiamine pyrophosphate (TPP), so that enzymic reactions were slowed down or stopped altogether with the resulting accumulation of reaction intermediates. This experimental approach has been adopted repeatedly when other metabolic systems have been investigated and it has been extended by the use of inhibitors. These compounds may act by combining with a coenzyme or with a chemical grouping needed for enzymic activity; or they may compete with a substrate for the catalytic site of the enzyme. Thus, in the glycolysis sequence iodoacetate inhibits the dehydrogenase of glyceraldehyde-3-phosphate: the inhibitor is known to form S-alkyl derivatives with sulfhydryl groups of cysteine residues in proteins; and fluoride ions inhibit enolase by forming a magnesium–fluorophosphate complex that displaces the divalent metal ion activator from the enzyme. Of the many competitive inhibitors used to study other systems we may mention malonate which competes with succinate for succinate dehydrogenase; fluorocitrate which is synthesized enzymically from monofluoroacetate and then competes with citrate for aconitase; and deoxypyridoxine which competes with pyridoxal phosphate for enzymes of amino acid metabolism.

Inhibitors: their use and limitations

When it is possible to identify the reaction products that accumulate by their action, inhibitors are valuable tools for revealing the individual reactions that constitute a sequence; but a mere reduction in overall reaction rate does not necessarily implicate a reaction known to be sensitive, for inhibitors do not always behave according to established precedent. For example, when the operation of the tricarboxylic acid cycle in bacteria was investigated it was found that sodium monofluoroacetate inhibited respiration and some citrate was formed as expected.

However, pyruvate also accumulated from some members of the cycle [1] and from later work it appears that in these cases fluoromalate may have been biosynthesized from the fluoroacetate added: fluoroacetyl coenzyme A can react with glyoxylate in the presence of malate synthase [2] and fluoromalate, like fluoropyruvate is a powerful inhibitor of pyruvate oxidation [3]. An experimental approach which is similar to the use of inhibitors, insofar as the enzyme system is exposed to 'unnatural' compounds, is that of studying the metabolism of a substrate of which the chemical structure has been modified: as a consequence the compound may be attacked by some of the enzymes present whereas the remainder may not be able to metabolize the products. Thus, in a study of bacterial aromatic metabolism, enzymes from bacteria oxidized benzoic acid; but catechol, which is now known to be a reaction intermediate in the degradative sequence, was decomposed so rapidly that its isolation was impossible. However, although *m*-chlorobenzoate was also oxidized, the enzyme that oxidized catechol was unable to oxidize 3-chlorocatechol; this compound accordingly accumulated and was identified [4]. Similarly, when a species of *Pseudomonas* grew with *p*-cresol as a source of carbon, no *p*-hydroxybenzoate was detected. Nevertheless, the probable conversion of *p*-cresol into *p*-hydroxybenzoate could be inferred from the observation that when an additional methyl group was introduced into the nucleus of the substrate, a methyl-substituted *p*-hydroxybenzoate accumulated in the culture fluid. That is, the enzyme system responsible for oxidizing the cresol was able to tolerate the presence of an additional methyl group in the molecule of the substrate, whereas oxidation of *p*-hydroxybenzoate was blocked by substitution [5]. In early studies of glycolysis it was possible to demonstrate the participation of 3-phosphoglycerate as a consequence of the enzymic formation of an 'unnatural' metabolite that could not serve as a substrate for later enzymes in the sequence. Thus, when added to yeast juice depleted of phosphate, arsenate could replace phosphate in the metabolism of 3-phosphoglyceraldehyde with the result that 1-arseno-3-phosphoglyceric acid was formed. However, this compound was then unable to react with ADP and instead it decomposed spontaneously to regenerate arsenate and produce 3-phosphoglycerate.

It is also possible to break a reaction sequence by adding a chemical reagent which is able to combine with a particular metabolite without at the same time interfering with the action of the enzymes that form it. Thus sodium bisulphite was used to 'trap' acetaldehyde formed in glycolysis; hydroxylamine combines with acyl coenzyme A compounds whilst they are being synthesized enzymically; and semicarbazide can be used to trap α-oxoacids. Such additions to reaction mixtures may, of course, also inhibit some of the enzymes that are present, and in consequence the trapping agent may have different effects from those anticipated: thus, the carbonyl reagents mentioned may be used to inhibit pyridoxal phosphate-dependent enzymes by combining with the coenzyme. Neither can it be assumed that, in crude enzyme mixtures or whole cells, the observed accumulations arise directly from the reaction sequence under investigation. Thus, arsenite is a useful inhibitor of oxidative decarboxylations of α-oxoacids and it may be used for this purpose with intact bacteria. Among the reactions so blocked are those of pyruvate and α-oxoglutarate in the tricarboxylic acid cycle and accordingly it is necessary to interpret with caution the significance of accumulations that may be observed. Thus, because these oxoacids are members of the cycle but only one of them, say α-oxoglutarate, accumulates when intact bacteria oxidize a given substrate, it may be tempting to place α-oxoglutarate on that part of the degradative pathway that leads from the substrate into the cycle. For it might be argued that if the accumulated α-oxoglutarate does in fact arise from the cycle, and not from a pathway that leads into it, why should not pyruvate also accumulate? However, the respective rates at which the two oxoacids are metabolized, and hence their concentrations in the metabolism fluid, may simply depend upon the conditions that obtained during the growth of the cells. For example, certain bacteria that grew with acetate as source of carbon excreted relatively large concentrations of α-oxoglutarate, and no pyruvate, when cell suspensions oxidized acetate [1]. In this case we can eliminate the possibility that the growth substrate is converted into α-oxoglutarate before it is oxidized. This observation, and no doubt others made for different growth substrates, may reflect the proportions of the various enzymes of the tricarboxylic acid cycle which were synthesized in response to the metabolic requirements of the bacteria when they were growing. One metabolite might therefore be readily identified and another escape detection although both occur in the same reaction sequence.

In 1940 D.D. Woods [2] discovered that the antibacterial action of sulphanilamide was reversed by the structurally-related compound *p*-aminobenzoic acid, which was later shown to be a naturally-occurring metabolite when it was isolated from yeast by K.C. Blanchard [3]. As a result of these observations, chemists were stimulated to synthesize a very large number of analogues of metabolites (antimetabolites) in the hope that they would possess therapeutic value. The action of an antimetabolite was considered to be exerted by blocking bacterial biosynthetic reactions through competition with a natural substrate for the active centres of its enzymes. We realize today that this interpretation of their action is oversimplified for many reasons: for example, metabolites may exert 'feed-back' inhibition of existing enzymes or may repress the synthesis of new ones. It is evident, and indeed it has been proved in certain cases, that an antimetabolite might mimic these actions as well as compete for an enzyme site. As a consequence of uncertainties about their precise sites of action, the high hopes of two decades ago have been disappointed: antimetabolites did not become universal tools for elucidating biosynthetic pathways. This type of investigation is best undertaken with mutant microorganisms in which metabolic blockages can be located with certainty. Nevertheless in a few areas the initial steps in delineating a biosynthetic sequence were greatly assisted by the use of bacteria poisoned with antimetabolites. For example, when the growth of *E. coli* was partially inhibited by sulphanilamide, 4-amino-5-imidazolecarboxamide was excreted into the culture. As its ribonucleotide, this compound is an intermediate in the biosynthesis of purines when it receives a 1-carbon fragment donated by N^{10}-formyl tetrahydrofolate. No doubt sulphanilamide interferes with the reactions of this coenzyme because its 'natural' counterpart, *p*-aminobenzoic acid, is a component of the folic acid molecule.

The use of inhibitors is not confined to breaking up metabolic sequences with a view to isolating each enzyme concerned. We may wish to understand the manner in which known enzymes function together as a group within the cell. Thus the use of inhibitors such as cyanide, carbon monoxide and narcotics has helped to reveal the order of assembly of the cytochromes in the electron transport system of mitochondria. Spectrophotometric measurements made when an inhibitor is added have enabled its point of action to be localized by the application of the 'cross-over theorem' of Chance and Williams [1] which states that if an inhibitor is added to an array of redox carriers in their steady state, those on the reduced side will become more reduced, while those on the oxidized side will become more oxidized. Oligomycin, 2,4-dinitrophenol and dicoumarol are reagents that prevent the coupling of the free energy released by the reactions of the respiratory chain to the synthesis of ATP. For a discussion of their use in attempts to localize the points of energy transfer the reader is referred to the essay of D.E. Griffiths [2].

Dismantling the catalytic apparatus of the cell

The success of the experimental approaches that were applied to the study of glycolysis make it clear that progress in elucidating metabolic pathways has depended on the development of methods by which chains of reactions can be broken up into their separate links. The ultimate goal of this approach, which has often been achieved, is to isolate and crystallize the enzymes that catalyse each reaction of a sequence. These methods are made necessary by the interweaving of enzymic pathways; but a paradoxical situation is introduced by the very fact that since reaction sequences are linked, the functioning of any one of them may depend upon the rest. This means that the physiological significance of a particular enzyme cannot be completely assessed without reference back to the intact organism from which it was isolated. To use the analogy of D.E. Green, cited by E. Baldwin [3]: 'A sufficiently ingenious mechanic could separate the parts of a baby Austin (a British car, now almost in the "vintage" class) and use them to make a perambulator or a pressure pump or a hair-dryer of sorts. If the mechanic was not particularly bright and was uninformed as to the source of these parts, he might be tempted into believing that they were in fact designed for the particular end he happened to have in view. The biochemist is presented with a similar problem in the course of his reconstructions. The materials of the cell offer unlimited possibilities of combinations and interaction, but only a few of these possibilities are realized in the cell under normal conditions. There is thus a grave element of risk in trying to reason too closely from reconstructed systems to the intact cell. The reconstruction can have no biological significance until some definite counterpart of these events is observed *in vivo*'.

Nevertheless, the investigation of a new metabolic pathway still begins with the intact organism and continues with the progressive dismantling of its enzymic apparatus. The rate of any scientific advance largely depends upon the techniques available for tackling current problems, and the components of cells may now be examined, separated and identified by means of the electron microscope, the ultracentrifuge and many types of column chromatography. In the days before radioactive isotopes were available, experiments with living animals could give little information about details of metabolic pathways. One exception of great value was a type of labelling experiment, performed by F. Knoop at the turn of the century, that led to the conception of β-oxidation of fatty acids. At the present time a labelling experiment would make use of 'heavy' or radio isotopes, but Knoop synthesized fatty acids in which a terminal carbon atom was joined to a phenyl group. The benzene nucleus remained intact when these substituted fatty acids were fed to animals, whereas the straight hydrocarbon chains were oxidized from the carboxyl ends. In consequence, those acids that contained an even number of carbon atoms were excreted as phenylaceturic acid, formed by condensation of phenylacetic acid and glycine *in vivo*, whereas those with an odd number of carbon atoms were excreted as the condensation product of benzoic acid and glycine, namely hippuric acid. These are the results to be expected from the theory that each fatty acid is degraded by the successive removal of two-carbon fragments from the parent molecule.

As the study of metabolism advanced, those biochemical sequences that are specialized functions of particular organs were located by removing an organ and studying the resulting derangement of the animal's metabolism. Thus, when a normal animal is fed with high quantities of protein there is an increase in the amount of urea excreted. When the liver is removed the animal will survive for several days if protein is withheld from its diet, but if protein is fed it soon dies. The kidneys remove nitrogen in the form of ammonia from amino acids; and since the liver in turn normally converts ammonia into urea, an animal without a liver dies from the toxic effects of the ammonia which accumulates. An extension of this approach was to bring an organ to the outside of an animal's body by means of a surgical operation: the organ was then kept functioning by providing it with an independent circulation of blood, plasma or a synthetic medium that imitated the composition of the normal blood supply. The action of the heart was replaced by mechanical pumps and these could also furnish the circulating medium with oxygen. Perfusion is still a valuable technique but is now applied more often to studies of metabolic control than to those concerned with metabolic pathways. However, it has been found recently that perfusion has certain advantages over the use of liver slices, described below, for the study of gluconeogenesis in rat liver [1]. Poor rates of synthesis of glucose from substrates such as succinate, malate, glutamate and aspartate were observed when slices were used, whereas rates in perfused livers were rather higher than the maximal rates for the living animal. From perfusion experiments it was shown that the liver converts ammonia quantitatively into urea and produces acetoacetate from fatty acids that contain even numbers of carbon atoms.

An extension of the procedure of taking the whole organism apart was the use of slices of an organ thin enough to permit the oxygen needed for its reactions to gain ready access to the sites where it is used. One of the notable biochemical advances which resulted was that of H.A. Krebs who used the liver-slice technique to establish the main reactions, involving citrulline, ornithine and arginine, by which urea is biosynthesized from ammonia. Liver slices also oxidize fatty acids, but for many years all attempts to isolate enzymes and reaction intermediates met with failure. For this reason it was believed until 1939 that oxidation of a fatty acid depended in some way upon the maintenance of the structure of the intact cell. In that year L.F. Leloir and J.M. Munoz dispelled this notion by making use of the homogenizer, developed three years earlier by V.R. Potter and C.A. Elvehjem, to make a 'brei' from rat liver. Although their suspensions of particles contained no intact cells they found that butyrate was oxidized and that its rate of oxidation was increased by additions of fumarate. This experiment was a starting point for one of three main lines of investigation that led to our present understanding of the sequence of reactions by which fatty acids are oxidized (Fig. 1).

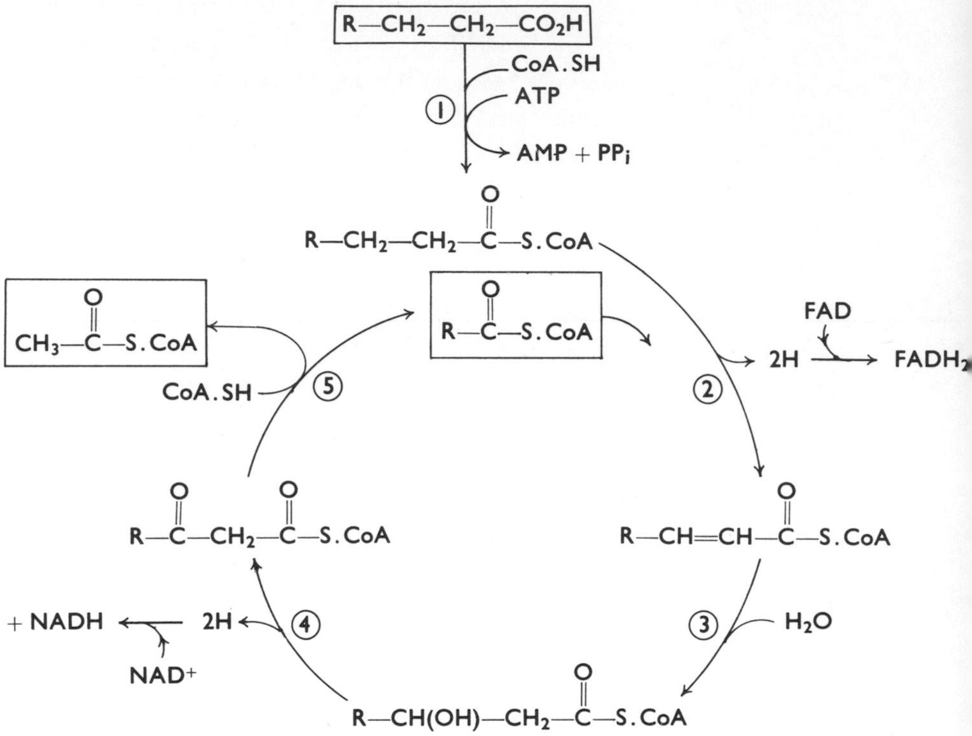

Fig. 1. The 'Fatty Acid Spiral'. A fatty acid $R.CH_2.CH_2.CO_2H$ is oxidized to $CH_3.CO{-}S.CoA$ (acetyl coenzyme A) and $R.CO{-}S.CoA$ from which further molecules of acetyl coenzyme A are formed by enzymes 2, 3, 4 and 5. These enzymes are reversible but are not those used for the synthesis of fatty acids.

Some lines of research that elucidated the 'fatty acid spiral'

The three separate lines of research that converged to provide a solution to this problem can be viewed in retrospect as follows. In one category were the studies of the enzymology of the process which stemmed from the work of Knoop, which were continued through the use of perfused organs and which led to experiments with cell-free preparations initiated by Leloir and Munoz. Secondly, there were the developments in cytology. Staining and microscopic examination revealed granules in the liver, and in particular, as techniques of staining and fixing developed, the existence of the structures now termed mitochondria were recognized. An authoritative account of these particles is given by A.L. Lehninger in his monograph *The Mitochondrion* [1]. Refinements in the use of the electron microscope, chiefly at the Rockefeller Institute, permitted a closer examination of mitochondria and aided in their isolation by G.H. Hogeboom, W.C. Schneider and G.E. Pallade [2] who prepared homogeneous suspensions by freeing the mitochondria from nuclei and microsomes by means of differential centrifugation in 0·88 M-sucrose. All this work had a direct bearing upon the problem of isolating the enzymes of the 'fatty acid spiral' when E.P. Kennedy and A.L. Lehninger [3]

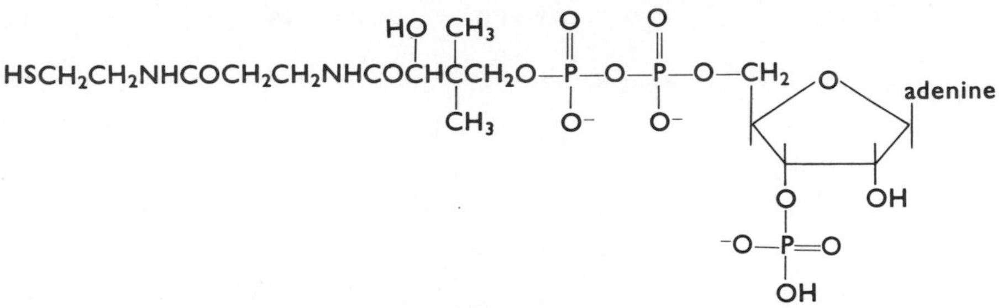

$$HSCH_2CH_2NHCOCH_2CH_2NHCOCH(OH)C(CH_3)_2CH_2O{-}P(=O)(O^-){-}O{-}P(=O)(O^-){-}O{-}CH_2\text{-ribose(adenine)}\text{-}3'{-}O{-}P(=O)(O^-)OH$$

Coenzyme A

$$HS{-}CH_2{-}CH_2{-}NH{-}CO{-}CH_2{-}H$$

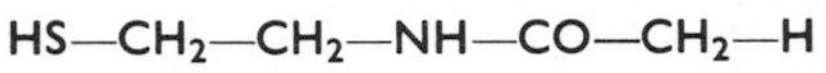

N-acetyl-thioethanolamine, an analogue of coenzyme A

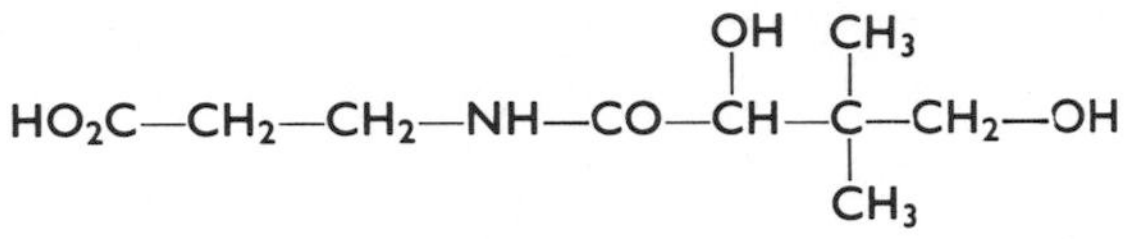

$$HO_2C{-}CH_2{-}CH_2{-}NH{-}CO{-}CH(OH){-}C(CH_3)_2{-}CH_2{-}OH$$

Pantothenic acid

$$HS{-}CH_2{-}CH_2{-}NH{-}CO{-}CH_2{-}CH_2{-}NH{-}CO{-}CH(OH){-}C(CH_3)_2{-}CH_2{-}OH$$

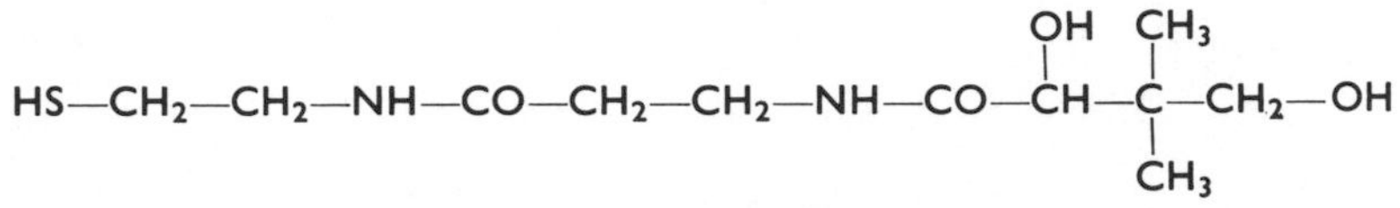

Pantetheine

Fig. 2. Coenzyme A and some related compounds.

applied the method used by the workers of the Rockefeller Institute to isolate rat-liver mitochondria and to show that they readily oxidized both fatty acids and compounds of the tricarboxylic acid cycle. G.R. Drysdale and H.A. Lardy [4] then showed that extracts of isolated rat-liver mitochondria provided a rich source of the enzymes concerned, far superior to extracts obtained by homogenizing whole cells. The stage was then set for the development of large-scale preparations of mitochondria from slaughterhouse material and hence for the isolation and purification of enzymes by D.E. Green, H.R. Mahler and their colleagues at the Institute for Enzyme Research, University of Wisconsin, and by other workers in various laboratories elsewhere.

The third main stream of research with a direct bearing upon the problem of fatty acid oxidation was that concerned with the chemistry and enzymology of coenzyme A. This coenzyme was discovered by F. Lipmann in 1945, and by D. Nachmansohn and M. Berman independently in 1946, as a requirement for enzyme systems by which acetate was converted by ATP into a form capable of reacting with acceptors such as sulphanilamide or choline. The work of F. Lipmann and others, which established the formula of coenzyme A shown in Fig. 2, was followed by that of F. Lynen and his colleagues [1] who recognized that it is the thiol group which functions in reactions of the coenzyme: acetyl coenzyme A is a thioester of the type CH_3CO—SCoA, where coenzyme A is represented as CoA—SH. Further progress was greatly facilitated when Lynen isolated acetyl coenzyme A from yeast and developed methods by which acyl derivatives of coenzyme A could be identified and determined. Another line of research was concerned with the entry of 'active acetate' into the tricarboxylic acid cycle. This work culminated in the preparation from pigs' hearts, by S. Ochoa, J.R. Stern and M.C. Schneider [2], of crystals of citrate synthase which catalyses the reaction:

$$\text{citrate} + \text{CoA} \rightleftharpoons \text{acetyl-CoA} + \text{oxaloacetate} + H_2O \quad \ldots\ldots\ldots \quad (8)$$

Meanwhile the participation of coenzyme A in the metabolism of fatty acids was shown by H.A. Barker, E.R. Stadtman and their colleagues [3] who worked with the obligate anaerobe *Clostridium kluyvei* which utilizes acetate and ethanol as sources of carbon. During growth, hydrogen atoms are transferred from ethanol and are accepted by reaction intermediates in the synthesis of fatty acids such as butyric and hexanoic acids which consequently appear as products of the fermentation. The somewhat unexpected observation was also made that soluble cell-free extracts of this anaerobe were able to catalyse the *oxidation* of fatty acids with four to eight carbon atoms. Oxygen was taken up, acetylphosphate was formed when inorganic phosphate was present and acetoacetate resulted when it was absent. However, the participation of the coenzyme A esters of the fatty acids in this process, rather than their phosphate esters, was indicated by the discovery of an enzyme that converts acetyl coenzyme A into acetylphosphate, which therefore appears as the end-product. This enzyme is phosphate acetyltransferase:

$$\text{acetyl-CoA} + \text{orthophosphate} \rightleftharpoons \text{CoA} + \text{acetylphosphate} \quad \ldots\ldots \quad (9)$$

It was shown later [4] that soluble cell-free extracts of an obligate aerobe (*Moraxella lwoffi*) catalysed the oxidation of hexanoic acid to give acetic acid as the end-product, and of nonanoic acid to give a mixture of acetic and propionic acids. Cofactors of the fatty acid spiral, namely coenzyme A, ATP, NAD and Mg^{2+} ions were needed for enzymic activity.

Progress in one particular area has often awaited developments in other fields. It is rarely possible to solve a central problem in science by rigid adherence to a programme, however plausible and detailed this may be. In devoting attention to an outline of some of the research on biological oxidation of fatty acids, we have tried to show how knowledge has grown from experiments that have ranged from cytological studies of mammalian cells to those concerned with the biochemistry of anaerobic bacteria; or from work on a different, though related, metabolic system (the tricarboxylic acid cycle) to studies of the chemistry of a cofactor required in the reactions.

The recognition by F. Lynen and his co-workers in Munich, that fatty acids react enzymically as their thioesters of coenzyme A enabled them to proceed with the purification of the enzymes before relatively large amounts of coenzyme A became available [1]. Instead of the coenzyme they used derivatives of *N*-acetylthioethanolamine which possess the same configuration of atoms next to the active thiol group as does coenzyme A itself (Fig. 2). Thioesters such as *S*-acetoacetyl- or *S*-crotonyl-*N*-acetylthioethanolamine were prepared as analytically pure compounds and shown to be substrates for the respective enzymes of Fig. 1. Higher concentrations were needed than those for the corresponding coenzyme A derivatives since the enzymes concerned had a lower affinity for the analogues, probably because they lacked certain groups that anchor coenzyme A itself to the enzymes. However, this was no serious handicap when they were used in assays performed during the course of enzyme purification in which spectroscopic methods were extensively employed. For example, acyl-CoA dehydrogenase could be assayed by following the bleaching of 2,6-dichlorophenol indophenol as it received hydrogen removed from the fatty acyl derivative of the analogue, a transfer mediated by flavin adenine dinucleotide; or alternatively the regenerating of colour from leucosafranin T could be measured on addition of *S*-crotonyl-*N*-acetylthioethanolamine. This work has been summarized by F. Lynen and S. Ochoa [2], and that on the purification of these enzymes using the acyl coenzyme A compounds themselves has been reviewed by D.E. Green [3]. Although our knowledge of the field has been extended since these reviews were written, they may still be read with profit for the account they give of progress made in the elucidation of a metabolic pathway.

Some reflections on the tricarboxylic acid cycle

A second reaction sequence that occupies a central position in the metabolism of most living tissues is the tricarboxylic acid cycle. In this case we shall not trace the converging lines of research which led to its discovery, although these were indeed instructive; but instead we shall consider the criteria which had to be satisfied before the cycle was universally adopted. Since the scheme is so familiar to us at the present time we do not readily appreciate that a sequence involving eight reactions for the complete oxidation of a molecule containing only two carbon atoms would be sceptically received by pure chemists less than three decades ago. First, it was not only necessary to show that the enzymes were present for the metabolism of all the reaction intermediates postulated, but even more important that they were found in amounts sufficient to account for the rates of complete oxidation of, say, pyruvate or acetyl coenzyme A. The need to account for observed metabolic rates in terms of the individual reactions proposed has not always been met in other instances, as we shall see. A compound cannot be established as a member of the sequence under investigation simply by showing that there are enzymes present in the cell which synthesize and degrade it. Second, in the case of the tricarboxylic acid cycle, this is a mechanism for the catalytic oxidation of metabolites and it was necessary to show that the addition of any of its various di- and tricarboxylic acids could stimulate the rate of endogenous respiration of the tissue preparation far in excess of any stoichiometric reactions of the added catalyst. Thus, in one experiment an amount of fumarate that would have been oxidized to completion with an uptake of 20 ml. of oxygen caused an increase in uptake of 151 ml. when added to minced pigeon breast muscle. Third, it was shown that additions of small amounts of malonate inhibit one step of the cycle, and one step only, namely the oxidation of succinate to fumarate. This enabled the participation of citrate, isocitrate, *cis*-aconitate and α-ketoglutarate to be demonstrated since these compounds were converted quantitatively into succinate. On the other hand the catalytic effect of a C-4 dicarboxylic acid such as fumarate was abolished by poisoning the tissue with malonate: instead, there was a stoichiometric conversion of the added compound to give succinate according to the equation:

$$\text{fumarate} + \text{pyruvate} + 2\,O_2 \rightarrow \text{succinate} + 3\,CO_2 + H_2O \quad \ldots\ldots \quad (10)$$

These experiments are reviewed by H.A. Krebs [1] who, with W.A. Johnson proposed the cycle in 1937 in a form that differed only slightly from the presently-accepted scheme. Finally, there is the evidence that the cycle is not merely a series of reactions which we ourselves can select from the metabolic network and so envisage acetate to be oxidized to carbon dioxide and water. That the enzymes do indeed function in concert in the living organism became evident when it was shown that they are organized together in one structure, the mitochondrion. In aerobic bacteria, which are too small to possess mitochondria, these enzymes appear to be located mainly in the cytoplasmic membrane.

Some difficulties in the use of ^{14}C-labelled metabolites

The use of ^{14}C-labelled metabolites has made it possible to trace their fates in intact organisms and hence to assess the physiological significance of the reaction sequences in which they participate. However, there are certain obstacles to making such assessments, as the following considerations show. First, as we have seen, a particular compound is likely to participate in several metabolic sequences besides the one under consideration; and second, metabolites of one pathway are often able to alter the reaction rates of another on account of the fact that certain enzymes exhibit 'allosteric' behaviour. In addition to the active site on the enzyme molecule where the substrate becomes attached, there may be another region where the enzyme is able to bind a second metabolite that does not serve as substrate. Accordingly, although such a compound is not decomposed, its attachment may cause an alteration in the configuration of the molecule of the enzyme which in turn can affect the rate at which the reaction is catalysed. This possibility cannot be ignored when accounting for the observed distribution of radioactivity after addition of a labelled metabolite to a complex enzyme system. We shall outline some of the difficulties which these considerations introduce to the use of radioisotopes, with examples chosen from systems discussed more fully by H.A. Krebs [2].

In the early 1960's experiments in several laboratories led to the conclusion that the formation and oxidation of amino acids in brain constituted a major pathway of glucose metabolism; and it appeared that in this tissue amino acids inhibited the oxidation of glucose and were themselves preferentially oxidized. These views were based on the following observations. When ^{14}C-glucose was injected into the tail veins of rats and the distribution of isotope examined at intervals of time it was found that 40% of the total radioactivity appeared in amino-acid fractions within two minutes, rising to 75% after half an hour. In other experiments, slices of brain cortex were incubated either with uniformly-labelled glucose alone or with uniformly-labelled glucose plus unlabelled glutamate or aspartate. The specific activity of the carbon dioxide evolved was then measured and expressed as a ratio to the specific activity of glucose carbon. The addition of either amino acid caused a substantial reduction of the above ratio, thus indicating that much of the carbon

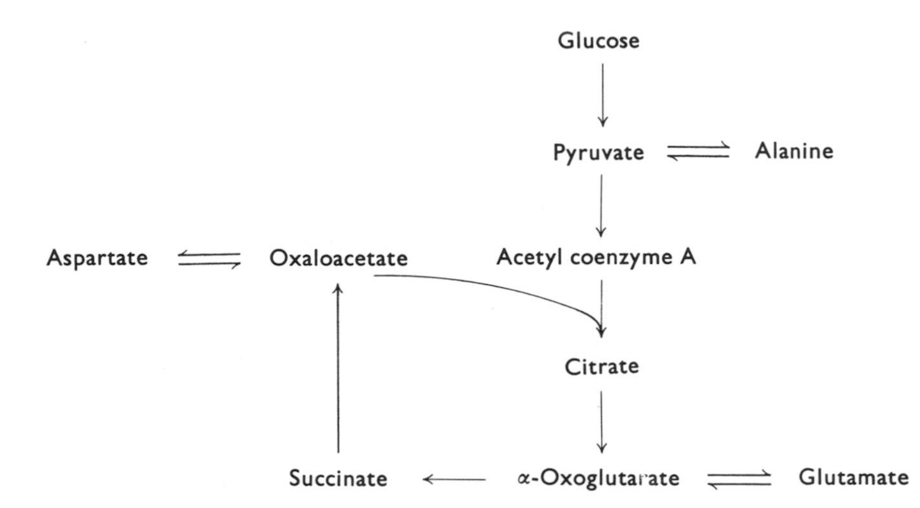

Fig. 3. Side reactions of α-oxoacids in the degradation of glucose.

dioxide was derived from precursors other than glucose. However, as Krebs [16.2] has shown, it is not necessary to conclude that amino acids lie on the main pathway of glucose catabolism in brain, or that they inhibit these reactions. Glucose is converted first into pyruvate and then into acetyl coenzyme A, and this is oxidized by the tricarboxylic acid cycle. The ketoacids of the cycle can now exchange their carbon skeletons with amino acids by the 'half reactions':

$$\text{amino acid} + \text{pyridoxal phosphate amino transferase} \rightleftharpoons \alpha\text{-ketoacid} + \text{pyridoxamine phosphate amino transferase}$$

These half reactions take place very rapidly: more so than those of the cycle in which the carbon skeletons are actually altered. As a consequence, although the main flow of carbon passes through α-oxoglutarate which is, in turn, resynthesized by the reactions of Fig. 3, much of the ^{14}C will appear as glutamate by a rapid side-step of these reactions, the carbon skeleton remaining unaltered. The following factors also contribute to the observed incorporation of isotope into amino acid fractions. (*a*) The size of the glutamate and glutamine pool which the isotope enters is substantially greater in brain than in other tissues. (*b*) The glucose pool is low, so that the isotope enters other metabolic processes rapidly. (*c*) The half reaction shown above also takes place in other major tissues and cannot by themselves account for the exceptionally rapid incorporation of isotope from glucose into glutamate. However, the rate of glycolysis in brain is very high and so, therefore, is the rate of formation of acetyl coenzyme A from which ^{14}C is incorporated into carbon skeletons of the α-ketoacids of the cycle. (*d*) The blood supply to the brain is relatively larger than that of muscle or liver, per unit of weight, so that more isotope reaches the brain.

In this system the interpretation of isotopic data must take into account the rapid transfer of a complete carbon skeleton from one compound to another. A second system, also discussed by Krebs [16.2] concerns the redistribution of isotope that occurs when two metabolic pathways 'cross over' through a common metabolite. In this case the synthesis of glucose from lactate (gluconeogenesis) by slices of rat kidney

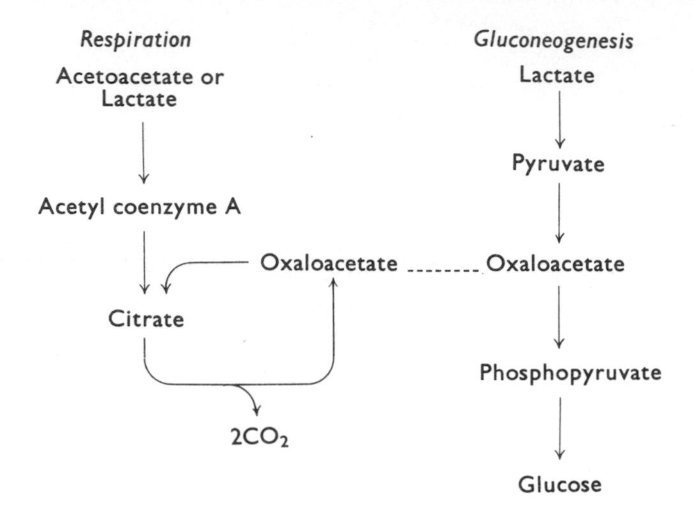

Fig. 4. Main reactions in kidney cortex incubated with acetoacetate and lactate. Oxaloacetate is an intermediate in both respiration and gluconeogenesis.

cortex was studied. It has long been known that the addition of acetoacetate to such a system causes a considerable increase in the yield of glucose, and from this it might be concluded that a net conversion of acetoacetate into glucose occurs. However, since there are no established reactions by which this conversion might be achieved an alternative explanation must be considered, namely that acetoacetate is metabolized as a preferred fuel of respiration of this tissue: an increased amount of lactate would then be 'spared' for gluconeogenesis. Acetoacetate is very rapidly oxidized by kidney cortex. This theory could be tested by the use of ^{14}C-acetoacetate when it might be expected that the preferred oxidation of this compound would release carbon dioxide having the same specific activity as acetoacetate carbon, whereas the glucose formed would incorporate only such label as would arise from carbon dioxide fixation. A corresponding experiment would involve using unlabelled acetoacetate, and ^{14}C-lactate which would give rise to labelled glucose. But these results were not obtained: on the contrary, ^{14}C-acetoacetate was metabolized to ^{14}C-glucose, the specific activity of the carbon dioxide being low, whereas the radioactivity of glucose arising from uniformly labelled lactate was much lower than expected.

The probable explanation of these results emerges when the reaction sequences in Fig. 4 are considered. It will be seen that (*a*) Oxaloacetate is an intermediate in both respiration and gluconeogenesis; moreover it is formed in the same intracellular compartment, the mitochondrion.

(*b*) Although the *amount* of glucose formed may correspond to the amount of lactate that disappears, *isotopic* carbon from lactate can find its way into the tricarboxylic acid cycle through oxaloacetate and so appear as $^{14}CO_2$. Likewise some of the carbon atoms originating in acetoacetate can appear in glucose without there being any *net* synthesis of glucose from acetoacetate. In fact, when acetyl coenzyme A enters the cycle its carbon atoms are not those released by the oxidation of isocitrate and α-oxoglutarate: instead, carbon dioxide arises from the portion of the citrate molecule into which oxaloacetate was incorporated. It therefore follows that isotopic carbon from acetyl coenzyme A (or acetoacetate) must be retained in oxaloacetate after one turn of the reactions of the cycle. On the other hand, ^{14}C-lactate when converted into oxaloacetate must give rise to $^{14}CO_2$ which is released by those reactions of the cycle that lie between isocitrate and succinate. This obligatory 'crossing over' of pathways is a consequence of the steric behaviour of citrate in the aconitase system to which A.G. Ogston [1] first drew attention; and it is for this reason that isotope from a fatty acid appears in glucose or glycogen although in mammals there is no overall synthesis of these compounds from the fatty acid. (*c*) The conversion of lactate into oxaloacetate involves the carboxylation of pyruvate. Now acetoacetate is converted into acetyl coenzyme A very rapidly, and this compound is known from the work of D.B. Keech and M.F. Utter [2] to be an allosteric effector for pyruvate carboxylase: accordingly its transitory accumulation from the acetoacetate added to the system assures a rapid conversion of lactate into glucose by accelerating the carboxylation of pyruvate. This is an example of how a compound that appears on one pathway can change the overall rate of a second, despite the fact that no additional metabolite is supplied to the enzymes of that pathway.

As H.A. Krebs has observed [16.2]: 'A general asset of the isotope method is its applicability to highly complex systems such as the whole animal and intact tissue where it can supply information under physiological conditions. But there is a price to be paid. What we gain by maintaining physiological conditions we may lose in part by the difficulties in interpreting results'.

On the other hand, such complications are minimal in studies of higher plants or fungi where biosynthetic pathways may lead to end-products that are not further metabolized and are therefore outside the metabolic network. Thus the labelling patterns of end-products such as orsellinic acid, 6-methylsalicylic acid, griseofulvin or terramycin that result from feeding labelled acetate to certain fungi are in agreement with those predicted by the hypothesis that benzene nuclei arise in certain natural products from condensation of acetate residues [3, 4].

An example of the use of ^{14}C: the pathway of biosynthesis of NAD in mice

A most instructive and successful application of radiotracer methods to a mammalian system was concerned with the biosynthesis of a metabolite NAD, which although participating in numerous metabolic pathways, does so simply as a carrier of hydrogen without undergoing any modification in its basic chemical structure. The experiments of H. Ijichi, A.

Ichiyama and O. Hayaishi [1] clarified a number of puzzling features associated with the biosynthesis of NAD in mammals. The pyridine nucleus that forms a part of its chemical structure might be contributed, in principle, by the operation of three different sets of reactions, all of which have been studied with isolated enzyme preparations. First, the pyridine nucleus might arise from tryptophan. This amino acid is degraded to 3-hydroxyanthranilic acid, the nucleus of which is opened by an oxygenase to give 2-acroleyl-3-aminofumarate; spontaneous ring-closure then occurs to give the pyridine derivative quinolinic acid. Second, the source of the pyridine nucleus may be nicotinamide, a component of the nucleotide structure; or third, it may arise from nicotinic acid in which case the carboxyl group must clearly be converted to an amide group at some stage in the synthesis of NAD.

It is not possible to decide how NAD is biosynthesized in the various organs of an animal simply by considering these reaction sequences in isolation (Fig. 5). The conversion of nicotinamide into nicotinamide ribonucleoside might be considered to be of little importance since it is catalysed at a slow rate by an enzyme from erythrocytes that shows little affinity for nicotinamide. Nevertheless, N.O. Kaplan and his co-workers [2] have established that substantial increases in the concentration of NAD in the livers of mice result from their injection with nicotinamide, whereas similar injections of nicotinic acid give rise to much smaller increases. The following observations were therefore in conflict: (*a*), experiments with intact animals implicated nicotinamide, rather than nicotinic acid, in the biosynthesis of NAD whereas (*b*), the activity of one enzyme possibly concerned in the metabolism of the amide was insufficient for its suggested function. The contradiction was resolved and accounted for by Ijichi, Ichiyama and Hayaishi [1] who measured the contributions made by nicotinic acid and its amide to the yields of NAD biosynthesized in the livers of mice (the possible contribution of tryptophan was not investigated). They used a technique that had the advantage of producing little or no disturbance in the metabolism of the living animal but which permitted the investigators, nevertheless, to disentangle the reactions they wished to study from those which might otherwise complicate the interpretation of their measurements. This was a modification of the technique of 'pulse labelling', similar in principle to that applied with such success to trace the pathway of carbon during photosynthesis of *Chlorella* [3] and to study various problems of bacterial metabolism discussed later. Small quantities (78 mμ moles) of either ^{14}C-nicotinic acid or ^{14}C-nicotinamide were injected into the portal veins of mice, and at intervals of time increasing from 20 seconds through 1 minute to several hours after the injection, the livers of the mice were removed and frozen and then extracted with 3% perchloric acid. The compounds to be studied dissolved in this solvent but the protein was precipitated and removed by centrifuging; and the perchloric acid was removed as its slightly soluble potassium salt by neutralizing with potassium hydroxide at 0°C. Nicotinic acid, nicotinamide and their ribonucleotides and also NAD and deamido-NAD were each separated by chromatography on a column of Dowex 1 (formate) so that the amount of each compound present in the sample

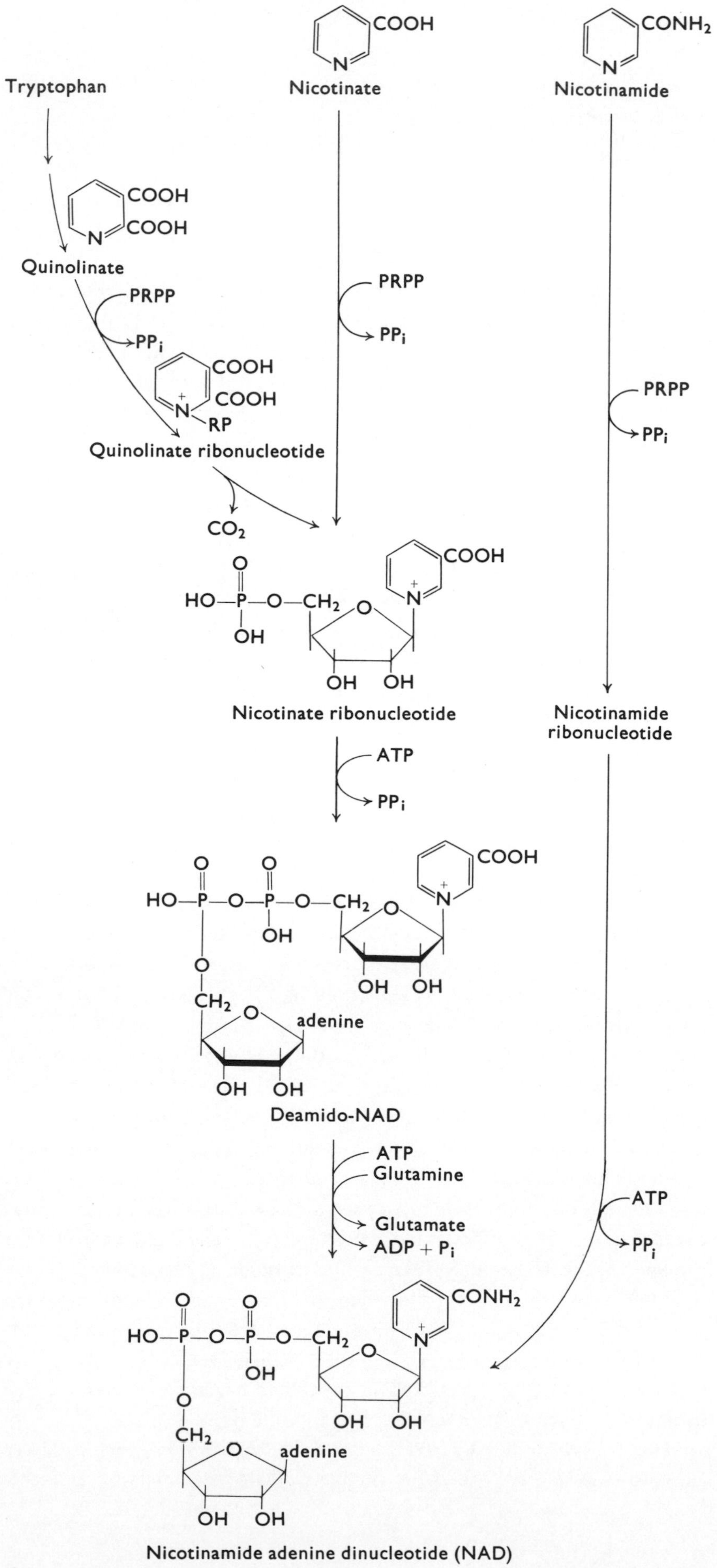

Tryptophan

Quinolinate

PRPP

PP_i

Quinolinate ribonucleotide

CO_2

Nicotinate

PRPP

PP_i

Nicotinate ribonucleotide

ATP

PP_i

Deamido-NAD

ATP
Glutamine
Glutamate
ADP + P_i

Nicotinamide adenine dinucleotide (NAD)

Nicotinamide

PRPP

PP_i

Nicotinamide ribonucleotide

ATP

PP_i

Fig. 5. Metabolic interrelationships in the biosynthesis of NAD.

and also the radioactivity it contained could be determined. It was found that both nicotinic acid and nicotinamide entered the liver so rapidly that about half of the radioactivity injected with each compound was found inside the organ within 20 sec. From 3 min. onwards, however, striking differences in behaviour of the two compounds became apparent. Thus, a large proportion of the radioactivity taken up by the liver from nicotinamide had disappeared after 3 min. whereas most of that from nicotinic acid was retained, and even after 4 hr. more than one fifth could still be recovered from the liver.

When the soluble compounds were extracted with perchloric acid and chromatographed it was possible to trace the fate of the administered ^{14}C-nicotinic acid. Three min. after injection most of the nicotinic acid itself had disappeared and the ^{14}C retained by the liver was largely due to NAD, deamido-NAD and nicotinic acid ribonucleotide. Then the amounts of the two last-named compounds decreased, so that after 10 min. they had also disappeared. However, at the end of 10 min. the radioactivity incorporated into NAD had attained its maximum value: this accounted for most of the radioactivity that entered the liver as nicotinic acid at the start of the experiment. In short, the movement of radioactivity was that expected from a reaction sequence in which nicotinic acid, nicotinic acid ribonucleotide and deamido-NAD are, in turn, the precursors of NAD (Fig. 5).

In contrast, very little NAD was biosynthesized over this period of time when ^{14}C-nicotinamide was injected and it is apparent that nicotinic acid and not nicotinamide is the metabolic precursor of NAD. On the other hand it is clear that the liver of a mouse can take up, but does not retain, nicotinamide. What, then, is the fate of this compound? In the course of experiments designed to answer this question Ijichi, Ichiyama and Hayaishi were able to account for the findings of previous workers who had also used whole animals but had reached a different conclusion. In these earlier experiments the amounts of NAD formed in the livers of mice had been determined at much longer intervals of time and the doses of nicotinic acid or nicotinamide had been about one thousand times greater. When the Japanese workers administered ^{14}C-nicotinamide in these larger doses the results they obtained entirely confirmed those of Kaplan and his colleagues, namely a ten-fold increase in concentration of liver NAD was observed at 8–10 hr. following the injection of nicotinamide whereas the increase due to nicotinic acid was much smaller. However, when they measured the total radioactivity in the liver at time-intervals of minutes rather than hours they found a rapid decrease over the first hour. Moreover, extraction and chromatography showed that at these high dosages nicotinamide entered the liver and then left it rapidly, just as it did when small amounts were injected; and further that little or no ^{14}C-NAD was found from nicotinamide during this period. After one hour the picture changed: both the total radioactivity of the liver and the NAD it contained began to increase so that substantial values were reached at the end of 8 hr.

These results indicated that the labelled nicotinamide which had been administered was excreted from the liver during the first hour and later returned in a changed form from which NAD could be biosynthesized. If this were so, at what site and into what precursor of NAD, had the administered nicotinamide been transformed? Various organs and tissues were analysed for the distribution of ^{14}C and it became evident that after leaving the liver, nicotinamide accumulated in the gastrointestinal tract where it was deamidated to give nicotinic acid. This compound was then reabsorbed by the liver where it served as a precursor for NAD over a prolonged period of time. One feature of these studies requires further elucidation, namely why it is that at high concentrations, nicotinic acid is converted only slowly into NAD. Doubtless the compound exerts an inhibitory action upon metabolism at these concentrations.

This work on the biosynthesis of NAD illustrates the value of kinetic experiments in aiding the investigator to decide which one, of various possible metabolic pathways, operates effectively within a living organism. On the other hand, NAD was particularly suited to these studies because its molecule, once formed, does not suffer rapid degradation: it plays its part in reaction sequences, but only as a coenzyme.

The study of a batch of closely related enzymes: fixation of carbon dioxide in heterotrophic organisms

However, in other instances the use of radioisotopes is of little value for making such decisions. As an example we may consider the role of phosphoenolpyruvate in metabolism. This compound is the starting point for the biosynthesis of hexoses such as glucose, for the pentoses that form parts of the structures of nucleic acids and for phenylalanine and tyrosine. It is formed during glycolysis and is converted into pyruvate by pyruvate kinase (equation 11); but this reaction is not readily reversed under physiological conditions for the following reasons: (*a*) Free energy is needed to convert pyruvate into phosphoenolpyruvate: that is, the equilibrium of the reaction lies to the left of the equation, (*b*) The affinity of the enzyme for pyruvate is low: that is, the Michaelis constant is high, well above the concentration of pyruvate we normally expect to find inside the cell, and (*c*) The amount of the enzyme present in the liver, where gluconeogenesis occurs, is also low.

$$\text{ATP} + \text{pyruvate} \rightleftharpoons \text{ADP} + \text{phosphoenolpyruvate} \quad \ldots\ldots\ldots \quad (11)$$

Accordingly, phosphoenolpyruvate is formed from pyruvate by other metabolic routes which are shown in the scheme of Fig. 6. Some of the reactions that are known to be involved are listed in Table 1 where it is seen that oxaloacetate may also take part in the interconversion. Indeed, the organism must also ensure that a concentration of this compound is maintained sufficient for the effective operation of the tricarboxylic acid cycle, and some of the reactions have this additional function. It is seen that pyruvate may be converted into phosphoenolpyruvate by the concerted action of two enzymes, namely (4) and (1) of Fig. 6, each of which uses one molecule of ATP. Alternatively, the conversion may be accomplished directly by reaction (5) of Fig. 6 when one molecule of ATP gives rise to adenosine monophosphate, AMP, instead of ADP. So far, this enzyme has been found only in bacteria. More free energy is released when ATP is hydrolysed to AMP and inorganic phosphate than when ADP is formed, and this additional energy ensures that reaction (5) will produce phosphoenolpyruvate from pyruvate when reaction (3) does not.

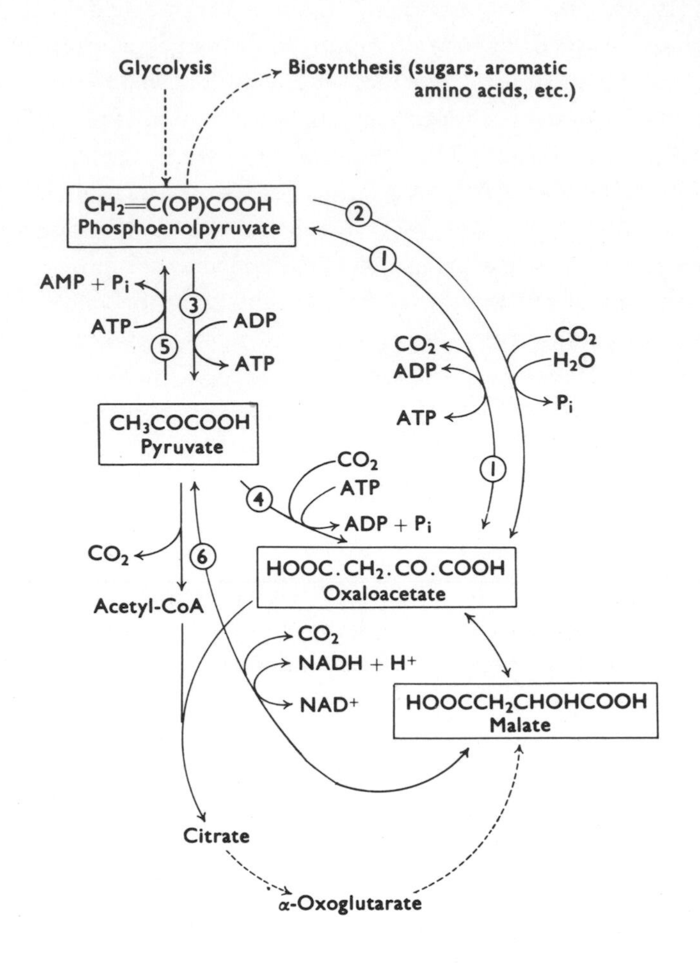

Fig. 6. Carbon dioxide fixation.

Table 1

Number	*Name of Enzyme*	*Reaction catalysed*
①	Phosphopyruvate carboxykinase	PEP + CO_2 + ADP(GDP, IDP) ⇌ OA + ATP(GTP, ITP)
②	Phosphopyruvate carboxylase	PEP + CO_2 + H_2O → OA + P_i
③	Pyruvate kinase	PEP + ADP → Pyruvate + ATP
④	Pyruvate carboxylase	Pyruvate + ATP + CO_2 + H_2O → OA + ADP + P_i
⑤	Phosphopyruvate synthase	Pyruvate + ATP → PEP + AMP + P_i
⑥	Malate enzyme	Pyruvate + CO_2 + NADH + H^+ ⇌ malate + NAD^+

Physiologically irreversible reactions are shown by single arrows. Enzymes I and 2 (4.1.1.32 and 4.1.1.31) are both correctly named 'phosphopyruvate carboxylase'. In order to distinguish between them without using their cumbersome systematic names, enzyme 1 has been referred to in the text and in this table under the name used by the various original investigators of this reaction.

The six enzymes of Fig. 6 serve to transfer an intact C_3-skeleton between three compounds one of which, oxaloacetic acid is formed simply by fixing carbon dioxide on to the skeleton. Malic acid, which is oxidized readily to oxaloacetate, can likewise be formed by the 'malic enzyme' (reaction 6) that reduces pyruvate and fixes carbon dioxide in one step.

It would be well nigh impossible to assess the physiological significance of one particular reaction in the presence of the others by a method that attempted to trace the flow of carbon. Progress can be made, however, if an enzyme system that converts pyruvate into phosphoenolpyruvate is taken and examined to find out which of the enzymes in Table 1 are absent, or are present in amounts too small to account for the observed conversion. Thus, mitochondria of chicken liver convert pyruvate into phosphoenolpyruvate despite the fact that they are essentially devoid of 'malic enzyme' (6), pyruvate kinase (3) or phosphopyruvate synthase (5) (a bacterial enzyme). Accordingly, D.B. Keech and M.F. Utter [1] sought for other reactions by which oxaloacetate, and hence phosphoenolpyruvate, could be formed; and in the course of their investigations they discovered pyruvate carboxylase (reaction (4) of Fig. 6). In crude extracts the specific activity of this enzyme was half that of phosphopyruvate carboxykinase, but after partial purification its amount increased to eight times that of the carboxykinase. Two distinct and separate enzymes were therefore present; but the fact that they tended to remain in association suggested that, in chicken mitochondria, phosphoenolpyruvate is formed by the concerted action of these two enzymes ((4) and (1) of Fig. 6).

By contrast, rat-liver mitochondria contain only one enzyme of the pair: pyruvate carboxylase. Although phosphoenolpyruvate is formed by the same route, the second enzyme involved in its formation is found in the cytosol that surrounds the mitochondria. This means that oxaloacetate, which cannot pass from the inside of a mitochondrion to the outside, must first be transaminated to aspartate or reduced to malate. These compounds can now move out into the cytosol where they are converted back again to oxaloacetate, the substrate of phosphopyruvate carboxykinase which is found there [2]. The high ratio of NAD: NADH in the cytosol favours the oxidation of malate by malate dehydrogenase, whereas this ratio is probably low inside the mitochondrion and there, the reduction of oxaloacetate is favoured.

An example of the use of bacterial mutants to decide the metabolic roles of closely related enzymes

The physiological significance of each reaction of Fig. 6 can perhaps be assessed most readily for bacteria, since mutants can be isolated that lack particular enzymes and the metabolic consequences of the deficiencies can be studied. This approach has been reviewed by H.L. Kornberg [3]. The tricarboxylic acid cycle serves to generate NADH and ATP and so provide energy for the cell. Aerobic bacteria also need to make members of the cycle, namely oxaloacetate and α-ketoglutarate from which cellular constituents are biosynthesized during growth. It is evident that when oxaloacetate or its precursors in the cycle are removed in this way they must be replenished by other reactions that Kornberg has termed *anaplerotic sequences* (from the Greek for 'filling up'). Were oxaloacetate not continually replenished the tricarboxylic acid cycle would cease to function since none would be available to react with acetyl coenzyme A to form citrate. The reactions of Fig. 6 and Table 1, by which pyruvate is converted into oxaloacetate, are therefore anaplerotic; and the question arises as to which of them are actually called into play by the growing cell.

In this work, mutants of Enterobacteriaceae (*S. typhimurium* or *E. coli*) were isolated that failed to grow with glucose unless small additions of tricarboxylic acid cycle compounds were made: this fact suggested that an enzyme which served an anaplerotic function was lacking. Indeed, these mutants differed from their wild type insofar as they lacked phosphopyruvate carboxylase (enzyme (2), Fig. 6) and therefore could not synthesize oxaloacetate from the phosphoenolpyruvate that is supplied from the degradation of glucose. T.S. Theodore and E. Englesberg [1] who had made this discovery with mutants of *S. typhimurium* determined the activity of enzyme (2) by assaying ^{14}C-oxaloacetate formed from ^{14}C-$NaHCO_3$. J.L. Cánovas and H.L. Kornberg [2] assayed oxaloacetate by its enzymic conversion either to malate or to citrate and they found that the latter was a much more effective trapping agent than the former. The reason for this emerged when it was shown that the acetyl coenzyme A, added to the reaction mixture that converts oxaloacetate into citrate, is also an allosteric effector of enzyme (2). The enzyme resembles enzyme (4) in this respect but differs from the phosphopyruvate carboxylase found in plants.

These same mutants of Enterobacteriaceae not only failed to grow with glucose as sole carbon source but they could not utilize pyruvate without supplements from the tricarboxylic acid cycle. This failure was not due to lack of enzyme (4), Fig. 6 because it was not present in wild type cells that grew with pyruvate. The missing enzyme was phosphopyruvate synthase (reaction (5)) which was discovered by Kornberg [2] and his colleagues as a result of investigating the mechanism by which pyruvate is converted into phosphoenolpyruvate by *E. coli*. From these experiments it is evident that when *E. coli* grows with pyruvate as a sole source of carbon, the oxaloacetate which they need is obtained by the concerted action of reactions (5) and (2) of Fig. 6. Of the other routes that could be suggested reaction (4) is not used because neither the wild type nor the mutants possess this enzyme. However, oxaloacetate might, in theory, be formed by enzymic oxidation of malic acid: this could be synthesized from pyruvate by 'malic enzyme' and, in turn, the phosphoenolpyruvate needed for various biosyntheses could be formed by reaction (1). It is remarkable that both wild type and mutants possess these two enzymes and yet they cannot adequately serve the requirements of the cells for growth with pyruvate. This example shows clearly that although we could construct, on paper, a feasible metabolic pathway from enzymes known to occur in the system which concerns us, we should not be justified in assuming that this would be a route of any significance for the economy of the cell. In short, although the enzymes are present they may not have the physiological function we assign to them. The cells possess phosphopyruvate carboxykinase but this enzyme does not serve to supply oxaloacetate. What, then, is its function? Mutants of *E. coli* have been isolated that lacked this enzyme and in consequence could not grow with tricarboxylic acid cycle compounds as sole sources of carbon, although unlike the mutants previously discussed they were able to grow with glucose or pyruvate [3]. Further, when the wild type organisms from which the mutants arose were grown with succinate the amounts of phosphopyruvate carboxykinase they contained were about ten times higher than when they grew with glucose. It therefore appears that the function of the enzyme is to furnish phosphoenolpyruvate from oxaloacetate when the latter is in abundant supply, as would be the case when succinate, fumarate, malate or α-ketoglutarate —but not pyruvate—serve as carbon sources for growth. In mammalian liver, as we have seen, its function also appears to be that of converting oxaloacetate into phosphoenolpyruvate, but in this case the oxaloacetate appears to be formed directly from pyruvate by reaction (4).

Advantages in the use of microorganisms for the study of metabolic pathways

Since it is relatively easy to isolate mutants, microorganisms have been of great value both for elucidating metabolic pathways and for assessing the physiological significance of particular biochemical reactions. Moreover, essentially the same metabolic sequences are used by a wide range of living organisms; for bacteria, birds, reptiles and mammals need to synthesize the same low molecular weight compounds. Further, the various species utilize essentially the same cellular machinery for accomplishing the incorporation of these compounds into certain macromolecules. Thus it appears that when amino acids are placed in their correct sequence in a protein, events of the same type occur whether the organism is a microbe or a mammal. The information required to solve certain general problems in biochemistry can therefore be obtained just as well from experiments with microbes as from any others, with the advantage that bacteria can be handled in ways that other organisms cannot be. As J.D. Watson [1] has said, *E. coli* is now the most intensively studied organism except for man. Bacteria may be exposed to radiation or mutagenic agents that damage their genetic apparatus, so that the biosynthetic pathways that their genes control are thereby interrupted at specific points: it then becomes possible to identify reaction intermediates which accumulate in cultures of the mutants. Whole bacteria may be frozen solid and then disrupted to release their enzymes. The application of such drastic methods to other species might be unprofitable and would certainly be unlawful. Finally, the degradative and synthetic capabilities of many bacteria are such that they are able to grow rapidly with a wide range of organic compounds as sole sources of carbon in simple mineral salts media. For this reason it is sometimes easier to follow the chemical transformations of a compound that serves as a nutrient for microbes than it is to follow those for the complex diets of higher organisms.

Some limitations to the use of intact bacteria

For many years the only methods available for the study of microbial metabolism were those which depended on painstaking analysis of fermentation products. Many valuable results were thereby obtained, notably for the alcoholic fermentations of glucose by yeast whereby the foundations were laid of our understanding of glycolysis; or by such studies as those of H.G. Wood and C.H. Werkman [2] who observed that propionic acid bacteria growing on glycerol actually used sizable amounts of carbon dioxide as substrate. This work established that heterotrophic as well as autotrophic organisms could fix carbon dioxide; a fact which, if not entirely unacceptable, was at least surprising at the time the observation was made. However, the general limitations of experiments which rely upon determining the overall conversion of nutrients into products of fermentation were well expressed in 1930 by M. Stephenson [1]. 'We are indeed in much the same position', she wrote, 'as an observer trying to gain an idea of the life of a household by a careful scrutiny of the persons or materials arriving at, or leaving the house: we keep an accurate record of the foods and commodities left at the door and patiently examine the contents of the dustbin (garbage can) and endeavour to deduce from such data the events occurring within the closed doors'.

Further, it is evident that reaction intermediates formed during the degradation of a source of carbon may not appear in metabolism fluids: they may never leave the cells because they cannot pass through membrane barriers, or because they are formed and then degraded so quickly that detectable amounts do not accumulate. If a compound can in fact be detected and is also a true intermediate in degradation it is clear that we can expect its concentration to increase to a maximum and then decrease as the substrate from which it originated disappears. The generalization might be made that the easier it is to isolate a compound from metabolism fluids, the greater the caution to be exercised before we assign to it the status of a degradative intermediate; for such a role implies rapid removal as well as rapid formation. For example, gluconic and 2- and 5-ketogluconic acids have been proposed as intermediates of a main pathway for glucose oxidation in *Acetobacter suboxydans*. This organism produces copious amounts of 5-ketogluconic acid that crystallize out of the medium as the calcium salt. However, J. De Ley and A.J. Stouthamer [2] have presented strong evidence that gluconic and ketogluconic acids are formed by a side pathway. When there is an abundance of reduced NADP, the specific and reversible dehydrogenases that these workers isolated no doubt effect a reduction of the ketoacids. The resulting gluconate is then phosphorylated and after conversion into ribulose-5-phosphate it is oxidized through the hexosemonophosphate cycle which appears to be the main oxidative process of *Acetobacter*.

Induction of enzymes in microorganisms: the technique of 'Simultaneous Adaptation'

Fortunately, the use of whole bacterial cells for studies of degradative pathways is not confined to chemical analyses of their metabolism fluids. When bacteria are removed from, say, a nutrient broth culture and are placed in a new mineral salts medium that contains a compound A not present in the broth, growth does not resume immediately even though the cells are able eventually to utilize A as a sole source of carbon. They must first synthesize enzymes that degrade A fast enough to supply the demands of growth for energy and material. If the bacteria are aerobic it is probable that A will be converted into metabolites of the tricarboxylic acid cycle and, as we have seen, this conversion will be accomplished through a reaction sequence that occurs in many enzymic steps. When growth at the expense of A eventually occurs we shall therefore expect the bacteria to contain not one 'new' enzyme but several that also catalyse the breakdown of reaction intermediates B, C and D in the sequence:

$$A \rightarrow B \rightarrow C \rightarrow D \rightarrow \rightarrow \text{(tricarboxylic acid cycle metabolites)}$$

We may say, therefore, that bacteria which have grown with A are also adapted to oxidize B, C and D. None of these compounds would be oxidized readily by cells taken from the nutrient broth before they had been exposed to A. Accordingly, if a washed suspension of bacteria grown with A as the sole source of carbon is shaken with A, B, C or D in a Warburg respirometer the rates of uptake of oxygen would probably be as shown in Fig. 7. Compounds B, C and D would be oxidized at about

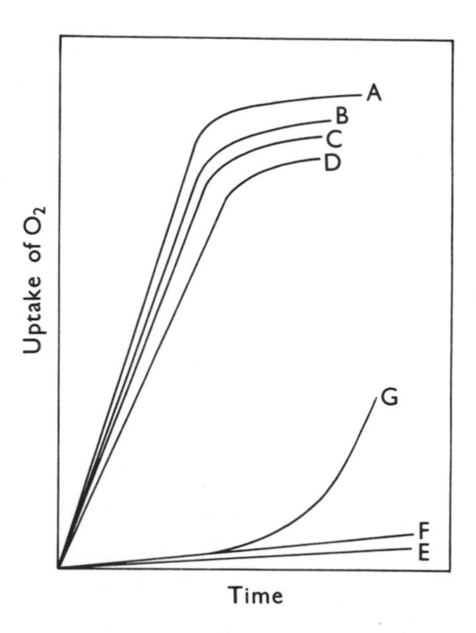

Fig. 7. The ability of a suspension of bacteria, grown with A as carbon source, to oxidize various compounds. E shows the uptake of oxygen when no substrate is added to the respirometer vessel. Compounds B, C and D may be intermediates in the breakdown of A; compounds G and F are probably not intermediates.

the same rate as the growth substrate A; but compounds F and G, which are not members of the metabolic sequence, would either not be oxidized any faster than the endogenous stores of material within the cells, or else a rapid rate of uptake of oxygen would only be achieved after additional enzymes have been synthesized in response to the added compound (curve G, Fig. 7). This experimental approach which was developed by R.Y. Stanier for a study of the bacterial degradation of benzenoid compounds enables us to write down a feasible reaction sequence for a metabolic pathway and to test the speculation by measuring initial rates of oxidation: compounds that are oxidized as readily as the growth substrate are probable intermediates in its degradation, and those that are poorly oxidized are not. The technique is subject to limitations which were clearly defined by R.Y. Stanier [1] but which have not always been heeded. Thus, a compound may be a reaction intermediate and yet whole cells may fail to oxidize it: the relevant enzymes are present inside the cells but the compound may fail to gain access to them because it cannot penetrate a cell membrane. Moreover, a compound may be oxidized readily when it is not an intermediate in the postulated reaction sequence: active enzymes for its metabolism may be present no matter what the growth substrate happens to be. Such a possibility can clearly be tested by studying rates of oxidation for suspensions of cells grown with substrates other than A as sources of carbon. This method was termed the technique of 'Simultaneous Adaptation' when its scope and limitations were first outlined by R.Y. Stanier. Later, the name 'Sequential Induction' was suggested [1] and has been adopted by many workers in the field. At the time when the change of name was proposed it was thought that the synthesis of a set of degradative enzymes took place as follows: the first enzyme was induced when the cells were exposed to A, so that B then accumulated; exposure to B led to induction of the second enzyme so that C accumulated; and so on, until all the enzymes needed for the degradation of A had been elaborated. However, we now know that enzymes are not always sequentially induced: they may be formed together, in batches, instead of one after the other. Systems are also known for which B is the inducer of the enzyme that attacks A; and further, it is generally agreed that the term 'derepressor' is often more apt than 'inducer' since the effect of exposing the genetic apparatus to B may be for the compound to remove a repressor that prevents the enzyme-synthesing machinery from functioning. The term 'Sequential Induction' is not satisfactory because it purports to describe the mechanism by which cells equip themselves to utilize a new growth substrate. 'Simultaneous Adaptation' merely indicates that when cells are able to oxidize A to completion, they are at the same time capable of oxidizing B, C and D. 'Adaptation' has an old and well-established biological meaning, denoting modifications of structure or function which increase fitness. It is true that enzymes can be induced by compounds that do not serve as metabolites, and that when this happens the organism is made less fit, rather than more fit, by synthesizing protein for which it has no immediate use. Nevertheless, despite exceptions which we shall mention, it is the general rule for bacteria to synthesize in large amounts only those enzymes that will be utilized when the cells are exposed to a new degradable metabolite. Since they are in consequence more fitted to their new environment, the term 'Simultaneous Adaptation' seems adequate to describe the technique by which the enzymic readjustment is investigated.

Difficulties in interpreting rates of oxidation of substrates by intact cells

Measurements of rates of oxidation have been a valuable aid in determining metabolic pathways in bacteria; but the need for cautious interpretation is evident when we consider the place assigned to the tricarboxylic acid cycle in bacterial metabolism over the last two decades. Although it is now assigned a central role in the metabolism of aerobes, in the late 1940's it was not believed to operate in these organisms. The evidence against its operation seemed impressive at that time. Thus, consider the reaction sequence:

$$\alpha\text{-ketoglutarate} \rightarrow \text{succinate} \rightarrow \text{fumarate} \rightarrow \text{malate} \rightarrow \text{pyruvate} \rightarrow \text{acetate}$$

It was found that *Azotobacter agilis*, for example when grown with α-ketoglutarate would readily oxidize all the compounds in this sequence. When grown with succinate, however, the progress of oxidation of all the compounds except α-ketoglutarate followed curves such as those for A, B, C and D of Fig. 7, whereas α-ketoglutarate was oxidized only after a lag period, as for compound G of Fig. 7. When acetate was the growth substrate, only acetate was oxidized readily and the oxygen uptake curves for succinate, fumarate and α-ketoglutarate all resembled curve G of Fig. 7. From this evidence it was eminently reasonable at the time (1948) to conclude that the tricarboxylic acid cycle did not operate in *A. agilis*.

Later work, however, provided an alternative explanation of these results [1]. When cell free extracts of *Azotobacter* were prepared, citrate and α-ketoglutarate were oxidized rapidly, although whole cells did not oxidize citrate at all and α-ketoglutarate was oxidized only after a lag period. Moreover, whereas acetate by itself was oxidized poorly by cell extracts, the rate was considerably increased when fumarate was added, the effect being much greater than could be accounted for by adding together the oxidation rates of the two compounds considered separately. It therefore appeared that the enzymes of the tricarboxylic acid cycle were present inside the cells but were inaccessible to certain metabolites placed outside them. Nevertheless it appeared that new enzymes were formed when α-ketoglutarate was presented to whole cells grown with succinate, since the rate of oxidation of the substrate accelerated during the course of the experiment. It is now considered that these enzymes are those concerned with the transport of the substrate across the cell membrane: they have been termed 'permeases', although it should be mentioned that the concept of specialized enzymes solely concerned with substrate transport has received criticism [2].

In the early 1950's further evidence was presented that some of the reactions of the tricarboxylic acid cycle did not operate in aerobic bacteria, and alternative reactions for oxidizing acetate were suggested, notably that proposed by T. Thunberg in 1920:

$$2 \text{ acetate} \rightarrow \text{succinate} + 2\text{ H}$$

If it were possible to initiate acetate oxidation in this manner, one molecule of succinate could then be converted by known reactions into one molecule of acetate, so that a 'dicarboxylic acid cycle' could operate in which one molecule of acetate is lost by each turn. Citrate and α-ketoglutarate would not be involved in this supposed cycle. Alternative oxidative pathways for acetate in microorganisms were indeed established later but they do not employ the 'Thunberg condensation', and there is no firm evidence that this reaction occurs. The experiments purporting to show that the tricarboxylic acid cycle did not operate were again performed with nonproliferating suspensions of bacteria. They were allowed to oxidize acetate labelled in the methyl group with ^{14}C: unlabelled α-ketoglutarate and succinate were then added, re-isolated and examined for radioactivity. It was argued that if these compounds were reaction intermediates then the added 'carrier' would mix with the labelled α-ketoglutarate (or succinate) formed as the radioactive acetate was oxidized and on re-isolation would themselves be radioactive. When *M. lysodeikticus* was used, both α-ketoglutarate and succinate were indeed found to be labelled on re-isolation, but the specific activities of the carbon atoms of α-ketoglutarate were lower than those of succinate. This result is not to be expected if α-ketoglutarate is the immediate precursor of succinate, as it is in the tricarboxylic acid cycle. Moreover, the specific activity of the respired carbon dioxide was many times greater than that of the carbon atoms in the carboxyl groups of both compounds; again, carbon dioxide originates from these groups in the reactions of the cycle. Results with *E. coli* also argued against the tricarboxylic acid cycle and favoured a 'Thunberg condensation' of acetate to give succinate. Very little label from acetate was incorporated into carrier α-ketoglutarate, thereby suggesting that this compound is not an obligatory intermediate in acetate oxidation; whereas there was good incorporation into the methylene groups of succinate as the 'Thunberg condensation' requires.

These experiments were shown by L.O. Krampitz and his colleagues [1, 2] to be misleading. The basic assumption that an added metabolite will mix freely with the same compound inside the cell was shown to be invalid. It is probable that succinate and α-ketoglutarate are bound to enzymes or coenzymes when they are formed and decomposed inside the cell, and accordingly their ability to mix with the added carrier compounds is extremely limited. This was shown by extracting the small amounts of metabolites actually present inside the bacteria. Large quantities of cells were naturally required for this purpose: for *E. coli* 80 g. (wet weight) were used in each experiment and for *M. lysodeikticus* amounts of lysed cells corresponding to 20 g. (dry weight). In these experiments the concentration of labelled acetate was sufficient to allow oxidation to proceed at half its maximum rate and the reaction time was kept short. In this way the free pool of isotopic acetate was reduced to a minimum and recycling of label was avoided. The cells of *E. coli* were disrupted by grinding with powdered glass, the enzymes and their substrates were thereby released and the labelled tricarboxylic acid cycle metabolites were separated by acidification and chromatography on columns of celite. The results were quite different from those obtained in the 'carrier' type experiments. Label was distributed quite uniformly between all the components of the tricarboxylic acid cycle examined, including α-ketoglutarate and succinate, whilst the specific activities of the carbon atoms of these compounds, and that of the respired carbon dioxide, were those expected from the operation of the tricarboxylic acid cycle. Similar results were obtained for suspensions of *M. lysodeikticus* which were submitted to enzymic lysis rather than mechanical disruption.

Some factors to be considered when using cell-free extracts

However, we should do well to remind ourselves of the analogy of D.E. Green (page 10) before we conclude that experiments with whole organisms are inherently faulty whereas those with isolated enzymes are automatically beyond reproach. From the examples discussed it is clearly prudent to supplement evidence supplied by Simultaneous Adaptation with observations on disrupted cells and the enzymes they contain. But caution must also be exercised in work with cell-free extracts for two reasons: (*a*) bacteria adapted to use one particular substrate will certainly contain relatively high concentrations of the enzymes concerned in its degradation, but those relating to other degradative pathways will not necessarily be absent, and (*b*) enzymes for other pathways may have been derepressed at the same time as those concerned with the degradation of the growth substrate. To illustrate, we shall consider the degradation of benzoic acid and *p*-hydroxybenzoic acid by *Pseudomonas putida*.

L.N. Ornston and R.Y. Stanier [1] have purified the enzymes for these pathways which are shown in Fig. 8. These metabolic routes are entirely separate and distinct. Thus, Ornston [2] crystallized enzymes 2′ and 3′ and established that they are specific for their substrate: enzyme 2′ lactonizes *cis-cis*-muconate but not β-carboxy-*cis-cis*-muconate, while enzyme 3′ isomerizes (+)-muconolactone but does not attack γ-carboxy-muconolactone. However, before this work was undertaken it was believed that (+)-muconolactone was a metabolite common to both pathways [3], β-carboxy-*cis-cis*-muconic acid being thought to decarboxylate and lactonize to give rise to this compound as shown by the dotted line of Fig. 8. The evidence was obtained using *Moraxella lwoffi* and rested upon the following observations: (*a*) heat-treated extracts of cells grown with *p*-hydroxybenzoate converted β-carboxy-*cis-cis*-muconic acid into (+)-muconolactone, and (*b*) untreated extracts converted (+)-muconolactone quantitatively into β-oxoadipate. The reason why these results were obtained despite the fact that (+)-muconolactone is not an intermediate in this sequence is now clear. Heat treatment completely destroys enzyme 4 but does not affect enzyme 2; hence γ-carboxy-muconolactone is still formed from β-carboxy-*cis-cis*-muconic acid by the treated preparation. Enzyme 3, although affected by the treatment, is still present in sufficient amounts to catalyse the slow conversion of

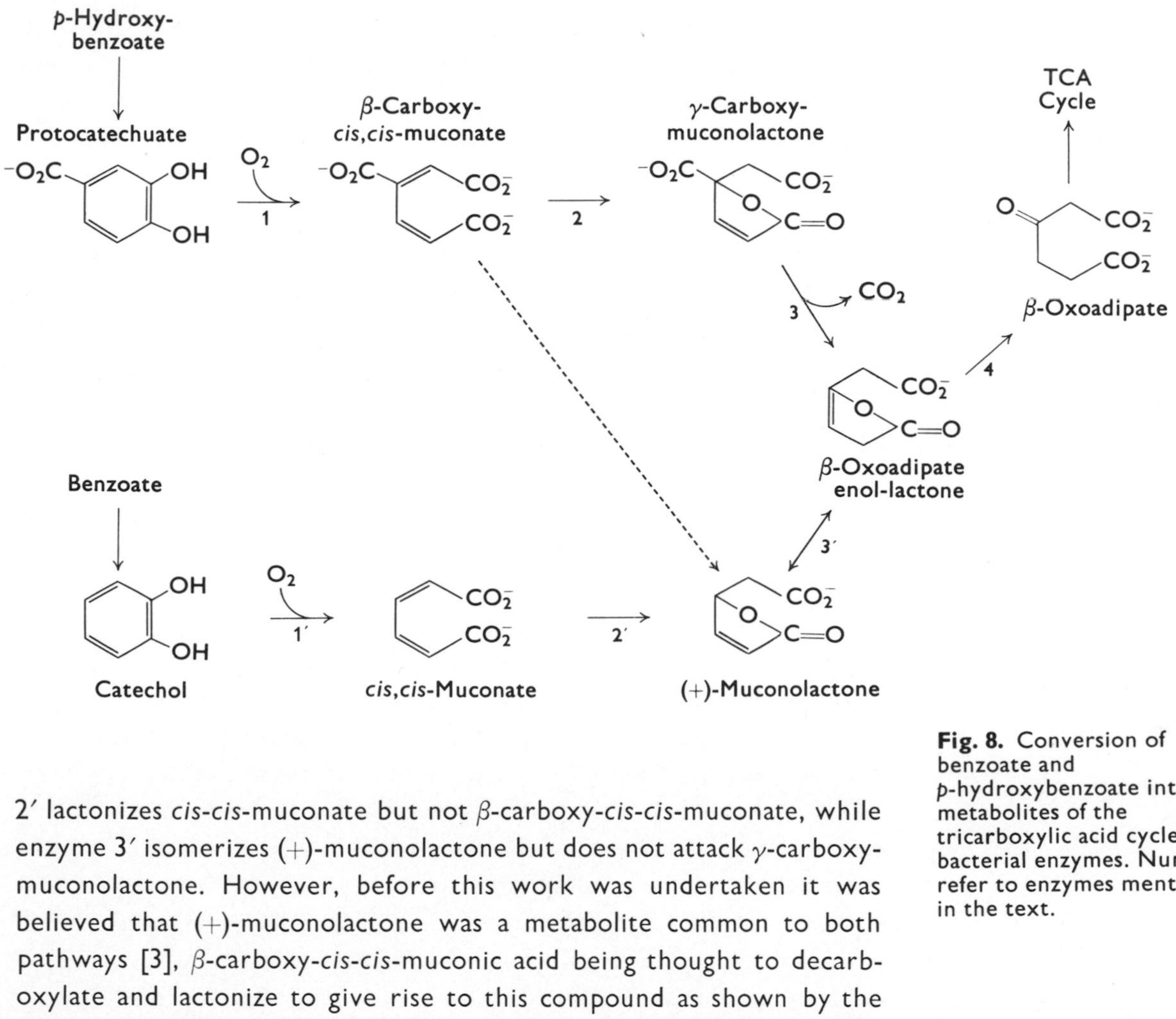

Fig. 8. Conversion of benzoate and *p*-hydroxybenzoate into metabolites of the tricarboxylic acid cycle by bacterial enzymes. Numbers refer to enzymes mentioned in the text.

γ-carboxymuconolactone to β-oxoadipate enol-lactone. The last-named compound cannot be hydrolysed to β-oxoadipate because the relevant enzyme 4 has been destroyed. Now it so happens that although the enzymes concerned in the degradation of protocatechuate are present in high concentration in *p*-hydroxybenzoate-grown cells, those concerned with catechol degradation are not entirely absent. Enzyme 3′ is present in sufficient amounts to catalyse the conversion of β-oxoadipate enol-lactone into (+)-muconolactone: the reaction is reversible and the equilibrium is strongly in favour of (+)-muconolactone formation. Accordingly this compound accumulates by the action of heat-treated extracts upon β-carboxy-*cis-cis*-muconic acid although it does not lie in the pathway of degradation. Furthermore, crude extracts of bacteria grown with *p*-hydroxybenzoic acid contain sufficient amounts of enzymes 3′ and 4 to catalyse its conversion into β-oxoadipate. However, one vital measurement was lacking before (+)-muconolactone could be placed with certainty on the pathway of degradation of protocatechuate. It is not sufficient to show that the compound is formed by extracts and is metabolized to β-oxoadipate. It must also be shown that the rates of these reactions are sufficient to account for the *overall* rate of conversion of β-carboxy-*cis-cis*-muconic acid into β-oxoadipate; and this is not the case: the overall conversion by crude extracts of *p*-hydroxybenzoate-grown cells proceeds much faster than the conversion of (+)-muconolactone into β-oxoadipate. Accordingly, (+)-muconolactone cannot be a reaction intermediate in this pathway.

Neither does the presence in a cell-free extract of an enzyme in large amounts necessarily prove its involvement in a particular sequence. It has been mentioned (page 30) that the product of enzymic action may serve as derepressor of that enzyme. Such is the case for enzyme 4 of Fig. 8 which, in *Ps. putida*, is derepressed by β-oxoadipate [1]. It might be objected that cells cannot decompose benzoate to β-oxoadipate unless enzyme 4 is present, and complete degradation of the growth substrate will therefore not be possible if the synthesis of one of the enzymes must await the accumulation of the end product of the sequence. But it will be recalled that induced enzymes are not entirely absent from the bacteria before they become adapted, and when they are exposed to benzoate there is a slow but significant formation of β-oxoadipate which is enough to start the derepression of enzyme 4. A further complication now arises. In *Pseudomonas putida* enzymes 2 and 3 are coordinately derepressed along with enzyme 4. As a consequence, when the bacteria synthesize the high concentration of enzyme 4 which they require to operate the benzoate pathway, they cannot help but synthesize large amounts of enzymes 2 and 3 at the same time. This means that enzymes for the 'wrong' pathway are induced by growth with benzoate, and without this knowledge one might well conclude that enzymes 2 and 3 function for the degradation of benzoate. It is possible that events of this type are more common than has been recognized hitherto: thus, catechol 2,3-oxygenase is derepressed when a strain of *Achromobacter* grows with β-phenylpropionate [2] although catechol is not an intermediate in the degradation of the growth substrate and the enzyme is presumably not used.

Incorporation of ^{14}C into actively-dividing bacteria: the technique of 'Isotope Competition'

Whereas non-proliferating suspensions of bacteria have been used for studying degradative pathways, cells that are actively dividing have proved valuable for tracing the flow of carbon into cell constituents and hence for elucidating biosynthesis. The early recognition that the tricarboxylic acid cycle functions for biosynthesis in *E. coli*, as well as for oxidation, owed much to a group of workers at the Carnegie Institution of Washington [1] who applied the technique of 'isotope competition' which is based on the following considerations. Consider the biosynthesis of an amino acid (X) by a culture growing at the expense of glucose (G)

$$G \rightarrow A \rightarrow B \rightarrow C \rightarrow \rightarrow \rightarrow X$$

If the growth substrate, or a product of its catabolism such as acetate, is labelled with ^{14}C, X will become labelled and will be incorporated into protein. If a non-radioactive amino acid A, which is a precursor in the pathway, is added to the growing culture the unlabelled carbon atoms will be incorporated into X and its specific activity will decrease. On the other hand, if unlabelled X is added, the specific activity of X will decrease but that of A or B and C will not. By making additions of ^{12}C-amino acids to a culture which is biosynthesizing protein from a source of ^{14}C it should therefore be possible to decide which amino acids are precursors and which are synthesized later. This reasoning is based upon an assumption which is known to be valid provided the bacteria are actively dividing, namely that the flow of carbon is essentially in one direction, from A to X, and is not reversed. That is, the ^{14}C arrives at X and stays there when X is incorporated into protein: the latter does not break down again into its constituent amino acids. This assumption would not be justified for non-proliferating bacteria, nor for other organisms where, as we have seen, the labelled carbon atoms and those from the added ^{12}C-amino acids would probably be distributed through the metabolic network so thoroughly as to introduce numerous complications.

The technique of isotope competition requires that the proteins of the bacteria be hydrolysed and their constituent amino acids be separated and radio-autographed. In consequence this method has been applied less extensively than the more direct procedures of 'rapid sampling' or 'pulse labelling' described later. Nevertheless the results published in 1955 [1] added considerably to our knowledge of biosynthetic processes in *E. coli*. In brief, it will be seen from Fig. 9 that the amino acids of this organism fell into two groups which were termed the 'aspartate family' and the 'glutamate family'. Thus, addition of ^{12}C-glutamate depressed the labelling

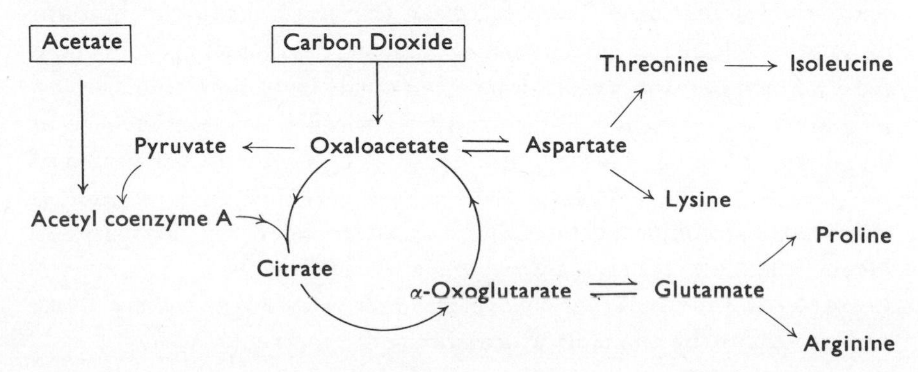

Fig. 9. Biosynthesis of some amino acids via the tricarboxylic acid cycle.

of proline and arginine in the protein of bacteria as they incorporated ^{14}C, while the addition of ^{12}C-aspartate depressed the specific activities of lysine and isoleucine. On the other hand, added ^{12}C-threonine depressed the specific activities of threonine and isoleucine but not that of aspartate, as would be expected for a reaction sequence in which aspartate is a precursor of threonine and threonine, in turn, is a precursor of isoleucine. Comparisons of the patterns of incorporation of ^{14}C-acetate and $^{14}CO_2$ accorded with the view that the tricarboxylic acid cycle operates to provide the source of protein precursors in *E. coli*. It has been pointed out (p. 19) that when acetate enters the cycle, both of its carbon atoms are incorporated into citrate and thence into α-ketoglutarate; whereas on this first turn of the cycle it is the oxaloacetate which provides the carbon dioxide released, and acetate which supplies two of the carbons of oxaloacetate used for the second turn. Accordingly when ^{12}C-glutamate was added to a culture growing with ^{12}C-glucose and incorporating ^{14}C-acetate, there was a rapid exchange of the carbon skeleton between glutamate and α-ketoglutarate with the result that radioactivity in the source of aspartate carbon (oxaloacetate) was diluted. That is, ^{12}C-glutamate depressed labelling in the aspartate family of amino acids whereas ^{12}C-aspartate had less effect upon the specific activities of members of the glutamate family. In contrast, addition of ^{12}C-aspartate caused labelling in the glutamate family to be depressed when the sole source of radioactivity was $^{14}CO_2$. In this case, $^{14}CO_2$ was fixed into oxaloacetate as it entered the cycle: oxaloacetate then exchanged its carbon skeleton with added aspartate before the label found its way into glutamate. Additions of ^{12}C-glutamate also depressed labelling in the aspartate family when the source of radioactivity was carbon dioxide.

The full significance of one aspect of this research was only appreciated later, namely the completeness with which additions of an amino acid such as threonine could remove radioactivity from the ^{14}C-threonine of the cell protein. The effect could not be ascribed entirely to dilution of the label, for in the presence of exogenous threonine the pathway from glucose to threonine appeared to be immediately and completely blocked. We now appreciate that the effect of a build-up of threonine in excess of the immediate requirements of the growing bacteria will be to inhibit allosteric enzymes that function in the biosynthetic pathway for threonine. Likewise an elevated concentration of isoleucine arising from the addition of this amino acid will exert end-product inhibition upon L-threonine deaminase: α-ketobutyrate will not be formed and biosynthesis of isoleucine will stop. It is clear, therefore, that the mechanism by which additions of certain amino acids depress labelling is not due simply to competition with isotope when unlabelled amino acid is drawn into a one-way stream of radioactivity. Rather, it is a consequence of feed-back inhibition of biosynthetic enzymes by which the cell is able to integrate metabolism with growth processes, so that when the concentration of, say, threonine increases then a brake is applied to its biosynthesis. Nevertheless, the technique provides valid information, because control is exerted in such a manner that reactions are slowed down only by the end-product or by one of its precursors in the metabolic pathway.

Studies of the incorporation of ^{14}C into growing cultures over short intervals of time

During the last decade attention has shifted from measurement of the incorporation of ^{14}C into end-products of growing microorganisms, such as proteins, and has concentrated upon the identification of transitory reaction intermediates in metabolic pathways. Much of the credit for the development of methods for identifying such intermediates is due to M. Calvin, J.A. Bassham and their colleagues as described in two authoritative and eminently readable monographs [20.3, 1]. Briefly, these compounds may be identified as spots on two-dimensional chromatograms by the procedure of A.J.P. Martin and R.L.M. Synge. Those spots that have become radioactive by incorporation of ^{14}C may be located by placing a thin sheet of X-ray film in contact with the two-dimensional chromatogram, for when the film is subsequently developed a black spot appears on the film at each site occupied by a radioactive compound. The amount of radiocarbon in each compound may also be determined by first placing a Geiger-Muller tube with a thin window over the radioactive compound on the paper and then counting the β-particles emitted. This technique, which is called 'radioautography', was also used by the group of workers at the Carnegie Institution of Washington, as was mentioned earlier. For accounts of the 'photosynthetic carbon reduction cycle' the general reader is referred to the monographs [20.3, 1] and to an article by J.A. Bassham [2]. We shall merely present an outline to facilitate a discussion of the methods as they have been applied to this metabolic pathway and others.

The cycle consists of four stages. (*a*) Ribulose-1,5-diphosphate is carboxylated to give two molecules of 3-phosphoglyceric acid (PGA). (*b*) PGA is reduced to triose phosphate (3-phosphoglyceraldehyde). (*c*) A series of reactions, which essentially entails the rearrangement of carbon atoms, converts five triose phosphate molecules into three pentose phosphate molecules. (*d*) The pentose phosphate molecules are phosphorylated to yield ribulose-1,5-diphosphate, which then re-initiates the cycle by combining with carbon dioxide as in (*a*). The reactions of stages (*b*) and (*c*) are encountered in glycolysis and the pentose phosphate cycle and they occur in many tissues; whereas the enzyme for (*a*), ribulose-diphosphate carboxylase, and that for (*d*), phosphoribulokinase, appear to function specifically for tissues that utilize carbon dioxide as their sole source of carbon for growth.

Most of the experiments of Bassham and Calvin were performed with the unicellular green algae, *Chlorella pyrenoidosa* and *Scenedesmus obliquus*. When a suspension of either organism in a solution containing the necessary mineral salts was illuminated and aerated, rapid photosynthesis occurred. The air stream usually contained 1% or 4% of carbon dioxide in equilibrium with bicarbonate ions in solution, and on addition of $H^{14}CO_3^-$ incorporation of radiocarbon into cell constituents began. At various times after the addition, samples of the culture were plunged into hot ethanol: the cells were killed, photosynthesis stopped immediately and the compounds extracted from the algae were submitted to two-dimensional chromatography and radioautography. When the algae were exposed to $H^{14}CO_3^-$ for a relatively long period, say 60 sec., many labelled compounds were identified including hexosemonophates, dihydroxyacetone phosphate, ribose phosphate and PGA; but when exposure occurred for a short time of 5 sec. or less, essentially all of the radioactivity appeared in PGA. This observation accords with the requirements of the cycle that the first stable intermediate arising from carbon dioxide fixation will be PGA. Further, when the light was turned off after the cells had photosynthesized for a relatively long period, it was found that label was rapidly lost from ribulosediphosphate and was gained by PGA just as quickly as it was lost. Turning off the light had stopped the production of ATP and NADPH required to reduce PGA to triose phosphate, while the disappearance of ribulosediphosphate was due to its conversion into PGA by the carboxylation reaction. The evidence for the photosynthetic carbon reduction cycle, obtained by the use of radiocarbon, was later reinforced by isolating the various enzymes, including ribulosediphosphate carboxylase, from spinach leaves and algae.

When we examine this approach to the determination of metabolic pathways by tracing the incorporation of radiocarbon into metabolites it becomes evident that certain experimental conditions should be observed, and possible limitations of the method are also apparent. During the steady and continuous operation of a metabolic pathway, the concentrations and amounts of the various biochemical intermediates adjust to constant values. The attainment of a constant 'pool size' for a particular metabolite is assured when its rate of formation is balanced by the rate or removal as a 'steady state' is established. Such a regulation of metabolic processes is achieved by microbes when cell division occurs at a steady rate in an environment that does not alter. When these conditions are observed, radioactivity will enter the first metabolic pool and the specific activity of the metabolite will rise until it is equal to that of the source of isotope to which the cells are exposed. Meanwhile, isotope will be entering the second pool and the same constant activity will shortly be attained, although the *amount* of radioactivity will depend upon the size of the pool. It is clear that measurements of radioactivity may reveal this sequence of events, but only when pool sizes remain constant. It is also evident that a compound may serve as a reaction intermediate and yet the kinetics of its formation and removal may be such as to provide a very small metabolic pool: in consequence, the amount of radioactivity that remains in the pool may be so small that the compound may escape detection on chromatograms. On the other hand, a labelled compound which is not a member of the reaction sequence under consideration might appear at early times because it has been formed from such a member by a very rapid reaction. Thus, oxaloacetate might be the direct intermediate and yet aspartate might be revealed by radioautography. A rapid exchange of its carbon skeleton with that of aspartate, present in a larger metabolic pool, might well have taken place. The evidence for a metabolic pathway provided by isotopic studies is usually supported by isolating the enzymes that are postulated, and by showing that their amounts are sufficient to account for the observed overall rates of reaction.

In a reaction sequence $A \rightarrow B \rightarrow C \rightarrow$ products, compound A contains initially all of the radioactivity entering the cells; but as the specific activity of A increases to its maximum value, radioactivity is also entering B and C in turn. Accordingly, if the radioactivity present in each compound is plotted as a *fraction* of the total radioactivity incorporated, a series of curves is obtained such as those shown in Fig. 10. Curves for the earliest compounds in the sequence will show negative slopes at early times, declining asymptotically to values determined by the pool sizes of the compounds concerned. The experimental methods used for bacteria are essentially those developed by Calvin and Bassham for algae: samples are removed from cultures which are dividing exponentially, and these are discharged into hot ethanol. The total radioactivity in the ethanol-soluble fraction is determined and shown to increase linearly over the short time of the experiment; and the radioactivity incorporated into each compound is then measured on chromatograms.

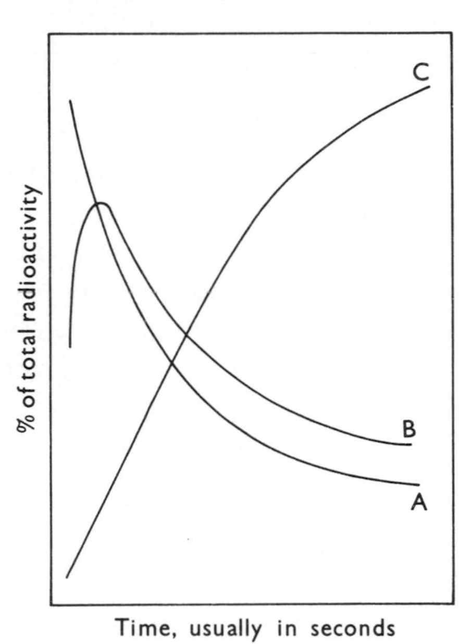

Fig. 10. Incorporation of radioactivity into cell constituents. Sampling for analysis is started as soon as possible after the growing culture has been exposed to the radioactive source of carbon. Compound A is the first detectable intermediate arising from the carbon source; B is also formed at early times, but later than A; and C arises later than B.

The importance of supplementing such experiments with observations of the enzymes involved in the sequence is illustrated with ribulosediphosphate carboxylase. This enzyme has a high Michaelis constant (a low affinity) for carbon dioxide and this has always been a disturbing feature in its postulated role in photosynthesis. Under natural conditions, plants are able to utilize the very low concentrations (0·03%) of carbon dioxide in the atmosphere for photosynthesis and growth, and accordingly this carboxylation reaction, if used, must operate at only a small fraction of its maximum velocity. M.D. Hatch and C.R. Slack [1] have re-examined the radioactive products formed from $^{14}CO_2$ under steady-state conditions for photosynthesis and at a light-intensity and concentrations of carbon dioxide as close to physiological conditions as possible. Under these conditions it does not appear that the initial carboxylation reaction is that catalysed by ribulosediphosphate carboxylase. Instead, the first product of carbon dioxide fixation is either malate or oxaloacetate, with aspartate probably formed by a side reaction from these products. Plots of the percentage of total radioactivity incorporated into metabolites, such as those of Fig. 10, indicate that the ^{14}C then moves from the C_4 dicarboxylic acid pool to PGA and subsequently to sucrose and a glucan via the hexose phosphates. One suggestion made to account for the prior fixation of carbon dioxide into C_4 acids is that oxaloacetate might be the donor of carbon dioxide to ribulose-1,5-diphosphate rather than the bicarbonate ion or carbon dioxide itself. Initially, carbon dioxide may be fixed on to pyruvate by the action of the 'malic enzyme' (Table 1), and the malate so formed may be oxidized to oxaloacetate; or alternatively oxaloacetate could be formed directly by fixation of carbon dioxide on to phosphoenolpyruvate by the enzyme phosphoenolpyruvate carboxylase (Table 1). Oxaloacetate is known from the work of H.G. Wood and his colleagues [1] to participate in transcarboxylation reactions, whereby a carboxyl group is inserted into an acceptor molecule without the release of free carbon dioxide. It is possible, but not proved, that ribulose-1,5-diphosphate may serve as an acceptor in such a reaction. The work of Hatch and Slack was performed with sugar-cane leaves, but it seems unlikely that the pathway they suggest is confined to this system. The physiological significance of ribulosediphosphate carboxylase is therefore not clear, at least in some organisms. The enzyme is present in various chemosynthetic autotrophs as well as in plants: the former organisms may encounter concentrations of carbon dioxide which permit the enzyme to operate at its maximum velocity. However, it might be mentioned that the concentration of carbon dioxide in air is not necessarily optimal for the growth of an organism; thus the growth rate of *Aerobacter aerogenes* in a mineral salts medium continues to increase at far higher concentrations than this [2].

Investigation of the glyoxylate cycle

A notable application of the technique of rapid sampling in isotopic studies was that of H.L. Kornberg [3] which led to the discovery of the glyoxylate cycle. Many species of microorganisms that employ the tricarboxylic acid cycle for biosynthesis of cell constituents are able to grow with acetate as sole source of carbon. The biochemical problem which this entails can be understood if we consider one complete assembly of enzymes for the cycle and imagine them to operate with a limited supply of oxaloacetate. If oxaloacetate is not removed, say by decarboxylation to pyruvate, no difficulties will be encountered in the *oxidation* of unlimited amounts of acetate to carbon dioxide and water because the intermediates of the cycle can operate in catalytic amounts. That is, each molecule of oxaloacetate used up when acetate is incorporated into citrate is ultimately regenerated. But when α-ketoglutarate is removed as glutamate to be used for biosynthesis, this regeneration does not occur:

$$\text{Acetyl coenzyme A} + \text{oxaloacetate} \rightarrow \text{citrate} \rightarrow \rightarrow \alpha\text{-ketoglutarate} \rightarrow \text{glutamate} \rightarrow \text{(biosynthetic reactions)}$$

When this sequence of reactions is repeated, the limited supply of oxaloacetate will become exhausted and the tricarboxylic acid cycle will cease to operate. However, this situation is avoided by the introduction of two enzymes, (*a*) isocitrate lyase and (*b*) malate synthase. As seen in Fig. 11, molecules of acetate can enter a cycle of reactions at the point indicated by 'Acetyl coenzyme A (1)' to provide a supply of glyoxylate without entailing a loss of oxaloacetate; whilst acetate from the growth medium can also enter at the point 'Acetyl coenzyme A (2)' to combine with glyoxylate, formed by the fission of isocitrate, and so provide malate and oxaloacetate. The net result of the reactions of Fig. 11 is therefore to convert two molecules of acetate into oxaloacetate and so replenish deficiencies arising from the withdrawal of α-ketoglutarate for biosynthesis. From Fig. 11 it is seen that acetate entering the sequence at point (1) will be incorporated into citrate, and at point (2) into malate. H.L. Kornberg found that when the percentages of isotope incorporated into citrate and malate were plotted against time, curves were given that had negative slopes, as for A of Fig. 10, whereas the curve for glutamate resembled curve C. Moreover, these pseudomonads growing with acetate incorporated radioactive substrate into malate before succinate: this is in accordance with the cycle of Fig. 11 but is the reverse of what would be

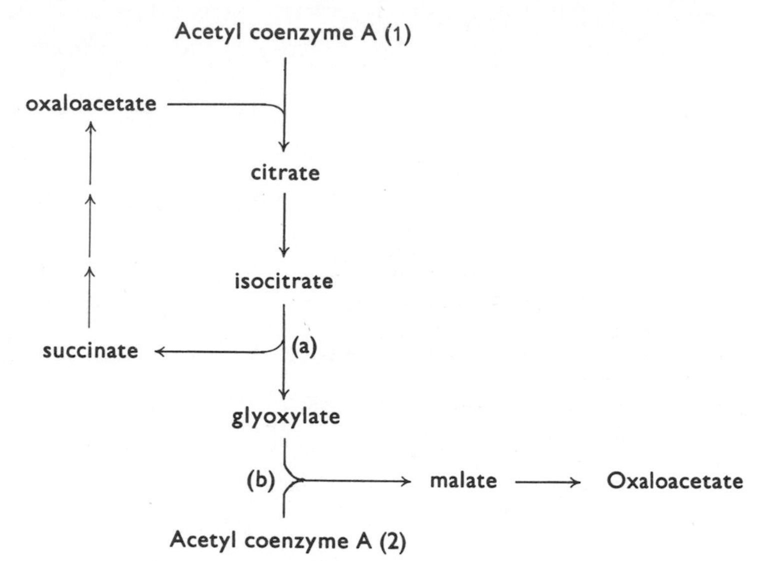

Fig. 11. Points of entry into the glyoxylate cycle (at (1) and (2)) for acetyl coenzyme A in the biosynthesis of oxaloacetate. The reactions indicated by (a) and (b) are respectively catalysed by isocitrate lyase and malate synthase.

predicted from the tricarboxylic acid cycle. Isocitrate lyase had been discovered in 1953 [1] and malate synthase in 1956 [2] but the full significance of these enzymes in metabolism had not been appreciated. It was now shown that their activities in extracts of acetate-grown cells were sufficient to account for the observed rates of growth. Extracts of bacteria grown on substances other than acetate (such as glucose or C_4-acids) contain greatly reduced amounts of isocitrate lyase. For some micro-organisms this change of substrate also results in a decrease in the amount of malate synthase, although for others the decrease is far less than that observed for isocitrate lyase. A good deal of later work has confirmed this 'anaplerotic sequence' (page 25), a description which is given to 'reactions that enable the provision of energy to be maintained under conditions in which the removal of biosynthetic precursors would otherwise interrupt the pathways of energy supply' [25.3]. These reactions are also employed by microbes that utilize long-chain fatty acids for aerobic growth: thus, *Moraxella lwoffi* converts octanoic and decanoic acids into acetate which therefore virtually serves as sole source of carbon [14.4].

Investigation of the glycerate pathway for the bacterial metabolism of glyoxylate

Since the difficulties of growth with acetate are surmounted by the operation of the cycle in Fig. 11 that converts acetate continuously into glyoxylate, the question then arises as to how bacteria can grow when they are supplied entirely with glyoxylate, but not with acetate. These are the circumstances that exist when bacteria grow with glycine [3] glycollate

[1] or oxalate [2] as sole sources of carbon because the initial steps in metabolism entail the conversion of each substrate into glyoxylate by deamination, oxidation or reduction. The metabolic pathway consists of the conversion of glyoxylate into phosphoenolpyruvate by the reactions shown in Fig. 12. Once phosphoenolpyruvate is formed, the anaplerotic requirements of the cell can be met: oxaloacetate can be generated by carbon dioxide fixation, or alternatively phosphoenolpyruvate may be metabolized to pyruvate and thence to acetyl coenzyme A which combines with glyoxylate to give malate. Evidence for the sequence of Fig. 12 was obtained by H.L. Kornberg and A.M. Gotto [1] who showed that when pseudomonads growing with glycollate were exposed to ^{14}C-glycollate for brief periods, all the intermediates of the tricarboxylic acid cycle

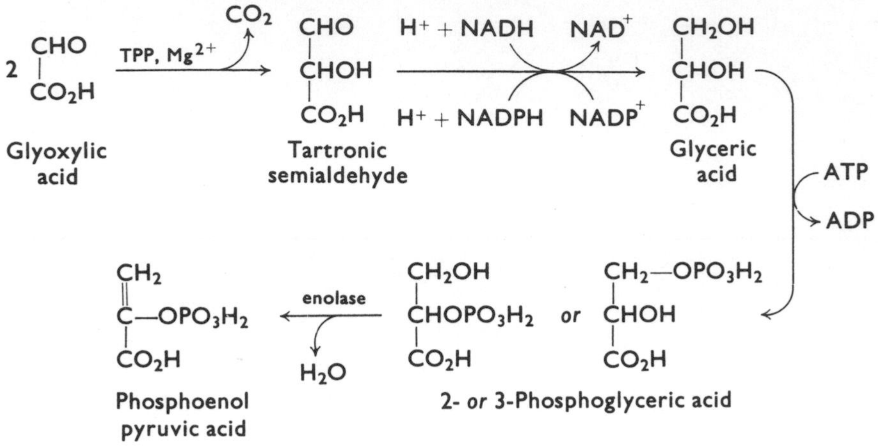

$2\ CHO{-}CO_2H$ (Glyoxylic acid) $\xrightarrow{TPP,\ Mg^{2+};\ -CO_2}$ $CHO{-}CHOH{-}CO_2H$ (Tartronic semialdehyde) $\xrightarrow{H^+ + NADH \to NAD^+ \text{ or } H^+ + NADPH \to NADP^+}$ $CH_2OH{-}CHOH{-}CO_2H$ (Glyceric acid) $\xrightarrow{ATP \to ADP}$ $CH_2OH{-}CHOPO_3H_2{-}CO_2H$ *or* $CH_2{-}OPO_3H_2{-}CHOH{-}CO_2H$ (2- *or* 3-Phosphoglyceric acid) $\xrightarrow{\text{enolase};\ -H_2O}$ $CH_2{=}C(OPO_3H_2){-}CO_2H$ (Phosphoenol pyruvic acid)

Fig. 12. The 'Glycerate Pathway' by which phosphoenol pyruvate is formed from glyoxylate. 2-Phosphoglycerate is the substrate for enolase, but it is not established whether this compound or its isomer arises directly from glycerate by phosphorylation.

acquired radioactivity but that label was incorporated most rapidly into glycine, phosphoglycerate and malate. Since glyoxylate was not detected by the chromatographic procedures used, the early labelling of glycine (which is readily formed from glyoxylate by amination) was taken to indicate the participation of glyoxylate in glycollate metabolism. As we have seen, the rapid labelling of phosphoglycerate might have suggested a metabolic pathway similar to that used by autotrophic organisms, but in this case the incorporation of $^{14}CO_2$ into PGA was negligible. The indications of the pathway obtained from rapid sampling of growing cultures were confirmed by showing that the activities of the enzymes present in cell-free extracts were adequate for the requirements of exponential growth. The enzyme tartronic semialdehyde reductase was later obtained in crystalline form from *Pseudomonas* grown with glycollate [3].

The researches of J.R. Quayle and his colleagues are particularly instructive in showing how the outlines of a metabolic pathway, as indicated by radiotracer studies, may guide the investigator to the discovery of new enzymes and metabolites. *Pseudomonas oxalaticus* rapidly incorporated label from ^{14}C-oxalate into glycine and phosphoglycerate and contained high concentrations of the enzymes required to operate the reaction sequence of Fig. 12 during growth. They also contained high concentrations of formate dehydrogenase. Now the conversion of oxalate into glyoxylate involves reduction of the substrate, and this suggests that formate dehydrogenase operates to provide the reducing power required.

Two problems are then apparent: first, the nature of this reduction and the way that it is coupled to formate oxidation; and second, the nature of the decarboxylation reaction by which oxalate is converted into formate. Untreated cell-free extracts readily decarboxylated oxalate when catalytic amounts of ATP were added, but when partially purified, the extracts showed rather poor and variable activity only when supplemented with ATP and coenzyme A. This observation suggested that an essential factor was lost in purification. The factor was shown to be succinate since extracts containing ATP, coenzyme A, thiamine pyrophosphate and Mg^{2+} ions rapidly decarboxylated oxalate to give formate when only catalytic amounts of succinate were supplied. Synthetic succinyl coenzyme A could

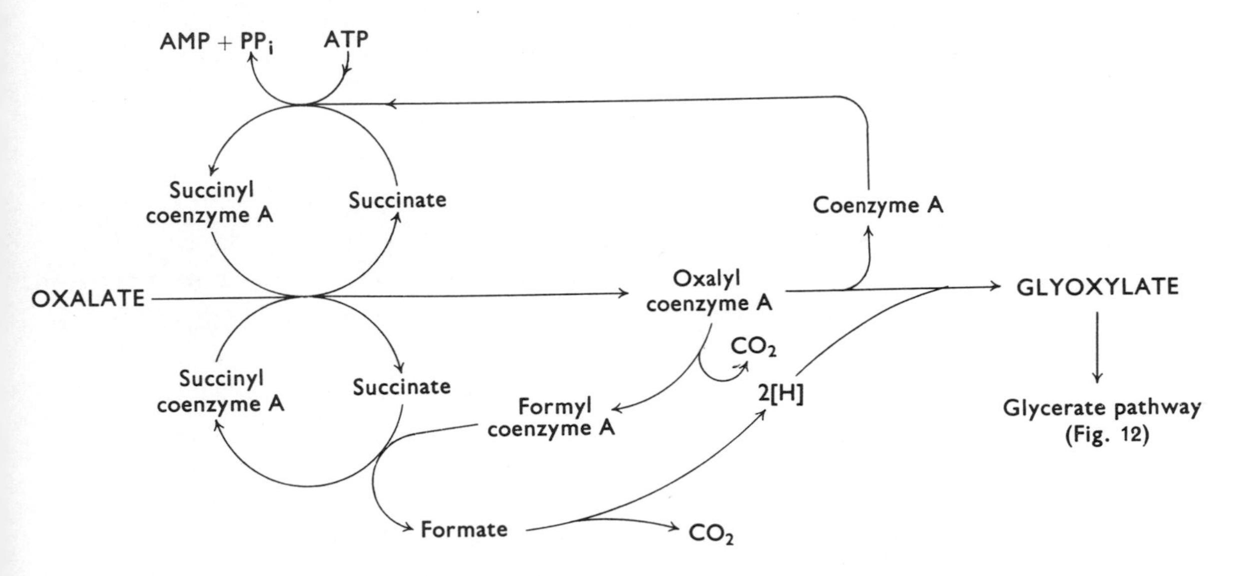

Fig. 13. Bacterial metabolism of oxalate. Hydrogen atoms from formate are transferred to NAD. Energy is provided by reoxidation of some of the NADH, and the remainder is used to reduce NADP to NADPH which initiates the biosynthetic sequence of reactions by reducing oxalyl-coenzyme A to glyoxylate. From Quayle J.R., Keech D.B. and Taylor G.A., (1961) *Biochem. J.* **78**, 225.

replace the mixture of ATP, coenzyme A and succinate. It then became apparent that the function of succinyl coenzyme A was to convert oxalate into oxalyl coenzyme A, the true substrate for the decarboxylation reaction. In this process only catalytic amounts of succinyl coenzyme A will be needed if it can be regenerated by the third of the three reactions shown below:

$$\text{oxalate} + \text{succinyl-SCoA} \rightarrow \text{oxalyl-SCoA} + \text{succinate}$$
$$\text{oxalyl-SCoA} \rightarrow \text{formyl-SCoA} + CO_2$$
$$\text{succinate} + \text{formyl-SCoA} \rightarrow \text{succinyl-SCoA} + \text{formate}$$

Later, Quayle synthesized oxalyl coenzyme A and showed that is was also the substrate for the reaction by which oxalate is reduced to glyoxylate. The enzyme, which was purified, catalyses the following reaction:

$$\underset{\text{Oxalyl coenzyme A}}{\begin{matrix} CO_2H \\ | \\ C{=}O \\ | \\ SCoA \end{matrix}} + NADPH + H^+ \rightleftharpoons \underset{\text{Glyoxylic acid}}{\begin{matrix} CO_2H \\ | \\ C{=}O \\ | \\ H \end{matrix}} + HSCoA + NADP^+$$

The manner in which these various reactions are coupled together by *Pseudomonas oxalaticus* when it utilizes oxalate for growth is shown by the scheme in Fig. 13. It is of interest that this organism can utilize formate as sole source of carbon, and when it does so the growth substrate is oxidized completely to carbon dioxide. This gas is then fixed by reactions apparently the same as those of the photosynthetic carbon reduction cycle, although in this case the bacteria utilize the energy released by the oxidation of formate and not light energy. This switch-over from a heterotrophic mode of existence with oxalate as carbon source to an autotrophic mode with formate was indicated by the marked difference in labelling patterns resulting from the transfer. *Pseudomonas oxalaticus* growing with formate incorporated most of the radioisotope of either $H^{14}CO_2H$ or $H^{14}CO_3^-$ into PGA at early times of sampling whereas most of the radioactivity of ^{14}C-oxalate passed into glycine when oxalate was the growth substrate. Enzyme patterns in cell-free extracts showed a corresponding change: from those of Fig. 12 when growth occurred with oxalate to those of the Calvin-Bassham cycle when the bacteria grew with formate [1].

Utilization of methane as a source of carbon

One further example of the value of these methods for elucidating metabolic pathways is again provided by the work of J.R. Quayle and his colleagues. It concerns the reactions by which methane gas is utilized as a growth compound by *Pseudomonas methanica*. The problem is particularly fascinating for two reasons. A century ago Baeyer [2] proposed that carbon dioxide is reduced to formaldehyde and then condensed to carbohydrate by the photosynthesizing organism; and although *Pseudomonas methanica* does not utilize light energy it appears to convert methane into formaldehyde, or into a compound like it, and then proceed to incorporate this compound into cell constituents by reactions closely resembling those of the Calvin-Bassham cycle. It is not, of course, proposed that formaldehyde should again be considered as an intermediate in carbon fixation by photosynthesizing organisms. Secondly, large quantities of methane are sometimes wasted in the petroleum industry and the research group of the 'Shell' company at Sittingbourne in Kent have shown that an edible and odourless white powder can be obtained by extracting the proteins from microorganisms grown with methane as a source of carbon. A thorough knowledge of the biochemistry of methane-utilizing microorganisms might conceivably assist the solution of world food problems.

P.A. Johnson and J.R. Quayle [3] exposed cultures of *Pseudomonas methanica* growing with methane to ^{14}C-methane or ^{14}C-methanol and found that 90% of the radioactivity was incorporated into phosphorylated compounds at the earliest times of sampling. Of this 90%, by far the largest proportion was in glucose and fructose phosphates: only about 10% was present in PGA. The origin of these compounds was further investigated by M.B. Kemp and J.R. Quayle [4] who incubated cell extracts with ribose-5-phosphate and either ^{14}C-methanol or ^{14}C-formaldehyde. Phosphorylated compounds were formed which were separated by chromatography and then submitted to the action of an acid phosphatase to remove phosphate groups; the dephosphorylated compounds were then rechromatographed. Two free sugars were identified: one was the familiar fructose and the other was the less familiar sugar, allulose. The incorporation of radioactivity from ^{14}C-formaldehyde was much stronger than that from ^{14}C-methanol. On the basis of these and other experiments it is suggested that methane is oxidized to the level of formaldehyde and that the following reaction then occurs:

	CHO H.C.OH H.C.OH H.C.OH $CH_2.O.PO_3H_2$		CH_2OH C=O H.C.OH H.C.OH H.C.OH $CH_2.O.PO_3H_2$		CH_2OH CO HO.C.H H.C.OH HC.OH $CH_2OPO_3H_2$
CHO +	Ribose-5-phosphate	⟶	Allulose-6-phosphate	⟷	Fructose-6-phosphate

The isomerization of allulose-6-phosphate to fructose-6-phosphate provides the starting material for the series of 'rearrangement' reactions encountered in the pentose cycle (Fig. 14). If we consider the reactions of

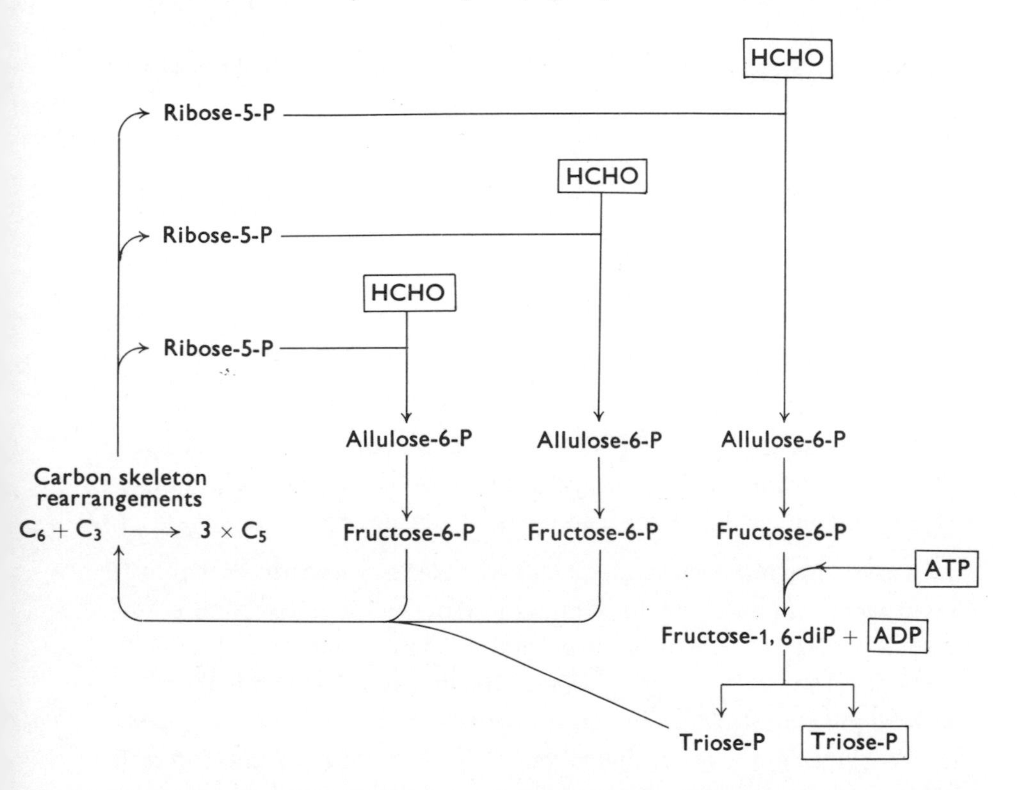

Fig. 14. Conversion of formaldehyde into triose phosphate. 3 HCHO + ATP → triose phosphate. P: ester phosphate. From Kemp M.B. and Quayle J.R. (1966) *Biochem. J.* **99**, 41.

three molecules of fructose-6-phosphate, one of them provides two molecules of triose phosphate. One molecule of triose phosphate participates in reactions with the other two molecules of fructose-6-phosphate, and rearrangements of the fifteen carbon atoms occur to give three molecules of the five-carbon compound, ribose-5-phosphate. These are now available to incorporate three more molecules of formaldehyde into three molecules of allulose-5-phosphate. If we count the number of molecules entering and leaving the cycle, shown in the rectangles of Fig. 14 we obtain the overall equation

$$3\ HCHO + ATP \rightarrow \text{triose phosphate} + ADP$$

That is, the oxidation product of methane, HCHO, has been converted into triose phosphate which can provide cellular constituents by well-established reactions.

The main difference between this cycle and the photosynthetic carbon reduction cycle of Calvin and Bassham lies in the conversion of ribose-5-phosphate into fructose-6-phosphate. In Fig. 14 these reactions are:

ribose-5-phosphate + HCHO → allulose-6-phosphate → fructose-6-phosphate

In the photosynthetic cycle the incorporated carbon dioxide must be reduced to the oxidation level of HCHO, and this is accomplished by the

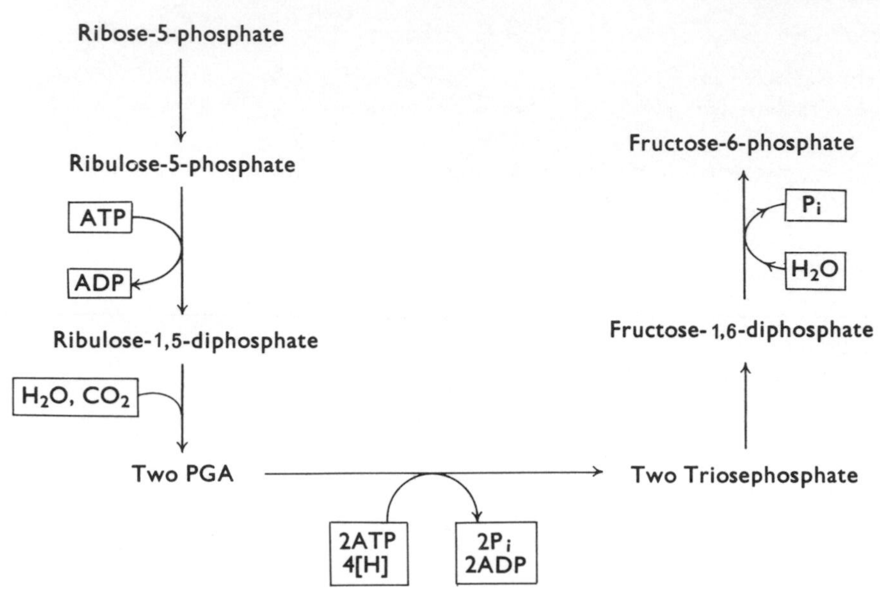

Fig. 15. Conversion of ribose-5-phosphate into fructose-6-phosphate by incorporation of carbon dioxide. P_i: inorganic orthophosphate.

$$\text{ribose-5-phosphate} + CO_2 + 3\,ATP + 4\,H + 2\,H_2O \rightarrow \text{fructose-6-phosphate} + 3\,P_i + 3\,ADP$$

reaction scheme outlined in Fig. 15 (details of reactions are omitted). Each molecule of carbon dioxide incorporated requires three molecules of ATP and two molecules of reduced pyridine nucleotide.

Nutritionally exacting microorganisms and the synthesis of vitamins

As we have seen, many bacterial species are able to synthesize all their cell constituents when placed in a mineral salts medium containing ammonium ions as source of nitrogen and a single organic compound as source of carbon and energy. However, many bacteria—like higher animals—require small additions of vitamins to their diet. In some cases these requirements are complex, but in others a single addition will suffice: thus, *Proteus vulgaris* can develop on a simple synthetic medium plus nicotinic acid. The use of such bacteria (natural auxotrophs) aided the discovery, assay and isolation of certain vitamins, particularly those of the B group which constitute portions of the molecules of coenzymes. Indeed, these studies not only provide some of the evidence that certain vitamins served this function but they also assisted the elucidation of the chemical structures of the vitamins themselves. Thus, the vitamin pyridoxine is a derivative of pyridine with methyl and hydroxyl groups at positions 2 and 3 respectively, and with two hydroxymethyl groups at 4 and 5. E.E. Snell showed that a strain of *Streptococcus faecalis* requiring this vitamin grew much more readily with pyridoxal or pyridoxamine than with pyridoxine: these two compounds carry —CHO and —CH_2NH_2 groups respectively at position 4 instead of hydroxymethyl. I.C. Gunsalus and W.W. Umbreit also showed that when these organisms grew in media deficient in pyridoxine their ability to decarboxylate tyrosine was impaired. This loss of activity was not restored by adding pyridoxine to the reaction mixture: a little stimulation was given by pyridoxal but strong decarboxylase activity was observed when ATP and pyridoxal were added together. This work led to the realization that the cofactor required for decarboxylations, transaminations and other reactions of amino acids is pyridoxal-5-phosphate.

Similarly, the recognition of lipoic acid as a cofactor involved in the oxidative decarboxylation of α-ketoacids stemmed from the observations that (*a*) a naturally occurring substance, called 'acetate-replacing factor', could take the place of acetate required for the growth of *Lactobacillus casei* and (*b*) that a 'pyruvate oxidation factor' was needed for the oxidation of pyruvate by *Streptococcus faecalis*. These 'factors' contained the same compound, lipoic acid: this is required for the production of acetyl coenzyme A from pyruvate by oxidative decarboxylation. Such examples could be multiplied. It is also clear that nutritionally exacting microorganisms will be valuable for determining the amounts of a cofactor present during various stages of its purification since the growth response of a culture may be measured and will be proportional to the amount of cofactor added. Thus, pantothenic acid and pantetheine are parts of the structure of coenzyme A (Fig. 2). Certain bacteria cannot synthesize pantothenic acid but are able to link it to β-mercaptoethylamine, giving pantetheine which they phosphorylate and join to 3′-adenosine-5′-phosphate to provide the coenzyme A they require. Other bacteria need to be supplied with pre-formed pantetheine for this synthesis. This information not only indicated which units composed the structure of coenzyme A, at a time when this was under investigation, but it also provided one of the methods of assay used for the cofactor and for the compounds involved in its synthesis. In a paper written in 1950, which is one of the milestones in the elucidation of the biochemistry of coenzyme A, G.D. Novelli and F. Lipmann [1] describe how they obtained yeast deficient in the coenzyme by growing the cells in a medium that lacked pantothenic acid. They divided the crop into two halves, one of which they aerated with glucose alone, and the other with glucose plus coenzyme A. The latter oxidized acetate twice as fast as the former; further, the yeast deficient in coenzyme A oxidized ethanol to acetate which then accumulated because the rate of oxidation was low in the absence of the cofactor. They also showed that the rate of synthesis of citrate from oxaloacetate and acetylphosphate, catalysed by cell-free extracts, responded to additions of coenzyme A. Novelli and Lipmann concluded that 'these results indicate involvement of coenzyme A in acetate activation for citrate synthesis'.

Induced auxotrophs and the biosynthesis of amino acids

Nowadays we do not have to rely upon Nature to furnish us with a microbe deficient in the particular biosynthetic pathway we wish to study. We can submit a 'wild type' organism, capable of growth in a simple mineral salts medium, to X-rays, ultraviolet radiation or chemical mutagenic agents. This treatment may damage the genetic apparatus of members of the microbial population and render them incapable of synthesizing particular cell constituents; growth can no longer occur in the simple medium unless it is supplemented with, say, a vitamin or an amino acid. Such mutants (or induced auxotrophs) can be distinguished from wild type by exposure to penicillin in the simple medium, since the antibiotic only acts upon actively dividing cells. Wild type organisms grow and are killed; mutants cannot grow without a supplement, and they survive. Induced auxotrophs have been used to study the biosynthesis of vitamins. In this section their use in elucidating the biosynthesis of amino acids will be illustrated with reference to the branched-chain amino acids valine and isoleucine.

Consider a sequence of reactions by which wild-type bacteria synthesize an amino acid (X) from a source of carbon, say glucose (G) with compounds A, B and C as intermediates in the reaction sequence:

$$G \rightarrow A \xrightarrow{(1)} B \xrightarrow{(2)} C \rightarrow \rightarrow X$$

Suppose a mutant is obtained that lacks the ability to make enzyme 1. It will not grow with either G or A as sources of carbon unless sufficient of X is added to the medium to satisfy its growth requirements. However, B or C or other intermediates can take the place of X since the enzymes that convert them to X are capable of functioning. Another mutant that requires X for growth may have its requirements met by C, as well as by X, but not by A or B: in this case the mutant is unable to synthesize enzyme 2. Clearly if a series of mutants is obtained, all of them requiring X for growth on glucose, it is then possible to suggest a route of biosynthesis. However, such data must be interpreted with caution. Thus, the requirements of mutants of *Aerobacter aerogenes* for aromatic amino acids could be satisfied by quinic acid as well as by 5-dehydroquinic acid. The latter is now accepted as an intermediate in the biosynthesis of tyrosine, phenylalanine and tryptophan from glucose, but the former is not. It appears that *A. aerogenes* can adapt to utilize quinate as a carbon source by converting it into 5 dehydroquinate: *E. coli* cannot do this, and accordingly its mutants that are blocked before 5-dehydroquinate do not respond to quinate. In our scheme, therefore, we must consider the possibility that when two compounds, A and F, are both able to satisfy the requirement for X, F may not be an obligatory intermediate in the reaction sequence G $\rightarrow$ X: it may be converted into A by a side reaction from the main pathway, the enzymes needed for the conversion being induced in the cells when they are exposed to F.

A further consequence of the lack of enzyme 1 is apparent. When the mutant grows in a glucose medium supplemented with X, compound A will accumulate and may be isolated and identified. Similarly, if enzyme 2 is blocked compound B will accumulate. Such studies will therefore complement those which identify reaction intermediates by their abilities to replace X in the growth medium.

The accepted metabolic pathways of biosynthesis of valine and isoleucine are shown in Fig. 16. In 1943 D.M. Bonner, E.L. Tatum and G.W.

Beadle [1] isolated mutants of the mould *Neurospora crassa* that required both valine and isoleucine for growth. Since genetic analysis showed that a particular mutant differed from the wild type by mutation in a single gene, this appeared to be at variance with the theory that one gene controlled the synthesis of one enzyme. It was therefore suggested that

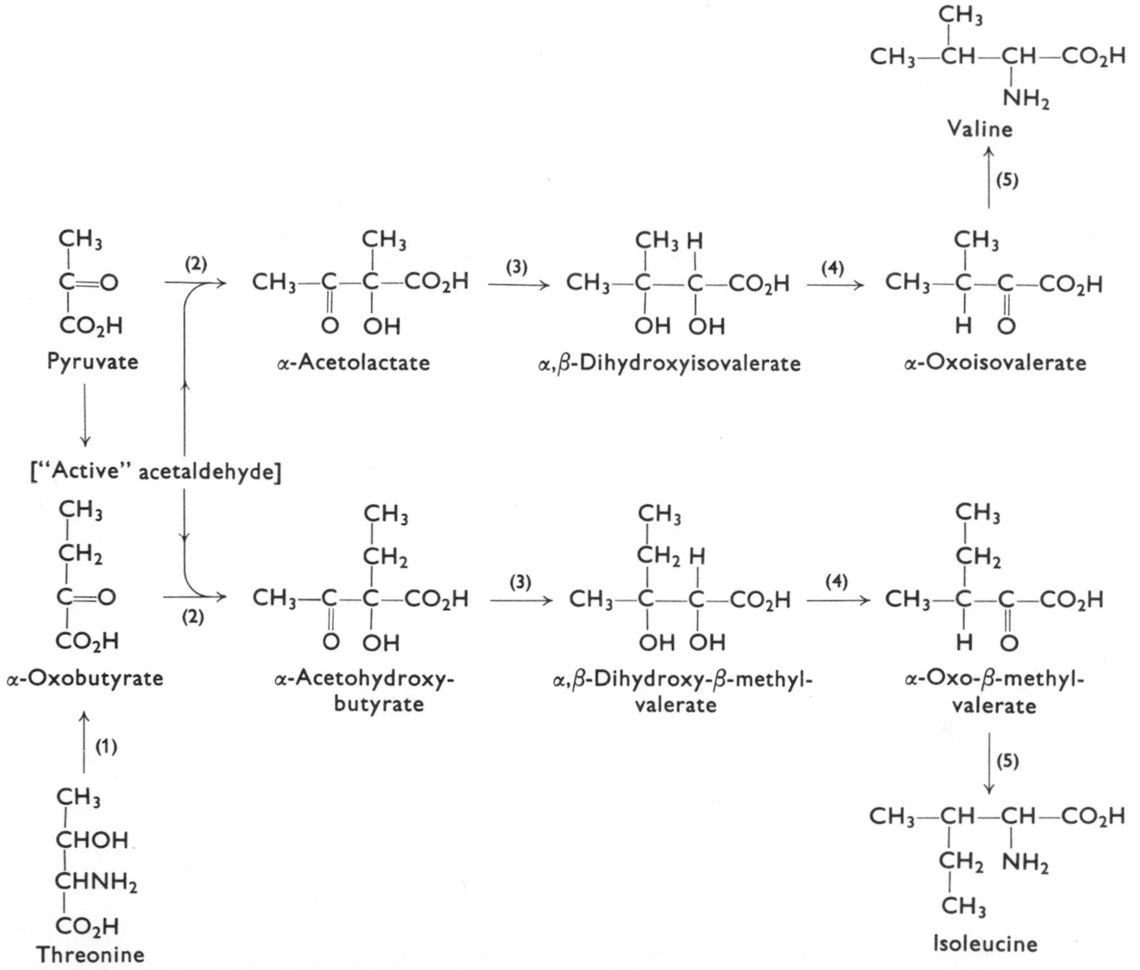

Valine: $CH_3-CH(CH_3)-CH(NH_2)-CO_2H$

↑ (5)

Pyruvate ($CH_3-CO-CO_2H$) →(2)→ α-Acetolactate ($CH_3-CO-C(CH_3)(OH)-CO_2H$) →(3)→ α,β-Dihydroxyisovalerate ($CH_3-C(CH_3)(OH)-CH(OH)-CO_2H$) →(4)→ α-Oxoisovalerate ($CH_3-CH(CH_3)-CO-CO_2H$)

Pyruvate ↓ ["Active" acetaldehyde] → (2)

Threonine ($CH_3-CHOH-CHNH_2-CO_2H$) →(1)→ α-Oxobutyrate ($CH_3-CH_2-CO-CO_2H$) →(2)→ α-Acetohydroxybutyrate ($CH_3-CO-C(C_2H_5)(OH)-CO_2H$) →(3)→ α,β-Dihydroxy-β-methylvalerate ($CH_3-C(C_2H_5)(OH)-CH(OH)-CO_2H$) →(4)→ α-Oxo-β-methylvalerate ($CH_3-CH(C_2H_5)-CO-CO_2H$)

↓ (5)

Isoleucine: $CH_3-CH(CH_2CH_3)-CH(NH_2)-CO_2H$

Fig. 16. Biosynthesis of valine and isoleucine. Reaction (1) is catalysed by threonine dehydratase (deaminase). Enzyme (3) is a reducto-isomerase, enzyme (4) a dihydroxyacid dehydratase.

the double requirement was due to the fact that metabolites accumulate just prior to a point in a reaction sequence at which an enzyme is deleted: in this case it was thought that the metabolite was the oxoacid of valine (or isoleucine) which would accumulate and compete with the oxoacid of isoleucine (or valine) for the enzyme that catalyses amination. However, attempts to prove this by isolating the oxoacids led instead to the isolation of enantiomers of α,β-dihydroxyisovalerate and α,β-dihydroxy-β-methylvalerate [2].

Evidence for the first step in the pathway from pyruvate to valine was obtained by M. Strassman, A.J. Thomas and S. Weinhouse [3] who isolated valine from yeast after growth with glucose plus trace quantities of labelled lactate. The carboxyl group of valine arose entirely from the carboxyl group of lactate but the distribution of the other two carbon atoms showed that the 3-carbon chain could not have remained intact. They suggested that pyruvate condensed with acetaldehyde, arising from pyruvate by decarboxylation, to give acetolactate; further, a migration of a methyl group must also have occurred. These changes, which are shown in Fig. 17, account for the distribution of ^{14}C in these experiments.

An enzyme that catalyses the formation of acetolactate from pyruvate was partially purified by E. Juni in 1952. This reaction is the first step in the formation of acetoin, a property long used by bacteriologists as a diagnostic test to distinguish *A. aerogenes* from *E. coli*.

$$2\,CH_3{-}\overset{\overset{\displaystyle O}{\|}}{C}{-}CO_2H \longrightarrow CH_3{-}\overset{\overset{\displaystyle O}{\|}}{C}{-}\underset{\underset{\displaystyle CO_2H}{|}}{\overset{\overset{\displaystyle OH}{|}}{C}}{-}CH_3 + CO_2$$

α-Acetolactic acid

$$CH_3{-}\overset{\overset{\displaystyle O}{\|}}{C}{-}\underset{\underset{\displaystyle CO_2H}{|}}{\overset{\overset{\displaystyle OH}{|}}{C}}{-}CH_3 \longrightarrow CH_3{-}\overset{\overset{\displaystyle O}{\|}}{C}{-}\underset{\underset{\displaystyle H}{|}}{\overset{\overset{\displaystyle OH}{|}}{C}}{-}CH_3 + CO_2$$

Acetoin

However, although the first of these two reactions is the same as that which occurs in the biosynthesis of valine, it is not catalysed by the same

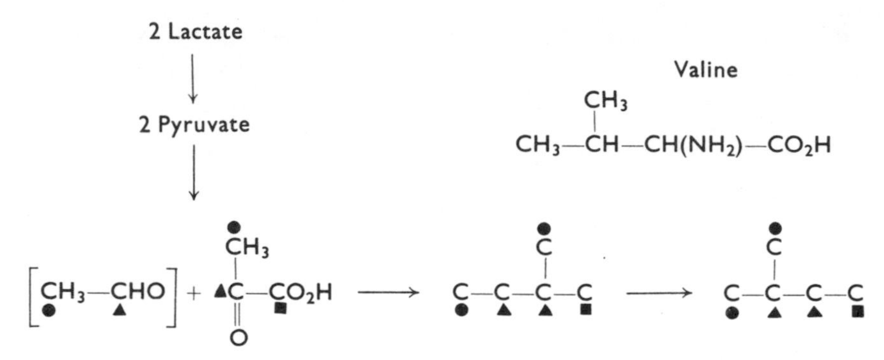

Fig. 17. Distribution of radiocarbon from lactate into valine. Details of the pathway are given in Fig. 16.

enzyme. Y.S. Halpern and H.E. Umbarger [1] showed that *A. aerogenes* synthesizes two enzymes that form acetolactate. One of them operates optimally near pH 6, is synthesized when the pH of the growth medium falls below 6 and functions to form the neutral product of glucose catabolism, acetoin, as shown above. The other operates near pH 8, is synthesized when the pH is greater than 6 and functions in the synthesis of valine. Most significantly, the second enzyme is not synthesized when valine is added to the growth medium and its catalytic activity is inhibited by valine. The formation and activity of the enzyme synthesized below pH 6 are not affected by valine; moreover, *E. coli* does not synthesize this enzyme, whereas the 'biosynthetic' enzyme has been isolated from this organism.

The same enzyme that forms α-acetolactate from pyruvate in *E. coli* also appears to form α-acetohydroxybutyrate from pyruvate and α-oxobutyrate. The last-named compound arises from the deamination of threonine which in turn is biosynthesized from aspartic acid. Just as the enzyme that forms acetolactate is inhibited and repressed by valine, so is 'biosynthetic' threonine deaminase inhibited and repressed by isoleucine. As we shall see, these are mechanisms whereby the cell is able to regulate the rates of formation of valine and isoleucine according to its requirements for growth.

We saw that a mutant of *Neurospora crassa* accumulated αβ-dihydroxy-β-methylvaleric acid. This compound is formed from α-acetohydroxybutyric acid by a remarkable reaction which appears to involve reduction of the carboxyl group (by NADPH) and simultaneous migration of an ethyl group (Fig. 16 reaction 3). One thing about the mechanism is certain: reduction does not occur first, because the (isomeric) compounds that might arise have been tested and shown not to be substrates for the enzyme. On the other hand, although cell extracts of *Salmonella typhimurium* will reduce the isomer that would be formed if rearrangement occurs first, F.B. Armstrong and R.P. Wagner [1] have concluded that enzyme preparations which only reduce but do not rearrange the molecule are probably altered forms of the reducto-isomerase. No preparation

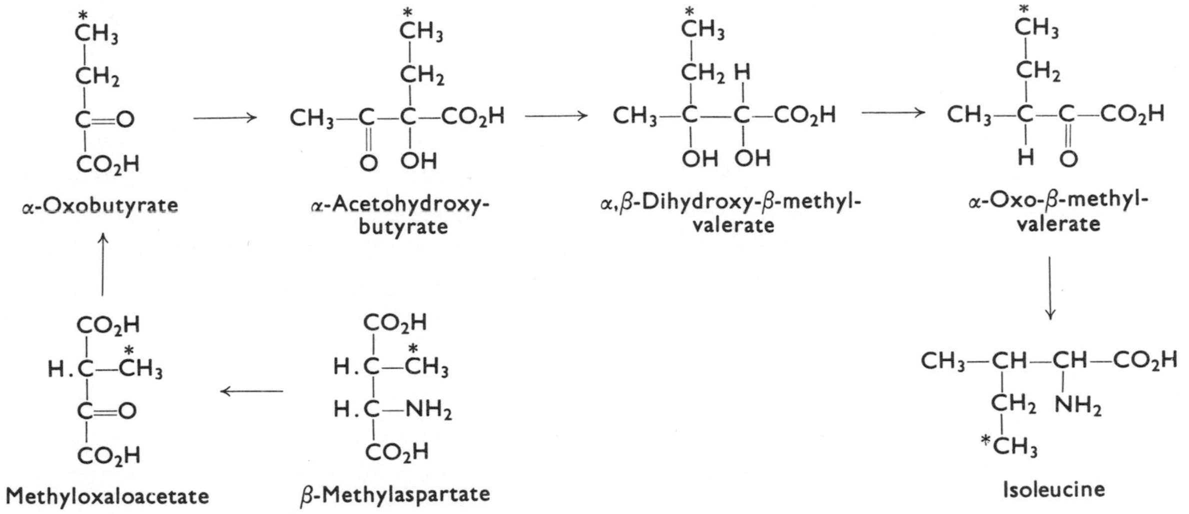

α-Oxobutyrate → α-Acetohydroxybutyrate → α,β-Dihydroxy-β-methylvalerate → α-Oxo-β-methylvalerate → Isoleucine

β-Methylaspartate → Methyloxaloacetate → α-Oxobutyrate

Fig. 18. Incorporation of ^{14}C from β-methylaspartate ($\overset{*}{C}H_3$—) into isoleucine.

has ever been obtained that can cause the ethyl group to migrate without reducing the carboxyl group at the same time. In short, it appears that reduction and isomerization occur at one fell swoop. R.L. Wixom and his colleagues [2] have shown that the reactions of Fig. 16 are widespread. They chose for special study the αβ-hydroxyacid dehydratase (Fig. 16 reaction 4). This enzyme was present in a wide range of microorganisms and plants that are able to synthesize valine and isoleucine, but it was absent from microbes that could not synthesize these amino acids and also from the livers and kidneys of mammals that need them in their diets. It appears that reactions 2, 3, 4, and 5 of Fig. 15 are catalysed by enzymes that are common to the valine and isoleucine pathways: for example, the αβ-dihydroxyacid dehydratase, enzyme 4, has been purified 120-fold from spinach leaves without any fractionation of activities between valine and isoleucine intermediates [3].

When we consider the development of our knowledge of the biosynthesis of valine and isoleucine we see how observations first made with induced auxotrophs of a bread mould (*Neurospora crassa*) have been extended to studies with *A. aerogenes*, *E. coli*, *S. typhimuriums*, higher plants and vertebrates. As in the other examples given earlier in this chapter attention has been focused upon the properties of individual enzymes, once the chemistry of the sequence has been established. No metabolic pathway is completely described without reference to the enzymes that catalyse its individual reactions. For example, it has been established that *E. coli* incorporates the methyl group of β-methylaspartic acid exclusively into isoleucine [1]. β-Methylaspartate, labelled with ^{14}C as shown, is converted into isoleucine by the reactions of Fig. 18 which provide an alternative route of biosynthesis that does not involve threonine. Indeed, an auxotroph that lacked threonine dehydratase responded to β-methylaspartate or α-oxobutyrate supplied for isoleucine biosynthesis. However, the isotope studies were performed with *E. coli* growing in a glucose medium containing additions of ^{14}C-β-methylaspartate, and as the authors point out [1] it is not yet clear whether these reactions occur only on the addition of β-methylaspartate to the growth medium. This will no doubt be decided by studies concerned with the repression and derepression of the enzymes used in the two pathways of isoleucine biosynthesis.

How complete is our knowledge of metabolic pathways?

From calculations based upon the DNA content of *E. coli* J.D. Watson [2] has concluded that one fifth—possibly one third—of the reactions that will ever be described in this organism are now known. As he says, 'The conclusion is most satisfying, for it strongly suggests that within the next 10 to 20 years we shall approach a state in which it will be possible to describe essentially all the metabolic reactions involved in the life of an *E. coli* cell.'

However, there is a vast area of metabolism relating to the biosynthesis of natural products which has so far been explored simply by establishing the isotopic patterns obtained by feeding compounds containing radiocarbon to higher plants and fungi. Little is known about the enzymic reactions involved: thus, few enzymes responsible for alkaloid biogenesis have been isolated. Moreover, whenever a complicated molecule is biosynthesized there are, so to speak, microbes waiting to degrade it in the soil. The exciting results obtained from work with *E. coli* over the last few decades may tend to obscure the fact that there exist microbes that are much more biochemically versatile. Indeed, a colleague who submitted for publication a manuscript which attempted an explanation for one aspect of the astounding versatility of pseudomonads received a comment from the editors of the journal, made in all seriousness, to the effect that *E. coli* could surely perform just as well but had not yet been tried! A study of the enzymes that accomplish the biosynthesis of complex molecules in plants, and of those that catalyse their degradation by microbes, will occupy the attention of biochemists concerned with metabolic pathways for several decades.

Regulation of Metabolism

In the previous chapter we saw that the chemical constituents of many living cells are in a dynamic state. Their biochemical stabilities show great differences so that, for example, the half-life of a molecule of ribosomal RNA is very much greater than that of a molecule of pyruvate, present only for a brief period inside a mitochondrion. Nevertheless, the overall impression is that of degradation nicely balanced by synthesis, with each one of the hundreds of different enzymes making its own particular and appropriate contribution to the steady state which has established itself within the cell. As the metabolic map of known enzymic reactions has been extended in scope and detail, biochemical effort has been directed increasingly towards achieving a better understanding of the ways in which these reactions are regulated and the balanced economy of the cell preserved.

In the case of a growing culture of bacteria, C.N. Hinshelwood and his colleagues [1] have stressed that the synthesis of any one component of the cell depends upon the functioning of others. For compounds of relatively low molecular weight a metabolic map consists, as we have seen, of an interwoven network of reactions. In macromolecules, for example, the synthesis of DNA depends upon the functioning of certain enzymes, and the synthesis of these depends upon DNA and other enzymes. The kinetics for a system such as this involving cellular components X_1, X_2, $X_3 \dashrightarrow X_n$, can be expressed in terms of (simplified) equations of the form $dX_1/dt = k_1X_2$; $dX_2/dt = k_2X_3$; $dX_3/dt = k_3X_4$ and so on. The series would be closed by denoting the dependency of X_n upon the first component X_1 by writing $dX_n/dt = k_nX_1$. This is a mathematical expression of the fact that a bacterium reproduces itself as a whole, and no part can function continuously without the others. In developing a mathematical description of bacterial growth, two further considerations are necessary. First, inflow of nutrients and outflow of waste products can only maintain a proper balance within certain limits of cell size, so that the cell divides when the upper limit is reached. Second, the division of a bacterium must be related to its growth processes and may begin when the amount of some cellular constituent has reached a certain critical value. Hinshelwood has developed systems of equations based on these principles and has shown that the operation of the laws of chemical kinetics are of themselves sufficient to account for the fact that a bacterial cell is able to make adjustments in rates of reactions, so that optimum growth rate is achieved in a given medium. In so doing it changes the proportions of its components, and when it is transferred from medium to medium it may abandon one pattern of chemical reactions for another. In this process it brings itself into an optimum relation with its environment.

It should be stressed that in this treatment Hinshelwood does not need to evoke the existence of mechanisms by which the activities of particular enzymes may be enhanced or decreased according to the needs of the cell. However, he does not, of course, deny their existence; and having in mind the perspective provided by the principle of 'total integration of cell function', we shall discuss some of the mechanisms by which bacterial, and other cells *do* appear to regulate the activities of specific enzymes. For example, when a compound such as isopropylthio-β-D-galactoside is added to a bacterial culture in a steady metabolic state the cells may be induced to synthesize relatively large amounts of an enzyme —in this case β-galactosidase. No other change in their metabolic condition may be detectable, and the enzyme serves no immediate purpose because the inducer is not a substrate for this, or any other enzymes in the cells. This is a situation which is not profitably discussed in terms of an overall adjustment in cellular economy; for a regulatory mechanism is called into play which is very rapid, specific, and in those particular circumstances, quite useless. On the other hand there are instances of specific regulatory responses so closely linked to function that the bacteria appear to be able to 'distinguish' between catabolic (energy-releasing) reactions on the one hand and reactions that are needed for synthesis on the other. Some examples are as follows:

Carbamoylphosphate + L-ornithine = L-citrulline + orthophosphate
L-Threonine = α-oxobutyrate + ammonia
2 Pyruvate = α-acetolactate + CO_2
L-Arginine = agmatine + CO_2
L-Ornithine = putrescine + CO_2

Each one of these reactions serves a biosynthetic function. However, in different circumstances each one may lie on a pathway of degradation, or may operate as a side reaction whose only function appears to be that of assisting the attainment of conditions more tolerable to the cells. In each case, as we shall see, the bacteria synthesize two enzymes, one used for biosynthesis and the other not so used. The last four reactions proceed from left to right whether or not they are used for biosynthesis. This is not the behaviour we should predict from purely chemical considerations. Here, the situation is not simply that the amount of an enzyme alters as the demand for its product changes. Rather, a second enzyme is brought into operation to catalyze the same reaction; and it is the metabolic fate of the product which has changed in this case.

Regulation due to availability of substrates or coenzymes

It is therefore evident that living matter has evolved specific mechanisms for enzyme control in addition to those predictable from the chemical kinetics of an integrated system. Before we discuss such mechanisms, however, we may note that the rate of a reaction sequence will depend upon the availability of substrates and cofactors for its enzymes. We saw that the early studies of glycolysis became possible when it was observed that the degradation of glucose by yeast juice soon slowed down and then stopped, but that fermentation resumed when orthophosphate was added to the reaction mixture. A similar limitation was imposed by ADP: glycolysis could proceed no further than the formation of fructose-1,6-diphosphate, and this was how the compound was first obtained by Harden and Young.

Consider a cell for which the total amount of ATP, ADP and orthophosphate is limited. ATP cannot be synthesized continuously by glycolysis unless this reaction sequence is linked with others which utilize the ATP it generates, and which furnish ADP in return. Degradation of glucose is therefore geared to synthesis; for when the former outstrips the latter and the ratio of ATP/ADP increases, those reactions of glycolysis that require ADP as substrate can be expected to slow down. Moreover, the interdependent reactions of certain multienzyme systems may occur in different compartments of the cell, and their effective coupling may depend upon the ease with which a linking metabolite is able to gain access to its enzymes, sometimes passing through a cellular membrane in order to do so. In some cases the transportation may itself involve an enzymic conversion. Thus, as mentioned in the previous chapter, the enzyme pyruvate carboxylase is situated inside rat liver mitochondria, whereas phosphoenol pyruvate carboxykinase is found outside. Accordingly, it appears that oxaloacetate, which is formed by the first of these enzymes and is required by the second, must be transported through the membrane of the mitochondrion. The compound is probably first reduced to malate, which unlike oxaloacetate can pass through the membrane, and when outside the mitochondrion, malate is oxidized to oxaloacetate. This interconversion of malate and oxaloacetate may serve a second purpose, namely to furnish NADH outside the mitochondrion.

Gluconeogenesis from pyruvate and other precursors involves a reductive step, in which glyceraldehyde phosphate is formed from diphosphoglycerate. This reaction occurs exclusively in the cytoplasm and raises the question of the origin of the required reducing agent, NADH. H.A. Krebs, T. Gascoyne and B.M. Notton [1] have presented evidence that the reducing power is generated inside the mitochondria as NADH, which again cannot pass through the membrane. However, the reduced cofactor passes its hydrogen over to oxaloacetate which, as malate, diffuses through the membrane and hands back the hydrogen to NAD in the cytoplasm.

Just as the degradation of glucose to pyruvate cannot proceed unless ADP is converted simultaneously into ATP, so it has been found that the flow of electrons along the electron transport chain in certain 'tightly-coupled' preparations of mitochondria cannot occur unless ADP is available for phosphorylation. In this case also, the rates of the chemical reactions that release the free energy to be harnessed will depend upon the ratio ATP/ADP which prevails in the system at a given time: when the ratio is low, and ATP is called for, the release of energy is accelerated. The overall equation for respiratory chain phosphorylation when NADH is oxidized is as follows:

$$\mathrm{NADH} + \mathrm{H}^{+} + 3\,\mathrm{P_i} + 3\,\mathrm{ADP} + \tfrac{1}{2}\,\mathrm{O_2} \rightarrow \mathrm{NAD}^{+} + 4\,\mathrm{H_2O} + 3\,\mathrm{ATP} \quad \ldots\ldots \quad (1)$$

where P_i denotes orthophosphate. The work of B. Chance and G.R. Williams [2] has shown that when both ADP and the substrate undergoing oxidation are present at high concentrations, then the proportion of reduced to oxidized form decreases as we pass from one component of the chain to the next, starting at NADH/NAD at one end and proceeding in the direction of oxygen. Under these conditions the rate limiting factor is the respiratory chain itself. If the concentration of ADP is now reduced to a low value, the respiratory rate is also lowered and the degree of reduction of the various components increases. As A.L. Lehninger [3] has said: 'The rate of respiration and the steady-state equilibrium of the electron carriers are thus exquisitely poised and are a function of concentration of respiratory substrate, ADP and oxygen'. In addition to the effect of ADP upon respiration in tightly-coupled mitochondria, ATP exerts an independent inhibitory effect of its own when added at physiological concentrations [1].

It is quite clear how the phosphorylation of ADP is coupled to the degradation of glucose, for ADP is a substrate for the two enzymes phosphoglycerokinase and pyruvate kinase by which the energy-rich compounds 1,3-diphosphoglycerate and phosphoenolpyruvate are converted respectively into 3-phosphoglycerate and pyruvate. However, oxidative phosphorylation is less clearly understood. It has been suggested that, at the sites where oxidation is coupled to phosphorylation, an electron carrier A combines with some endogenous substance X. This occurs when the reduced form of A becomes oxidized by transferring hydrogen (or electrons) to the next electron carrier B:

$$AH_2 + B + X \rightleftharpoons A \sim X + BH_2 \quad (2)$$

The symbol ~ denotes that AX is an energy-rich compound, the bond between A and X being very readily broken by hydrolysis. Accordingly, sufficient free energy can be released in the following reaction to accomplish the phosphorylation of ADP:

$$A \sim X + P_i + ADP \rightleftharpoons A + X + ATP \quad (3)$$

If reaction (3) cannot occur, oxidation will cease as well as phosphorylation, for A must be freed from X before it can accept the next pair of hydrogens (or accept an electron) so that oxidation can continue. This appears to happen when the antibiotic oligomycin is added: phosphorylation and electron transport both stop. When an 'uncoupling agent' such as dinitrophenol is added then the hydrolysis of AX—which is energetically favoured—is catalysed. Oxidation can proceed since A is freed to accept electrons, but phosphorylation stops. In fact the addition of dinitrophenol is found to relieve the inhibition of respiration that oligomycin causes.

However, it must be said that the experimental evidence for the involvement of energy-rich intermediates such as AX is weak. Despite a great concentration of effort, no compounds have been isolated that would serve the functions assigned to them in mitochondria. At the time of writing these intermediates appear to be entirely hypothetical and it is possible that they will be dispensed with in future descriptions of oxidative phosphorylation. Indeed P. Mitchell [2] has contended for some years that when we take into account the orientation which membranes must impose upon enzymes attached to them, such compounds become irrelevant to the description.

The ability of mitochondria to oxidize substrates survives various treatments more readily than their capacity to convert ADP into ATP. This may be due to the fact that the phosphorylation site is damaged or removed during the treatments employed. Nevertheless, fragments of mitochondria have been isolated that can still phosphorylate; moreover, soluble heat-labile components have also been isolated in various laboratories that can restore phosphorylating activity to submitochondrial particles. It is possible that these preparations contain enzymes that are needed when ATP is formed from energy-rich compounds such as the hypothetical AX of equation (3). Alternatively they may act by removing an endogenous compound that is able to stop phosphorylation. In either event, 'coupling factors' of this type might be involved in the regulation of oxidative phosphorylation.

In order to function continuously a tightly-coupled electron transport system and the glycolytic sequence both require the presence of ADP. As the ratio ADP/ATP within a cell is lowered, the system that has the lower affinity for ADP will be expected to slow down first. Since the Michaelis constants for ADP are far larger for glycolytic enzymes than for the respiratory chain, it may be predicted that glycolysis would slow down and then stop when conditions are aerobic and when ADP is efficiently converted into ATP by oxidative phosphorylation. Inhibition of fermentation by air was, in fact, first observed by Pasteur. Alternative explanations of the Pasteur effect have been put forward, however. Thus, orthophosphate is required for oxidative phosphorylation and is also a substrate for the glycolytic enzyme glyceraldehyde-3-phosphate dehydrogenase, so that when phosphate is drained away by the first process, the second (glycolysis) would be brought to a halt. Another interpretation of the Pasteur effect stems from an attempt to answer the question: why do most malignant tissues form lactate aerobically in appreciable quantities, whereas normal tissues do not? This is a case of a breakdown of the normal regulatory mechanisms we have discussed. The explanation resembles those based upon ADP and orthophosphate acting as common metabolites for the two processes, but in this case the linking metabolite is NADH. Since the enzymes of glycolysis are in the cytoplasm which contains only catalytic amounts of NAD, the continuous breakdown of glucose depends upon the continuous reoxidation of NADH, formed when glyceraldehyde-3-phosphate is oxidized to 1,3-diphosphoglycerate. As we have seen, NADH itself cannot pass through the mitochondrial membrane to be oxidized, but pyruvate produced by glycolysis is able to accept hydrogen to form lactate, which therefore accumulates. This process normally occurs under anaerobic conditions. G.E. Boxer and T.M. Devlin [1] have presented evidence that when aerobic conditions are established, the hydrogen atoms from NADH find their way into the mitochondria where they are oxidized to water despite the permeability barrier mentioned. They suggest that in the cytoplasm the hydrogens are transferred to the glycolytic intermediate dihydroxyacetone phosphate by the action of the enzyme α-glycerophosphate dehydrogenase. The α-glycerophosphate so formed is then able to diffuse into the mitochondria where the transferred hydrogen atoms are removed. This 'shuttle' for hydrogen atoms resembles that described for malate earlier, but differs insofar as reduction of oxaloacetate in that case was postulated as taking place inside the mitochondria and oxidation outside. The explanation accounts for the fact that glycolysis and lactate formation are favoured when respiration is low, and it also predicts that lactate would still accumulate even under aerobic conditions, were it not for the α-glycerophosphate shuttle. Boxer and Devlin [59.1] showed that α-glycerophosphate dehydrogenase is present in normal cells but is lacking in malignant tumours. As a consequence these tumours accumulate lactate aerobically and so waste the energy (15 mole equivalents of ATP) which would otherwise be liberated by oxidation of the pyruvate taken out of circulation.

Regulation of specific enzymes

We saw in the preceding chapter that studies of bacterial metabolism have greatly extended our knowledge of the range of chemical reactions that occur in living cells in general. These studies have been even more successful in revealing the ways in which these reactions are regulated. The reasons for this success were most clearly and prophetically expressed in 1949 by the late Marjory Stephenson as she wrote the final passage for the third edition of her famous monograph on bacterial metabolism [1]. 'It is impossible to exaggerate', she wrote, 'the importance of the variability of the bacterial cell or the desirability of studying the laws regulating it. Biochemically, bacterial cells are the most plastic of living material . . . The bacterial cell, by reason of its small size and consequently relatively large surface, cannot develop by maintaining a constant chemical environment, but reacts by adapting its enzyme systems so as to survive and grow in changing conditions. It is immensely tolerant of experimental meddling and offers material for the study of processes of growth, variation and development of enzymes without parallel in any other biological material'. She concluded the book with the words: 'Bacterial studies pay the highest dividends on biochemical investigation'.

Bacteria are able to regulate their enzymic activities by two mechanisms which H.L. Kornberg [2] has described as 'fine control' and 'coarse control' respectively. Their operation may be illustrated as follows. Suppose a bacterial culture is growing exponentially at the expense of ^{14}C-glucose, so that radioisotope is flowing into all the cell constituents, including the amino acids of the various proteins. Then, as was shown by R.B. Roberts and his colleagues [3] the addition of, say, ^{12}C-isoleucine to the medium has the immediate effect of stopping the incorporation of ^{14}C into isoleucine, but not into other amino acids. This is an example of 'fine control' when the amounts of the enzymes present in the cell remain unchanged but those enzymes that catalyse the synthesis of isoleucine rapidly cease to function. As a result of this type of regulation, the cells do not continue the useless activity of making isoleucine when that amino acid is in abundant supply. Suppose now that these bacteria are capable of growing in a medium that furnishes isoleucine as the sole source of carbon. When they are transferred to such a medium they will not grow immediately, but after a time certain enzymes that previously were present in barely detectable amounts begin to increase in concentration. Eventually the degradation of isoleucine will be catalysed at a rate that can satisfy the metabolic requirements of exponential growth. This is an example of 'coarse control'. It is a much slower adjustment than that which follows from a change in the rates of pre-existing enzymes; for the composition of the bacterial cell itself has altered as new enzymes needed for isoleucine degradation have been biosynthesized. Fine control depends upon changes in properties of proteins that are caused by the presence of new factors in their environment; whereas coarse control is exercised through a change in the rate of biosynthesis of the enzymes themselves when the environment alters.

This example also serves to illustrate another feature of metabolic regulation which is found to be general, namely that a pathway of biosynthesis leading from, say, the tricarboxylic acid cycle to a particular substrate is not the one used when that substrate is degraded. In the case of glycolysis it is true that some of the enzymes are used in reverse for the synthesis of glucose, and only at certain points are the pathways different. In other cases the whole metabolic route is altered, so that the reactions for degrading isoleucine are entirely different from those used in its biosynthesis. It is evident that when synthesis and degradation of a cell constituent occur side by side in the same organism—as so often happens—it is an advantage for the cell to be able to regulate each pathway separately. Adjustments of metabolism do not have to wait until such time as reverse reactions are favoured by a build-up of reaction intermediates.

Regulation of rates of enzymic reactions by particular metabolites

In 1940 Z. Dische made the discovery that the phosphorylation of glucose in erythrocytes was prevented by addition of phosphoglycerate [1]. This was perhaps the first demonstration that a member of a metabolic sequence, formed after several reaction steps, could depress the rate of the initiating reaction of the sequence. However, the full significance of this phenomenon was not appreciated until 1956 when H.E. Umbarger pointed out in a short article [2] that when the end-product of a sequence strongly inhibits an earlier enzyme, a mechanism is established by which the concentration of that product can be maintained at a suitable physiological value. This type of interaction has been called 'end-product inhibition' and has been compared with negative feedback-control devices employed in technology. The phenomenon was clearly illustrated by Umbarger in the case of isoleucine which strongly and specifically inhibits threonine dehydratase (enzyme 1, Fig. 16, page 49). As we saw in the previous chapter, *E. coli* elaborates two different threonine dehydratases, namely the enzyme concerned with biosynthesis which is inhibited by isoleucine and a catabolic enzyme which is not. The second enzyme, which also deaminates serine, is induced only under rather specialized conditions which include the presence of serine or threonine and the absence of air or fermentable carbohydrate. A similar situation exists with regard to the enzymes that form acetolactate from pyruvate. The biosynthetic enzyme (enzyme 2, Fig. 16, page 49) is inhibited by valine, is formed above pH 6, has a high pH optimum and is found in both *E. coli* and *A. aerogenes*. The catabolic enzyme is not inhibited by valine, is only formed below pH 6, has a low pH optimum and is found in *A. aerogenes* but not in *E. coli*. About the same time as these investigations by H.E. Umbarger and his colleagues, R.A. Yates and A.B. Pardee [3] discovered that the biosynthetic enzyme aspartate carbamoyltransferase is inhibited by cytidine triphosphate, one of the end-products of pyridine biosynthesis (Fig. 21, page 65). The later investigations by J.C. Gerhart and A.B. Pardee have shed much light upon the nature of this inhibition exerted by cytidine triphosphate.

Our understanding of this mode of regulation of biological activity stems from the realization that protein molecules are flexible and may alter their shapes or their sizes, by joining subunits together. A previous theory of enzyme catalysis was based upon the diametrically opposite view. A concept that had proved fruitful for over sixty years regarded the active centre of an enzyme as being so constructed that only the substrate molecule, or a few other molecules of like size and shape, could be accommodated at that point in the surface. Those molecules that could fit but did not react were competitive inhibitors. The analogy of a key fitting a lock has often been used; but that of a piece fitting into a jig-saw puzzle may emphasize the basic concept of a rigid region of protein structure whose contours would not yield to the intrusion of molecules differing from the substrate in shape or size.

This concept, the 'template hypothesis', may still be applied to many enzymes, but D.E. Koshland [1] has examined several systems for which it is not adequate and has developed the theory of 'induced fit'. Thus, the enzyme 5′-nucleotidase can liberate the phosphate group from adenylic acid and also from nicotinamide ribonucleotide in which adenine has been replaced by nicotinamide. The consequence of this replacement was a reduction, though not a dramatic one, in the maximum velocity of the reaction and also in the binding of the substrate to the enzyme, as judged by an increase in the Michaelis constant. Both effects might be predicted from the template hypothesis because nicotinamide is smaller than adenine and also differs in shape. If the nicotinamide moiety is now replaced by a hydrogen atom, to give ribose phosphate, the substrate would still be expected to gain access to the active site: moreover, the bond that the enzyme is required to break remains the same. In fact, it was found that whereas ribose phosphate was bound as strongly as was nicotinamide ribonucleotide, the maximum velocity attained was extremely small in comparison with the other substrates. Something other than ability to reach the active site, and to be bound there, was clearly involved. Koshland has given many examples of substrates which, after chemical modification to reduce their size, have still been tightly bound to an enzyme but have not reacted. In other cases substrates have been made more bulky, and although they have not been degraded they have still been bound to the enzyme—a situation difficult to accommodate on the template hypothesis.

The induced-fit hypothesis retains that particular feature of the template theory that accounts for specificity of action, namely that there must be an intimate relation in space between the substrate and its enzyme. However, Koshland has modified the theory in two respects. First, it is suggested that the substrate fits into place only when a change in the shape of the enzyme has occurred; and second, it is assumed that more than one catalytic group is involved in enzyme action. Two such groups, A and B, are shown in Fig. 19(*a*). They are envisaged as being particular amino acid residues of the protein molecule and it is seen that when the substrate is bound there is a change in the configuration of the enzyme molecule that brings the catalytic groups together at the point where the substrate is to be attacked. In Fig. 19(*b*) it is seen that a larger compound than the substrate could still be bound, but it may not react because the necessary alignment of A and B no longer occurs.

This view of enzyme action provides an explanation of the fact that a second compound which bears no structural resemblance to the substrate, and is not bound to the enzyme site, may be capable, nevertheless, of altering profoundly the rate of the reaction. If there exists a different site of attachment for this compound the configuration of the enzyme may be altered during binding, just as such an alteration occurs when the substrate is bound. This change of configuration may in turn affect the active

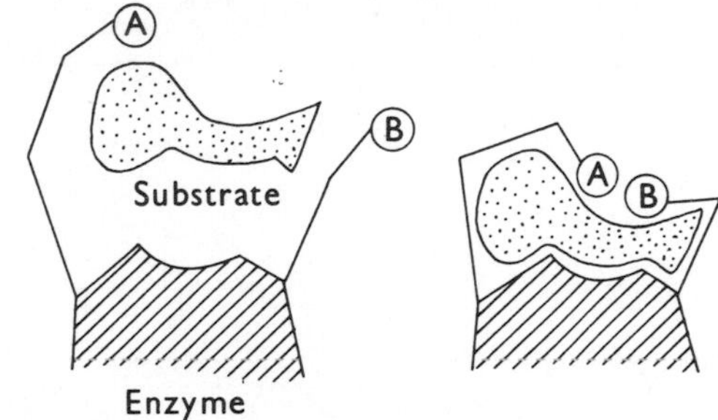

(*a*) A substrate induces a conformational change that results in a catalytic reaction.

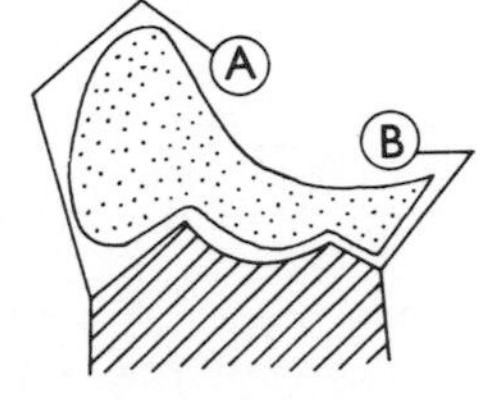

(*b*) A more bulky compound is bound to the enzyme but the reaction is not effectively catalysed.

Fig. 19. The theory of induced fit for enzyme catalysis. Redrawn from Koshland D.E., Jr. (1963) *Cold Spring Harbor Symposia on Quantitative Biology*, vol. 28, page 473.

site, so that the catalytic groups can no longer be brought into the positions they need to take up for effective catalysis to occur. This is the probable situation when isoleucine interacts with 'biosynthetic' threonine dehydratase, so that isoleucine inhibits the reaction. Alternatively an enzyme may be rendered capable of catalysing a reaction that did not occur previously, as when pyruvate carboxylase is activated by additions of acetyl coenzyme A. These effects of metabolites that bind at different sites from those involved in catalysis were discussed in a classical article by J. Monod, J-P. Changeux and F. Jacob [1] who described them as 'allosteric'. Isoleucine is a negative allosteric effector of threonine dehydratase, acetyl coenzyme A a positive effector of pyruvate carboxylase.

J. Monod, J. Wyman and J-P. Changeux [2] have extended the conception of allosterism by proposing models that take into account two experimental observations: first, that allosteric enzymes show unusual kinetics and second, that their molecules (oligomers) appear to consist of associations of smaller units (monomers) and for this reason are often difficult to purify. As far as their kinetics are concerned we may note that for many enzymes a hyperbola is obtained when substrate concentration is plotted against reaction velocity, as first-order Michaelis kinetics require. By contrast, similar plots for allosteric enzymes give sigmoid-shaped curves, which is the behaviour to be expected when more than one molecule of substrate interacts in a rate determining way with each molecule of enzyme (Fig. 20). When a positive effector (activator) is present the curve is less markedly sigmoid and is shifted to the left, although the maximum velocity finally attained is not altered. A negative effector (inhibitor) shifts the curve to the right.

The following is a simplified and qualitative description of some possible consequences arising from the ability of enzyme monomer molecules to associate. Suppose that P is a positive and N a negative effector of an allosteric protein that catalyses a reaction of a substrate S. Suppose also

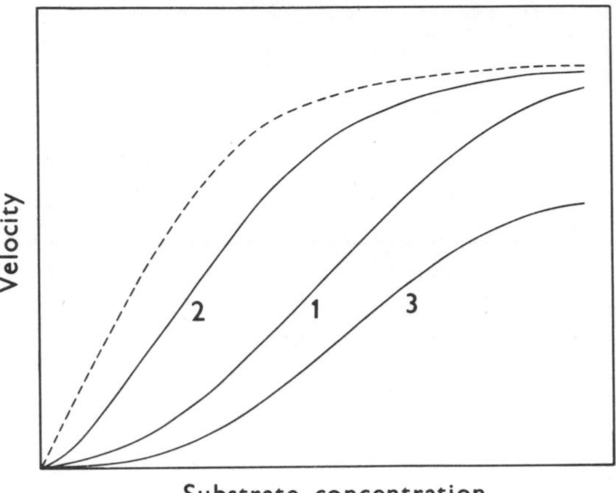

Fig. 20. Kinetics of allosteric enzymes. 1, Sigmoid relationship between velocity and substrate concentration; 2, the same enzyme with a positive effector (activator) present; 3, negative effector (inhibitor) present. Dotted curve: Michaelis-Menten kinetics given by an enzyme that does not show allosteric behaviour.

that each monomer possesses one active site for S and one site at which P can combine. One type of interaction between monomers may give dimers that still retain these sites for S and P: a dimer would therefore have two sites for each molecule and would be catalytically active. On the other hand a somewhat different distribution of bonding and energy might occur during the association, and a dimer might be formed for which the sites of S and P are distorted so that these molecules are no longer accommodated readily. This second type of dimer would not be catalytically active, and it may also possess sites for N that do not function in active dimers. If there exists initially an equilibrium between active and inactive dimers, then addition of N would displace this equilibrium in favour of inactive enzyme. Conversely, S would cause the equilibrium to be displaced in favour of active dimers, and as each molecule of S was added, more sites for S would become available. In contrast to the model used by Michaelis theory, which requires that addition of S increasingly saturates active sites, we have a situation in which, initially, the more substrate we add the more catalytic sites we form. Clearly, when P is added the equilibrium is also disturbed in the direction of forming more active sites.

At the present time a great deal of research is being directed to the study of allosteric enzymes and space permits discussion of only two outstanding investigations. They have been chosen to illustrate two main features of these enzymes: (1) their possession of binding sites for effectors, in addition to the usual catalytic sites and (2) their tendency to exist in the form of large molecules composed of subunits.

The enzyme aspartate carbamoyltransferase (ACTase; also named 'aspartate transcarbamylase' and abbreviated to ATCase) initiates the metabolic pathway of biosynthesis of pyrimidines and nucleotides (Fig. 21, below). As long as nucleoside phosphates are converted into nucleic acids, ACTase continues to catalyse the formation of the first product, carbamoyl aspartate. However, if one of the end products, cytidine triphosphate (CTP) accumulates, ACTase is inhibited and pyrimidines and nucleotides are no longer synthesized. This enzyme is an exception to the general rule that allosteric enzymes are very difficult to purify. It was

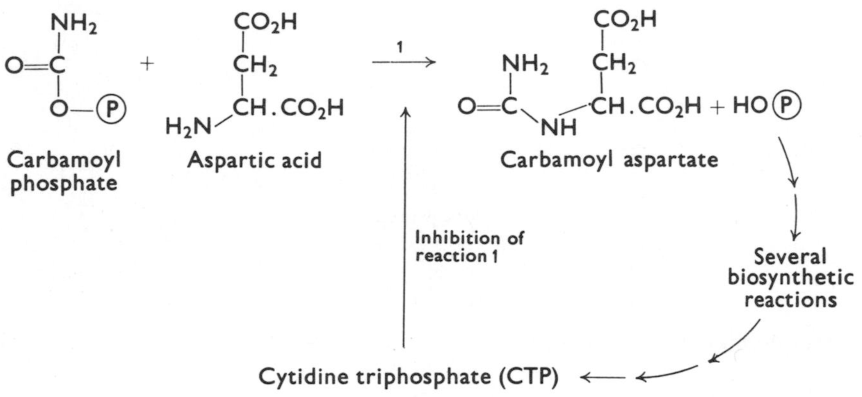

Carbamoyl phosphate + Aspartic acid $\xrightarrow{1}$ Carbamoyl aspartate + HOⓅ

Carbamoyl aspartate → Several biosynthetic reactions → Cytidine triphosphate (CTP)

Cytidine triphosphate (CTP) → Inhibition of reaction 1

Fig. 21. Reaction 1: aspartate carbamoyltransferase, the first enzyme of the biosynthetic pathway from aspartate to pyrimidines and nucleotides. Ⓟ denotes the group $—PO_3H_2$

examined in the analytical ultracentrifuge by J.C. Gerhart and H.K. Schachman [1] and shown to be a globular protein having a molecular weight of about 3×10^5. When the enzyme was treated with mercurials, or subjected to mild heating, its catalytic activity was retained, and indeed increased; and ultracentrifugal analysis showed that the native protein had yielded two subunits of molecular weights $9{\cdot}6 \times 10^4$ and 3×10^4 respectively [2]. These proteins could be separated by chomatography on DEAE-Sephadex. The larger protein catalysed the formation of carbamoyl aspartate but was not inhibited by CTP; the second protein did not catalyse this reaction; but when the two were mixed together and incubated with the sulphydryl compound mercaptoethanol, the proteins re-associated and showed catalytic activity which was now susceptible to inhibition by CTP. The presence of a site for attachment of CTP to the smaller unit was shown by sedimentation studies with 5-bromocytidine triphosphate, a compound chosen because its extinction at 298 mμ enabled it to be located during sedimentation when protein was present. These studies showed, therefore, that ACTase is composed of two separate proteins which together confer upon the enzyme the two functions of catalysis and metabolic control. One protein, the catalytic subunit, possesses the entire catalytic activity and the other, the regulatory subunit, takes up CTP. There appear to be two catalytic subunits and four regulatory subunits per molecule of native protein. Each catalytic unit loses some of its activity when it enters this complex, although all the activity is regained at high concentrations of aspartate. The regulatory unit can therefore be regarded as a proteinaceous inhibitor which acts in a reversible fashion; and CTP can be regarded as a co-inhibitor that increases the inhibitory action of this protein. It is interesting that, as the model requires, the catalytic subunit by itself gives kinetics of the Michaelis type since interactions between monomers have been abolished. Sigmoid curves are obtained when velocities of reaction catalysed by *native* protein are plotted against concentration. Gerhart and Schachman suggest that allosteric enzymes such as ACTase could have evolved from two separate proteins, one of

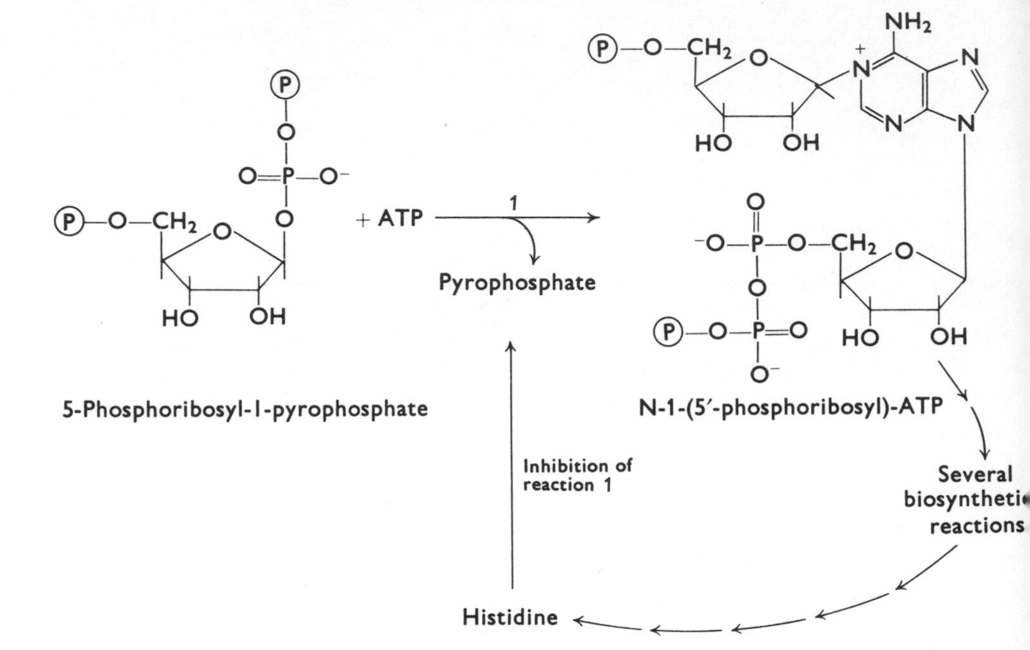

Fig. 22. Reaction 1: Phosphoribosyl-ATP-pyrophosphorylase, the first enzyme of the biosynthetic pathway of histidine.

them descended from a primitive enzyme that bound CTP as a substrate or product. The mutations most important for the appearance of a regulatory enzyme were those that allowed the association and specific interaction of the two proteins.

The enzyme phosphoribosyl-ATP-pyrophosphorylase (Fig. 22, above) which catalyses the first reaction of histidine biosynthesis is strongly inhibited by histidine but not by other purified intermediates in the metabolic pathway. R.G. Martin [1] isolated this enzyme from *S. typhimurium* and showed that, like ACTase, the pyrophosphorylase is rendered insensitive to its inhibitor by heating whilst its catalytic activity is retained. Desensitization to histidine is also achieved by treatment with mercurials, and the addition of mercaptoethanol reverses this effect. These experiments are in agreement with the view that histidine does not bind to the catalytic site of the enzyme. This conclusion was supported by the observation that whereas the substrates, 5-phosphoribosyl-1-pyrophosphate and ATP both protect the enzyme against digestion with trypsin, histidine made it more susceptible to attack. However, the enzyme differed from ACTase insofar as histidine was found to be bound to enzyme in its histidine-insensitive, as well as its histidine-sensitive form. Moreover, although this enzyme is a much larger protein than any other in the histidine biosynthetic pathway, and indeed can be dissociated into subunits by various means, there is no evidence that any dissociation took place when feedback control was lost. From this it would appear that inhibition by histidine is caused by a conformational change in the protein that occurs after histidine has been bound. These differences between the pyrophosphorylase and ACTase suggest that it may be too early to apply a general theory to all allosteric proteins. However, as R.G. Martin [17] pointed out, this enzyme is difficult to obtain pure; and if and when this is achieved, further light may be shed upon its behaviour.

Further examples of bacterial allosteric enzymes will not be given here; many have been listed and discussed in an excellent review by G. Cohen [18]. It may be mentioned, however, that the regulation of pathways of biosynthesis involving branch points presents a problem. Thus, the conversion of aspartate into aspartyl phosphate, which is catalysed by aspartokinase, is the first step in the biosynthesis of lysine, methionine, threonine and isoleucine—the latter being synthesized directly from threonine, as we have seen, whereas lysine and methionine lie on separate pathways. To take one example by way of illustration, threonine regulates the rate of its own formation by its ability to inhibit aspartokinase. However, were this enzyme inhibited completely the cell would be deprived of lysine which, unlike threonine, might not be readily available. This difficulty is surmounted by the ability of *E. coli* to elaborate three aspartokinases, one of which is inhibited noncompetitively by L-lysine, another competitively by threonine, and the third repressed by methionine. In *Rhodopseudomonas capsulata* the same end is achieved, not by two enzymes that catalyse the same reaction but by a single enzyme which is only inhibited when the two substrates, threonine and lysine, are present together. Inhibition is never total, so that the biosynthesis of methionine from aspartyl phosphate is still permitted. These aspects of regulation are discussed more fully by G. Cohen [1].

Glycolysis and gluconeogenesis

Much of the definitive work on enzyme regulation has been performed with bacteria, but these mechanisms are also important for the control of mammalian metabolism. Although substrate limitation, as we have seen, is a factor in regulating the release of energy, it has become increasingly evident that many of the enzymes of the central metabolic processes are able to receive *direct* 'signals' from AMP, ADP, ATP and acetyl coenzyme A, with the result that the rates of synthesis of glucose, lipids and other materials are coordinated with each other and also with their rates of degradation.

Three of the enzymes concerned with glucose degradation catalyse one-way reactions that are not reversible under normal physiological conditions. Denoting them by the numbers used in Fig. 23 (page 68) these enzymes are (1) glucokinase (hexokinase), (2) phosphofructokinase and (3) pyruvate kinase. When glucose is synthesized, these enzymes are bypassed by: (4) glucose-6-phosphatase, (5) fructose-1,6-diphosphatase and (6) and (7) which denote pyruvate carboxylase and phosphopyruvate carboxykinase respectively. Consider first the relationship of AMP, ADP and ATP to the key enzymes concerned with degradation. Phosphofructokinase (2) can be activated by AMP, ADP and several other compounds; and it is inhibited by citrate and ATP. Clearly, when the ratio ATP/ADP is low in the cell, further breakdown of glucose will be favoured; and when it is high breakdown will be opposed. Pyruvate kinase from yeast [1] and liver [2] is stimulated by fructose-1,6-diphosphate, a phenomenon which, by comparison with feedback, could be called 'feed-forward'.

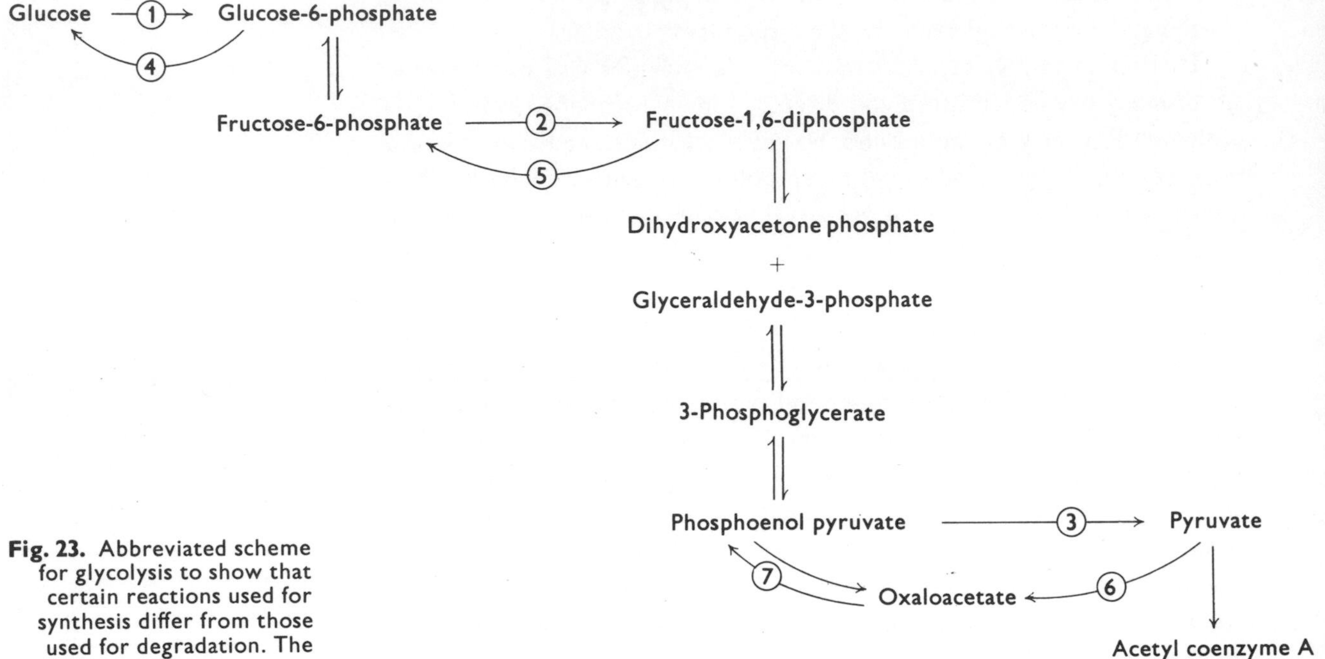

Fig. 23. Abbreviated scheme for glycolysis to show that certain reactions used for synthesis differ from those used for degradation. The numbers refer to reactions discussed in the text.

In glucose biosynthesis, fructose-1,6-diphosphatase (5) is a key reaction and appropriately, the effects of AMP and ATP are the reverse of those for phosphofructokinase; that is enzyme (5) is stimulated by ATP and is inhibited by AMP. Acetyl coenzyme A is a positive effector for pyruvate carboxylase (6). This is also appropriate since its effect is to favour carbohydrate synthesis when acetyl coenzyme A is in abundant supply. Here we have a compound which is also product of lipid degradation giving a 'signal' for carbohydrate metabolism. Enzyme (6) is also responsive to the ATP/ADP ratio since it is inhibited by ADP.

Glucose is stored as glycogen and starch in mammalian and plant tissues respectively, and the reactions by which it is converted into these reserve materials differ from those by which it is released. Liberation of glucose is catalysed by the enzyme glycogen phosphorylase according to the equation

$$(\text{Glucose})_n + H_3PO_4 \rightleftharpoons (\text{glucose})_{n-1} + \text{glucose-1-phosphate}$$

This reaction is reversible and indeed under certain favourable conditions it may be used for synthesis of the polysaccharide. The enzyme glycogen phosphorylase is subject to complex regulation which will be outlined here only briefly; the subject is reviewed by E.G. Krebs and E.H. Fischer [3] to whom much of our knowledge of the system is due. The enzyme exists in two forms, one of which, phosphorylase *a* is active while the other, phosphorylase *b*, is not normally active. Both forms can be dissociated into subunits of molecular weight 125,000: the *b* form consists of two subunits and the *a* form of four. When *b* is converted into *a*, the protein is phosphorylated according to the equation:

$$2\text{ Phosphorylase } b + 4\text{ ATP} \rightarrow \text{phosphorylase } a + 4\text{ ADP}$$

Clearly, the activity of the kinase which catalyses this reaction will determine the amount of active enzyme available for glycogenolysis. It transpires that this kinase in turn can exist in both active and inactive forms, the latter being converted into the former in the presence of adenosine-3′,5′-phosphate (cyclic AMP). The last-named compound is formed from ATP by an enzyme that is stimulated by the hormones epinephrine (adrenaline) and glucagon. By this complex chain of events, an increase in concentration of adrenaline leads to the conversion of phosphorylase *b* into phosphorylase *a* and hence to the liberation of glucose from reserve polysaccharide. It may be mentioned that the reconversion of phosphorylase *a* into phosphorylase *b* is catalysed by a phosphatase. Recent work has added another complication to this system with the demonstration that both phosphorylases appear to be allosteric enzymes for which AMP is the positive effector. Activation by AMP alone could therefore signal the release of glucose-1-phosphate from glycogen when the ratio ATP/ADP falls and the cells require energy to be generated.

When glycogen is biosynthesized, glucose units are transferred to the non-reducing end of the polysaccharide from uridine diphosphate glucose. The reaction is catalysed by glycogen synthetase (UDPG-glycogen glucosyl transferase). J. Larner and co-workers [1, 2] have shown that the enzyme can exist in a form that requires glucose-6-phosphate for activity (D form) and also in a form that does not (I form). Evidently glycogen storage by the D enzyme will be favoured by those factors, which we have discussed, that promote the conversion of pyruvate into glucose-6-phosphate. Further, the I form shows greater activity than does the D form and it is probably significant that D can be converted into an inactive protein by warming, and that this protein in turn is readily transformed to I in the presence of glucose-6-phosphate. Insofar as it exists in two forms, glycogen synthetase resembles glycogen phosphorylase; however a notable difference is that the more active I form is converted into the less active D form by a kinase that is stimulated by cyclic AMP.

The synthesis of fatty acids and its relation to the tricarboxylic acid cycle

A living cell does not synthesize fatty acids by merely reversing the enzymes of the 'fatty acid spiral'. First, the β-ketothiolase reaction (5, Fig. 1, page 12) reaches an equilibrium strongly in favour of cleavage. This reaction is the last one of the 'spiral' and would need to be reversed in order to initiate the biosynthetic pathway. Accordingly synthesis is initiated by a reaction that brings in ATP to provide the free energy needed, malonyl coenzyme A being formed from acetyl coenzyme A by fixation of carbon dioxide:

$$\text{ATP} + \text{acetyl-CoA} + CO_2 + H_2O \rightarrow \text{ADP} + \text{orthophosphate} + \text{malonyl-CoA} \quad (4)$$

This reaction is catalysed by acetyl coenzyme A carboxylase, an enzyme that contains biotin which becomes carboxylated in the first step of the reaction. When the malonyl thioester condenses with the fatty acid carbon chain to be lengthened, two carbon atoms are added and carbon dioxide is lost. Another difference between synthesis and degradation is that NADPH is used to donate hydrogen atoms and NAD is used to receive them. But perhaps of greater interest is the fact that during the biosynthetic reactions, malonate and fatty acid are bound to a protein as their thioesters: in plants and bacteria this is a small, soluble 'acyl carrier protein', whereas in yeast it is not free but is attached to a complex formed by the association of the six enzymes involved in the process of lengthening the carbon chain by addition of two more atoms. F. Lynen and his colleagues have isolated this multienzyme complex as a particle of molecular weight 2·3 millions. The fatty-acid chain becomes attached to one end of the panthotheine molecule through a thioester linkage whilst the other end of pantotheine is thought to be anchored (through a phosphate group) to protein at the centre of the complex. The pantotheine molecule is thus envisaged as an 'arm' that can revolve over each catalytic site in turn, taking the fatty acid from one enzyme to the next as two-carbon additions are made to the chain and successive reductions and hydrations are catalysed [1]. The experimental basis for this model emphasises the basic differences between biosynthesis and degradation of fatty acids in yeast; for the processes differ not only in their biochemistry but in the nature and location of the enzymic structures used.

In 1952, R.O. Brady and S. Gurin [2] obtained a cell-free extract from pigeon liver that converted ^{14}C-acetate into long chain fatty acids at a rate that was greatly enhanced by additions of citrate. In more recent years various workers have shown that citrate exerts its effect upon acetyl coenzyme A carboxylase (equation 4) and that activation by citrate is accompanied by an increase in the sedimentation velocity of the carboxylase. Acetyl coenzyme A carboxylase has now been isolated in pure form by investigators at the Department of Biochemistry of the New York University School of Medicine [3] and its physicochemical properties, including its behaviour with citrate, have been correlated with its catalytic activity. Studies using the analytical ultracentrifuge showed that the enzyme is a very large asymmetric molecule of molecular weight about 8 millions. By dialysis against a solution of sodium chloride in tris buffer the protein was dissociated into subunits of molecular weight 400,000. These subunits were not catalytically active, but when they were dialysed against phosphate buffer, or against tris buffer containing citrate or isocitrate, large and catalytically active forms of the enzyme were regained. The molecular weight of the enzyme 'reconstituted' using phosphate was 8 millions and that using citrate or isocitrate was 4 millions, so that apparently the former contained 20, and the latter 10, subunits. Under the electron microscope the inactive small forms appeared as particles of which the largest were 100–300 Å in size, whereas the active enzyme particles obtained by treating them with isocitrate were in the form of filaments 70–100 Å wide and up to 4,000 Å in length. The model proposed for these findings was:

$$\text{SMALL FORM} \rightleftharpoons \begin{bmatrix}\text{MODIFIED} \\ \text{SMALL FORM}\end{bmatrix} \rightleftharpoons \text{LARGE FORM}$$
(subunit.) (filaments)

The 'modified' small form, for which there is at present no direct experimental evidence, is suggested to be in equilibrium with the subunit and to have a conformation that permits linear aggregation to give the large form. Isocitrate and citrate are among the factors favouring displacement of the equilibria to the right; ATP and malonyl coenzyme A favour a shift towards the small form. Whatever the significance of these effects of citrate and isocitrate might be as far as metabolic regulation is concerned, it seems likely that the long, fibrous active protein may also serve some structural purpose.

In order for citrate (or isocitrate) to be solely responsible for regulating the activity of this enzyme within the cell, it must accomplish the conversion of inactive to active form at physiological concentrations. F. Lynen and his colleagues [70.1] have presented evidence that the concentrations required for activation are too high for citrate to exert sole control, but they have studied the inhibition of activity due to palmitoyl coenzyme A and other acyl coenzyme A compounds. These inhibitions are exerted at much lower concentrations and are competitive with citrate, the enzyme activator. It appears, therefore, that this is one more example where the first enzyme of a biosynthetic sequence is inhibited by an end product. Raised concentrations of fatty acids in the blood are associated with starvation and diabetes, conditions in which fatty acid synthesis is known to be almost fully blocked. This can also be accomplished in normal animals when they are fed on a diet rich in fat or when injected with chylomicrons. Feedback inhibition of acetyl coenzyme A carboxylase is a regulatory mechanism for fat biosynthesis that would account for many observations made with intact animals.

Acetyl coenzyme A is widely encountered in metabolism, but there are three main channels through which it is utilized. It may be used, as we have seen for fat biosynthesis; or it may be oxidized by the tricarboxylic acid cycle. Alternatively, condensation reactions may occur to form the coenzyme A derivatives of acetoacetate and β-hydroxy-β-methylglutarate; and from the latter, steroids may be biosynthesized or the molecule may cleave to give acetoacetic acid. It has been known for a very long time that acetoacetate and β-hydroxybutyrate accumulate in the blood in certain states of carbohydrate deficiency such as starvation or diabetes. However, these compounds are now recognized to be important metabolites under normal conditions for, as H.A. Krebs and his colleagues [1] have shown, acetoacetate constitutes over 90% of the fuel of respiration in the heart whilst other tissues consume smaller but still substantial quantities. Acetoacetate and β-hydroxybutyrate may be regarded, in fact, as transport forms by which acetyl coenzyme A is distributed through the blood stream from the liver to other tissues where they serve as readily available sources of energy.

We have discussed one mechanism which affects the distribution of acetyl coenzyme A through the three main channels of its metabolism, namely the inhibition of fatty-acid biosynthesis by long-chain acyl coenzyme A derivatives. It now appears from the work of a group of investigators in the University of Munich that a similar mechanism may operate to regulate its oxidation through the tricarboxylic acid cycle [1]. The enzyme Citrate synthase from liver, which condenses acetyl coenzyme A with oxaloacetate, also has been found to be strongly inhibited by fatty acid thioesters of coenzyme A. The kinetics of inhibition suggest that allosteric interaction occurs. Thus, sigmoid plots were given when reaction rates were plotted against concentrations of stearyl coenzyme A, and the enzyme lost sensitivity to the inhibition, but little catalytic activity, when treated with *p*-chloromercuribenzoate. Moreover, the analytical ultracentrifuge showed that the single peak given by the crystalline enzyme was split into two components when stearyl coenzyme A was added, and that this dissociation was reversed by adding oxaloacetate. Clearly, the concentration of oxaloacetate within the tissue must also play an important part in regulation, not only because of this effect upon the stability of citrate synthase but also because the compound is a substrate for the enzyme. Thus, when the rate of gluconeogenesis is high, the concentration of oxaloacetate is lowered because of its rapid conversion into phosphoenol pyruvate. Under these circumstances, the liver may be regarded as compensating for a diminished rate of the tricarboxylic acid cycle by an increased rate of oxidation outside the cycle, namely by converting fatty acids into acetoacetate which is then distributed to the various tissues and oxidized.

Two other examples of probable means by which the tricarboxylic acid cycle may be regulated are provided by the work of J.H. Hathaway and D.E. Atkinson [2]. They have shown that citrate synthase from yeast is inhibited by ATP as well as by palmityl coenzyme A. In this case, however, ATP competes at physiological concentrations with acetyl coenzyme A and not with oxaloacetate; neither do changes in molecular weight accompany this inhibition. From the point of view of metabolic regulation it appears that when energy is available in abundance, that is when the ratio ATP/ADP is high, the metabolism of acetyl coenzyme A may be directed towards biosynthesis rather than oxidation. These authors also found that AMP is a positive effector for yeast isocitrate dehydrogenase so that when the ratio ATP/ADP is low the operation of this step in the tricarboxylic acid cycle is speeded up.

In this section we have illustrated the fruitful concept of metabolic regulation mediated through allosteric effects by choosing a few examples from an extensive literature. In the course of time many more examples will be found of enzymes that are inhibited by a range of metabolites, none of which bear any close structural resemblance to the substrates. In such cases it is tempting to invent an equal number of ingenious mechanisms by which the enzymes could, in principle, be regulated. However, it should be borne in mind that in order to serve this function these metabolites must powerfully inhibit the action of the enzymes at physiological concentrations. For example, isocitrate lyase is inhibited by fumarate, α-oxoglutarate, malate, succinate and glycollate all of which are substrates for metabolically related enzymes. Nevertheless it appears from the work of

H.L. Kornberg that regulation is not imposed by these compounds, but by phosphoenol pyruvate which inhibits more strongly than any of them.

Regulation of enzyme synthesis in bacteria

A bacterial cell exercises 'coarse control' of its metabolism by altering the rates of synthesis of particular enzymes in response to changes in environment. This phenomenon may be illustrated by the following experiments. First, a species of *Pseudomonas* when grown in a nutrient broth may contain barely detectable amounts of the enzymes of Fig. 8, page 33. When the cells are transferred to a mineral salts medium there will be no growth initially, but during this period of lag, considerable synthesis of the relevant enzymes occurs, sufficient to support the requirements of exponential growth at the expense of benzoate as sole source of carbon. The synthesis of these catabolic enzymes was 'induced', or 'derepressed', by benzoate and by certain compounds arising from its degradation. If the mineral salts medium had contained succinate as well as benzoate, much smaller amounts of these enzymes would have been induced, and would then be said to show catabolite repression by succinate. But a compound need not be metabolized in order to act as an inducer. Thus, when lactose is used as a carbon source for the growth of *E. coli*, the sugar undergoes transgalactosidation inside the cells and then induces the enzyme β-galactosidase. However, the most potent inducers of this enzyme are methyl-thio-β-D-galactoside and isopropylthio-β-D-galactoside; and neither compound serves as substrate for the enzyme. These experimental conditions which are very favourable for studies of induction have been called *induction gratuit* by J. Monod and M. Cohn [1]. The control of this enzyme has been investigated more thoroughly than any other, mainly thanks to the work of J. Monod and his colleagues: it has been shown that this, and other, induced enzymes are formed directly from amino acids and not from polypeptide precursors; and when fully induced an *E. coli* cell may contain 5% of its protein in the form of this single enzyme, β-galactosidase.

In these examples, 'coarse control' was exerted over catabolic sequences. A second type of experiment concerns anabolic, or biosynthetic, reactions. We shall illustrate from the early work of B.N. Ames [2] and of A.B. Pardee [3] and their colleagues whose subsequent investigations of the biosynthesis of histidine and pyrimidines respectively have contributed so much to our knowledge of the processes of enzyme repression and derepression. A mutant of *Salmonella typhimurium* required histidine in order to grow rapidly but could grow slowly without an external supply of this amino acid. During growth with histidine present in the medium, the bacteria contained only low concentrations of the enzymes of the histidine biosynthetic pathway; but as soon as the histidine was exhausted, and the cells grew at the slower rate, they began to synthesize relatively large amounts of all four of the enzymes that were assayed. By contrast, the amount of glutamate dehydrogenase remained unaltered: this enzyme is not directly concerned with histidine metabolism. Whereas catabolic enzymes in our previous examples were induced by addition of benzoate or lactose, in this case anabolic enzymes were repressed by adding histidine and derepressed when it was removed.

A difference in meaning of the words 'inhibition' and 'repression' might be emphasized at this point. Histidine *inhibits* the enzyme phosphoribosyl-ATP-pyrophosphorylase when it reduces its rate of catalysis. The amino acid (or, as now appears likely, its acyl-sRNA derivative) *represses* when it reduces the rate at which the enzyme itself is synthesized. Just as the enzymes of histidine—requiring mutants of *S. typhimurium* were repressed by histidine, so mutants of *E. coli* that needed a pyrimidine for growth also began to synthesize large amounts of the enzymes of this metabolic pathway when the cells had removed the added pyrimidine from the growth medium. Advantage was taken of this behaviour when aspartate carbamoyltransferase (ACTase) was first obtained in crystalline form. One particular mutant was grown in a mineral salts plus glycerol medium, supplemented with arginine and a limiting concentration of uracil; and after growth had ceased due to uracil depletion, dihydroorotate (an intermediate in the pathway) was added to permit slow growth. Under these conditions the specific activity of ACTase increased one-thousandfold over the basal level of the enzyme.

Our current conceptions of some of the events that may occur when proteins in general are synthesized have been enlarged by a careful scrutiny of the particular phenomenon of repression and derepression in bacteria. In the previous chapter we referred to the reactions by which *Pseudomonas putida* converts benzoate and *p*-hydroxybenzoate into metabolites of the tricarboxylic acid cycle (see Fig. 8, page 33 and also the summary given below).

Protocatechuate $\xrightarrow{1}$ CMA $\xrightarrow{2}$ $\searrow^{3}$

β-Oxoadipate enol-lactone $\xrightarrow{4}$ β-Oxoadipate $\xrightarrow{5}$ Succinate + Acetyl-CoA

$\nearrow^{3'}$

Catechol $\xrightarrow[1']{}$ MA $\xrightarrow[2']{}$

When *Pseudomonas putida* is exposed to benzoate, this substrate induces an oxygenase that converts it into catechol. By a separate event, catechol 1,2-oxygenase (enzyme 1′) is next induced, probably by *cis-cis*-muconate (MA); and then enzymes 2′ and 3′ are coordinately induced by MA. Enzyme 4 is induced either by β-oxoadipate or by its thioester with coenzyme A. When the bacteria are exposed to protocatechuate, the growth substrate induces the dioxygenase (enzyme 1) that opens the benzene nucleus and gives rise to *cis-cis*-β-carboxymuconate (CMA); but in contrast to enzymes 2′ and 3′, enzymes 2, 3 *and* 4 are coordinately derepressed. In this case β-oxoadipate is the inducer of enzymes 2 and 3 (and 4), and the product of fission of the benzene nucleus is not their inducer. There are several ways of deciding whether a particular enzyme is induced as an individual, or whether it is induced along with several others so that the ratio of the amounts of the various enzymes is maintained constant, although their absolute amounts may vary. One method, used by L.N. Ornston [1] consisted in growing the bacteria on benzoate —or *p*-hydroxybenzoate—with various amounts of succinate or glucose present in order to exert catabolite repression to different degrees. In this way it was possible to study different batches of cells that contained amounts of these enzymes varying over a wide range. It was then shown that the specific activities of enzymes 2 and 3, for example, could be plotted against those of enzyme 4 to give a straight line in each case, whereas no relationship existed between the amounts of enzyme 4 and those of enzyme 1: the synthesis of the latter is not regulated by β-oxoadipate, but by protocatechuate, and is not coordinate with enzymes 2, 3 and 4.

The synthesis of enzymes for a particular metabolic pathway is not necessarily regulated in the same manner by bacteria of different genera. Thus *Moraxella calcoacetica* catalyses exactly the same set of reactions for protocatechuate catabolism as those described for *Pseudomonas putida*,

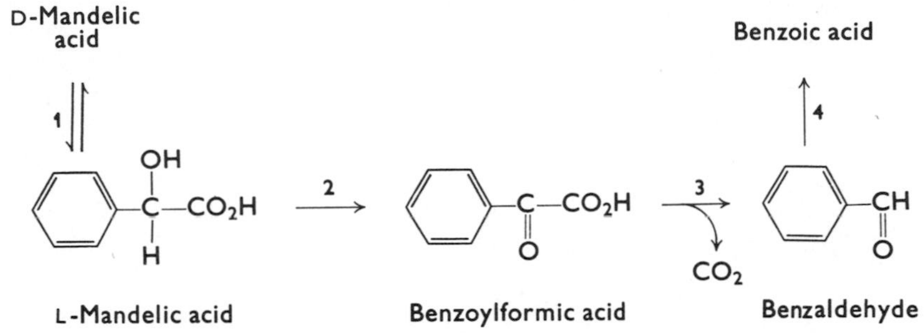

D-Mandelic acid ⇌ (1) L-Mandelic acid →(2) Benzoylformic acid →(3, $-CO_2$) Benzaldehyde →(4) Benzoic acid

Fig. 24. Conversion of D-mandelic acid into benzoic acid by *Pseudomonas*.

but in this case all the enzymes, from protocatechuate oxygenase (1) to β-oxoadipate succinylcoenzyme A transferase (5), are synthesized in one coordinate block [1]. This raises the question of how enzymes 4 and 5 are induced when the cells metabolize benzoate and do not encounter protocatechuate. It transpires that a pair of enzymes, which we can denote as 4′ and 5′, are elaborated by benzoate-grown cells. They catalyse the same respective reactions as enzymes 4 and 5, that is they are isofunctional with them, but one pair differs from the other in respect of stability towards heat. Moreover, just as enzymes 1 to 5 are induced by protocatechuate, so are enzymes 2′ to 5′ induced in one coordinate block in *Moraxella calcoacetica* by *cis-cis*-muconic acid.

R.Y. Stanier and his colleagues have also used mutants to reveal the existence of coordinate blocks of certain enzymes engaged in catabolism. Thus, D-mandelic acid is converted into benzoate by the four reactions of Fig. 24, above. Five enzymes are actually involved since reaction 4 can be catalysed by two enzymes, one requiring NAD and the other NADP. A mutant of *Pseudomonas fluorescens* was isolated that lacked the racemase (enzyme 1) which interconverts the two optical isomers. This mutant could not grow with D-mandelate as a source of carbon on account of its genetic block, but nevertheless the enzymes for reactions 2, 3 and 4 were fully induced when it was incubated with D-mandelate. No other enzymes were induced. Those that degrade benzoate and the compounds that follow it in the sequence require benzoate, *cis-cis*-muconate and β-oxoadipate for their induction, and these catabolites were not formed by the cells because D-mandelate could not be metabolized. The first two enzymes of the pathway of degradation of tryptophan by *Pseudomonas* species are also coordinately derepressed. A mutant lacking enzyme 1 of

Fig. 25, below, synthesized enzyme 2 when it was exposed to L-kynurenine, proving that L-tryptophan was not the inducer. Similarly, enzyme 1 was induced by L-kynurenine in another mutant that lacked enzyme 2. These and other observations, including the induction of the two enzymes by non-metabolizable analogues of L-kynurenine, were used to establish coordinate induction in this pathway.

There are several examples of coordinate repression by end products of enzymes in biosynthetic sequences. Thus, five enzymes are involved in biosynthesizing uridine phosphate from carbamoyl phosphate and aspartate. The last four of these enzymes are coordinately repressed, but the first one (ACTase), although repressed by pyrimidines as we have seen,

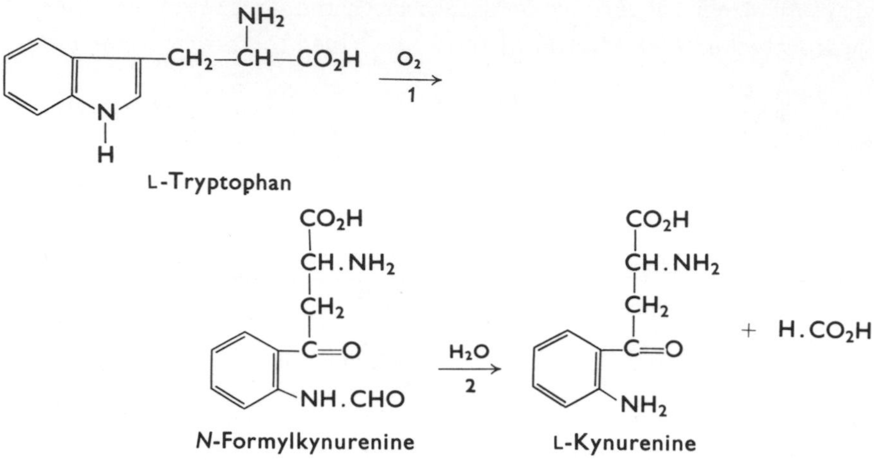

L-Tryptophan $\xrightarrow[1]{O_2}$ *N*-Formylkynurenine $\xrightarrow[2]{H_2O}$ L-Kynurenine + $H.CO_2H$

Fig. 25. Conversion of L-tryptophan into L-kynurenine by *Pseudomonas*.

is not coordinate with the remainder. It will be recalled that for the repression of a group of enzymes to be coordinate, the ratio of the amount of any enzyme to that of any other in the group must be constant, regardless of the extent of the repression or induction. The earlier work of B.N. Ames and his colleagues, which showed that four of the enzymes in the pathway of biosynthesis of histidine were coordinately repressed, has since been extended. All ten enzymes that convert phosphoribosyl pyrophosphate and ATP into histidine have now been shown to be coordinate. Moreover, several hundreds of histidine-requiring mutants of *Salmonella typhimurium* have been submitted to genetic analysis by P.E. Hartman and co-workers, and it has been shown that the structural genes for the enzymes of histidine biosynthesis are in a cluster, the histidine operon, on the *Salmonella* chromosome. This combination of biochemical and genetic analysis is the most extensive carried out for bacteria and, as we shall see, it has had a considerable impact upon theories of enzyme biosynthesis and regulation.

We have given examples of metabolic routes that serve for the biosynthesis of two different compounds by branching out from one reaction step or sequence that is common to both pathways. This situation presents problems for regulation which, in the case of feedback inhibition, may sometimes be solved by the elaboration of two separate enzymes to catalyse the same reaction step, one being subject to inhibition by the first metabolite and the other by the second. Clearly this raises the possibility that the synthesis of the two enzymes might also be repressed or derepressed independently. However, there appears to be another solution to the problem. Thus, four of the enzymes that catalyse the biosynthesis of valine from pyruvate are the same as those that catalyse the synthesis of isoleucine from pyruvate and threonine (Fig. 16, reactions 2, 3, 4 and 5, page 49). It is clear that should additions of valine prevent the formation of these enzymes, then the apparatus for isoleucine biosynthesis would also be lost in consequence. Moreover, since α-oxoisovalerate is an obligatory intermediate in the biosynthesis of leucine (not shown in Fig. 16, page 49) the organism would also lose some of the enzymes that operate in this biosynthetic pathway. The solution to the problem has been arrived at by a mechanism that H.E. Umbarger and his colleagues [1] have called 'multivalent repression'. It appears that there is no repression

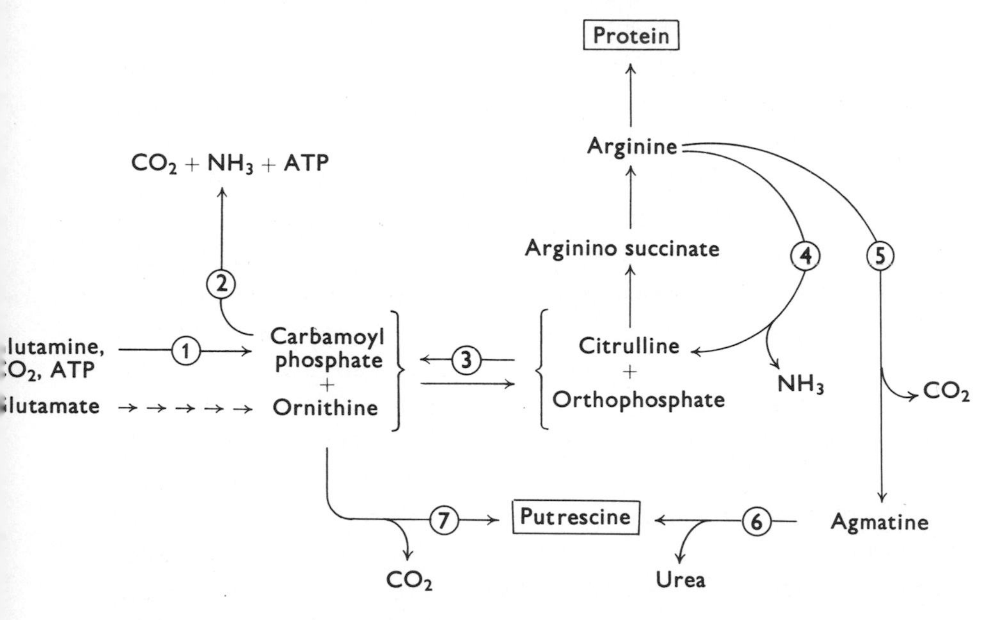

Fig. 26. Biosynthesis and degradation of arginine by bacteria. The numbers refer to reactions discussed in the text. In *E. coli*, putrescine and polyamines are synthesized by reactions 5, 6 or 7. Redrawn from Stalon V., Ramos F., Piérard A. and Wiame J. (1967) *Biochim. biophys. Acta*, **139**, 91.

of the common enzymes in these pathways unless each of the ultimate end products, valine, isoleucine and leucine are simultaneously present in excessive concentrations. No single amino acid can therefore shut off the synthesis of the enzymes required by the others.

One further example may be given to illustrate how bacteria are able to regulate their enzymic activities according to the functions these enzymes are required to perform. Some of the metabolic interrelationships of arginine, citrulline and ornithine are shown in Fig. 26, above. Bacteria, like other organisms, incorporate arginine into protein and they synthesize this amino acid by way of ornithine, citrulline and argininosuccinate. They also make the carbamoyl phosphate they need by means of reaction 1. By contrast, carbamoyl phosphate is usually degraded by a different route, reaction 2. Species of *Pseudomonas* that can utilize arginine as a source of carbon and nitrogen develop enzyme 4, arginine deiminase, and obtain energy through the ATP released in reaction 2. However, when this pathway is used, reaction 3 must operate from right to left and this is not accomplished by reversing the action of 'biosynthetic' ornithine carbamoyltransferase. V. Stalon, F. Ramos, A. Piérard and J.M. Wiame [2] have shown that *Pseudomonas* uses another enzyme. The synthesis of this 'catabolic' ornithine carbamoyltransferase, which exhibits maximal activity at neutrality, is induced by arginine, whereas the amino acid strongly represses the anabolic enzyme which has a pH optimum at 8·5. R.W. Bernlohr [1] has also shown that this enzyme exists in two forms in *Bacillus licheniformis*. The other biosynthetic product of Fig. 26, page 77, besides arginine, is putrescine which, in *E. coli* at least, probably serves an important metabolic purpose along with the polyamines spermidine and spermine. These compounds appear to help in stabilizing cellular organelles, membranes and ribosomes; they also stimulate DNA-directed RNA polymerase [2]. A. Raina and S.S. Cohen [3] showed that when an arginine-requiring mutant of *E. coli* was incubated in a medium lacking arginine, additions of spermidine caused a marked stimulus in synthesis of RNA; putrescine, from which spermidine is derived, could reverse this effect. If we assume, then, that *E. coli* needs to synthesize putrescine, a problem in enzyme regulation immediately arises. The most direct route to putrescine is by decarboxylation of ornithine (reaction 7), but when arginine is present in the growth medium it represses the enzymes of its own biosynthetic pathway, and these include the enzymes that form ornithine. However, there exists an alternative route by way of reactions 5 and 6; that is, arginine is decarboxylated to give agmatine which is then hydrolysed to urea and putrescine. D.R. Morris and A.B. Pardee [4] have shown that both of these decarboxylases are present in *E. coli* when grown at neutrality, and their existence appears to provide an unusual solution to the regulatory problems encountered in a branched biosynthetic pathway. It is also interesting that two other enzymes that catalyse reactions 5 and 7 can be induced in *E. coli* by arginine and ornithine respectively under specialized conditions. They were discovered in 1940 by E.F. Gale and their sole function appears to consist in decarboxylating these amino acids when cultures grow increasingly acidic. The physical properties of these two catabolic enzymes differ from those of their biosynthetic counterparts in several respects.

The biosynthesis of proteins

In order to understand modern views as to how the synthesis of an enzyme may be repressed or induced, it is first necessary to consider the events that are believed to take place when the amino acids of any protein are assembled in their correct sequence. This subject has been reviewed on many occasions and at all levels of sophistication in recent years. Accordingly we shall attempt only a brief outline, and the general reader is referred to a particularly lucid account by J.D. Watson in *Molecular Biology of the Gene* [5].

DNA is the material of the genes. A strand of DNA consists of a great many molecules of deoxyribose joined in a linear fashion through phosphate diester linkages at the 3′ and 5′ positions of the sugar molecule; and each deoxyribose carries either a purine or a pyrimidine attached at the 1′ position. The polynucleotide chain, therefore, consists of a very long *backbone*, consisting of sugar and phosphate groups, which carries at perfectly regular intervals the purine bases adenine (A) and guanine (G) and the pyrimidines cytosine (C) and thymine (T). DNA is not usually encountered as single strands, however, but as double strands in which the backbones are held at a constant distance apart throughout their lengths, due to the fact that A must always be paired with T and G with C. This feature is dictated by the sizes and shapes of the bases constituting each pair, since only the pairing stated can permit the ready formation of the hydrogen bonds that are required for structural stability. However, as the famous model of Watson and Crick requires, the two backbones are not held at a constant distance apart like the lines of a railroad track. Instead, the two long chains are twisted about each other in the form of a regular double helix, with the pairs of bases stacked one above the other in planes perpendicular to the axis of the helix. The random assembly of the four bases along a very long strand of DNA would permit an astronomical number of possible sequences; but only one particular sequence is consistent with the biological functions of the molecule. Moreover, the model of Watson and Crick demands that since A must always face T, and G must face C, the two strands are complementary: a strand with a given sequence can only pair with this complementary sequence and with none of the millions of alternatives. When replication occurs, the two strands separate and each one guides the formation of its partner, so that when replication is finished there will be two completely identical molecules. The principal requirement for a molecule to serve as genetic material is therefore satisfied, namely that of maintaining an unaltered pattern when it replicates.

A strand of DNA is also copied as a complementary molecule of RNA when the nucleotides of the latter are assembled by DNA-directed RNA polymerase, although uracil (U) takes the place of T in RNA. When a protein is biosynthesized the first step is to form a molecule of messenger RNA (m-RNA): this will contain a base sequence which is complementary to the genetic material from which it was copied. The copying of DNA by m-RNA is termed *transcription*. The process of assembling the twenty or so different amino acids in their correct sequence in the finished molecule of protein is called *translation* because the order of the bases in m-RNA is translated into the order of the amino acids. Translation is accomplished at the ribosomes. These are particles of ribonucleoprotein composed of two subunits, of molecular weights 0·9 and 1·8 million respectively, associated to give a functioning particle of molecular weight 2·7 millions. Several ribosomes can be attached at different points to the same strand of m-RNA to constitute a polysome; and each amino acid in turn is brought to a site on the ribosome near its point of attachment to the m-RNA. Each amino acid was previously prepared for the occasion by enzymic reactions that attached it to its own particular species of transfer RNA which is denoted by s-RNA (or alternatively as t-RNA). This s-RNA contains at some point in its polynucleotide chain a sequence of three bases (CCA in the case of tryptophan s-RNA); and this triplet of bases is able to pair with the three complementary bases (GGU, for example), and *only* with these three bases, when they become available on the m-RNA at its point of contact with the ribosome. Each ribosome is envisaged as moving along, or rolling along, the strand of m-RNA so that one triplet at a time is placed opposite the point to which an amino acid is brought. In this way, each molecule of s-RNA can be regarded as an *adaptor* that brings its own particular amino acid to one unique place on the m-RNA strand. The triplet that directs the placing of the amino acid, or 'codes' for it, is called a *codon*. A codon for tryptophan is thus GGU. After being freed from its s-RNA, the amino acid is then joined enzymically, through its amino group, by peptide bond formation with the carboxyl of the amino acid at one end of the growing polypeptide chain. Each ribosome has two adjacent sites, one for the growing polypeptide chain and the other for the incoming s-RNA with its amino acid attached. The biosynthesis of a polypeptide appears to be initiated by the attachment to the ribosome of one particular type of s-RNA, namely that for *N*-formylmethionine, which recognizes a codon, AUG, GUG or GUA [1] on the m-RNA. It is then bound to the 'polypeptide' site and initiates polypeptide synthesis. Since *N*-formylmethionine cannot form a peptide bond through nitrogen, it is not incorporated into protein and appears to have the specific function of initiating protein biosynthesis.

When the polypeptide chain is being synthesized, a particular ribosome is first attached at the 5′ end of the m-RNA and is thought to move along the whole length, translating the sequence of triplets into a sequence of amino acids as they appear in the finished protein molecule. A strand of m-RNA may contain the information for several different polypeptides, and it is believed that each one is finished off when the ribosome encounters a 'nonsense' triplet, UAA or UAG, which does not code for any amino acid. When the ribosome reaches the 3′ end of the m-RNA one imagines that it falls off, and so do the other ribosomes in its wake, when their turns come. In bacteria at least, m-RNA that is not attached to ribosomes is quickly degraded by enzymes, so that new molecules are produced continuously. As for the ribosomes, there are stable structures and there is growing evidence that they are something more than inert 'work benches' on which proteins are made, for they also appear to be active in pulling m-RNA molecules from the templates at which they are formed.

Research into protein biosynthesis expands rapidly and continuously and no attempt has been made to do full justice to the subject in this brief outline; but it may suffice as a basis for discussing the regulation of enzyme biosynthesis.

Regulation of the biosynthesis of enzymes

In 1961, on the basis of their extensive genetic and biochemical analyses of β-galactoside formation in *E. coli*, F. Jacob and J. Monod [2] put forward the Operon Theory of regulation of gene activity. This theory, and the Watson-Crick model for the structure of DNA, have been amongst the most fruitful concepts in molecular biology.

For several systems we have discussed, blocks of functionally related enzymes can be induced or repressed coordinately. It is therefore reasonable to assume that one region of the chromosomal DNA would be set aside for coding this group of enzymes, and that one strand of m-RNA would be used for their translation. Such a region of the chromosome is called an *operon*, although it may be observed that an operon need not necessarily contain a number of genes: it could contain only one, which would govern the synthesis of one individual enzyme independently of all the others. The functioning of the operon is controlled by an *operator*, a sequence of nucleotides situated at one end of the operon. The existence of operators was first revealed by genetic analysis, and their function is essentially negative: that is, if an operator combines with a repressor the adjacent genes are not transcribed; but if it is free, the transcribing enzyme, possibly DNA-directed RNA polymerase, can begin to assemble the constituent nucleotides of m-RNA. Jacob and Monod proposed that in addition to the *structural* genes—that part of the chromosome where transcription occurs—there is another region, the *regulator* gene, which controls the synthesis of such a repressor. To date, it has emerged from independent studies of two different systems that their repressors are proteins (P in Fig. 27, below). From the functions assigned to repressors

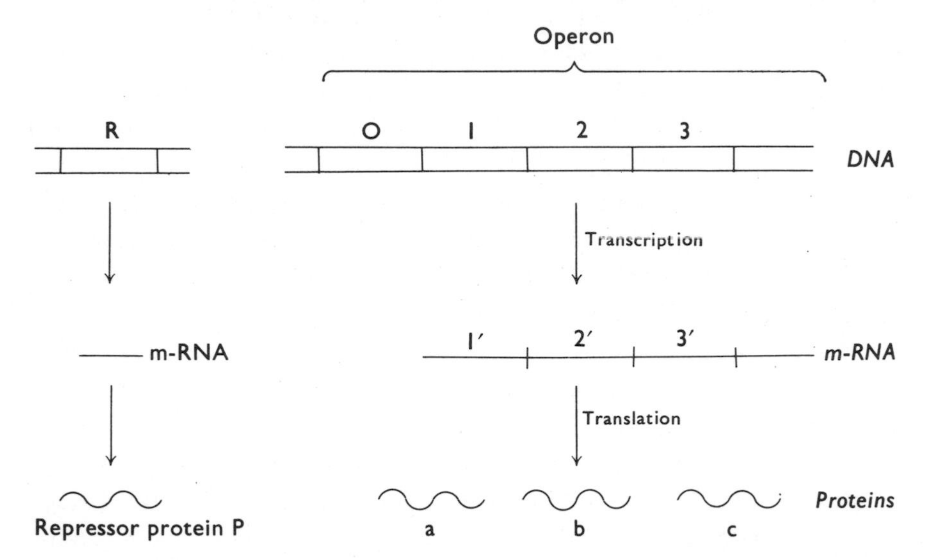

Fig. 27. The operon model for regulation of protein biosynthesis. R and O are the regulator and operator genes respectively. 1, 2, 3 ... are structural genes. The product of the regulator gene, P, may combine directly with O and may be prevented from doing this if it first combines with an inducer. Alternatively it may not be able to combine with O unless it first binds to a co-repressor.

we can expect them to exhibit allosteric properties. Thus, if the proposed mechanism operates for cells whose proteins *a*, *b*, *c*, etc. are capable of being induced, then P is considered to be able to combine with the operator site, O, before they are exposed to the inducer but not afterwards. The inducer combines with P, modifies its properties, so that its affinity for O is abolished: repression is therefore lifted and transcription proceeds. The theory could be enlarged to embrace the repression observed when the concentration of a catabolite is raised: this compound would combine with P and reverse the effect of the inducer. Turning from catabolic to anabolic sequences, the regulator gene for a series of biosynthetic enzymes would control the synthesis of a repressor that would have little or no affinity for O until it combined with the end product of the sequence. In genetic analysis the regulator gene is termed the R gene (i in the case of the *lac*-system), the O gene refers to the operator site which has affinity for the product of the R gene, and the structural genes are those that control the synthesis of the proteins *a*, *b*, *c*, etc., required for the metabolic pathway.

We shall outline the methods used to isolate two repressor proteins. This work is clearly of the greatest importance because it opens up many possibilities for the future investigation of control mechanisms. When a compound has existed hitherto simply as a necessary postulate in a satisfying theory, its isolation increases confidence; but more important, it enables assessments to be made of the relative merits of various alternative proposals. Thus the two proteins that were isolated showed affinity for DNA rather than RNA; therefore their action is exerted at the stage of transcription as Jacob and Monod first postulated, and not during translation as others have suggested later. W. Gilbert and B. Müller-Hill [1] isolated a mutant of *E. coli* which bound very tightly a nonmetabolizable inducer of β-galactosidase, namely isopropylthio-β-D-galactoside (IPTG). This exceptional tightness of binding was a necessary property for a component of the cell to possess if, as in this case, the only physical property that could be used for assay and isolation was its capacity to bind inducer. A cell-free extract of the mutant was prepared and fractionated to give a protein solution that was able to draw ^{14}C-IPTG into a dialysis sac to give a concentration about twice that outside the sac. When the preparation was sedimented through a glycerol gradient in the presence of a solution of ^{14}C-IPTG of uniform concentration, the radioactivity became redistributed to give a high concentration in one particular zone. From the position of this peak of radioactivity it was calculated that the material binding IPTG had a molecular weight of 150,000–200,000. The compound behaved as a protein since it was destroyed by protease but was not attacked by enzymes that hydrolyse DNA and RNA; moreover its ability to bind IPTG was lost at 50°.

M. Ptashne [2] has isolated a repressor from λ phage. This repressor blocks the expression of the other genes of the phage and so keeps the phage dormant within its host, *E. coli*. Cells of the host organism were first irradiated with ultra-violet light to destroy their DNA, thereby causing a drastic reduction in cellular protein synthesis. The irradiated cells were then infected with many λ phages which, under these conditions, synthesized little or no phage protein except the repressor. By feeding the infected cells with radioactive amino acids, therefore, a single labelled protein could be separated on columns of DEAE-cellulose. This protein was identified as the product of the regulator gene by showing that it was not produced by a λ phage which, from genetic analysis, was known to have suffered a mutation in this region of the chromosome. Finally, the isolated, labelled protein was mixed with DNA from λ phage, sedimented through a sucrose gradient and shown to bind to this DNA. It did not bind to DNA from a closely related phage that differed only insofar as it was not sensitive to the repressor material from the λ phage used in these experiments. The *lac* repressor isolated by Gilbert and Müller-Hill likewise binds specifically to *lac* DNA and is removed by IPTG (experiments of W. Gilbert quoted in reference [2]).

Modulation and polarity

If a ribosome starts its journey on the m-RNA from the end of the strand that corresponds to the operator gene, and if it moves *steadily* to the other end, then we should expect equal numbers of proteins *a*, *b*, *c*, etc., to be formed (Fig. 26, page 77). However, there are two lines of evidence that this is not so. Consider the ordering of the genes in the *lac* operon, which is known to be as follows:

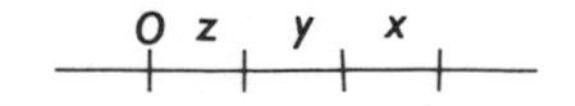

Here, *O* is the operator region, *z* is the structural gene for β-galactosidase, *y* is the gene for galactoside permease and *x* is the gene for thiogalactoside transacetylase. The regulator gene *i*, it might be mentioned, maps outside of the *lac* operon. One indication of the limitations of this simple model was the crystallization of the enzyme thiogalactoside transacetylase and the determination of its molecular weight by I. Zabin [1]. The value he obtained enabled him to calculate that, far from these enzymes being formed in equal numbers, about 25 molecules of the subunit of β-galactosidase are made for each one molecule of the transacetylase. A second indication was the discovery of polar mutants [2, 3] in which a mutation in the *z* gene not only affected the formation of β-galactosidase but also led to a fall in the rates at which the products of the *y* and *x* genes were made. This phenomenon of polarity, as well as the finding of different rates of polypeptide synthesis for the products of one operon, both receive an explanation through an extension of the operon theory proposed by B.N. Ames and R.G. Martin [4] and later modified in some details [5]. A strand of m-RNA can accommodate several ribosomes, the whole constituting a polysome, and each ribosome begins its translation at the operator end. If the messenger is polycistronic, that is, coding for several polypeptides, each ribosome is supposed to encounter a modulating codon when it has finished translating for each polypeptide. A modulating codon appears to be the same as a 'nonsense' codon [5]. After this encounter the ribosome will need to be reorientated on the m-RNA before it can resume translation, and it may be assumed that this happens when it makes contact with a triplet whose function is to initiate formation of the next polypeptide. As to the nature of this triplet, it is established that codons exist for *N*-formylmethionine whose only known function, as we have seen, is that of an initiator. However, synthetic polyuridylic acid can direct polyphenylalanine synthesis and mixed synthetic polynucleotides can direct incorporation of all amino acids into peptides. These observations suggest that many different codons might serve as initiators, although their effectiveness for this purpose might differ very considerably. In the model of R.G. Martin and his colleagues [5] these factors are taken into account as determining whether the ribosome delays for a long time when it encounters a modulating codon, or whether it gets off the mark again quickly; that is, whether or not it is likely to drop off, or be bumped off, the m-RNA. In any event there will be a greater frequency of translation of cistrons near the operator, before the ribosomes encounter too many of these hazards, than there will be for cistrons at a distance from the operator. More molecules will be synthesized of those polypeptides that are formed earlier than those formed later. Now the sequence of genes in an operon need not correspond to the sequence of their enzymes in the metabolic pathway. Indeed, for the ten enzymes of the histidine biosynthetic pathway, although the gene for enzyme 1 is nearest the operator, that for enzyme 10 is next and that for enzyme 2 is last [4]. The theory of modulation therefore predicts that fewer enzyme molecules are made as the ribosomes approach the end of the operon, and as a result it may transpire that the least efficient catalyst is synthesized first and in greatest amounts, and the most efficient are made last and in small amounts. The significance of the order of the genes might be, therefore, that this is the selection that was arrived at, on the basis of efficiency, in the course of evolution. In the case of polarity mutants, where it appears that a nonsense triplet has been introduced into the gene so that it does not function, the ribosome will have a greatly increased chance of leaving the m-RNA in this region and the numbers of enzyme molecules formed beyond that point will therefore be decreased, although they will remain co-ordinate.

Catabolic repression

It has been known since the 19th century that the biochemical activities of bacteria are changed when fermentable carbohydrate is added to their growth media: thus, indole was no longer found to be produced in putrefying meat broth when sucrose was present. The first clear-cut studies of what we now call repression were probably those of F.C. Happold and L. Hoyle in 1936 who found that although glucose did not inhibit the action of tryptophanase, the formation of the enzyme was considerably repressed in *E. coli* when glucose was added to broth media. In 1942, H.M.R. Epps and E.F. Gale showed that several other amino-acid deaminases were similarly affected, and J. Monod discovered that glucose prevented the formation of enzymes needed to degrade other sugars. Many other examples, in yeasts as well as bacteria, were later discovered and the phenomenon was called 'the glucose effect'. Much of our present understanding of this effect is due to B. Magasanik and his colleagues (for a review see reference [1]). Thus, F.C. Neidhardt and B. Magasanik pointed out that all the enzymes repressed by glucose have one thing in common, namely that they function to provide metabolites for growth which can be obtained more rapidly from glucose: the deamination of amino acids, for example, provides oxoacids that arise from glycolysis, the tricarboxylic acid cycle and related reaction sequences. In contrast, the ultimate products of glucose-insensitive enzymes cannot be obtained independently by separate pathways from glucose: thus, glucose cannot furnish the products of tetrathionate reductase or nitrate reductase, and these enzymes are not repressed by glucose. However, glucose does not merely serve as an alternative source of metabolites obtainable from other growth substrates. It is usually degraded more rapidly, often by alternative pathways that may operate simultaneously, such as glycolysis and the pentose-phosphate cycle. It appears, in fact, that bacteria can make intermediary metabolites from glucose faster than they are able to use them for biosynthesis, and we might expect the cells to experience an *embarassante richesse* when glucose is added to a culture growing relatively slowly at the expense of another carbon compound. Accordingly, an adjustment of metabolism will surely follow this condition. This adjustment will ensure that the metabolites, accumulating under the changed conditions, will be used more effectively. Further, we may expect that the cells will no longer gear their protein-synthesizing machinery to the task of making enzymes whose sole functions are to convert a growth substrate into those metabolites which glucose is able to provide much more rapidly. Whatever the detailed explanation of this metabolic adjustment might be, it is evident that Magasanik's description *catabolite repression* is to be preferred to 'glucose effect', particularly since compounds other than glucose can cause repression provided they are rapidly metabolized.

V. Moses and his colleagues [1, 2] have shown that catabolite repression of β-galactosidase in *E. coli* consists of two phases (Fig. 28, below). They found that when glucose was first added to cells that were synthesizing β-galactosidase in the presence of inducer (IPTG), no further enzyme was made over a period of time which they called the phase of *transient* catabolite repression. Afterwards, synthesis of β-galactosidase was resumed at a constant but slower rate than that observed before glucose was added. It seems that this permanent but less severe second phase may be one in which a reapportionment of the total protein occurred. The

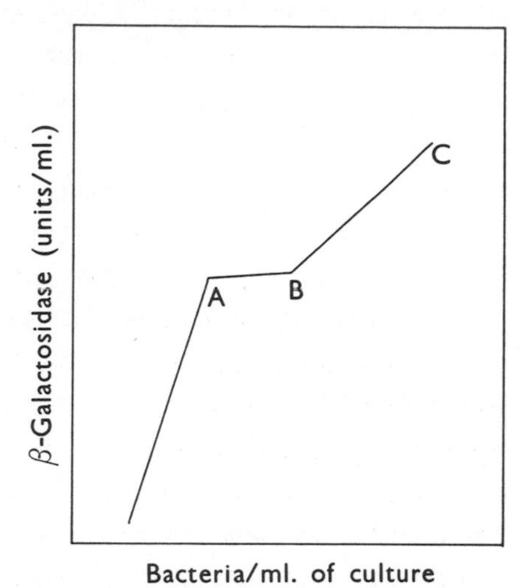

Fig. 28. Transient catabolite repression. *E. coli* grew in a glycerol medium with a strong inducer of β-galactosidase present, so that this enzyme formed a definite fraction of the protein being synthesized by the cells at a particular moment. When data are plotted as shown, the slope of the line gives a measure of this fraction. Addition of glucose at A was followed by a phase of acute transient repression AB. After B, repression was permanent but less severe. Redrawn from data of Palmer J. and Moses V. (1967) *Biochem. J.* **103**, 358.

cells were induced to make larger amounts of the 'profitable' enzymes of glucose metabolism; and these would automatically form a greater proportion, than was the case before glucose was introduced. On this view it is scarcely appropriate to describe the second phase as one of repression; for an enzyme which, in these conditions, is useless has merely suffered dilution by increased synthesis of the enzymes needed to catalyse essential reactions. By contrast, degradation of glucose provided the cells with concentrations of metabolites in excess of their immediate growth requirements during the transient phase (AB of Fig. 28, above), and it appears that one or more of these metabolites strongly repressed the synthesis of β-galactosidase. These experiments raise two questions which Moses and his colleagues have attempted to answer, namely: which metabolites cause repression? and, how is repression exerted? As regards the first question, two strains of *E. coli* were studied, the first of which exhibited acute transient repression of β-galactosidase when supplied with glucose during growth on glycerol whereas the second strain did not. The two strains did not differ genetically in their *lac* operons. However, the first used twice as much glucose per unit of cell mass as the second, and also showed much stronger pentose-phosphate cycle activity. The pool sizes of glucose-6-phosphate, 6-phosphogluconate, fructose-1,6-diphosphate and NADPH all increased when glucose was added to cultures of the first strain, and the concentration of each compound fell when the cells recovered from the transient phase of repression. These changes in concentration did not occur in the second strain which was only weakly repressed by glucose. Accordingly it was suggested that repression of β-galactosidase by glucose requires the rapid operation of the pentose-phosphate cycle and is mediated by one of the four compounds mentioned.

V. Moses and J. Palmer [85.2] have presented evidence for the genetic regulation of transient catabolite repression. They compared the behaviour of mutants 'constitutive' for β-galactosidase with strains that regulated the production of this enzyme in the normal manner, requiring the presence of an inducer before they showed activity. A strain may be constitutive if there has been a mutation either in the regulator or the operator gene. In principle, regulator mutants (i^-) may give rise to an altered repressor (P in Fig. 27, page 83) which is no longer able to combine with the operator, or alternatively the i gene may produce no repressor at all. Such a mutant will not require an inducer in order to synthesize β-galactosidase: it will be constitutive because it cannot repress the synthesis of this enzyme. The same behaviour will be shown when a mutant has a defective operator; but in this case, even though the i gene has synthesized a normal repressor, this molecule is no longer able to combine with the defective operator. Palmer and Moses found that i^+ (normal) strains and i^- mutants both showed transient catabolite repression when IPTG was used as inducer. By contrast, a mutation in the operator gene abolished transient repression, although the milder and permanent phase of 'repression' (BC, Fig. 28, page 85) was still given. These and other experiments suggested that i^- mutants did in fact make a repressor, but that this molecule was not biologically active until it combined with a co-repressor—presumably one of the four compounds previously implicated in transient catabolite repression. When it combined with the powerful inducer IPTG the product of the i^+ gene was likewise inactive until further combination occurred with the co-repressor. However, a functional operator gene was essential before transient catabolite repression could occur, for otherwise the repressor molecule from either strain would fail to combine with the operator, even though it had been activated by co-repressor.

The general phenomenon of catabolite repression is not completely understood and remains the subject of active research. Indeed, recent studies [1] suggest that the severe transient repression we have described is a phenomenon quite distinct from catabolite repression: for example, it may be elicited by analogues of glucose that are not catabolized. On the other hand, a nutritional situation analogous to that encountered in the experiments of V. Moses and his colleagues occurs in the so-called 'step-up' conditions when bacteria are suddenly provided with nutrients in excess of their immediate requirements. In these conditions a rich supply of nutrients—a mixture of amino acids, for example—is added to a culture of bacteria growing in a mineral salts medium. Although the cells grow more rapidly in due course, they are not able to do so immediately the amino acids are added, because the rate of protein biosynthesis is limited by the number of ribosomes available. More ribosomes are needed for a higher rate of growth, and in fact it is found that the RNA required is made much faster over a short initial period during which synthesis of DNA and protein shows little change. By contrast with transient catabolite repression, we have here a situation in which a repression (on RNA synthesis) is *lifted* by the conditions of 'luxury' provided for the bacteria. According to one explanation the various uncombined s-RNA species repress RNA synthesis, but do not do this when combined with their respective amino acids. When the latter are present in abundance, little or no s-RNA remains uncombined, so that synthesis of RNA is thereby derepressed. We shall not discuss the regulation of macromolecular synthesis further: the present status of the subject has been well reviewed by O. Maaløe and N.O. Kjeldgaard [1]. However, we may note that a stimulating alternative theory to regulation of RNA synthesis by s-RNA has been proposed in which it is suggested that nascent RNA serves as repressor, and a fraction of ribosomal RNA as the inducer, of the biosynthesis of RNA of ribosomes [2].

Induction and repression in mammals

Much of our present knowledge of induction and repression has been obtained from experiments with bacteria. The striking advances of the last decade have justified the observations of the late Marjory Stephenson (page 60) concerning the wisdom of using bacteria to study the regulation of enzymes. She wrote them at a time before molecular biology was conceived, or at least when it was no more than a twinkle in the eye of W.T. Astbury who gave the new science its name.

The concept of allosteric enzymes, developed mainly from studies with bacteria, has already proved fruitful, as we have seen, for understanding the regulation of enzymic activities in mammals. Nevertheless, it is far from established that mammals and bacteria regulate the *biosynthesis* of their enzymes in similar ways. The very factors that make bacteria such convenient experimental material are those which raise doubts about extrapolating our concepts to embrace mammalian systems. A multicellular organism may maintain a remarkably constant environment for its various tissues. These tissues may have highly specialized functions, with metabolic activity coordinated by hormonal and nervous mechanisms. In contrast many bacteria need to make rapid adjustments to a changing environment which they have little or no ability to modify; but they do have an enormous capacity for synthesizing new enzymes which enable them not merely to tolerate but to take advantage of the changes that occur. Since they are unicellular they do not need hormones and their metabolism is focused upon cell division; and whilst it is true that protein turnover occurs when they are not dividing, this is not so during exponential growth: protein is synthesized but is not degraded in this phase. The situation in an adult multicellular organism is quite different. Mitosis in many organs is rare, and protein synthesized in excess of requirements must be removed, so that protein turnover is common and indeed necessary. When bacteria use the new enzymes they have synthesized in response to a change in environment, those enzymes that have become redundant are rapidly diluted out during the course of cell division. The amounts of certain enzymes in the various organs of a mammal can be shown to alter with nutritional states, but it is clearly more difficult to decide whether this is due to an increased rate of synthesis or a decreased rate of breakdown, or both; and if synthesis has increased, whether this is due to induction or derepression along the lines of the operon model used with such success to interpret the behaviour of bacteria.

Thus it appears from present experimental evidence that the induction of tryptophan pyrrolase in rat livers may occur by a mechanism quite different from those we have described for bacteria. This oxygenase cleaves the indole nucleus of L-tryptophan to give *N*-formylkynurenine and it is the first enzyme on the pathways of degradation of tryptophan by mammals and bacteria (Fig. 25, page 76). The amount of the enzyme found in the livers of rats may be increased by including L-tryptophan in their diets. When the protein is isolated, it is not active enzymically until incubated with L-tryptophan and a reducing agent such as ascorbic acid, a process which has been shown to consist of two steps. First, the inactive protein (apoenzyme) must be conjugated with its prosthetic group, hematin, to form the holoenzyme. This reaction needs L-tryptophan (or an analog) as well as the source of hematin, such as methemoglobin. Second, the holoenzyme so formed is in its oxidized state, and must be reduced before it can catalyse the oxidation of tryptophan; and the reduction itself requires tryptophan at low concentrations:

$$\xrightarrow{\text{Biosynthesis}} \begin{array}{c}\text{Apoenzyme}\\ \text{(inactive)}\end{array} \xrightarrow{\text{L-Tryptophan}} \begin{array}{c}\text{Oxidized}\\ \text{Holoenzyme}\\ \text{(inactive)}\end{array} \underset{\begin{array}{c}O_2\\ \text{No Tryptophan}\end{array}}{\overset{\begin{array}{c}\text{L-Tryptophan +}\\ \text{asorbic acid}\end{array}}{\rightleftharpoons}} \begin{array}{c}\text{Reduced}\\ \text{holoenzyme}\\ \text{(active)}\end{array}$$

M. Piras and W.E. Knox [1] have shown that there are two sites on the apoenzyme molecule, namely a catalytic site and a site at which conjugation with hematin occurs. They also showed that certain homologues of tryptophan, as well as tryptophan itself promote the reaction by which the apoenzyme is conjugated. Moreover, those analogues—and only those—that promoted this conjugation were effective *inducers* of the enzyme in adrenalectomized rats, irrespective of whether the compounds had any affinity for the catalytic site or not. These workers are inclined to interpret their findings on a model [2] published more than twenty years earlier than the operon model. The enzyme is visualized as being in dynamic equilibrium with a precursor. A substrate, or a tightly bound coenzyme could then cause an increase in the amount of active enzyme simply by combining with the protein to give a form less readily degraded. Accordingly, equilibrium would be displaced towards active enzyme formation because synthesis would continue but degradation would not. In the system discussed, tryptophan pyrrolase would be 'induced' by tryptophan or its analogues because the apoenzyme would be stabilized by conjugation.

Other workers have used the operon model to interpret the changes in enzyme concentrations that accompany alterations in the metabolic states of animals. G. Weber and his colleagues [3] showed that when rats were starved there were profound decreases in the concentrations of three 'key' enzymes of glycolysis, namely enzymes 1, 2 and 3 of Fig. 23, page 68. The amounts of these enzymes increased again on re-feeding, but if the animals were first given inhibitors of protein biosynthesis, no such increases were found. This suggests that the rise in glucokinase, phosphofructokinase and pyruvate kinase activities was due to the *de novo* biosynthesis of these enzymes. In contrast to the glycolytic enzymes, the key enzymes of gluconeogenesis (4, 5, 6, 7 of Fig. 23, page 68) were preferentially increased or maintained in the normal range on starvation. These workers have suggested that synthesis of the two groups of key enzymes for glycolysis and gluconeogenesis respectively may be controlled in each case by genes on a single operon. These observations have been extended to explain one of the many effects of insulin which is to increase the concentration of enzymes 1, 2 and 3 and to decrease those of enzymes 4, 5, 6 and 7 (Fig. 23, page 68). In the terms of the operon theory, insulin would activate the repressor of the operon for the second batch of enzymes and would inactivate that for the first batch.

The action of hormones will not be discussed further except to say that one hypothesis, of which the foregoing is an application, proposes that they stimulate the synthesis of specific messenger RNA molecules. There are many experimental findings which suggest that this may be so, but there are also many which the hypothesis cannot accommodate. A vigorous review is that of A. Korner [1] who, whilst discussing the many instances of stimulation of protein biosynthesis by hormones, has concluded that we do not yet know how this occurs.

Metabolic Pathways

Structure

$NAD(P)^+$ $\underset{-2H}{\overset{+2H}{\rightleftharpoons}}$ $NAD(P)H + H^+$

Function

Hydrogen transfer.

Co-enzymes for oxidoreductases (dehydrogenases).

Typical Reactions

The ability of NAD and NADP to accept hydrogen depends on the tendency of $=\overset{+}{N}=$ to withdraw electrons and so create an electrophilic site at C_4 of the pyridine nucleus. A hydrogen atom then appears to be transferred from the substrate to this site in the form of a hydride ion (a proton with an electron pair). At the same time the substrate loses a second hydrogen as a proton.

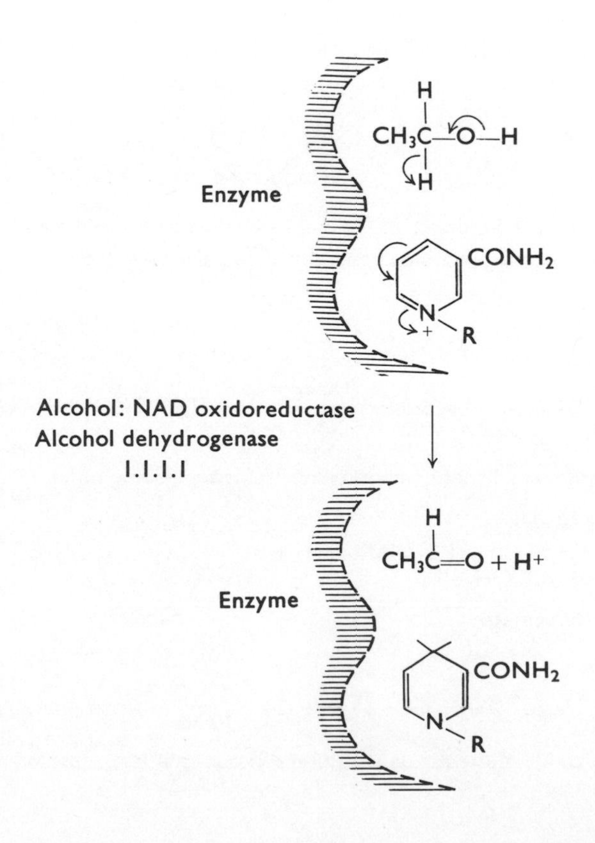

Structure

Riboflavin

Flavin mononucleotide (FMN)

Flavin adenine dinucleotide (FAD)

FMN and FAD are prosthetic groups of a wide variety of enzymes, and are often associated with other co-factors such as metals, non-heme iron compounds, sulphur compounds etc. in highly complex complexes. Two or more molecules of one or of both of the flavins may be present in the same enzyme system.

Function

Dehydrogenation and hydrogen transfer.

Oxidized $\underset{-H}{\overset{+H}{\rightleftharpoons}}$ Semiquinone $\underset{-H}{\overset{+H}{\rightleftharpoons}}$ Reduced

The transfer of two hydrogen atoms from donor to flavin may involve a two-fold step in which a semiquinone is formed as an intermediate. This contains a nitrogen atom possessing an unpaired electron.

Typical Reactions

1 *Oxidation of reduced NAD(P)*

$$\text{e.g. NAD(P)H} + \text{H}^+ + \text{acceptor} \xrightarrow[\text{1.6.99.1—3}]{\text{Flavoprotein}} \text{NAD(P)}^+ + \text{reduced acceptor}$$

2 *Removal of two hydrogen atoms* from adjacent carbon atoms to form a double bond.

$$\text{e.g.}\ \begin{matrix} CH_2COOH \\ | \\ CH_2COOH \end{matrix} + \text{E—FAD} \xrightarrow[\text{1.3.99.1}]{} \begin{matrix} HCCOOH \\ \| \\ HOOCCH \end{matrix} + \text{E—FADH}_2$$

Succinate → Fumarate (see 123 BC)

3 *Transfer of hydrogen to oxygen*

$$\text{e.g. Xanthine} + H_2O + O_2 \xrightarrow[\text{1.2.3.2}]{} \text{Urate} + H_2O_2$$

(See 164 DF)

The enzyme xanthine: oxygen oxidoreductase or xanthine oxidase is a flavoprotein containing molybdenum.

Structure

```
     H2
    .C.
H2C     CH CH2CH2CH2CH2COOH
 |       |
 S ----- S
```

(1,2-Dithiolane-3-valeric acid)

May be protein bound through an amide link formed by condensation with the ε-amino group of a lysine residue.

```
                                    :
                                    NH
                                    :
     H2                             CO
    .C.                             :
H2C     CH(CH2)4CONH(CH2)4CH
 |       |                          :
 S ----- S                          NH
                                    :
                                    CO
                                    :
```

Function

Acyl transfer (in the reduced form).

Hydrogen transfer—from reduced form to regenerate oxidized form required for further acyl transfer.

Typical Reaction

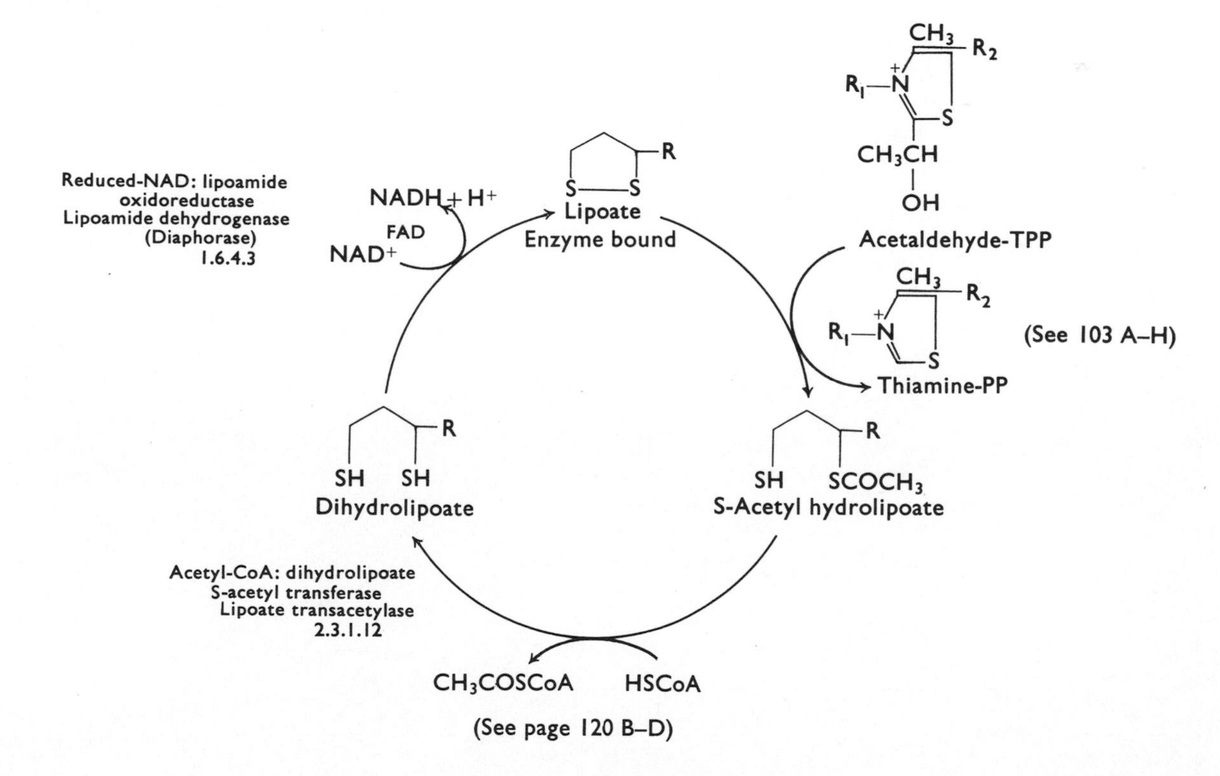

(See page 120 B–D)

Ubiquinone

Structure

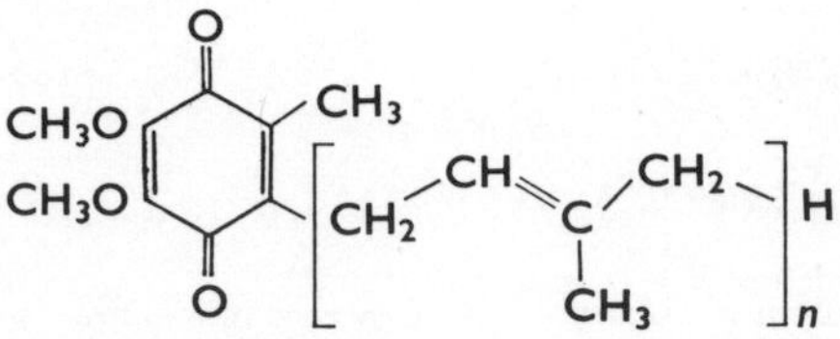

O
CH_3O CH_3
CH_3O CH_2 CH CH_2 H
O C
CH_3 n

Ubiquinone-10
Co-enzyme Q_{10}

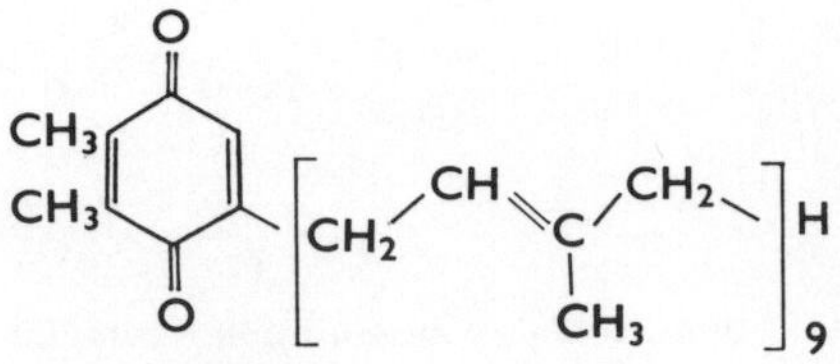

O
CH_3
CH_3 CH_2 CH CH_2 H
O C
CH_3 9

Plastoquinone

Ubiquinone occurs in mitochondria chiefly as ubiquinone-10 (i.e. with 10 isoprenoid units in the side chain) but the number can vary from 6 to 10.
Plastoquinone is found in plant chloroplasts.

Function

Hydrogen carrier between flavins and cytochromes (109).

Mechanism

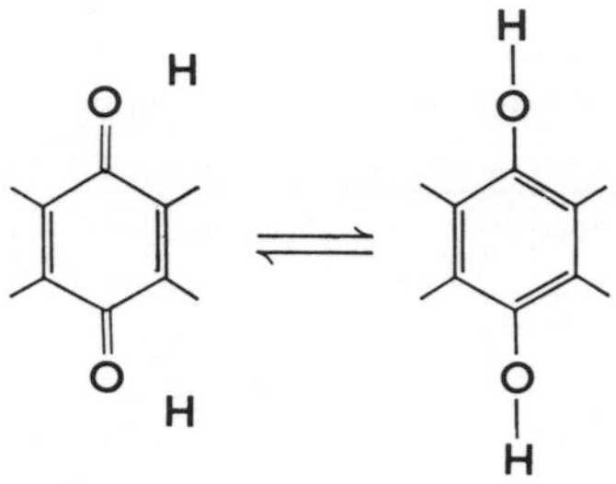

H H
O O
O O
H H

Quinone ⇌ Hydroquinone

Structure

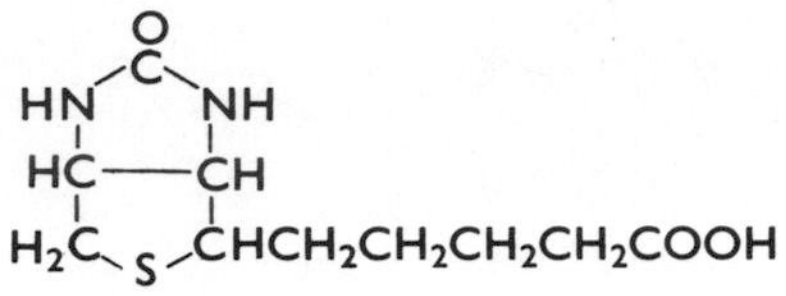

```
        O
        C
   HN       NH
   HC ----- CH
   H2C  S   CHCH2CH2CH2CH2COOH
```

Biotin

```
        O
        C     O
   -O       C
          N     NH
          HC -- CH
          H2C S CH(CH2)4COOH
```

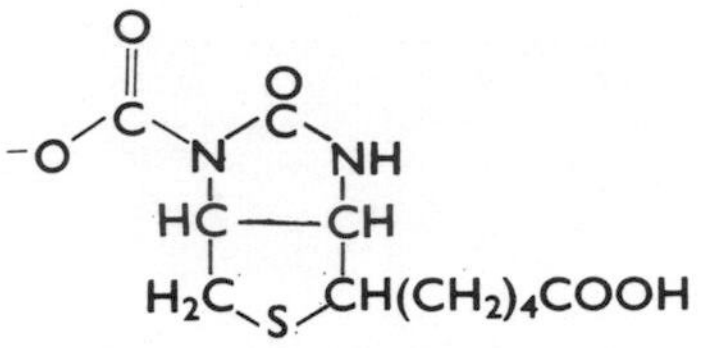

Carboxybiotin

Enzyme-bound by condensation with ε-amino group of a lysine residue.

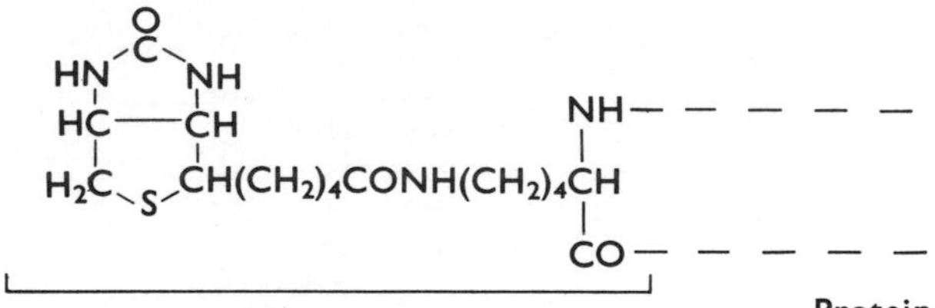

```
        O
        C
   HN       NH
   HC ----- CH                       NH— — — — —
   H2C  S   CH(CH2)4CONH(CH2)4CH
                                     CO— — — — —
```

ε-N-Biotinyl-L-lysine
Biocytin

Protein

Function
Carboxylation

Mechanism
Carboxylations require 'activated' carbon dioxide (or carbonic acid). Free energy of cleavage of carboxybiotin enzymes is sufficient to make carboxylations possible. The formation of carboxybiotin therefore requires the participation of ATP.

$$HCO_3^- + \text{biotin-enzyme} + ATP \rightleftharpoons \text{carboxybiotin-enzyme} + ADP + P_i$$

$$\text{Carboxybiotin enzyme} + \text{acceptor} \rightleftharpoons \text{carboxylated acceptor} + \text{biotin}$$

Typical Reaction

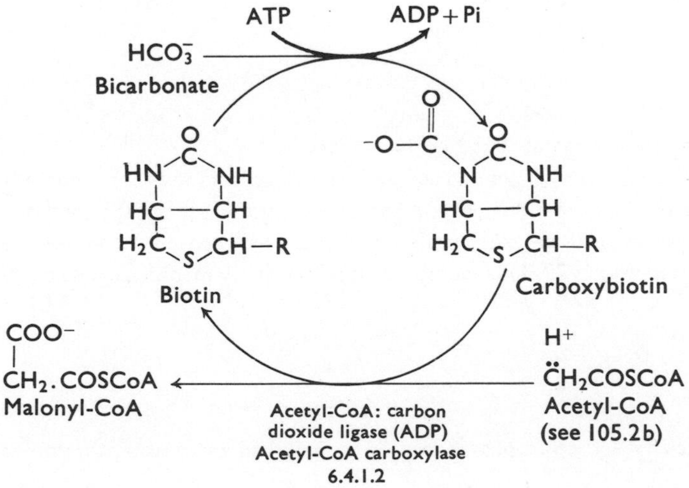

ATP

ADP + Pi

HCO_3^-

Bicarbonate

```
        O                     O
        C                -O—C    O
   HN       NH                 N   C   NH
   HC ----- CH                 HC ---- CH
   H2C  S   CH—R               H2C  S  CH—R
```

Biotin

Carboxybiotin

H^+

COO^-

$CH_2.COSCoA$

Malonyl-CoA

$\ddot{C}H_2COSCoA$

Acetyl-CoA

(see 105.2b)

Acetyl-CoA: carbon dioxide ligase (ADP)
Acetyl-CoA carboxylase
6.4.1.2

Adenosine triphosphate (ATP)

Structure

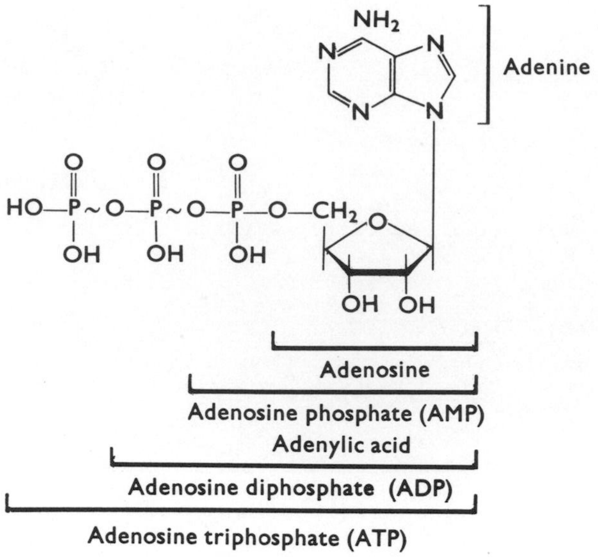

$$\mathrm{HO{-}\overset{O}{\underset{OH}{P}}{\sim}O{-}\overset{O}{\underset{OH}{P}}{\sim}O{-}\overset{O}{\underset{OH}{P}}{-}O{-}CH_2{-}ribose(OH,\,OH){-}adenine}$$

Adenine

Adenosine

Adenosine phosphate (AMP)
Adenylic acid

Adenosine diphosphate (ADP)

Adenosine triphosphate (ATP)

The pyrophosphate bond(s) of ATP are hydrolysed with the liberation of more free energy than the bonds of phosphate esters such as glycerol-1-phosphate or 3-phosphoglycerate (about 8 kcal as against 3 kcal). They are often (though inaccurately) called 'high-energy bonds' and are denoted by the sign ~ (the squiggle).

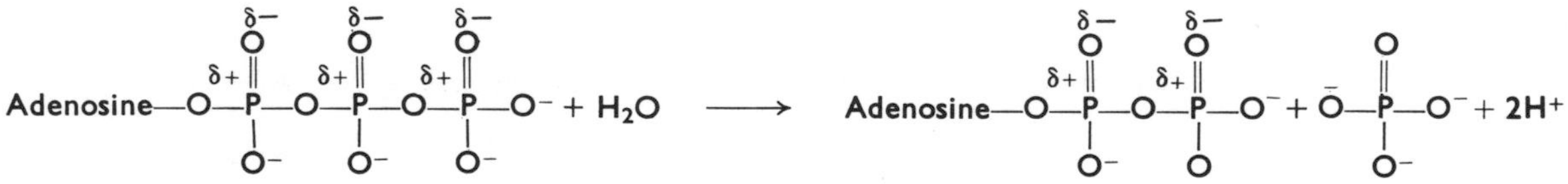

$$\mathrm{Adenosine{-}O{-}\overset{\overset{\delta-}{O}}{\overset{\delta+\|}{\underset{O^-}{P}}}{-}O{-}\overset{\overset{\delta-}{O}}{\overset{\delta+\|}{\underset{O^-}{P}}}{-}O{-}\overset{\overset{\delta-}{O}}{\overset{\delta+\|}{\underset{O^-}{P}}}{-}O^- + H_2O \longrightarrow Adenosine{-}O{-}\overset{\overset{\delta-}{O}}{\overset{\delta+\|}{\underset{O^-}{P}}}{-}O{-}\overset{\overset{\delta-}{O}}{\overset{\delta+\|}{\underset{O}{P}}}{-}O^- + \bar{O}{-}\overset{O}{\overset{\|}{\underset{O^-}{P}}}{-}O^- + 2H^+}$$

$$\mathrm{ATP + H_2O \longrightarrow ADP + P_i + 8{,}000\ cals.}$$

The high free energy of hydrolysis depends upon two main factors

1 The close proximity of the negative charges on the oxygen atoms places the molecule under a stress which is relieved by hydrolysis. The forces of electrostatic repulsion contribute to the energy released on hydrolysis.

2 When the orthophosphate ion separates, the number of possible resonance forms is increased: that is, the products of the reaction are more stable than ATP. This increased stabilisation by resonance also contributes towards the energy released.

Function

Energy transfer

Part of the molecule (especially orthophosphate) may be transferred to an acceptor molecule to increase its energy content (Reactions A-D).

Typical Reactions

The molecule of ATP can be cleaved in several places and at least four fragments can be transferred to acceptor molecules.

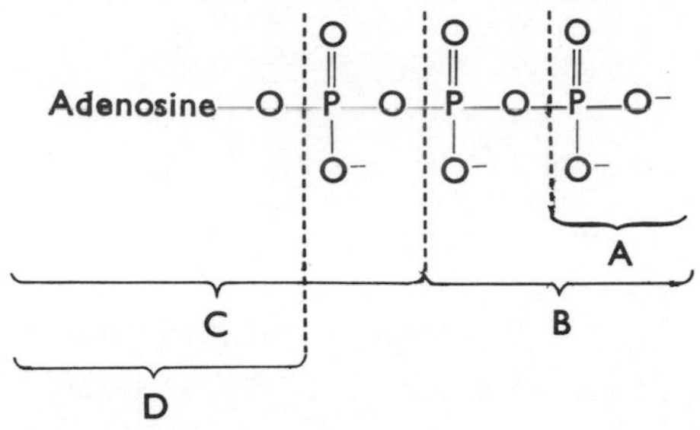

Adenosine—O—P(=O)(O⁻)—O—P(=O)(O⁻)—O—P(=O)(O⁻)—O⁻

A B C D

A Transfer of the terminal orthophosphate group by a large number of phosphotransferases or kinases (2.7.1).
e.g. ATP + Glucose → ADP + Glucose-6-phosphate (2.7.1.2). (See 112 AB)

B Transfer of pyrophosphate group by pyrophosphotransferases (pyrophosphokinases) (2.7.6).
e.g. ATP + D-ribose-5-phosphate → AMP + 5-phosphoribosyl pyrophosphate (2.7.6.1). (See 160 BC)

C Transfer of adenylic acid (adenosine phosphate), leaving pyrophosphate, utilized by nucleotidyltransferases (2.7.7).
e.g. ATP + nicotinate ribonucleotide → pyrophosphate + desamino-NAD (2.7.7.1). (See 239 CD)

D Transfer of adenosine, leaving orthophosphate and pyrophosphate.
e.g. ATP + L-Methionine + H_2O → Orthophosphate + pyrophosphate + 5-adenosyl-methionine (2.5.1.6).
(See 205 AC)

E The free energy of hydrolysis of ATP may be utilised by ligases or synthetases (6.2) to link other molecules together. A phosphorylated reaction intermediate is usually formed which is not shown in the overall equation:

ATP + acetate + HSCoA → AMP + pyrophosphate + acetyl-SCoA (6.2.1.1).
or ATP + succinate + HSCoA → ADP + orthophosphate + succinyl-SCoA (6.2.1.5). (See 153 FE)

Adenosine-3′,5′-cyclic phosphate (cyclic AMP)

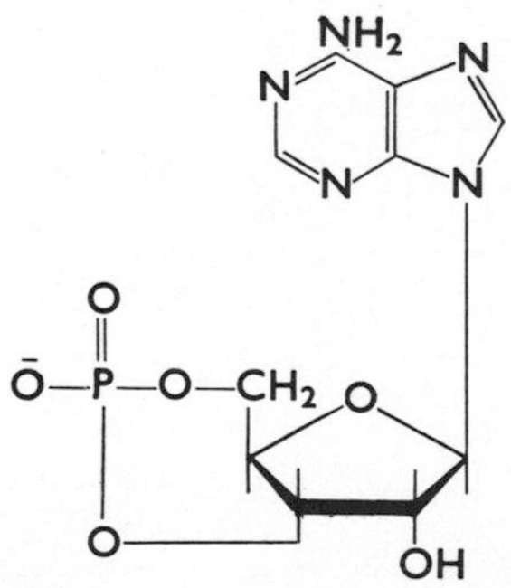

NH₂ N N N N O Ō—P—O—CH₂ O O OH

Function

Intermediate in the action of various hormones. By serving as the activator for one of the enzymes in a series, cyclic AMP takes part in a 'cascade' effect by which the action of a few molecules of hormone is tremendously amplified.

Typical Reactions

F Adrenaline stimulates adenyl cyclase by means of which cyclic AMP is formed from ATP, and cyclic AMP then activates the kinase which is needed for the conversion of inactive phosphorylase *b* into active phosphorylase *a* (page 71).

G Active glycogen synthetase I (also called '*a*') is converted into a less active form D (also called '*b*') by a kinase that is stimulated by cyclic AMP. The effects of adrenaline or glucagon are therefore to suppress the synthesis of glycogen as well as to accelerate the breakdown of glycogen (page 71).

H The effect of thyroxine and catecholamines in stimulating lipolysis in adipose tissue may be due to the fact that adenyl cyclase is first activated and the cyclic AMP formed by its action is then used to activate a kinase by which inactive lipase is phosphorylated and converted into an active form.

Pyridoxal Phosphate

Structure

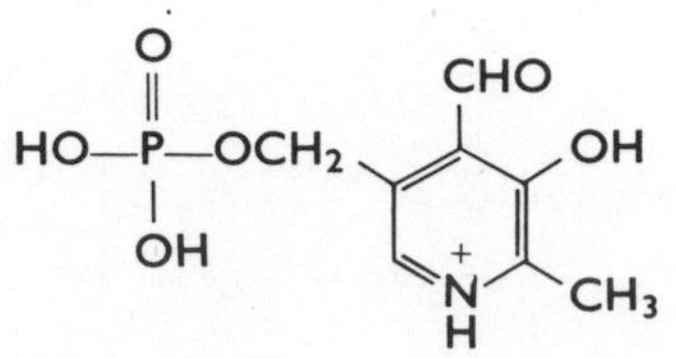

O
HO—P—OCH$_2$ CHO OH
OH
$\overset{+}{N}$ CH$_3$
H

Active form

As a Schiff's base formed by the condensation of the aldehyde group with the ϵ-amino group of a lysine residue of an enzyme. This then reacts with an amino acid with which it forms a Schiff's base and liberates a free ϵ-amino group on the lysine residue.

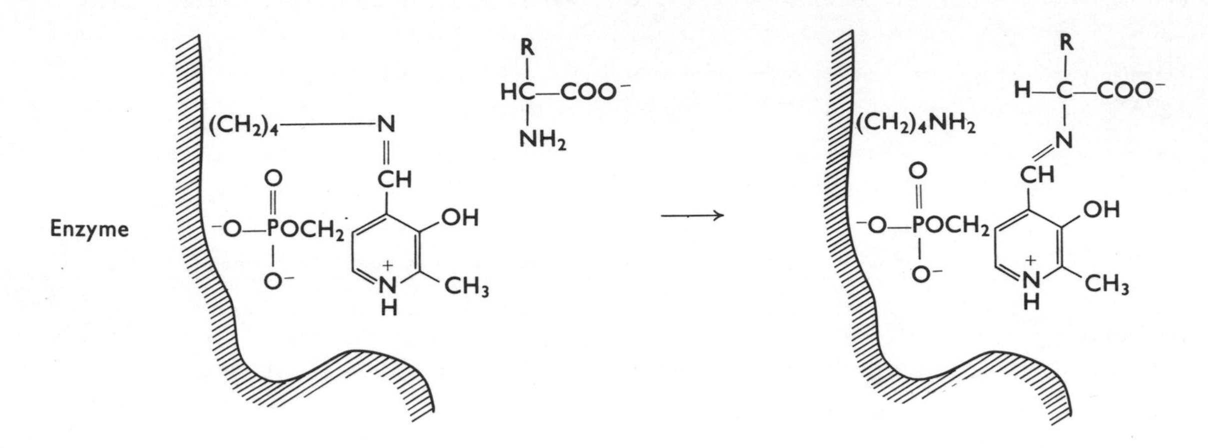

Function

Transamination of aminoacids
Decarboxylation of amino acids
Racemization of amino acids
Also involved in condensation of indole and serine to form tryptophan, the α,β-dehydration of serine, the β,γ-desulphuration of homocysteine, and the formation of serine from glycine.

In each reaction the starting point is the Schiff's base formed between the aldehyde group of pyridoxal phosphate and the amino group of the amino acid

Decarboxylation

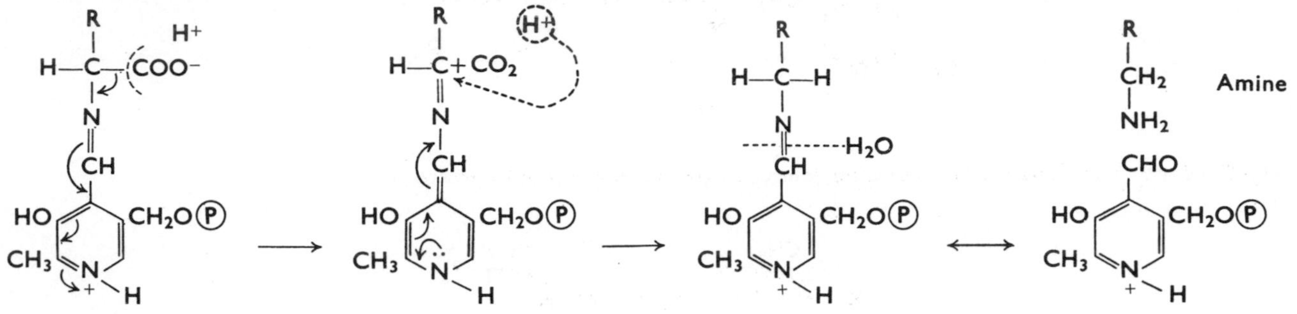

Amine

A

Net result

$$RCH(NH_2)COOH \longrightarrow RCH_2NH_2 + CO_2$$

Transamination

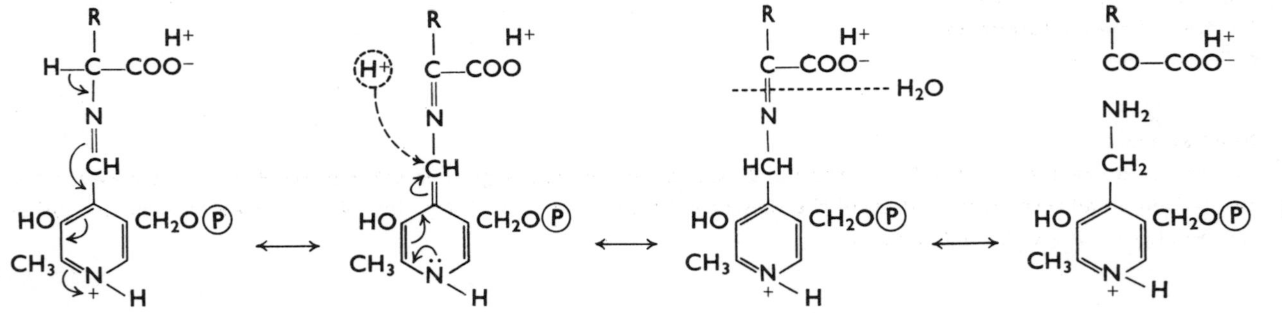

B

Donor amino acid + pyridoxal phosphate Schiff's base ⟶ Product oxo acid + Pyridoxamine phosphate

Product amino acid + Pyridoxal phosphate Schiff's base ⟵ Donor oxo-acid + Pyridoxamine phosphate

Racemization

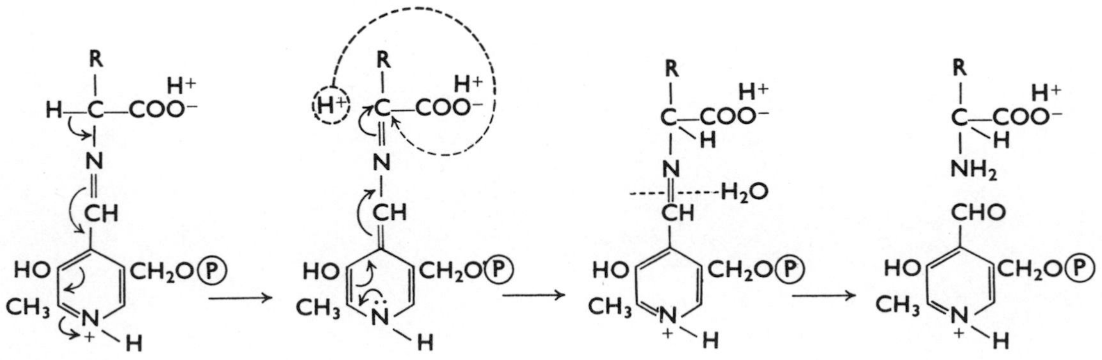

C

Net result

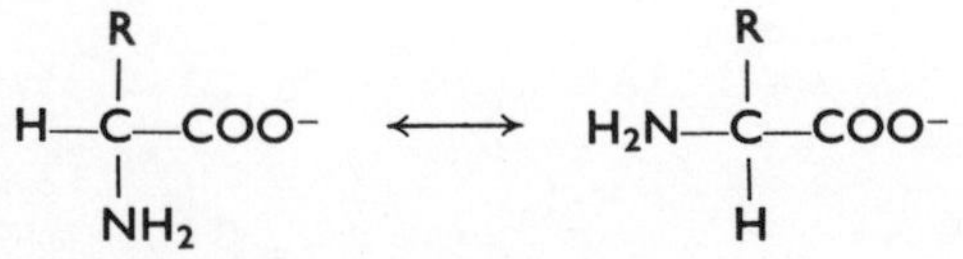

$$H{-}\underset{NH_2}{\overset{R}{C}}{-}COO^- \longleftrightarrow H_2N{-}\underset{H}{\overset{R}{C}}{-}COO^-$$

Thiamine Pyrophosphate (TPP)

Structure

The C_2 of the thiazole part of the molecule, ionises to give a carbanion and a proton.

Function

1 Decarboxylation.
2 Oxidative decarboxylation.
3 α-Ketol (acyloin) formation.
4 Transketolation.

Mechanism

The carbanion reacts with the (δ+) carbon of a carbonyl group. Cleavage then takes place of the bond between this carbon and an adjacent carbon atom of the substrate. This results in the formation of an 'active' aldehyde and either carbon dioxide or a stable aldehyde.

$$\text{TPP} + \text{R—}\underset{\underset{\text{O}}{\|}}{\text{C}}\text{—COO}^- + \text{H}^+ \longrightarrow \text{TPP—}\overset{\overset{\text{R}}{|}}{\underset{\underset{\text{OH}}{|}}{\text{C}}}\text{—H} + \text{CO}_2$$

or

$$\text{TPP} + \text{R—}\underset{\underset{\text{O}}{\|}}{\text{C}}\text{—}\overset{\overset{\text{H}}{|}}{\underset{\underset{\text{OH}}{|}}{\text{C}}}\text{—R}' + \text{H}^+ \longrightarrow \text{TPP—}\overset{\overset{\text{R}}{|}}{\underset{\underset{\text{OH}}{|}}{\text{C}}}\text{—H} + \text{R}'\text{CHO}$$

Typical Reactions

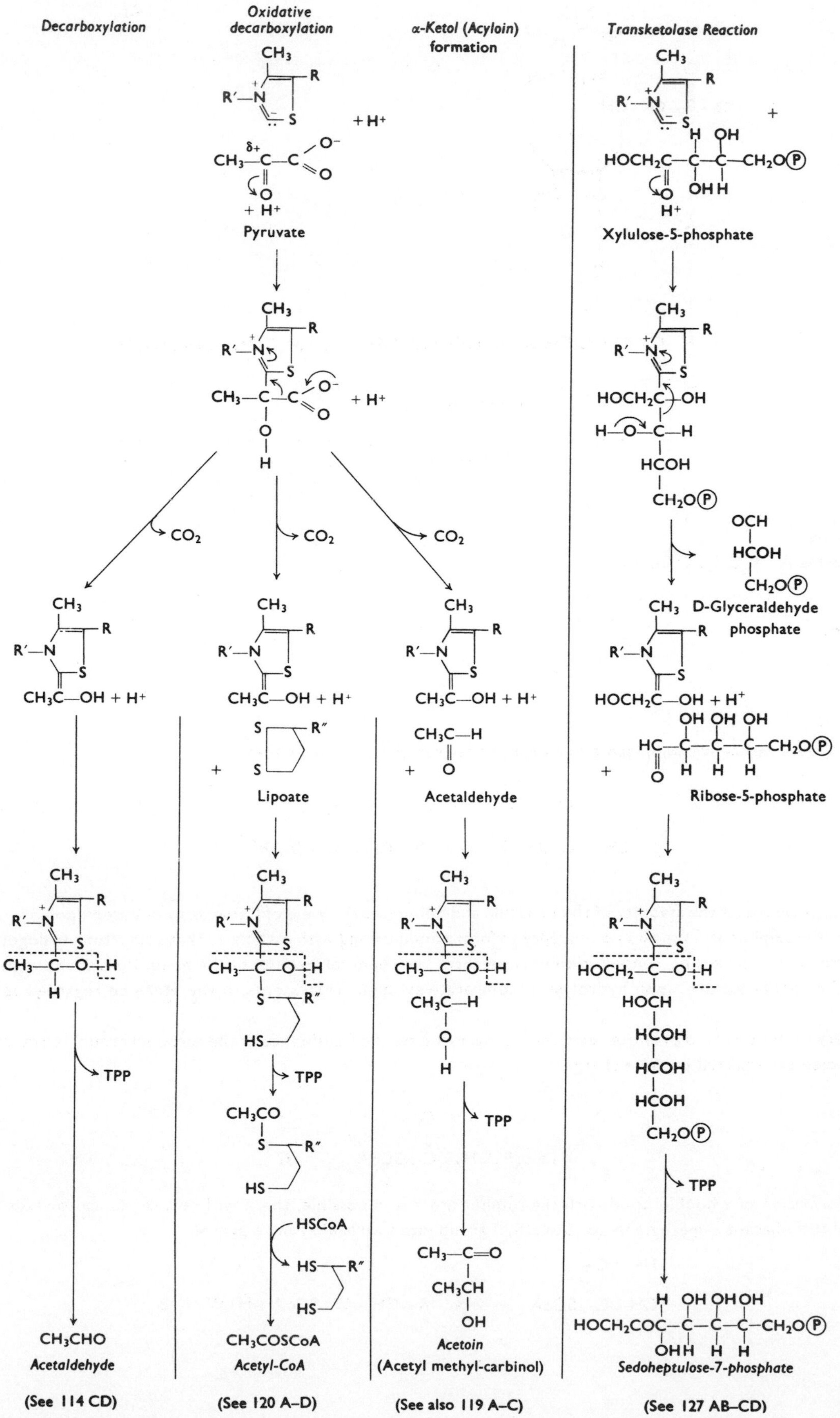

Decarboxylation

Oxidative decarboxylation

α-Ketol (Acyloin) formation

Transketolase Reaction

AB

Pyruvate

+ H^+

Xylulose-5-phosphate

C

CO_2

D-Glyceraldehyde phosphate

DEF

Lipoate

Acetaldehyde

Ribose-5-phosphate

GHJK

TPP

HSCoA

CH_3CHO
Acetaldehyde

$CH_3COSCoA$
Acetyl-CoA

Acetoin
(Acetyl methyl-carbinol)

Sedoheptulose-7-phosphate

(See 114 CD)

(See 120 A–D)

(See also 119 A–C)

(See 127 AB–CD)

Structure

NH_2

$^{-}O-\overset{\overset{O}{\|}}{P}-O-CH_2$

OH

$O=P-O^{-}$

O^{-}

$^{-}O-\underset{\underset{O}{\|}}{P}-O-CH_2C(CH_3)_2-CHOH.CO.NH.CH_2.CH_2CONHCH_2CH_2SH$

Pantothenic acid

Pantetheine

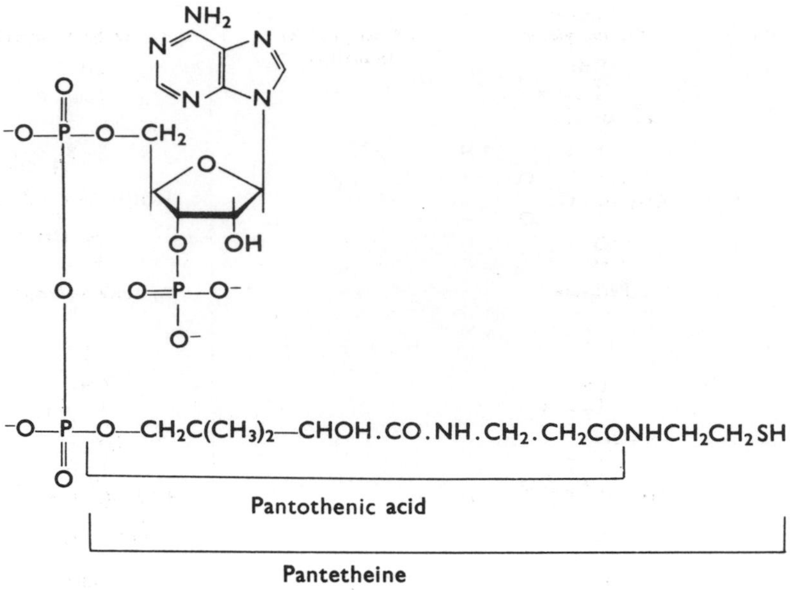

Active form

Acyl coenzyme A R CH_2CO-SCoA

Function

Acyl transfer.

Mechanism

Normal (oxygen) esters are stabilised by resonance between the two forms shown:

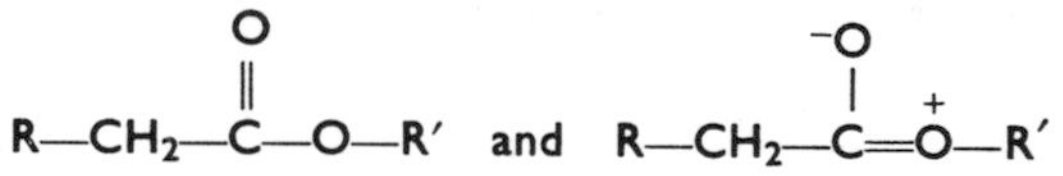

$$R-CH_2-\overset{\overset{O}{\|}}{C}-O-R' \quad \text{and} \quad R-CH_2-\overset{\overset{^{-}O}{|}}{C}=\overset{+}{O}-R'$$

This resonance increases the stability of the molecule and minimizes the yield of free energy released upon hydrolysis. In *thioesters* the sulphur atom shows no tendency to form a double bond with carbon, so that a structure analogous to the second form (above) makes no contribution to resonance. The chemical stability of the molecule is thus comparatively low, and the energy yielded upon hydrolysis is comparatively high. Thioesters can therefore be regarded as 'energy rich'.

This inability of thioesters to produce resonance forms increases the significance of the carbonyl group, the carbon atom of which possesses a partial positive charge.

$$\text{i.e.} \quad R.CH_2-\overset{\overset{\delta- \; O}{\|}}{\overset{\delta+}{C}}-SCoA \qquad \textbf{A}$$

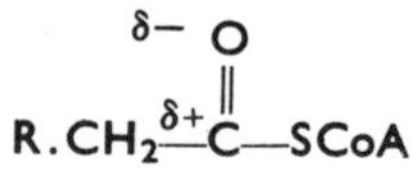

Since the formation of a double bond with the sulphur atom is impossible, there will be a tendency towards the dissociation of the adjacent α-methylene- (or α-methyl) group into a carbanion and a proton.

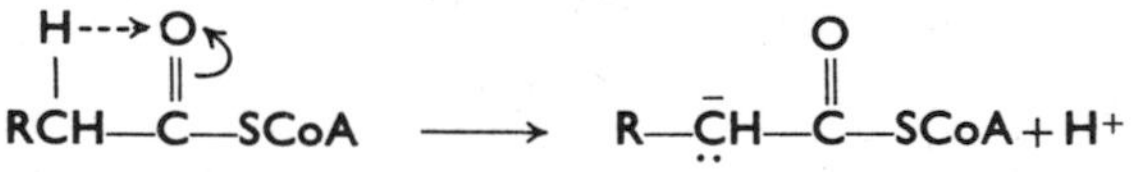

$$RCH(H)-\overset{\overset{O}{\|}}{C}-SCoA \longrightarrow R-\overset{-}{\underset{\cdot\cdot}{C}}H-\overset{\overset{O}{\|}}{C}-SCoA + H^{+} \qquad \textbf{B}$$

Acyl-CoA can therefore exist in two forms:

1 *An electrophilic form* (A) in which the positively charged carbonyl carbon atom is open to attack by nucleophilic (negatively charged) compounds.

2 *A nucleophilic form* (B) in which the negatively charged methylene group is open to attack by electrophilic (positively charged) compounds.

Typical Reactions

1 *As an electrophile*

a

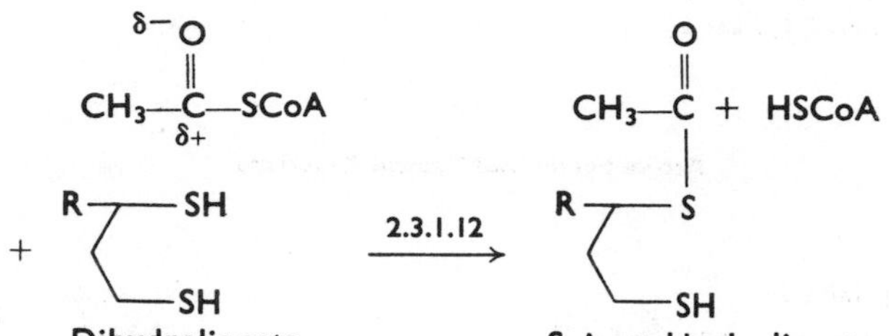

$$CH_3{-}\overset{\delta+}{C}(=\overset{\delta-}{O}){-}SCoA + \text{Dihydrolipoate } (R(SH)(SH)) \xrightarrow{2.3.1.12} CH_3{-}C(=O){-}S{-}R(SH) + HSCoA$$

Dihydrolipoate → S-Acetyl hydrolipoate

(See 120 DC)

b

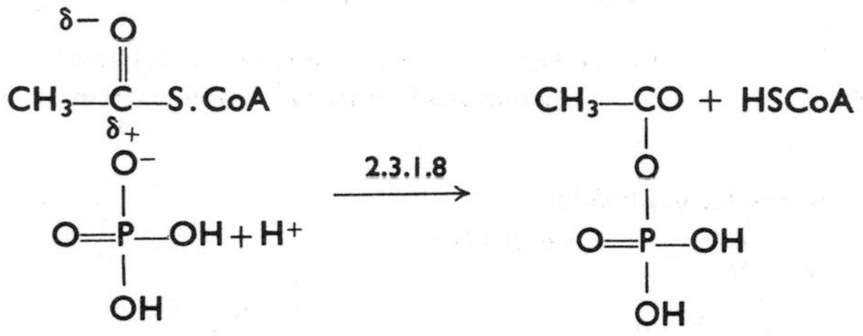

$$CH_3{-}\overset{\delta+}{C}(=\overset{\delta-}{O}){-}S.CoA + {}^{-}O{-}P(=O)(OH){-}OH + H^+ \xrightarrow{2.3.1.8} CH_3{-}CO{-}O{-}P(=O)(OH){-}OH + HSCoA$$

Orthophosphate → Acetyl phosphate

(See 118 FE)

2 *As a nucleophile*

a

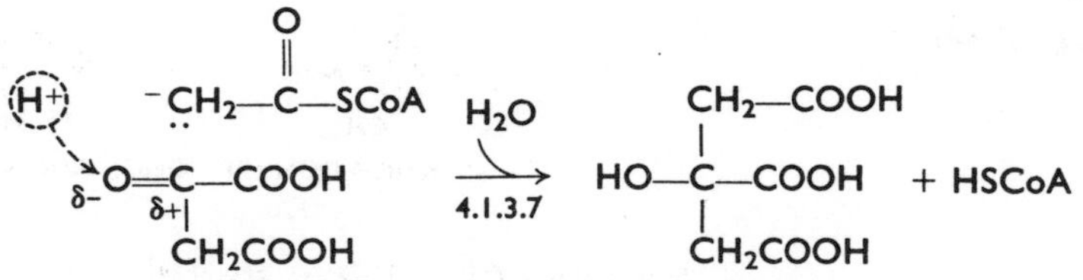

$$H^+ \quad {}^{-}\ddot{C}H_2{-}C(=O){-}SCoA + {}^{\delta-}O{=}\overset{\delta+}{C}(CH_2COOH){-}COOH \xrightarrow[4.1.3.7]{H_2O} HO{-}C(CH_2{-}COOH)(CH_2COOH){-}COOH + HSCoA$$

Oxaloacetate → Citrate

(See 122 AB–C)

b

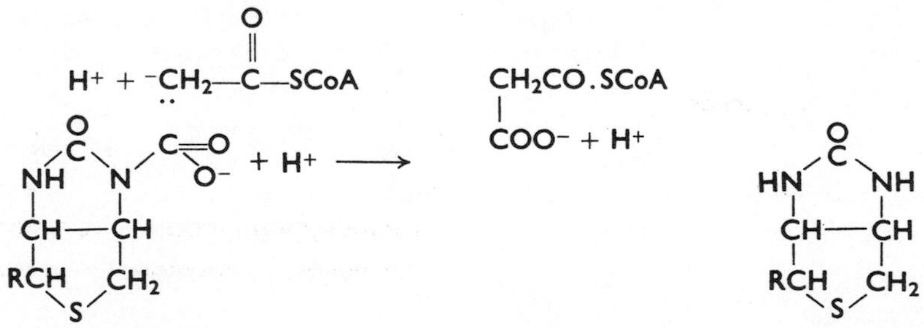

$$H^+ + {}^{-}\ddot{C}H_2{-}C(=O){-}SCoA + \text{Carboxybiotin} + H^+ \longrightarrow CH_2CO.SCoA(COO^-) + H^+ + \text{Biotin}$$

Carboxybiotin: O=C(NH)(N–C(=O)O⁻); CH—CH; RCH, CH₂, S

Biotin: O=C(HN)(NH); CH—CH; RCH, CH₂, S

Carboxybiotin — Malonyl CoA — Biotin

(See 174 AB)

3 *As an electrophile and a nucleophile*

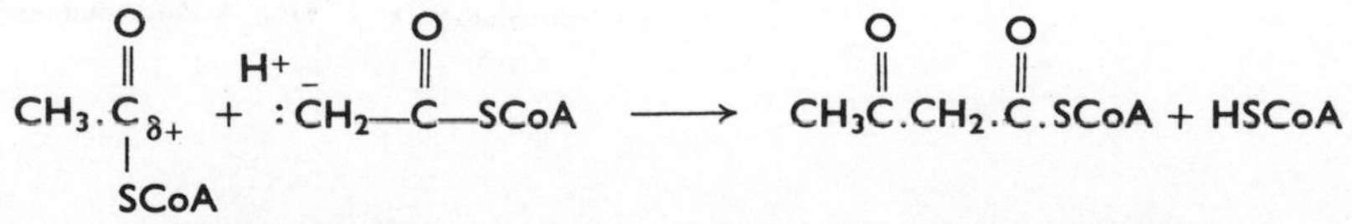

$$CH_3.\overset{\delta+}{C}(=O){-}SCoA + H^+ \; {}^{-}{:}CH_2{-}C(=O){-}SCoA \longrightarrow CH_3C(=O).CH_2.C(=O).SCoA + HSCoA$$

2 Acetyl-CoA → Acetoacetyl-CoA

(See 177 JK–GH)

Structure of Folic Acid

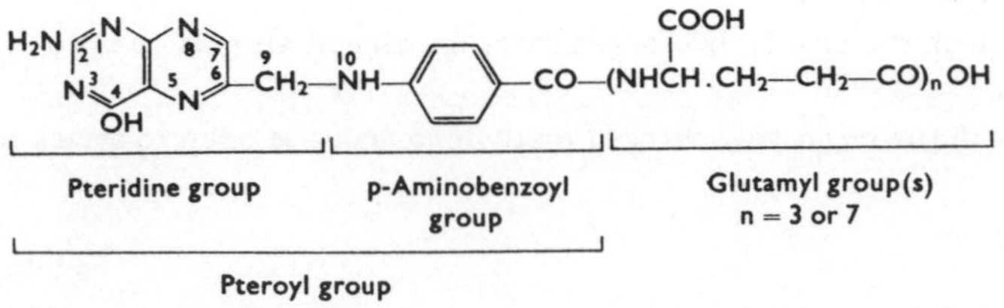

H_2N — pteridine (positions 1–8), OH — CH_2(9) — NH(10) — C₆H₄ — CO — $(NHCH(COOH).CH_2—CH_2—CO)_nOH$

Pteridine group | p-Aminobenzoyl group | Glutamyl group(s) n = 3 or 7

Pteroyl group

Dihydrofolate (FH_2) is reduced at positions 7 and 8.
Tetrahydrofolate (FH_4) is reduced at positions 5, 6, 7 and 8.

Function

1 Carbon Transfer

Active Forms and Typical Reactions

A

10-Formyltetrahydrofolate

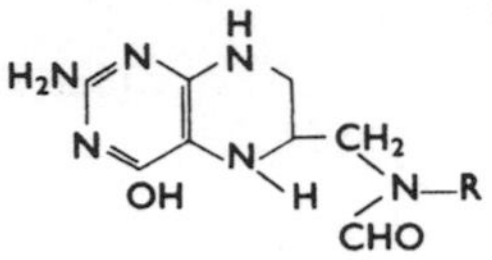

(e.g. 160 CD)

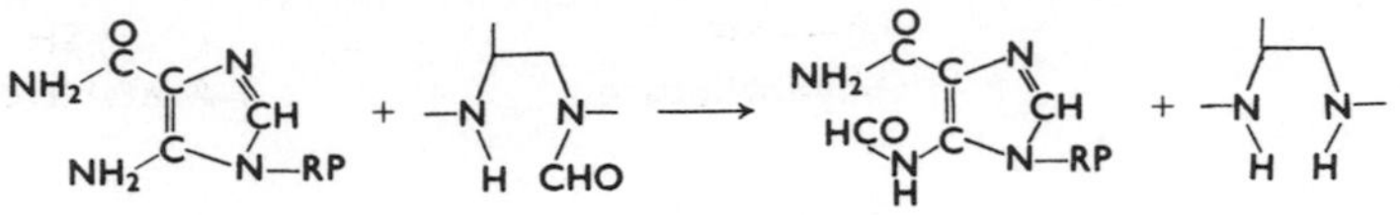

5′-Phosphoribosyl-5-aminoimidazole-4-carboxamide + 10-formyl FH_4 →
5′-phosphoribosyl-5-formamidoimidazole-4-carboxamide + FH_4

B

5,10-Methylenetetrahydrofolate or 10-Hydroxytetrahydrofolate

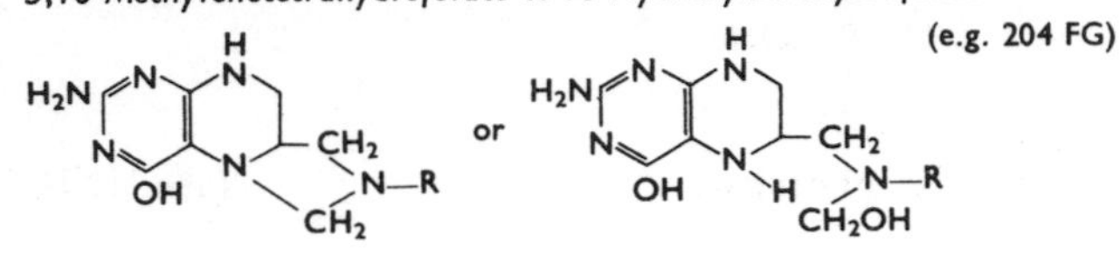

(e.g. 204 FG)

$CH(NH_2)COOH$ → $HOCH_2CH(NH_2)COOH$

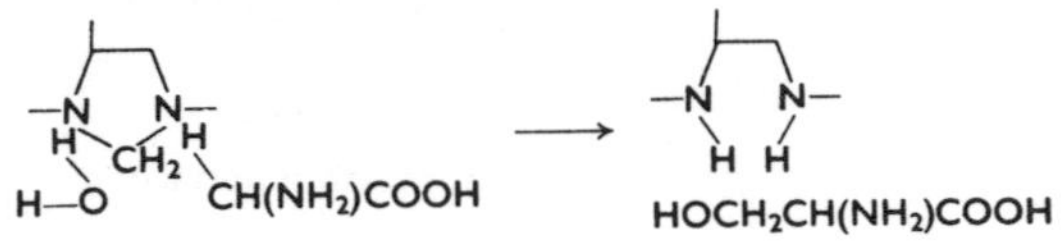

5,10-Methylene-FH_4 + H_2O + Glycine ↔ FH_4 + serine

C

5,10-Methenyltetrahydrofolate

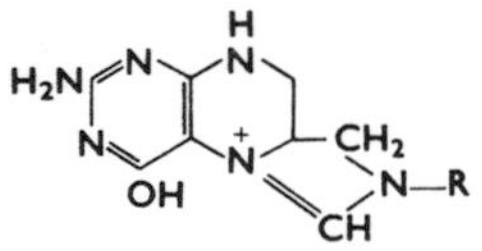

(e.g. 159 AB)

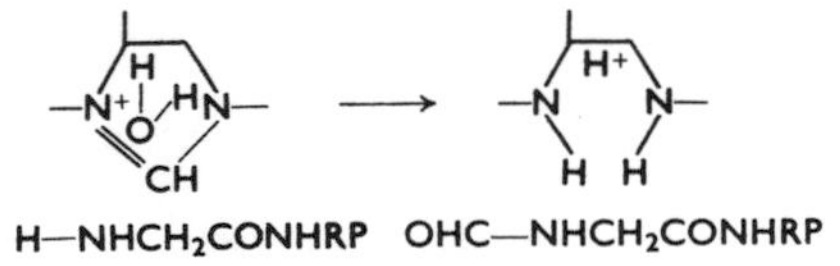

H—$NHCH_2CONHRP$ → OHC—$NHCH_2CONHRP$

5,10-Methenyl-FH_4 + H_2O → FH_4 + H^+
+ 5′-phosphoribosyl glycineamide → + 5′-phosphoribosyl-N-formylglycineamide

D

5-Methyltetrahydrofolate

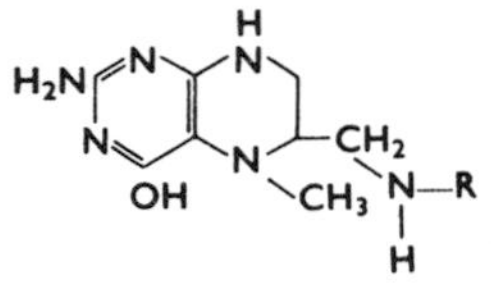

(e.g. 201 DE)

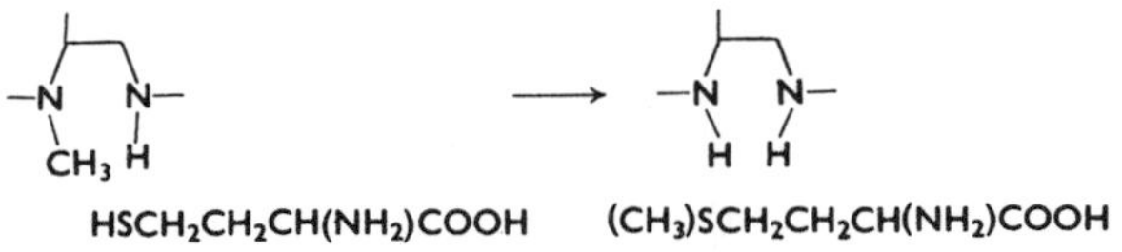

$HSCH_2CH_2CH(NH_2)COOH$ → $(CH_3)SCH_2CH_2CH(NH_2)COOH$

5-Methyl-FH_4 + homocysteine → FH_4 + methionine

E

5-Formiminotetrahydrofolate

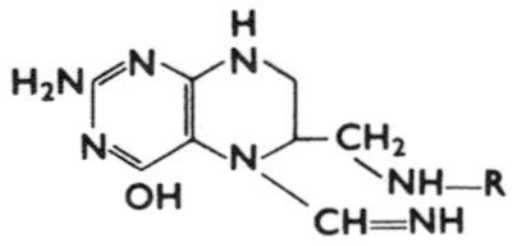

(e.g. 204 DE)

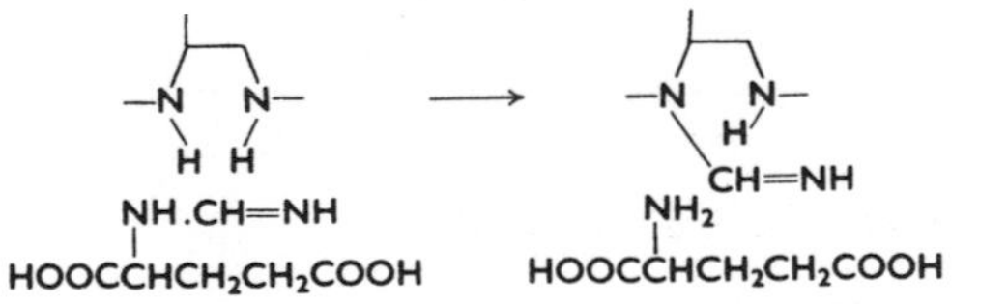

NH.CH=NH / $HOOCCHCH_2CH_2COOH$ → NH_2 / $HOOCCHCH_2CH_2COOH$

Formiminoglutamate + FH_4 → 5-formimino-FH_4 + glutamate.

Interrelationships between Active Forms of Tetrahydrofolate

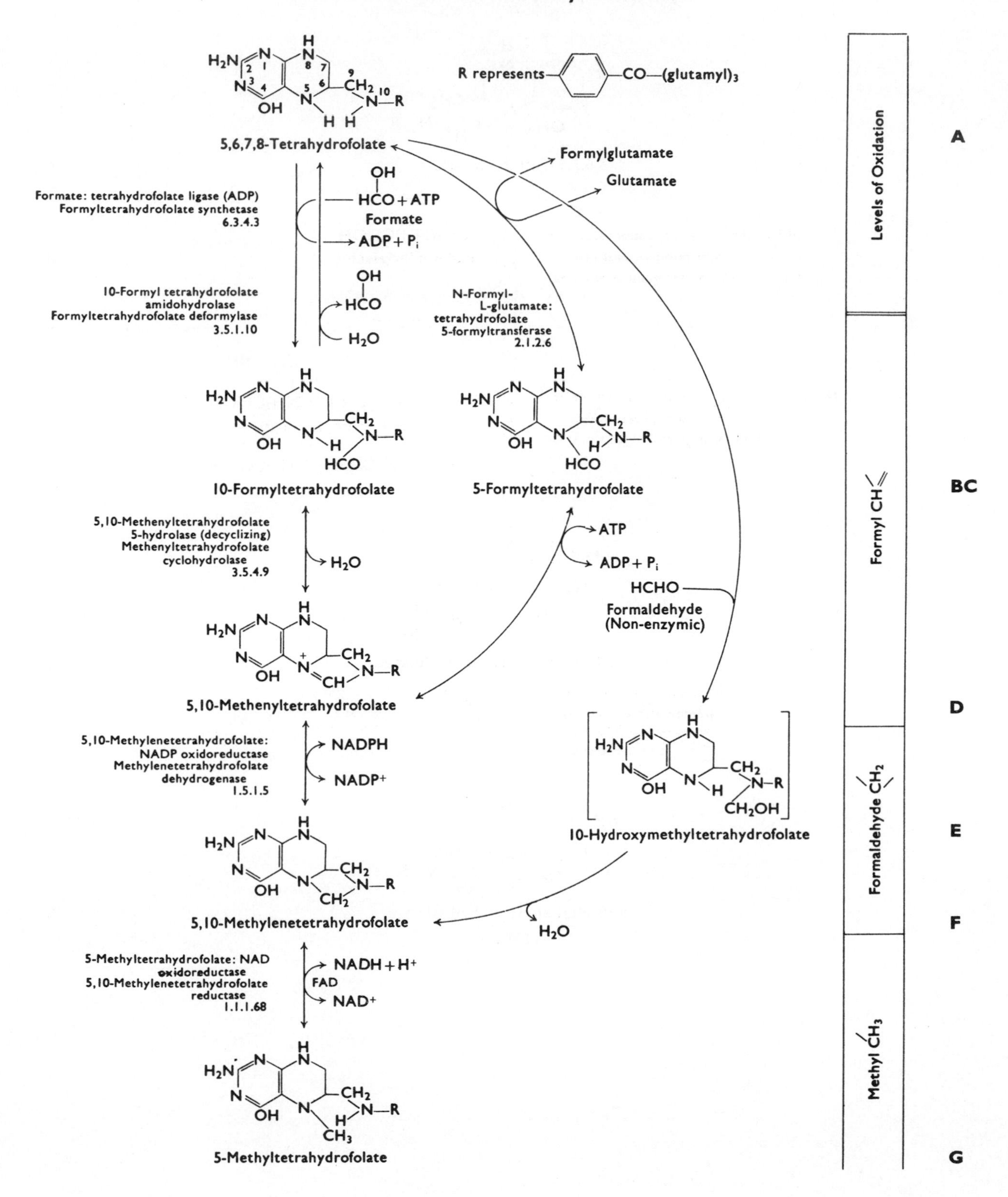

R represents—C6H4—CO—(glutamyl)3
5,6,7,8-Tetrahydrofolate
Formylglutamate
Glutamate
Formate: tetrahydrofolate ligase (ADP)
Formyltetrahydrofolate synthetase
6.3.4.3
HCO+ATP
Formate
ADP+Pi
10-Formyl tetrahydrofolate amidohydrolase
Formyltetrahydrofolate deformylase
3.5.1.10
HCO
H2O
N-Formyl-L-glutamate: tetrahydrofolate 5-formyltransferase
2.1.2.6
10-Formyltetrahydrofolate
5-Formyltetrahydrofolate
5,10-Methenyltetrahydrofolate 5-hydrolase (decyclizing)
Methenyltetrahydrofolate cyclohydrolase
3.5.4.9
H2O
ATP
ADP+Pi
HCHO
Formaldehyde
(Non-enzymic)
5,10-Methenyltetrahydrofolate
5,10-Methylenetetrahydrofolate: NADP oxidoreductase
Methylenetetrahydrofolate dehydrogenase
1.5.1.5
NADPH
NADP+
10-Hydroxymethyltetrahydrofolate
5,10-Methylenetetrahydrofolate
H2O
5-Methyltetrahydrofolate: NAD oxidoreductase
5,10-Methylenetetrahydrofolate reductase
1.1.1.68
NADH+H+
FAD
NAD+
5-Methyltetrahydrofolate
Levels of Oxidation
Formyl CH
Formaldehyde CH2
Methyl CH3
A
BC
D
E
F
G

Interrelationships between Active Forms of Tetrahydrofolate

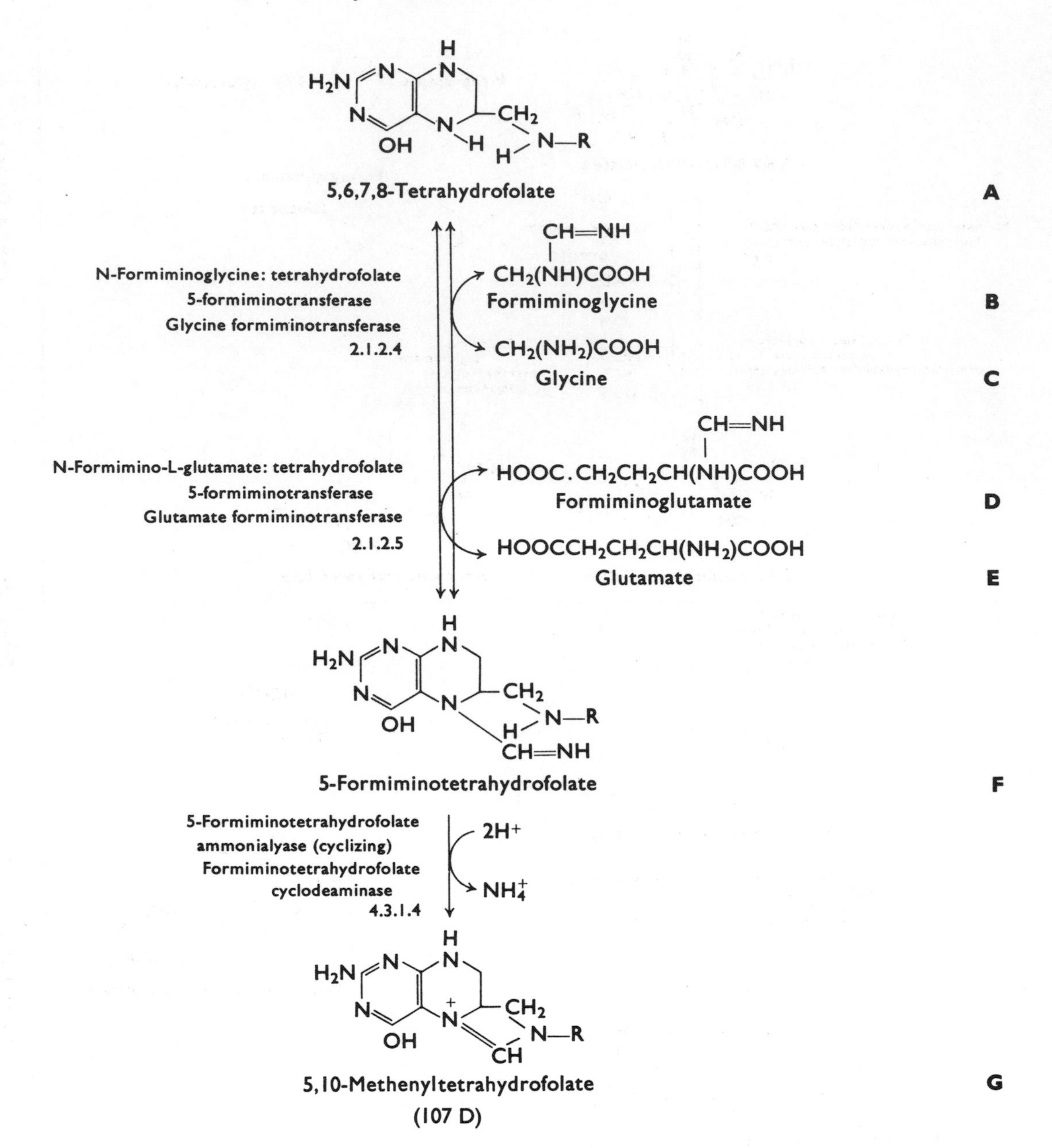

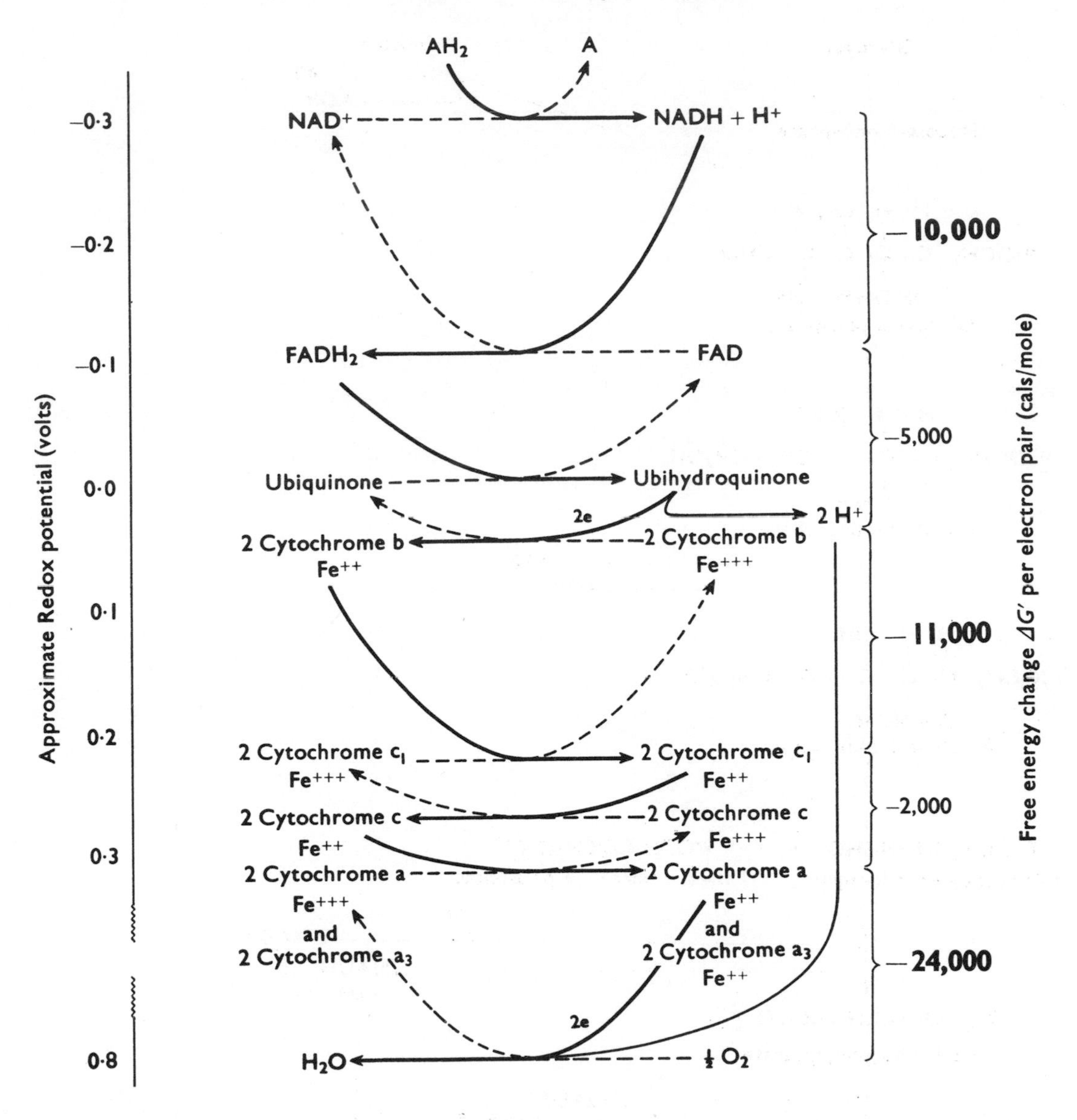

Heavy lines show flow of electrons (with protons in the first 3 steps).
Dotted lines show utilization and regeneration of oxidized co-enzymes.

Loss of electrons (oxidation) results in the release of free energy, the amount of which is related to the charge in redox potential by the equation

$$\Delta G' = -nF\Delta E_0'$$

where n is the number of electrons transferred, F is a faraday (—23,063 calories/volt/equivalent), and ΔE is the difference in redox potentials of the reacting carriers.

The formation of ATP from ADP and P_i requires at least 8,000 calories and this is released in 3 of the above steps, shown in heavy figures.

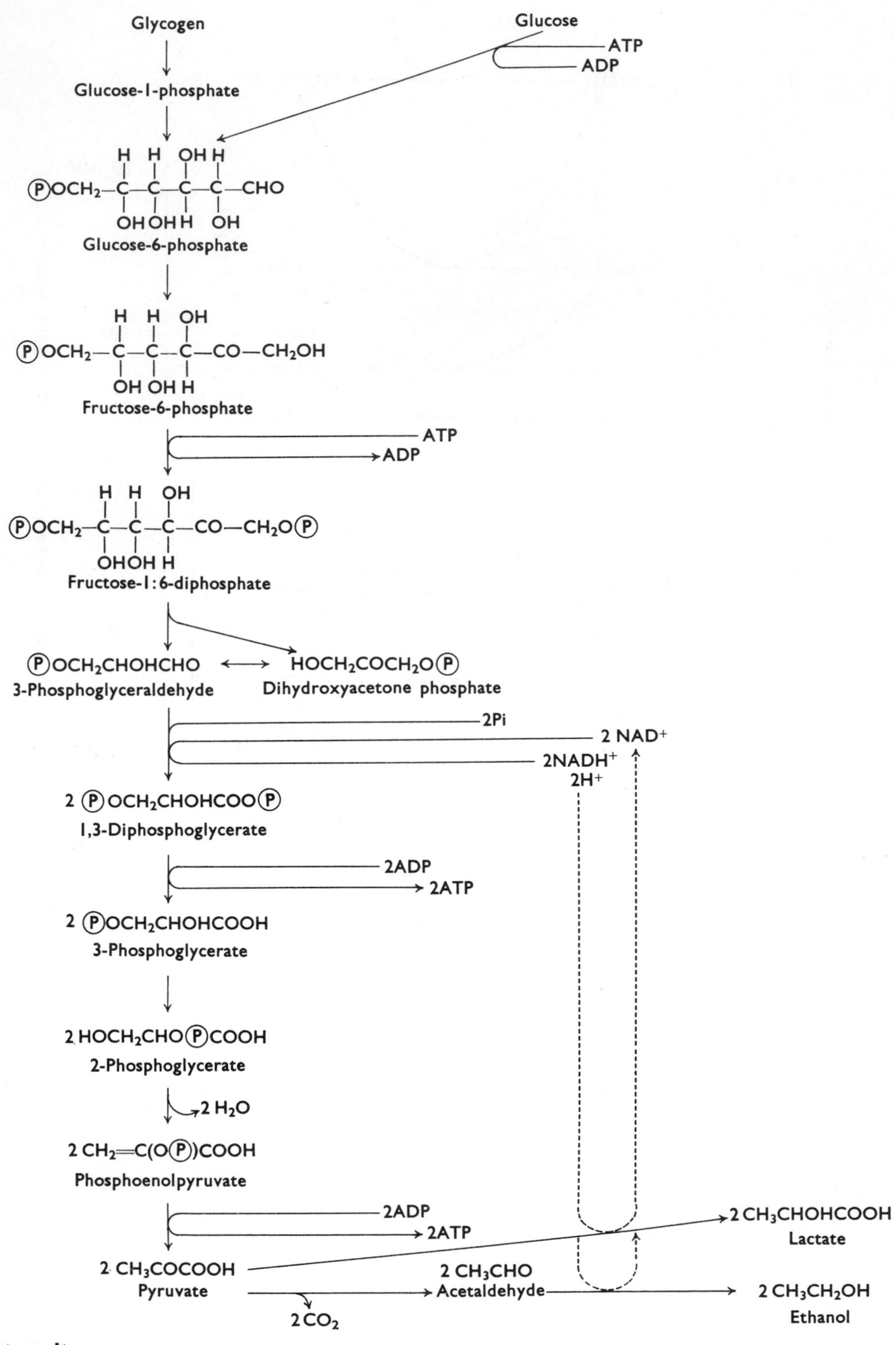

Net result

From glucose $C_6H_{12}O_6 + 2Pi + 2ADP \longrightarrow 2CH_3CHOHCOOH + 2ATP + 2H_2O$

From glycogen $[\alpha\text{-}1,4\text{-glucosyl}]_n + 3P_i + 3ADP \longrightarrow 2CH_3CHOHCOOH + 3ATP + 2H_2O + [\alpha\text{-}1,4\text{-glucosyl}]_{n-1}$

Starting from glycogen

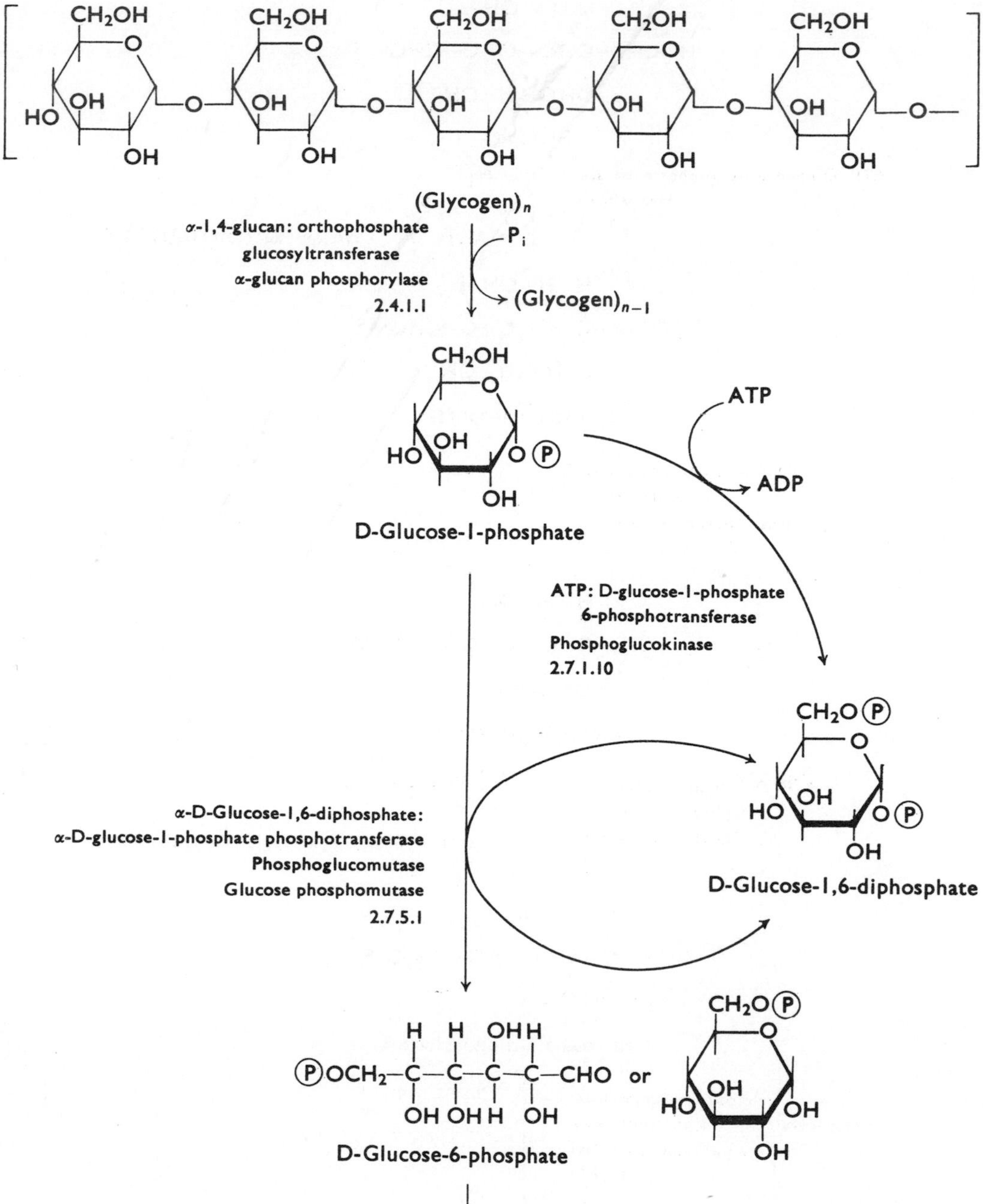

CH_2OH CH_2OH CH_2OH CH_2OH CH_2OH

HO OH OH OH OH OH —O—

OH OH OH OH OH

$(Glycogen)_n$

A

α-1,4-glucan: orthophosphate glucosyltransferase
α-glucan phosphorylase
2.4.1.1

P_i

$(Glycogen)_{n-1}$

CH_2OH

HO OH O(P)

OH

D-Glucose-1-phosphate

B

ATP

ADP

ATP: D-glucose-1-phosphate 6-phosphotransferase
Phosphoglucokinase
2.7.1.10

CH_2O(P)

HO OH O(P)

OH

D-Glucose-1,6-diphosphate

α-D-Glucose-1,6-diphosphate:
α-D-glucose-1-phosphate phosphotransferase
Phosphoglucomutase
Glucose phosphomutase
2.7.5.1

H H OH H

(P)OCH_2—C—C—C—C—CHO or

OH OH H OH

CH_2O(P)

HO OH OH

OH

D-Glucose-6-phosphate

C

To 112 B

Starting from glucose

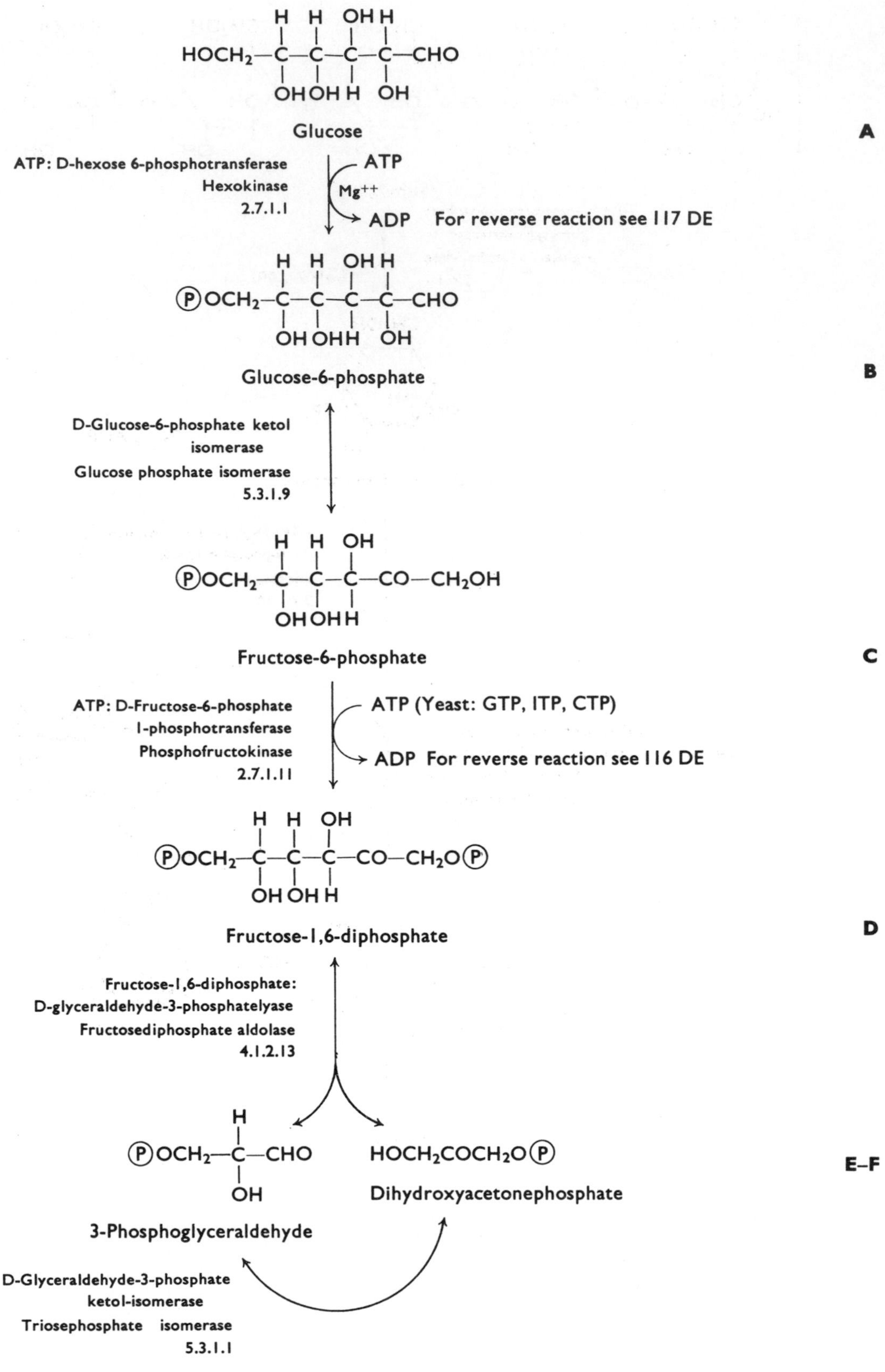

$$HOCH_2-\underset{OH}{\overset{H}{C}}-\underset{OH}{\overset{H}{C}}-\underset{H}{\overset{OH}{C}}-\underset{OH}{\overset{H}{C}}-CHO$$

Glucose **A**

ATP: D-hexose 6-phosphotransferase
Hexokinase
2.7.1.1

ATP → (Mg^{++}) → ADP For reverse reaction see 117 DE

$$ⓅOCH_2-\underset{OH}{\overset{H}{C}}-\underset{OH}{\overset{H}{C}}-\underset{H}{\overset{OH}{C}}-\underset{OH}{\overset{H}{C}}-CHO$$

Glucose-6-phosphate **B**

D-Glucose-6-phosphate ketol isomerase
Glucose phosphate isomerase
5.3.1.9

$$ⓅOCH_2-\underset{OH}{\overset{H}{C}}-\underset{OH}{\overset{H}{C}}-\underset{H}{\overset{OH}{C}}-CO-CH_2OH$$

Fructose-6-phosphate **C**

ATP: D-Fructose-6-phosphate 1-phosphotransferase
Phosphofructokinase
2.7.1.11

ATP (Yeast: GTP, ITP, CTP) → ADP For reverse reaction see 116 DE

$$ⓅOCH_2-\underset{OH}{\overset{H}{C}}-\underset{OH}{\overset{H}{C}}-\underset{H}{\overset{OH}{C}}-CO-CH_2OⓅ$$

Fructose-1,6-diphosphate **D**

Fructose-1,6-diphosphate: D-glyceraldehyde-3-phosphatelyase
Fructosediphosphate aldolase
4.1.2.13

$$ⓅOCH_2-\underset{OH}{\overset{H}{C}}-CHO$$

3-Phosphoglyceraldehyde

$$HOCH_2COCH_2OⓅ$$

Dihydroxyacetonephosphate **E–F**

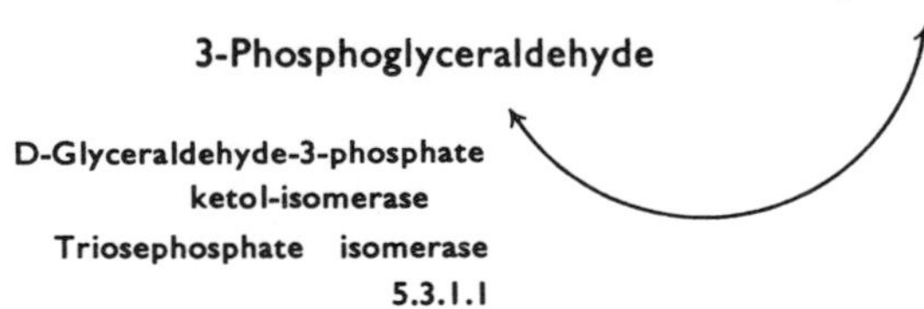

D-Glyceraldehyde-3-phosphate ketol-isomerase
Triosephosphate isomerase
5.3.1.1

The two kinase reactions (**A-B** and **C-D**) are irreversible under normal physiological conditions.

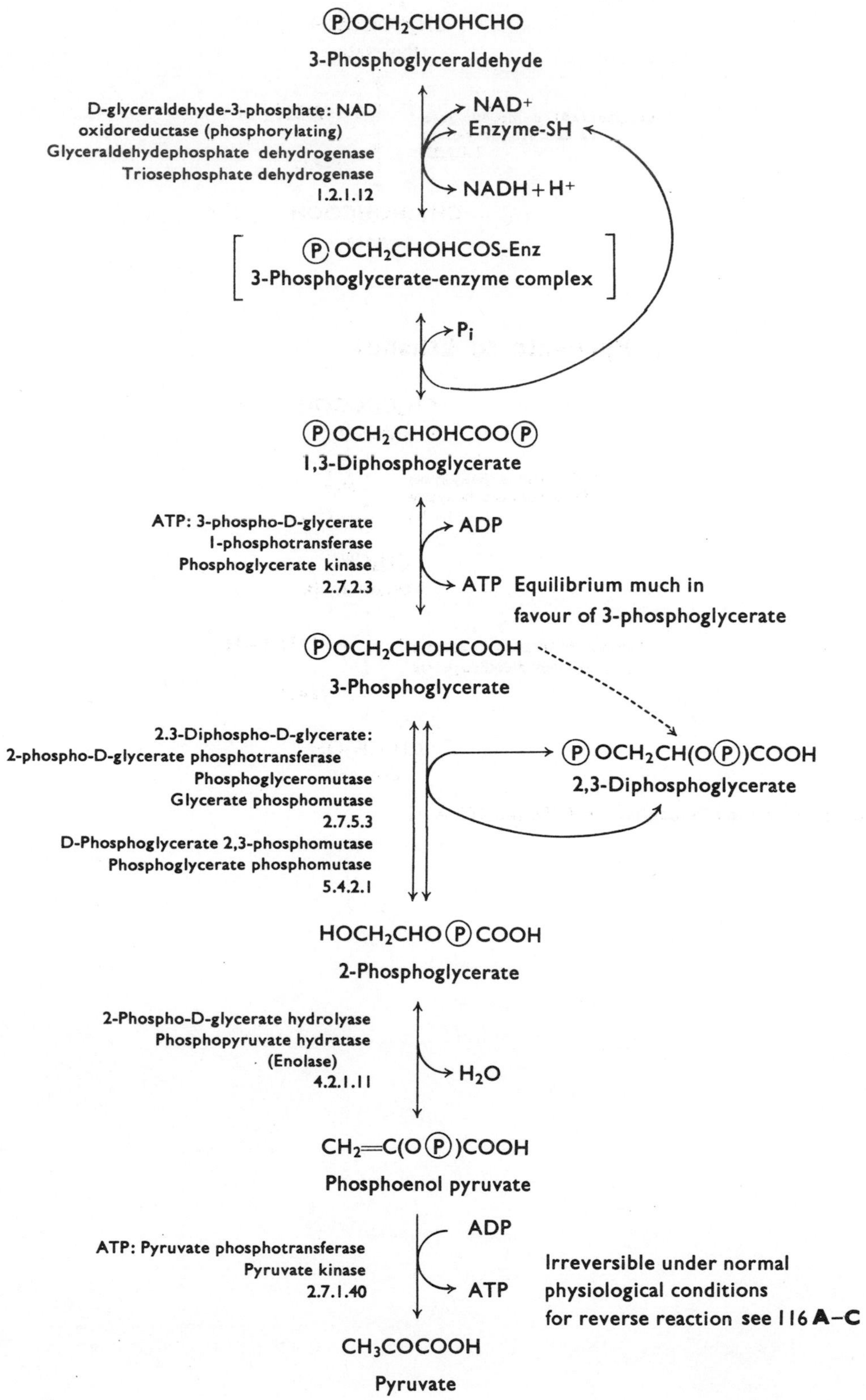

Ⓟ$OCH_2CHOHCHO$
3-Phosphoglyceraldehyde **A**

D-glyceraldehyde-3-phosphate: NAD oxidoreductase (phosphorylating)
Glyceraldehydephosphate dehydrogenase
Triosephosphate dehydrogenase
1.2.1.12

NAD^+
Enzyme-SH
$NADH + H^+$

[Ⓟ $OCH_2CHOHCOS$-Enz
3-Phosphoglycerate-enzyme complex] **B**

P_i

Ⓟ$OCH_2CHOHCOO$Ⓟ
1,3-Diphosphoglycerate **C**

ATP: 3-phospho-D-glycerate 1-phosphotransferase
Phosphoglycerate kinase
2.7.2.3

ADP
ATP Equilibrium much in favour of 3-phosphoglycerate

Ⓟ$OCH_2CHOHCOOH$
3-Phosphoglycerate **D**

2.3-Diphospho-D-glycerate: 2-phospho-D-glycerate phosphotransferase
Phosphoglyceromutase
Glycerate phosphomutase
2.7.5.3
D-Phosphoglycerate 2,3-phosphomutase
Phosphoglycerate phosphomutase
5.4.2.1

Ⓟ $OCH_2CH(O$Ⓟ$)COOH$
2,3-Diphosphoglycerate

$HOCH_2CHO$Ⓟ$COOH$
2-Phosphoglycerate **E**

2-Phospho-D-glycerate hydrolyase
Phosphopyruvate hydratase
(Enolase)
4.2.1.11

H_2O

$CH_2{=}C(O$Ⓟ$)COOH$
Phosphoenol pyruvate **F**

ATP: Pyruvate phosphotransferase
Pyruvate kinase
2.7.1.40

ADP
ATP

Irreversible under normal physiological conditions
for reverse reaction see 116 **A–C**

$CH_3COCOOH$
Pyruvate **G**

Pyruvate to Lactate

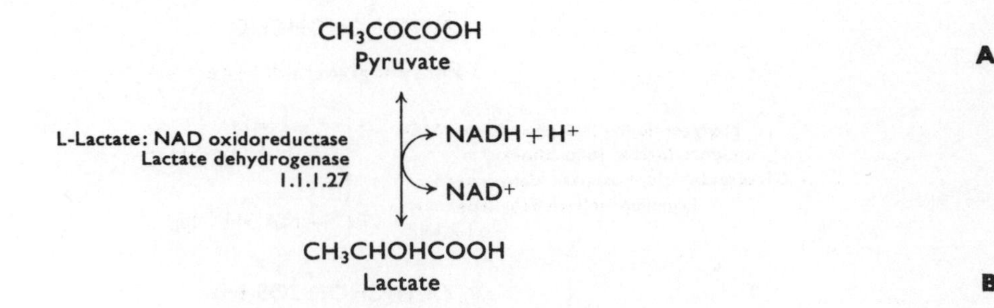

Pyruvate to Ethanol

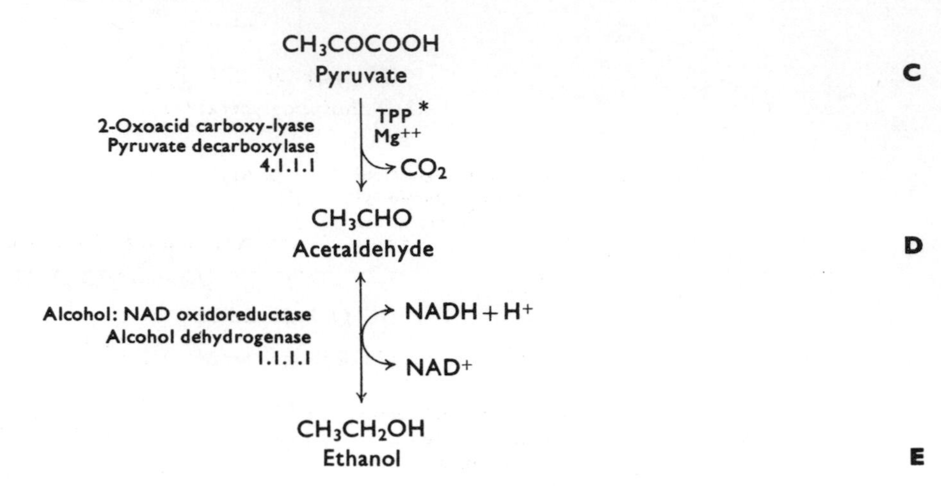

* For function of thiamine diphosphate in **C-D**, see 103 A–G.

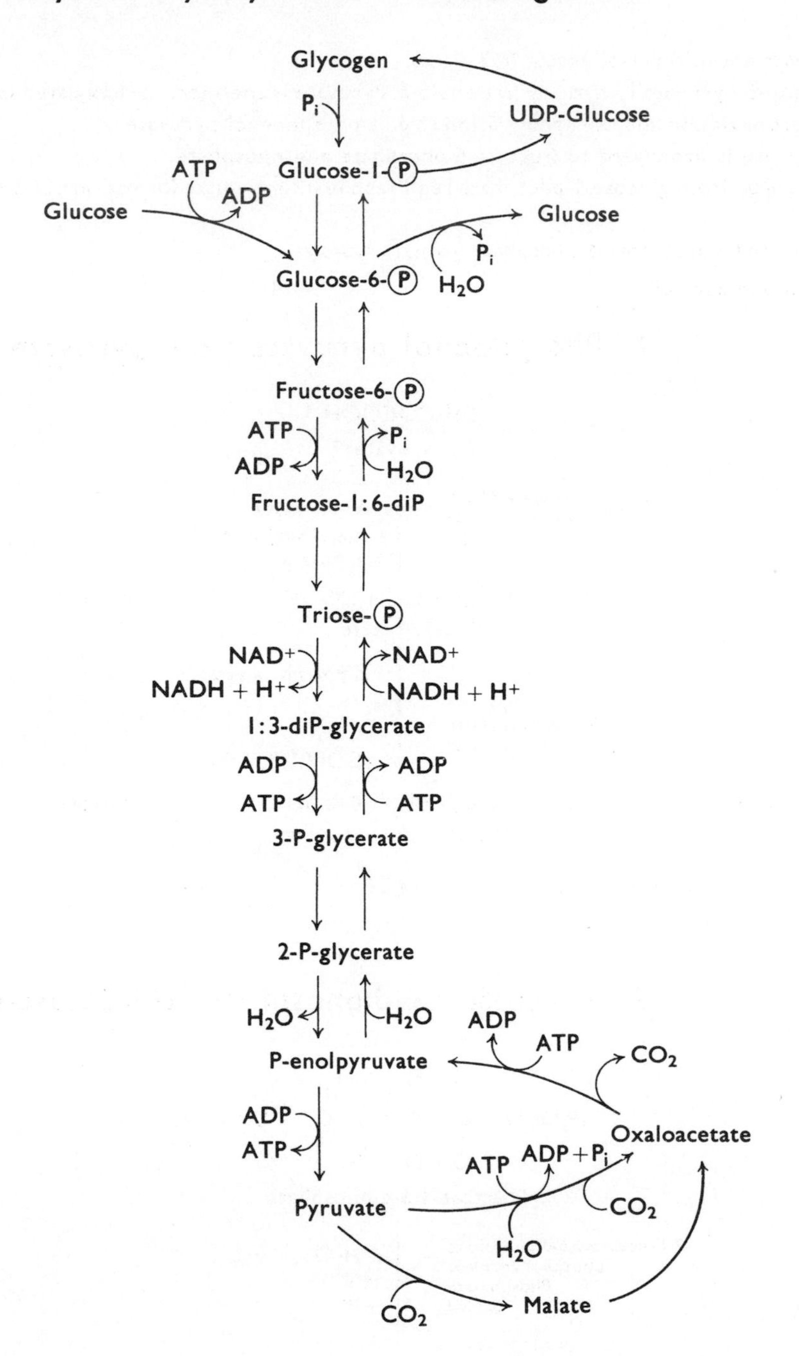

For discussion see pages 69–71.

These are glycolytic reactions in reverse, except in 3 places.

1 Phosphoenol pyruvate to pyruvate is virtually irreversible. Pyruvate is therefore carboxylated to oxaloacetate which then undergoes decarboxylation and phosphorylation to give phosphoenol pyruvate.

2 Fructose-1,6 diphosphate is *hydrolyzed* to fructose-6-phosphate and phosphate.

3 The formation of glycogen from glucose-1-phosphate requires the intermediate formation of UDP-glucose

or

The formation of glucose from glucose-6-phosphate involves *hydrolysis*.

For general discussion see page 69.

I. Phosphoenol pyruvate from pyruvate

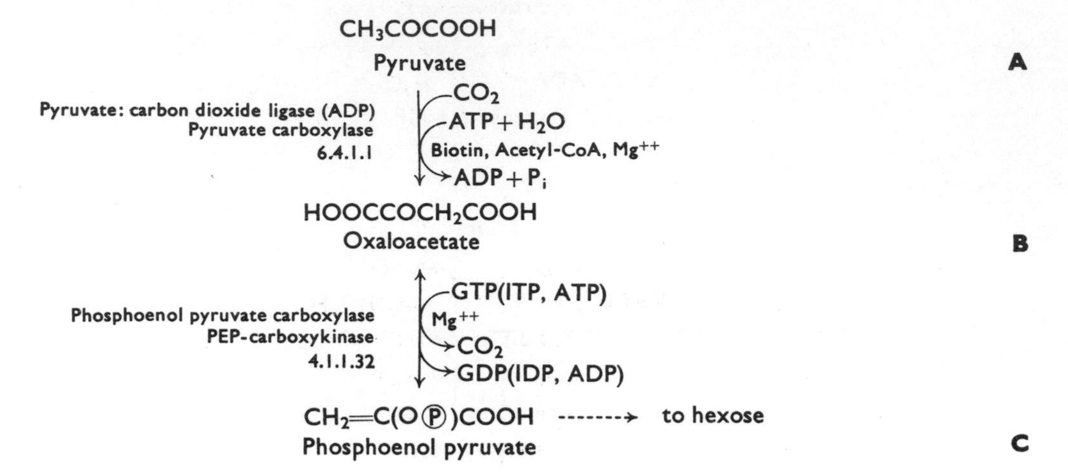

For further discussion on carboxylation reactions see page 24.

2. Fructose-1,6-diphosphate to fructose-6-phosphate

ⓅOCH_2—C(H)(OH)—C(H)(OH)—C(OH)(H)—CO—CH_2OⓅ

Fructose-1,6-diphosphate **D**

D-Fructose-1,6-diphosphate phosphohydrolase
Phosphatase
3.1.3.1

H_2O, M^{++} → P_i

ⓅOCH_2—C(H)(OH)—C(H)(OH)—C(OH)(H)—CO—CH_2OH

D-Fructose-6-phosphate **E**

3. Glucose-1-phosphate to Glycogen

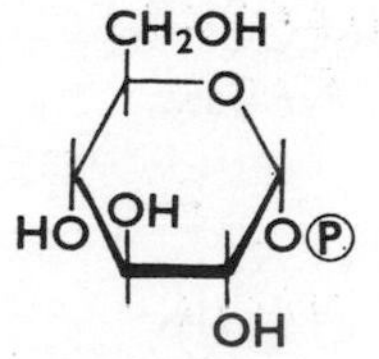

CH₂OH
HO OH OⓅ
OH

Glucose-1-phosphate

A

UTP: α-D-glucose-1-phosphate uridylyltransferase
Glucose-1-phosphate uridylyltransferase
UDPG-pyrophosphorylase
2.7.7.9

UTP → PP_i

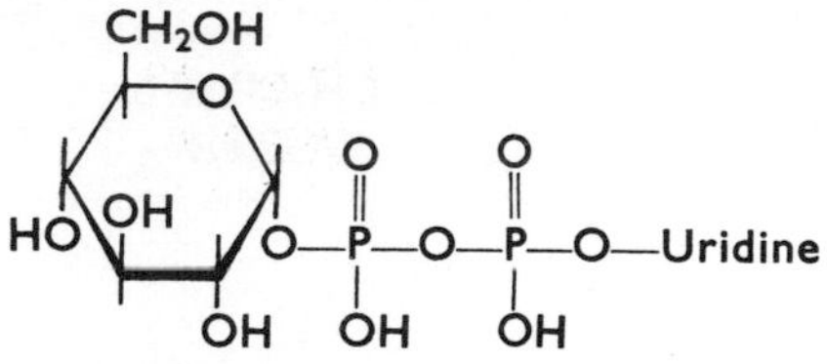

Uridine diphosphoglucose

B

UDP glucose: glycogen α-4-glucosyltransferase
UDP glucose-glycogen glucosyltransferase
Glycogen-UDP glucosyl transferase
2.4.1.11

(α-1,4-glucosyl)$_n$ or (glycogen)$_n$
Mg^{++}s
UDP

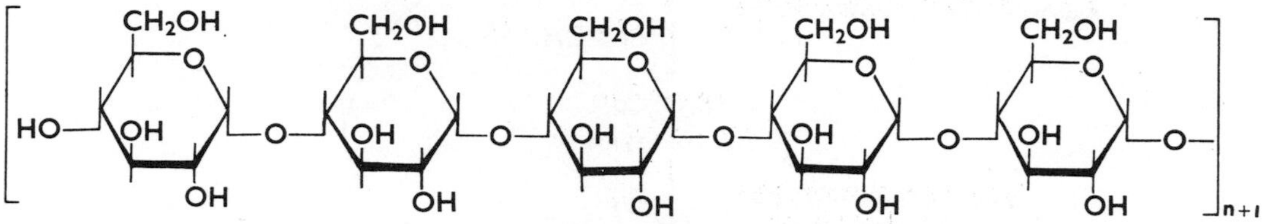

$n+1$

Glucosyl added at this end

Glycogen fragment

C

3a. Glucose-6-phosphate to glucose

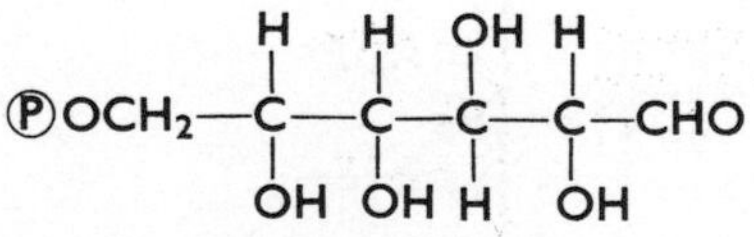

Glucose-6-phosphate

D

D-Glucose-6-phosphate phosphohydrolase
Glucose-6-phosphatase
3.1.3.9

H_2O → Pi

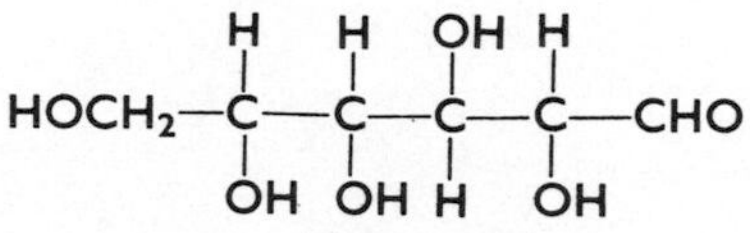

Glucose

E

Acetate Formation

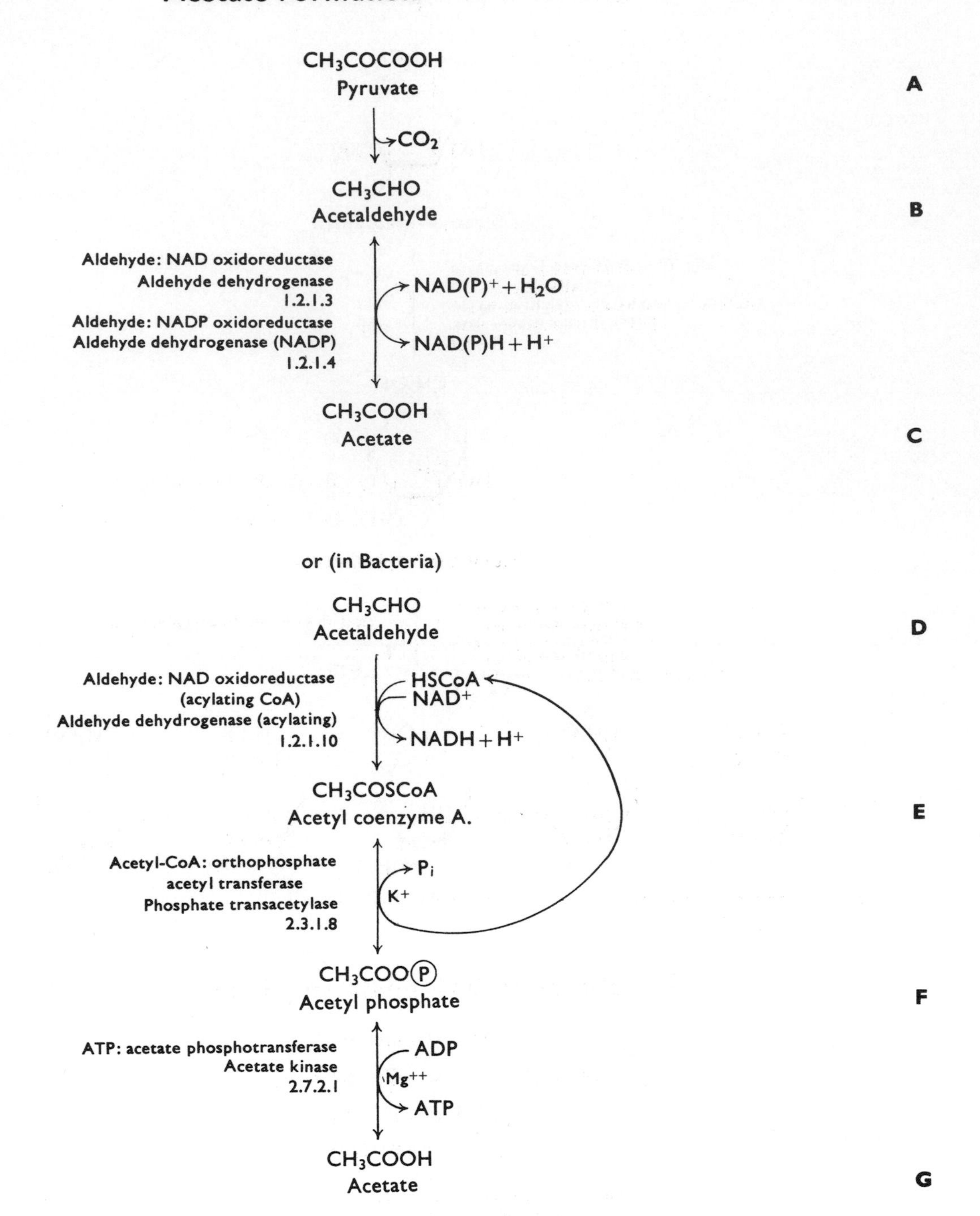

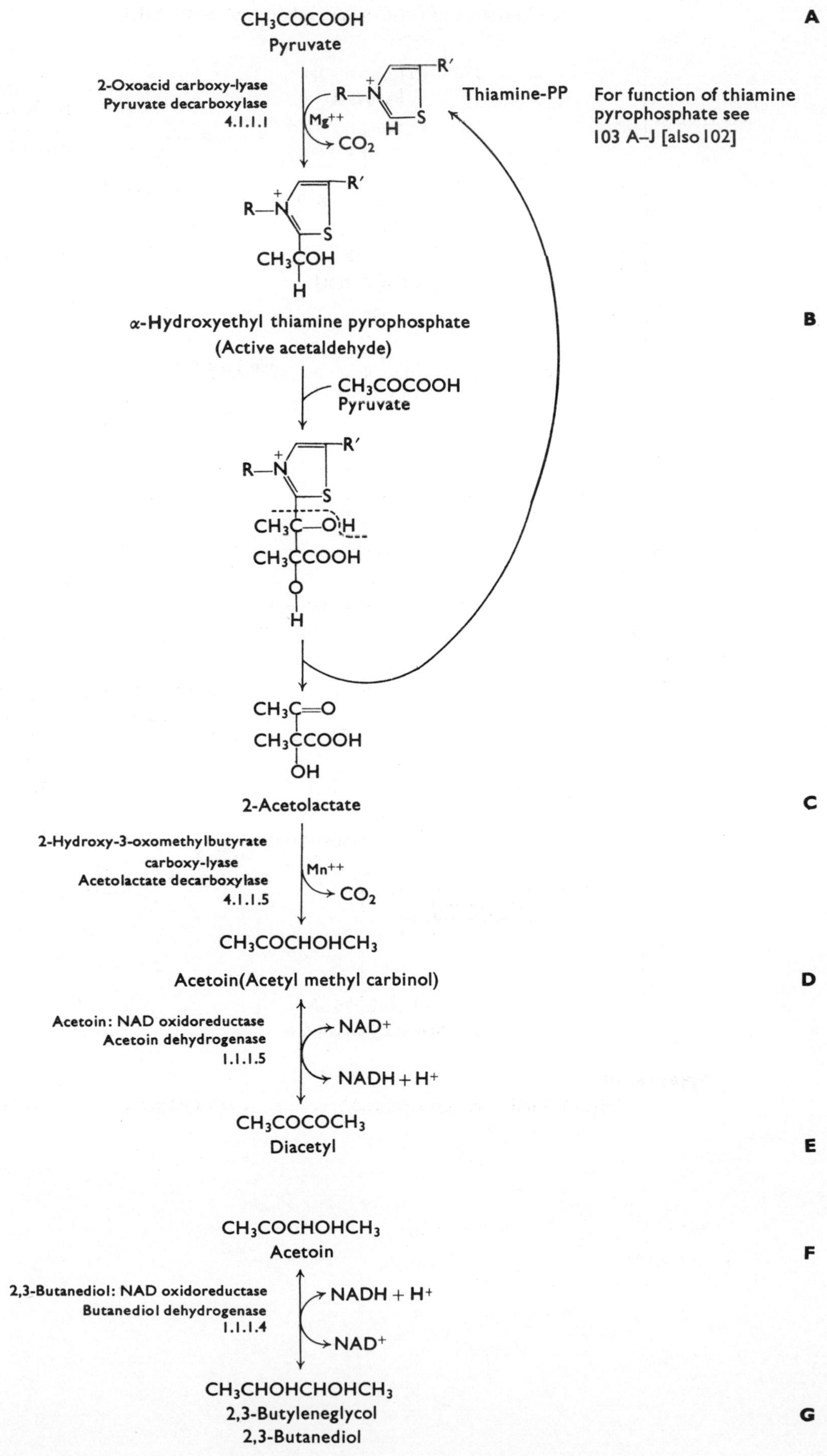

A

$CH_3COCOOH$
Pyruvate

2-Oxoacid carboxy-lyase
Pyruvate decarboxylase
4.1.1.1

Mg^{++} → CO_2

Thiamine-PP

For function of thiamine pyrophosphate see
103 A–J [also 102]

B

α-Hydroxyethyl thiamine pyrophosphate
(Active acetaldehyde)

$CH_3COCOOH$
Pyruvate

C

2-Acetolactate

2-Hydroxy-3-oxomethylbutyrate carboxy-lyase
Acetolactate decarboxylase
4.1.1.5

Mn^{++} → CO_2

D

$CH_3COCHOHCH_3$
Acetoin(Acetyl methyl carbinol)

Acetoin: NAD oxidoreductase
Acetoin dehydrogenase
1.1.1.5

NAD^+
$NADH + H^+$

E

$CH_3COCOCH_3$
Diacetyl

F

$CH_3COCHOHCH_3$
Acetoin

2,3-Butanediol: NAD oxidoreductase
Butanediol dehydrogenase
1.1.1.4

$NADH + H^+$
NAD^+

G

$CH_3CHOHCHOHCH_3$
2,3-Butyleneglycol
2,3-Butanediol

(See Reactions of Thiamine pyrophosphate p. 103 A–K)

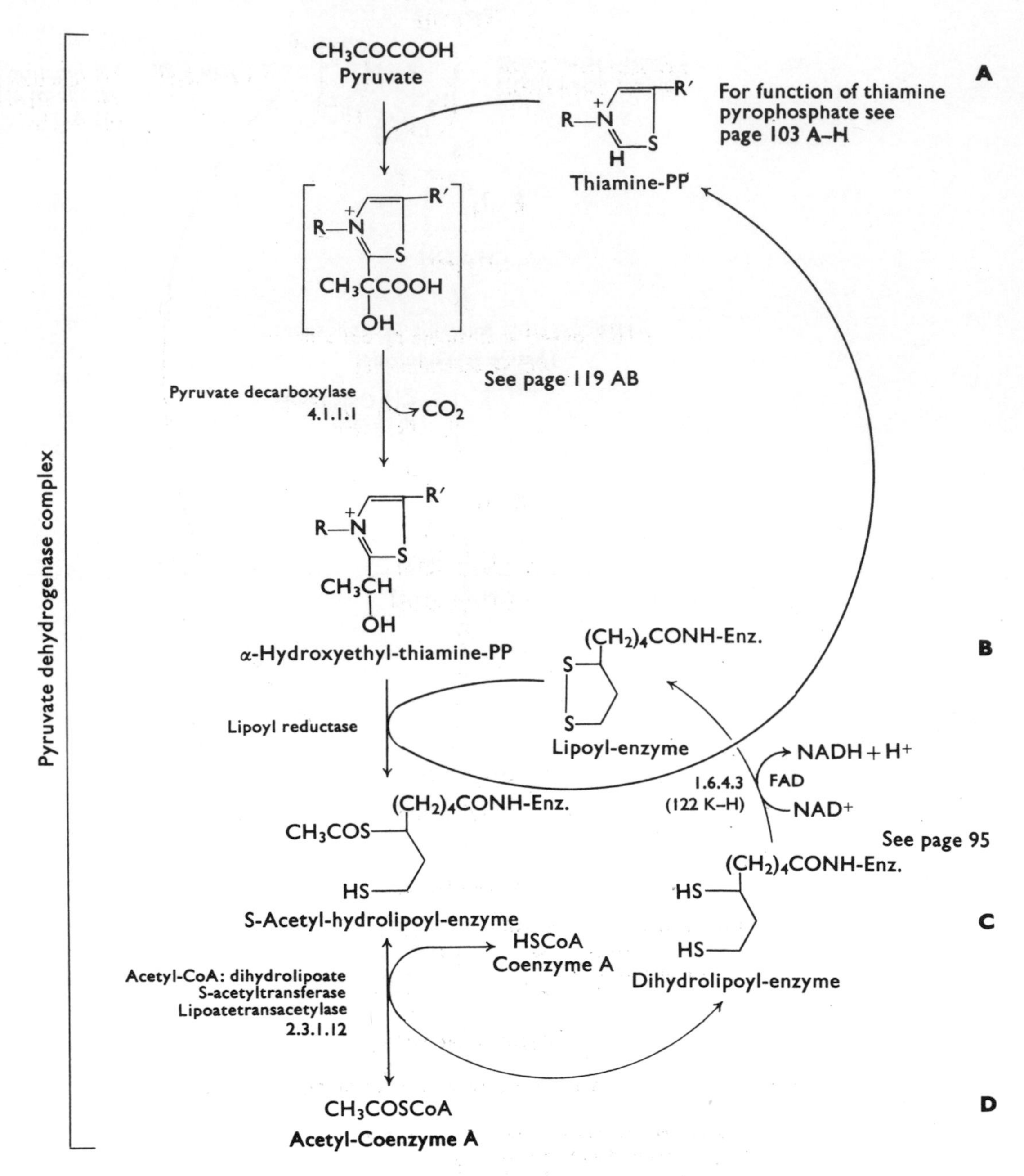

Net result

$$CH_3COCOOH + NAD^+ + HSCoA \longrightarrow CH_3COSCoA + CO_2 + NADH + H^+$$

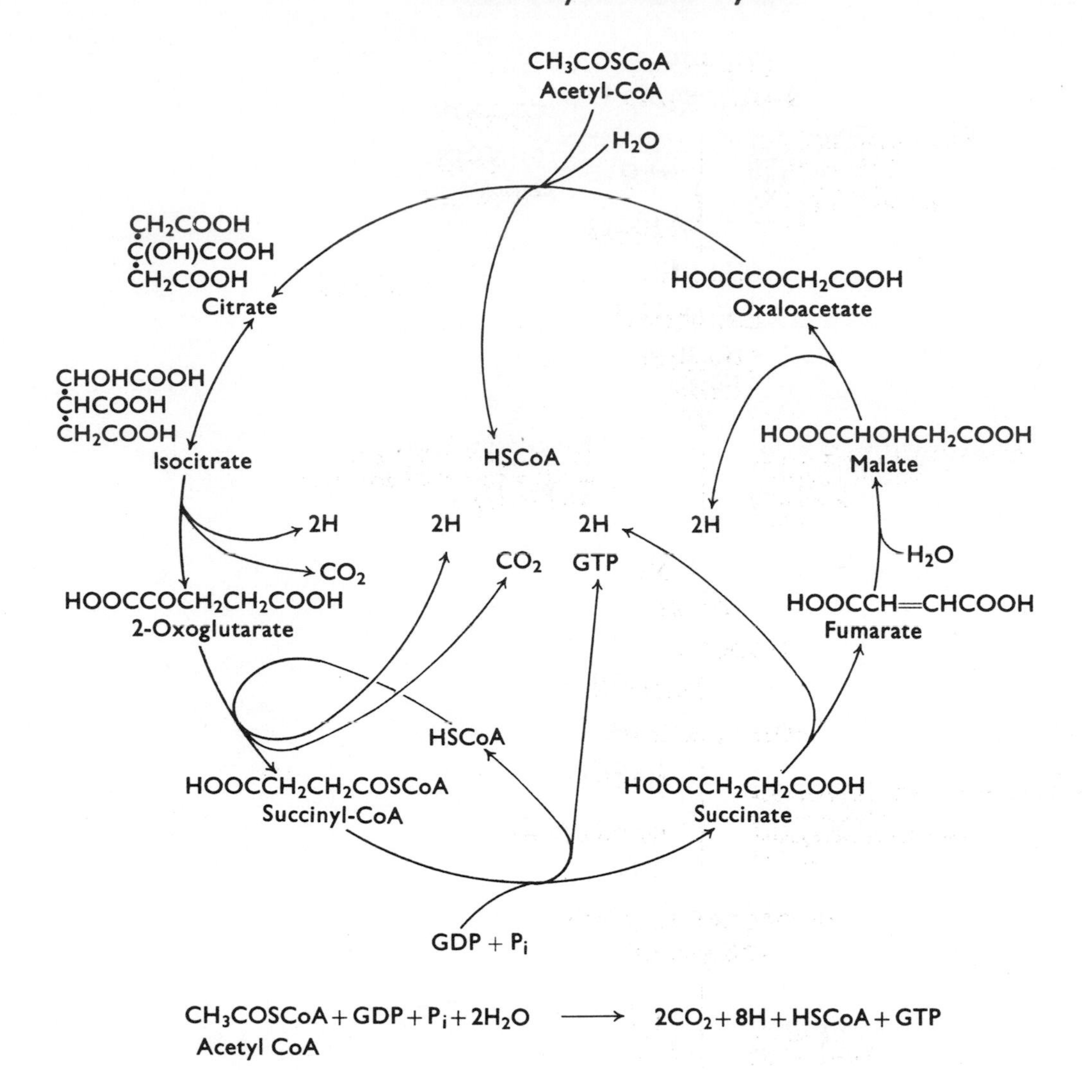

$$\underset{\text{Acetyl CoA}}{CH_3COSCoA} + GDP + P_i + 2H_2O \longrightarrow 2CO_2 + 8H + HSCoA + GTP$$

Of the 8H, 3 pairs pass through NAD to the cytochrome electron transport chain yielding 3 ATP each. 1 pair passes to FAD and hence to the electron transport system yielding 2 ATP
Total—11 ATP.

$$8H + 2O_2 + 11ADP + 11P_i \longrightarrow 4H_2O + 11H_2O + 11ATP$$

Sum Total

$$CH_3COSCoA + GDP + 11ADP + 12P_i + 2O_2 \longrightarrow 2CO_2 + 13H_2O + GTP + 11ATP + HSCoA$$

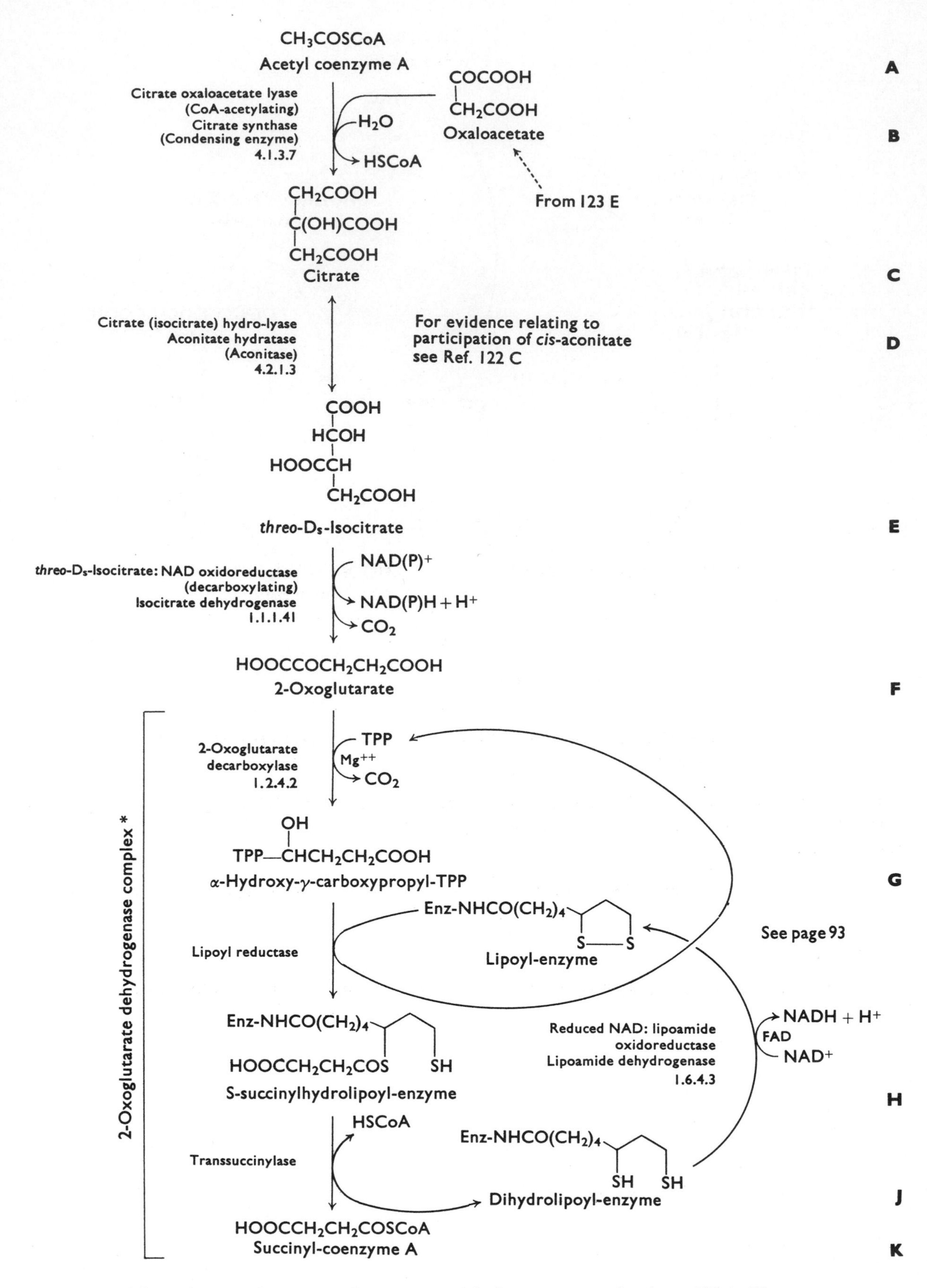

CH3COSCoA
Acetyl coenzyme A
A
COCOOH
CH2COOH
Oxaloacetate
Citrate oxaloacetate lyase
(CoA-acetylating)
Citrate synthase
(Condensing enzyme)
4.1.3.7
H2O
HSCoA
B
From 123 E
CH2COOH
C(OH)COOH
CH2COOH
Citrate
C
Citrate (isocitrate) hydro-lyase
Aconitate hydratase
(Aconitase)
4.2.1.3
For evidence relating to
participation of cis-aconitate
see Ref. 122 C
D
COOH
HCOH
HOOCCH
CH2COOH
threo-Ds-Isocitrate
E
threo-Ds-Isocitrate: NAD oxidoreductase
(decarboxylating)
Isocitrate dehydrogenase
1.1.1.41
NAD(P)+
NAD(P)H + H+
CO2
HOOCCOCH2CH2COOH
2-Oxoglutarate
F
2-Oxoglutarate dehydrogenase complex *
2-Oxoglutarate
decarboxylase
1.2.4.2
TPP
Mg++
CO2
OH
TPP—CHCH2CH2COOH
α-Hydroxy-γ-carboxypropyl-TPP
G
Lipoyl reductase
Enz-NHCO(CH2)4
S—S
Lipoyl-enzyme
See page 93
Enz-NHCO(CH2)4
HOOCCH2CH2COS SH
S-succinylhydrolipoyl-enzyme
Reduced NAD: lipoamide
oxidoreductase
Lipoamide dehydrogenase
1.6.4.3
NADH + H+
FAD
NAD+
H
Transsuccinylase
HSCoA
Enz-NHCO(CH2)4
SH SH
Dihydrolipoyl-enzyme
J
HOOCCH2CH2COSCoA
Succinyl-coenzyme A
K

* See references for very similar pyruvate dehydrogenase complex (page 120 A–D).

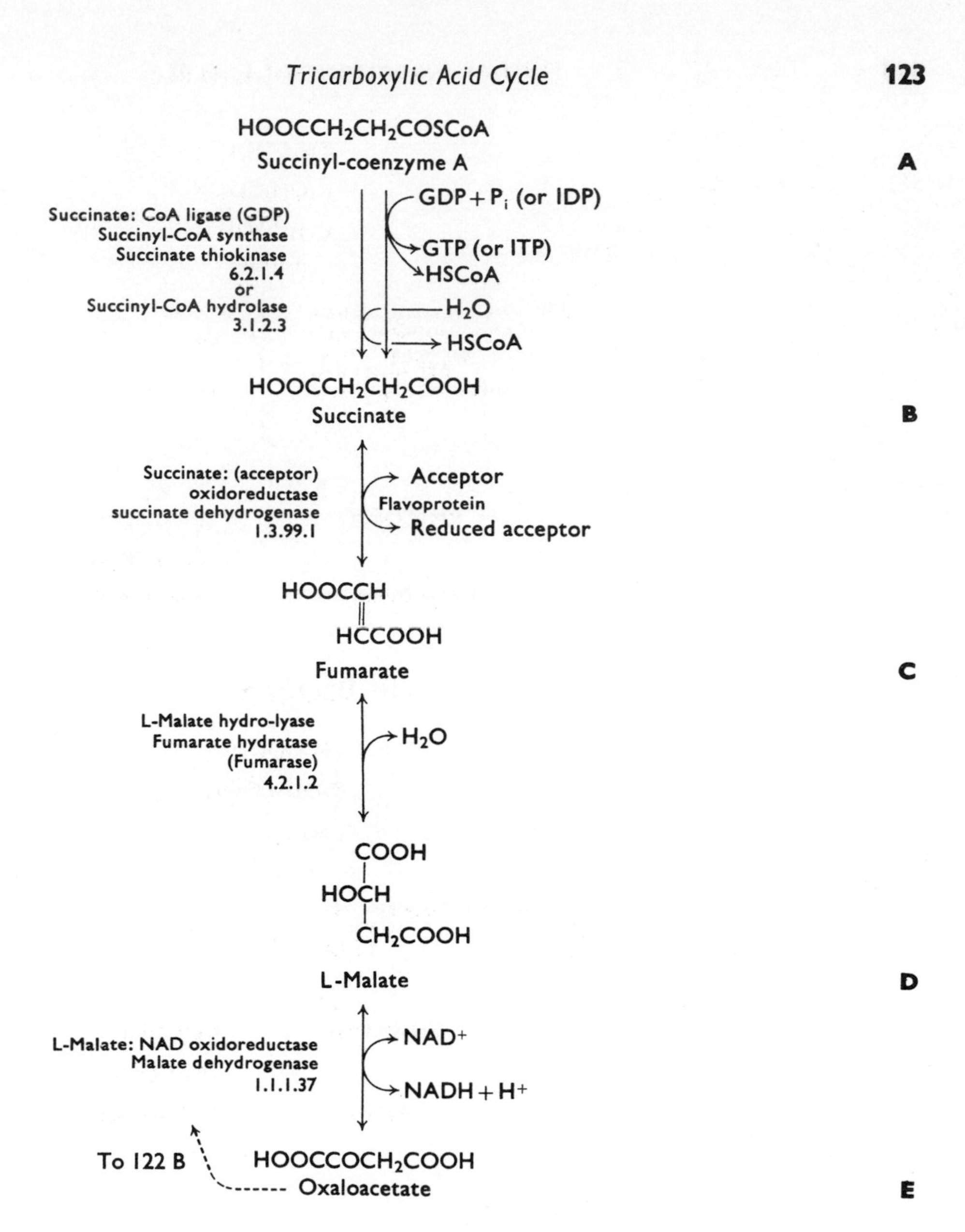
HOOCCH₂CH₂COSCoA
Succinyl-coenzyme A
A
Succinate: CoA ligase (GDP)
Succinyl-CoA synthase
Succinate thiokinase
6.2.1.4
or
Succinyl-CoA hydrolase
3.1.2.3
GDP + Pi (or IDP)
GTP (or ITP)
HSCoA
H2O
HSCoA
HOOCCH2CH2COOH
Succinate
B
Succinate: (acceptor)
oxidoreductase
succinate dehydrogenase
1.3.99.1
Acceptor
Flavoprotein
Reduced acceptor
HOOCCH
HCCOOH
Fumarate
C
L-Malate hydro-lyase
Fumarate hydratase
(Fumarase)
4.2.1.2
H2O
COOH
HOCH
CH2COOH
L-Malate
D
L-Malate: NAD oxidoreductase
Malate dehydrogenase
1.1.1.37
NAD+
NADH + H+
To 122 B
HOOCCOCH2COOH
Oxaloacetate
E

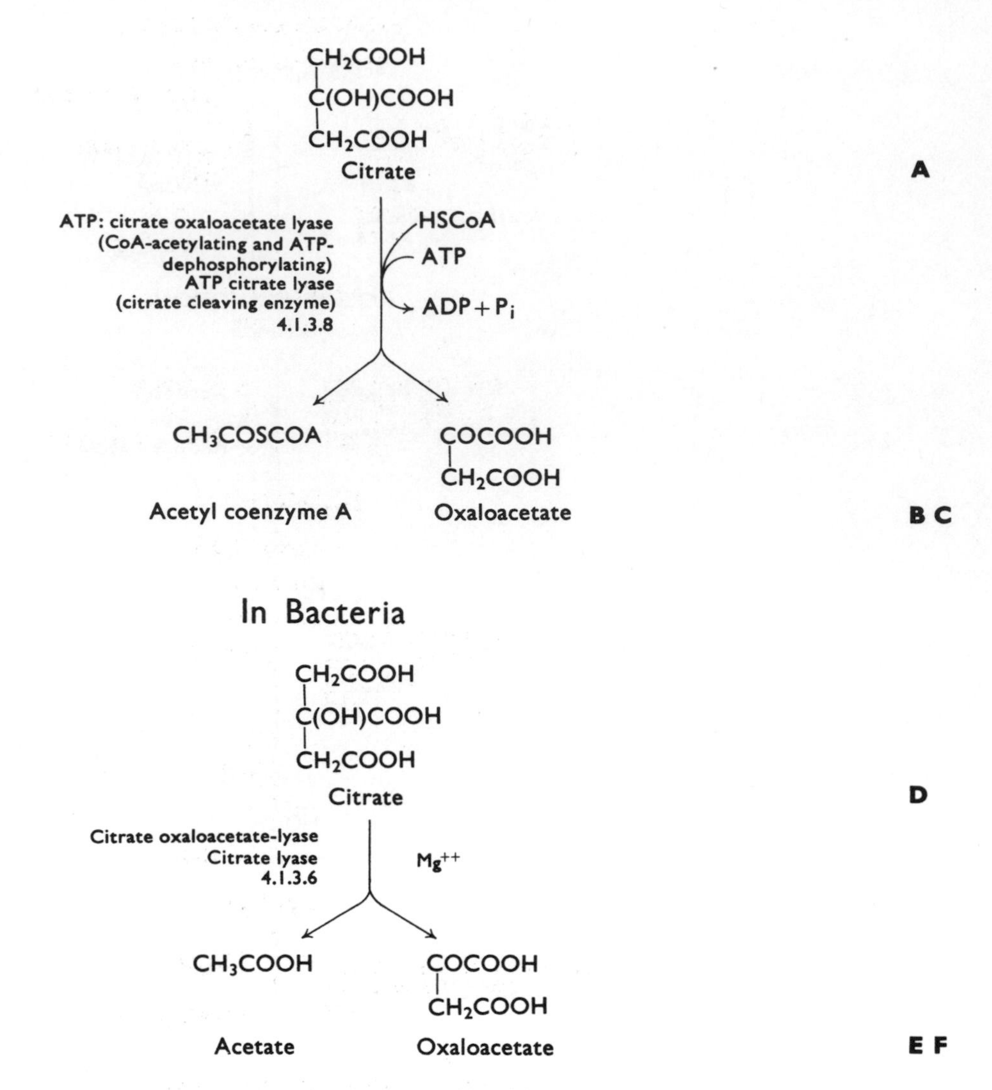

CH_2COOH
$C(OH)COOH$
CH_2COOH
Citrate
A
ATP: citrate oxaloacetate lyase (CoA-acetylating and ATP-dephosphorylating)
ATP citrate lyase
(citrate cleaving enzyme)
4.1.3.8
HSCoA
ATP
ADP + P_i
$CH_3COSCOA$
$COCOOH$
CH_2COOH
Acetyl coenzyme A
Oxaloacetate
B C
In Bacteria
CH_2COOH
$C(OH)COOH$
CH_2COOH
Citrate
D
Citrate oxaloacetate-lyase
Citrate lyase
4.1.3.6
Mg^{++}
CH_3COOH
$COCOOH$
CH_2COOH
Acetate
Oxaloacetate
E F

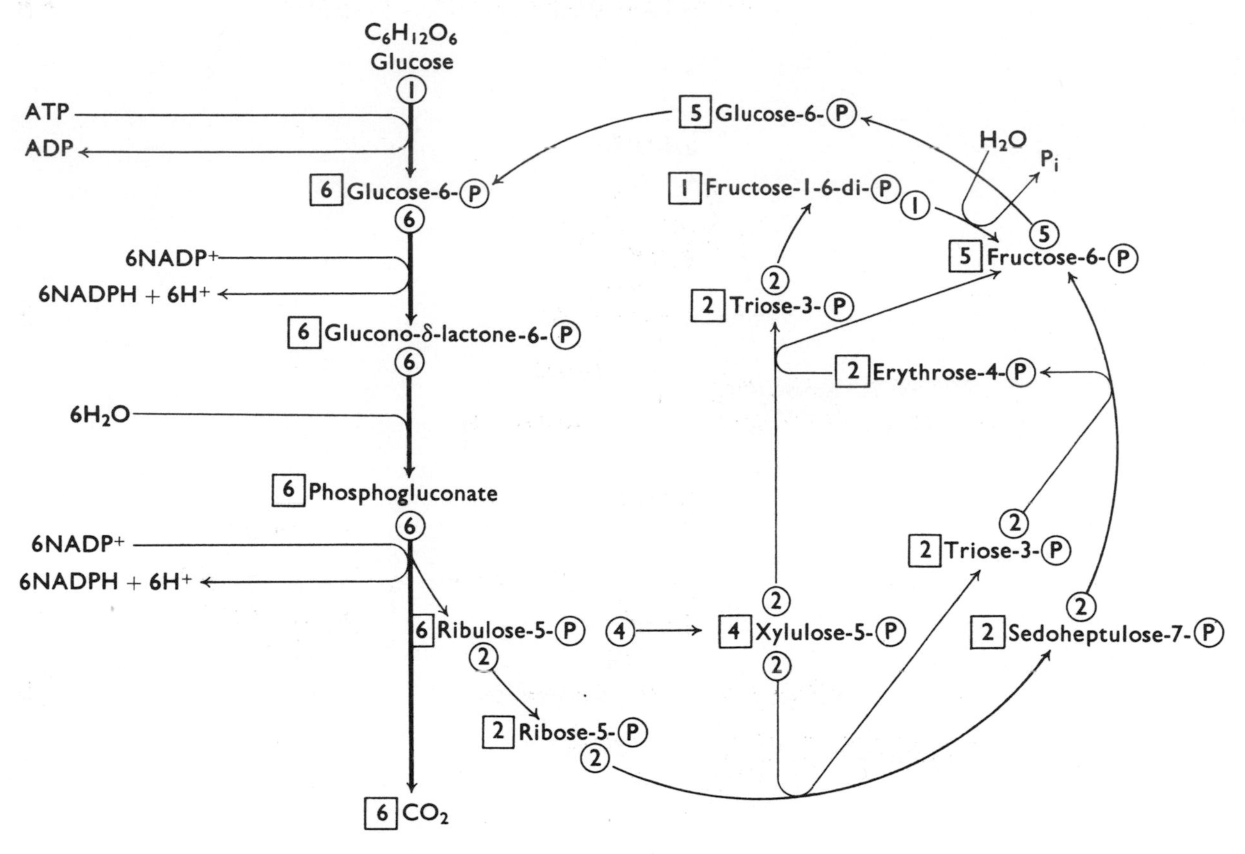

Net result

$$C_6H_{12}O_6 + 7H_2O + 12NADP^+ + ATP \longrightarrow 6CO_2 + 12NADPH + 12H^+ + ADP + P_i$$

Functions of Cycle

1 To provide reduced NADP for synthetic purposes such as the synthesis of fatty acids.

2 To provide energy. If all the reduced NADP is completely oxidized, 36 molecules of ATP are produced for each molecule of glucose oxidized. However, it should be noted that, in general, energy is provided by oxidation of NADH, whereas NADPH is used in biosynthetic reactions.

3 Certain reactions provide a source of pentoses for nucleotides.

4 To make possible the interconversion of hexoses and pentoses.

Mechanism of Cycle

Regenerate 5 molecules of hexose from 6 molecules of pentose ($6C_5 \rightarrow 5C_6$).

Note

The molecule of glucose entering the cycle is not itself completely oxidized. The six CO_2 molecules arise from the C_1 groups of the six molecules of glucose-6-phosphate.

The figures in squares denote the number of molecules resulting from previous reaction(s). The figures in circles denote the number of these molecules utilized in different pathways.

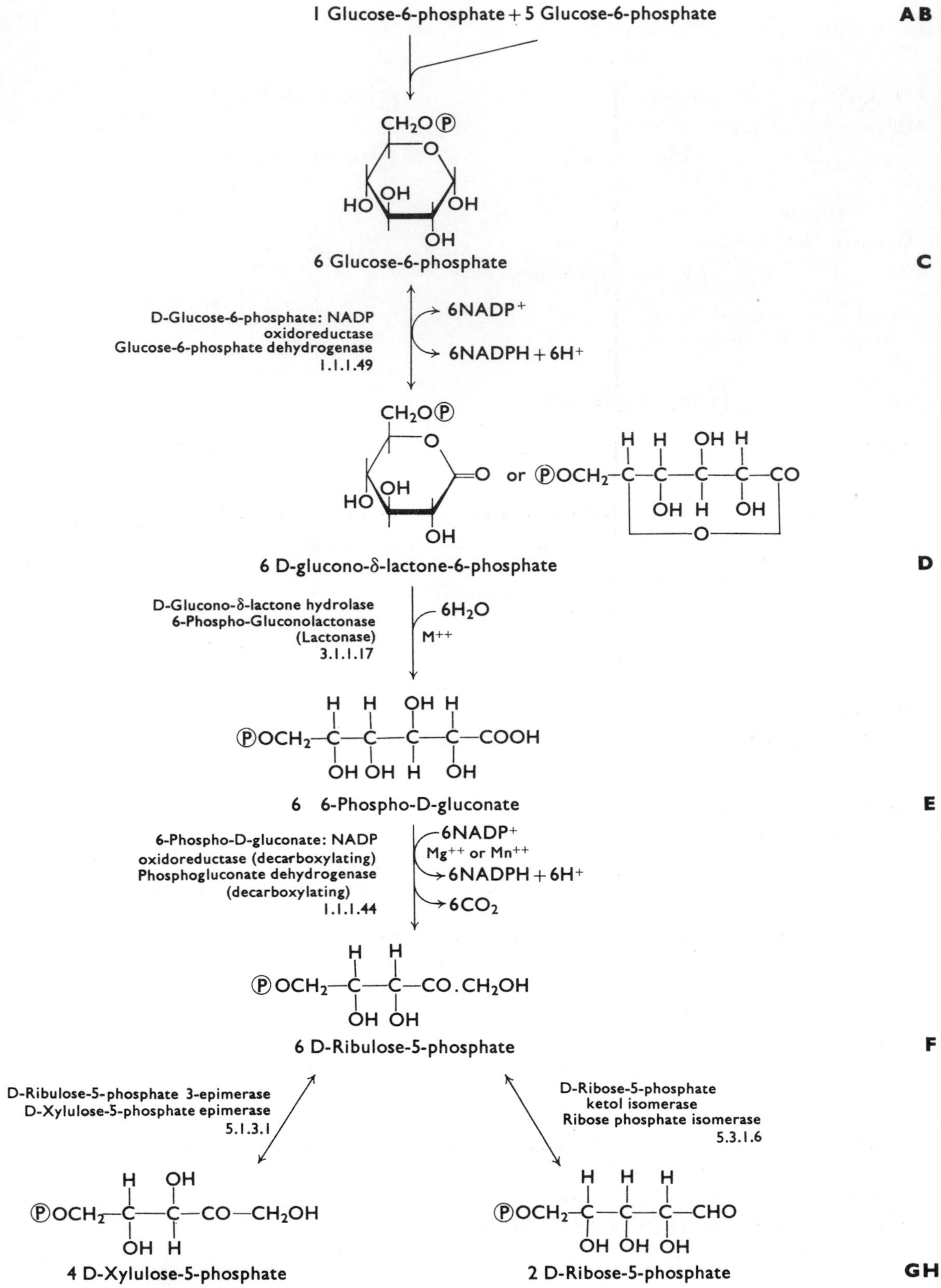

1 Glucose-6-phosphate + 5 Glucose-6-phosphate **AB**

6 Glucose-6-phosphate **C**

D-Glucose-6-phosphate: NADP oxidoreductase
Glucose-6-phosphate dehydrogenase
1.1.1.49

$6NADP^+$ → $6NADPH + 6H^+$

6 D-glucono-δ-lactone-6-phosphate **D**

D-Glucono-δ-lactone hydrolase
6-Phospho-Gluconolactonase
(Lactonase)
3.1.1.17

$6H_2O$, M^{++}

6 6-Phospho-D-gluconate **E**

6-Phospho-D-gluconate: NADP oxidoreductase (decarboxylating)
Phosphogluconate dehydrogenase (decarboxylating)
1.1.1.44

$6NADP^+$, Mg^{++} or Mn^{++} → $6NADPH + 6H^+$, $6CO_2$

6 D-Ribulose-5-phosphate **F**

D-Ribulose-5-phosphate 3-epimerase
D-Xylulose-5-phosphate epimerase
5.1.3.1

D-Ribose-5-phosphate ketol isomerase
Ribose phosphate isomerase
5.3.1.6

4 D-Xylulose-5-phosphate

2 D-Ribose-5-phosphate **GH**

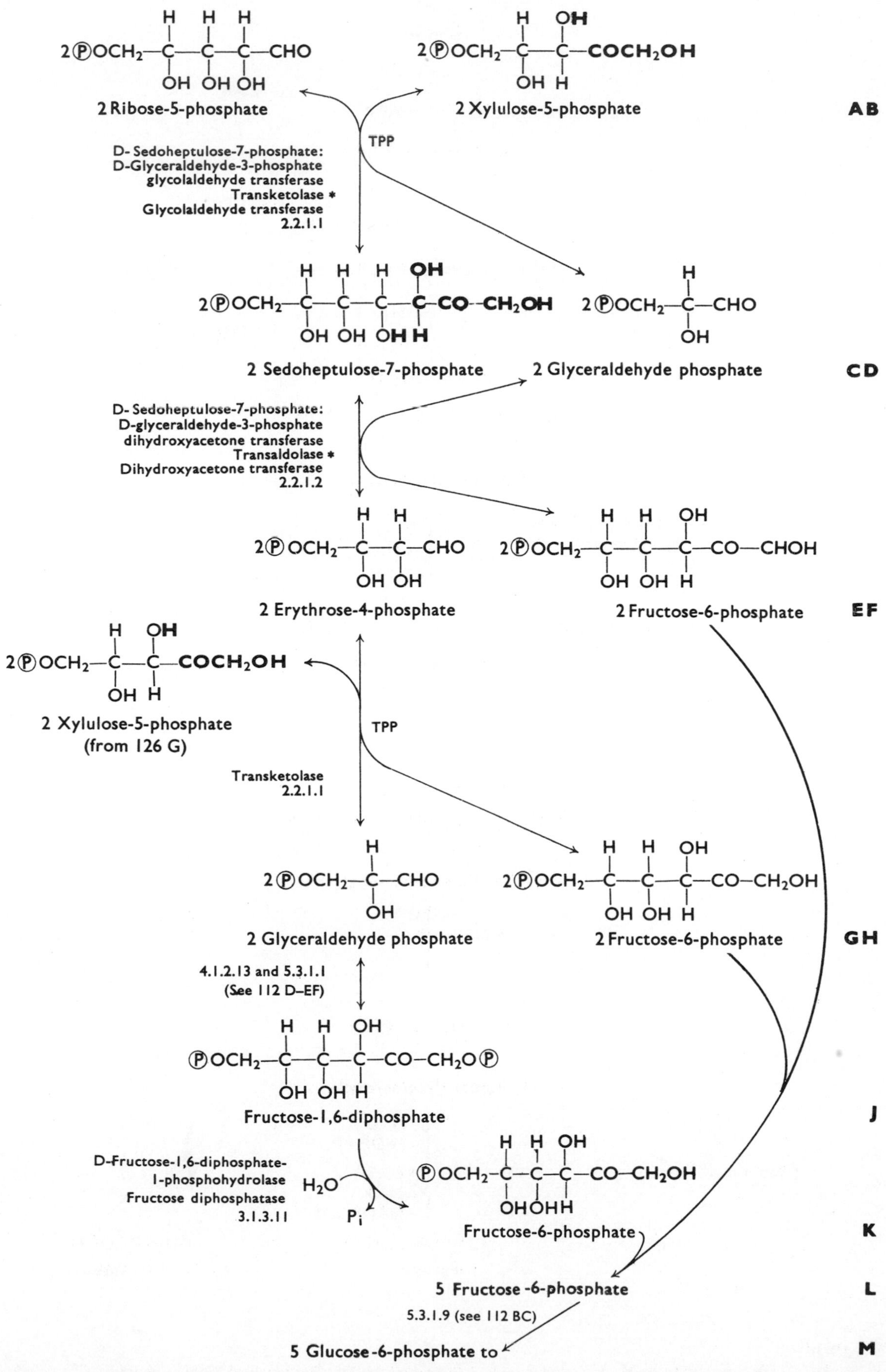

* Trans*ketol*ase transfers 'active' glycol *aldehyde* whereas trans*aldol*ase transfers the *ketone* moiety dihydroxyacetone. The transferred moieties are shown in heavy type.

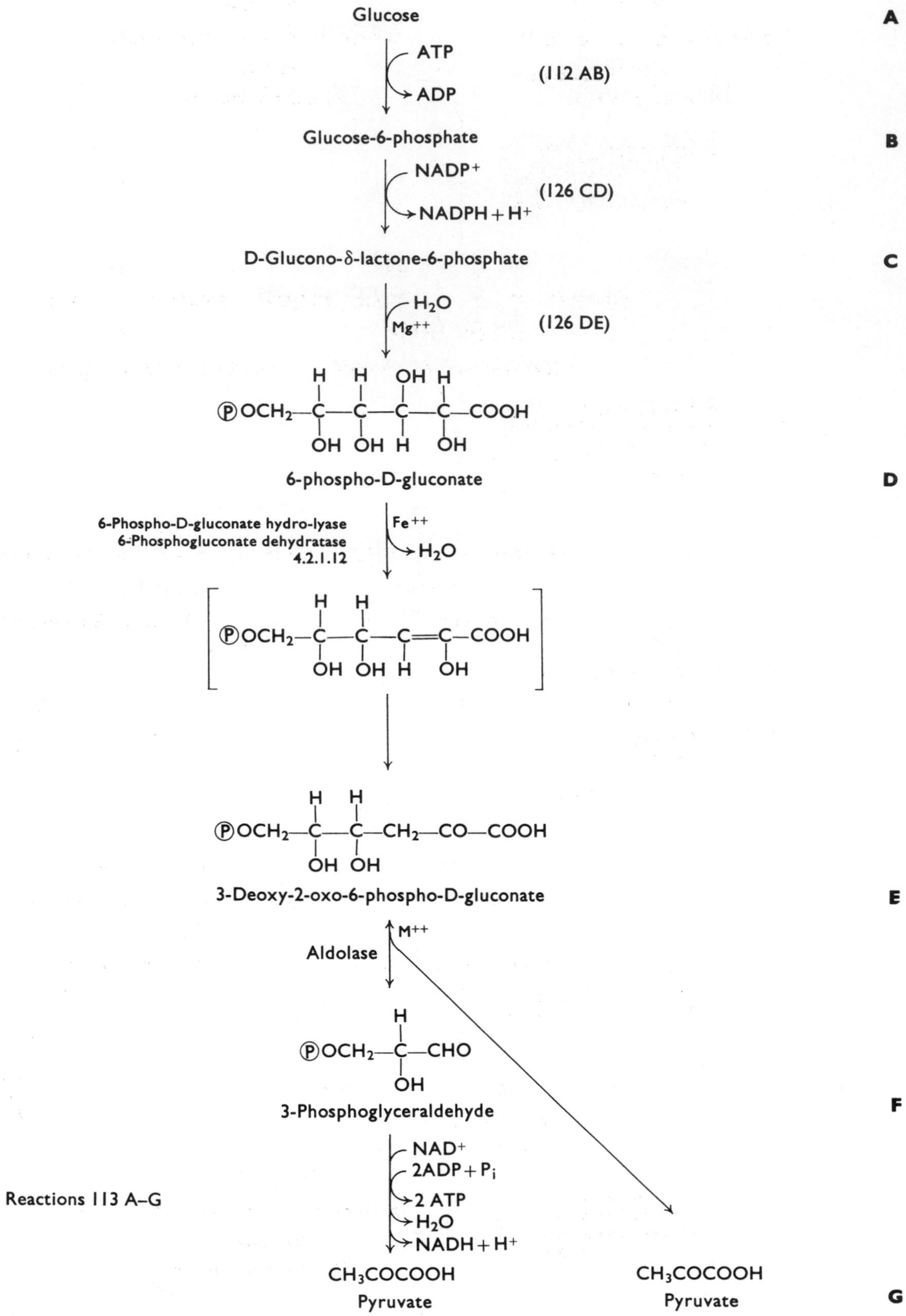

Entner-Doudoroff Pathway

Glucose — A

$\xrightarrow{ATP \rightarrow ADP}$ (112 AB)

Glucose-6-phosphate — B

$\xrightarrow{NADP^+ \rightarrow NADPH + H^+}$ (126 CD)

D-Glucono-δ-lactone-6-phosphate — C

$\xrightarrow{H_2O,\ Mg^{++}}$ (126 DE)

$$\text{Ⓟ}OCH_2{-}\underset{OH}{\overset{H}{C}}{-}\underset{OH}{\overset{H}{C}}{-}\underset{H}{\overset{OH}{C}}{-}\underset{OH}{\overset{H}{C}}{-}COOH$$

6-phospho-D-gluconate — D

6-Phospho-D-gluconate hydro-lyase
6-Phosphogluconate dehydratase
4.2.1.12

$\xrightarrow{Fe^{++}}$ H_2O

$$\left[\text{Ⓟ}OCH_2{-}\underset{OH}{\overset{H}{C}}{-}\underset{OH}{\overset{H}{C}}{-}\underset{H}{C}{=}\underset{OH}{C}{-}COOH\right]$$

$\downarrow$

$$\text{Ⓟ}OCH_2{-}\underset{OH}{\overset{H}{C}}{-}\underset{OH}{\overset{H}{C}}{-}CH_2{-}CO{-}COOH$$

3-Deoxy-2-oxo-6-phospho-D-gluconate — E

Aldolase $\rightleftarrows$ M^{++}

$$\text{Ⓟ}OCH_2{-}\underset{OH}{\overset{H}{C}}{-}CHO$$

3-Phosphoglyceraldehyde — F

Reactions 113 A–G

NAD^+, $2ADP + P_i$ $\rightarrow$ $2\ ATP$, H_2O, $NADH + H^+$

$CH_3COCOOH$ Pyruvate — $CH_3COCOOH$ Pyruvate — G

Net result

$$C_6H_{12}O_6 + ADP + P_i + 2NAD(P) \longrightarrow 2CH_3COCOOH + ATP + 2NAD(P)H + 2H^+ + H_2O$$

Glucose — Pyruvate

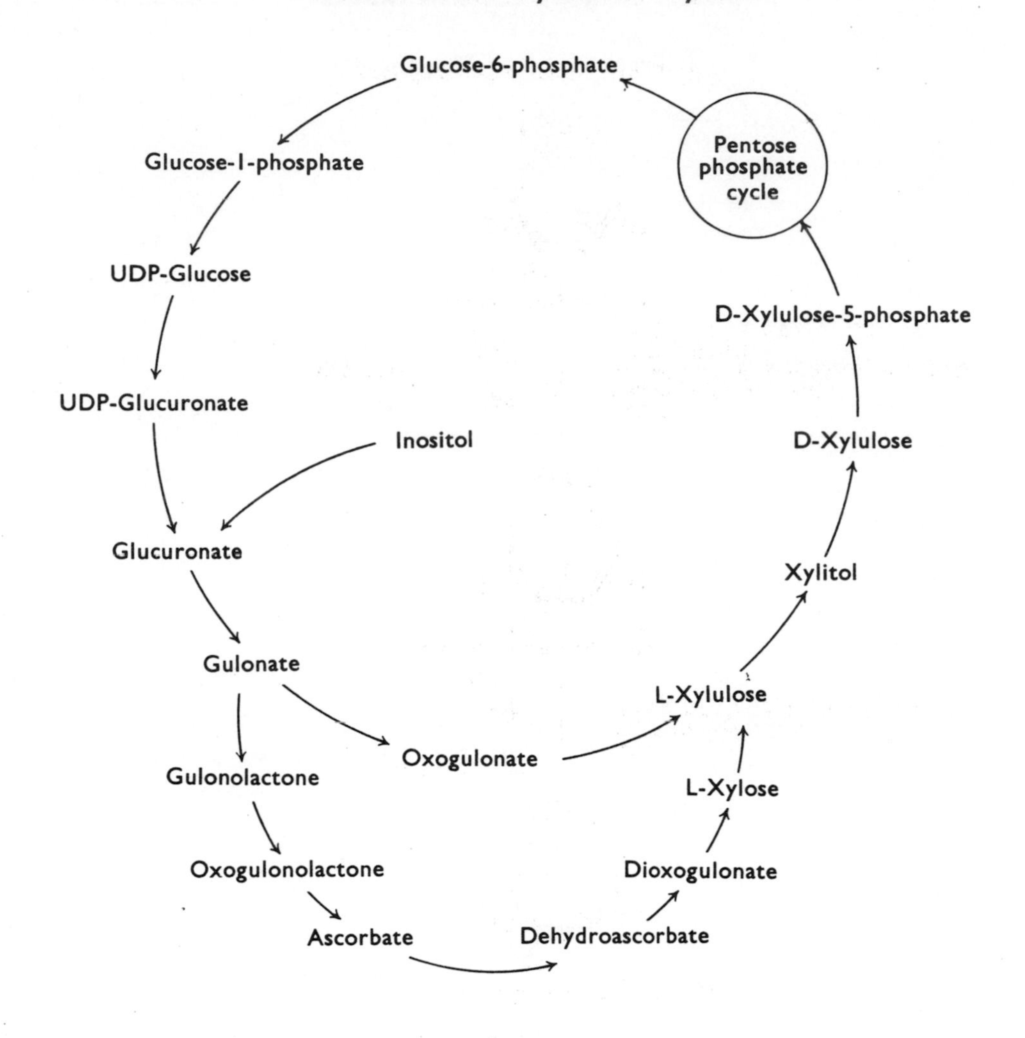
Glucose-6-phosphate
Glucose-1-phosphate
UDP-Glucose
UDP-Glucuronate
Inositol
Glucuronate
Gulonate
Gulonolactone
Oxogulonolactone
Ascorbate
Oxogulonate
Dehydroascorbate
Dioxogulonate
L-Xylose
L-Xylulose
Xylitol
D-Xylulose
D-Xylulose-5-phosphate
Pentose phosphate cycle

1. From Glucose-1-Phosphate

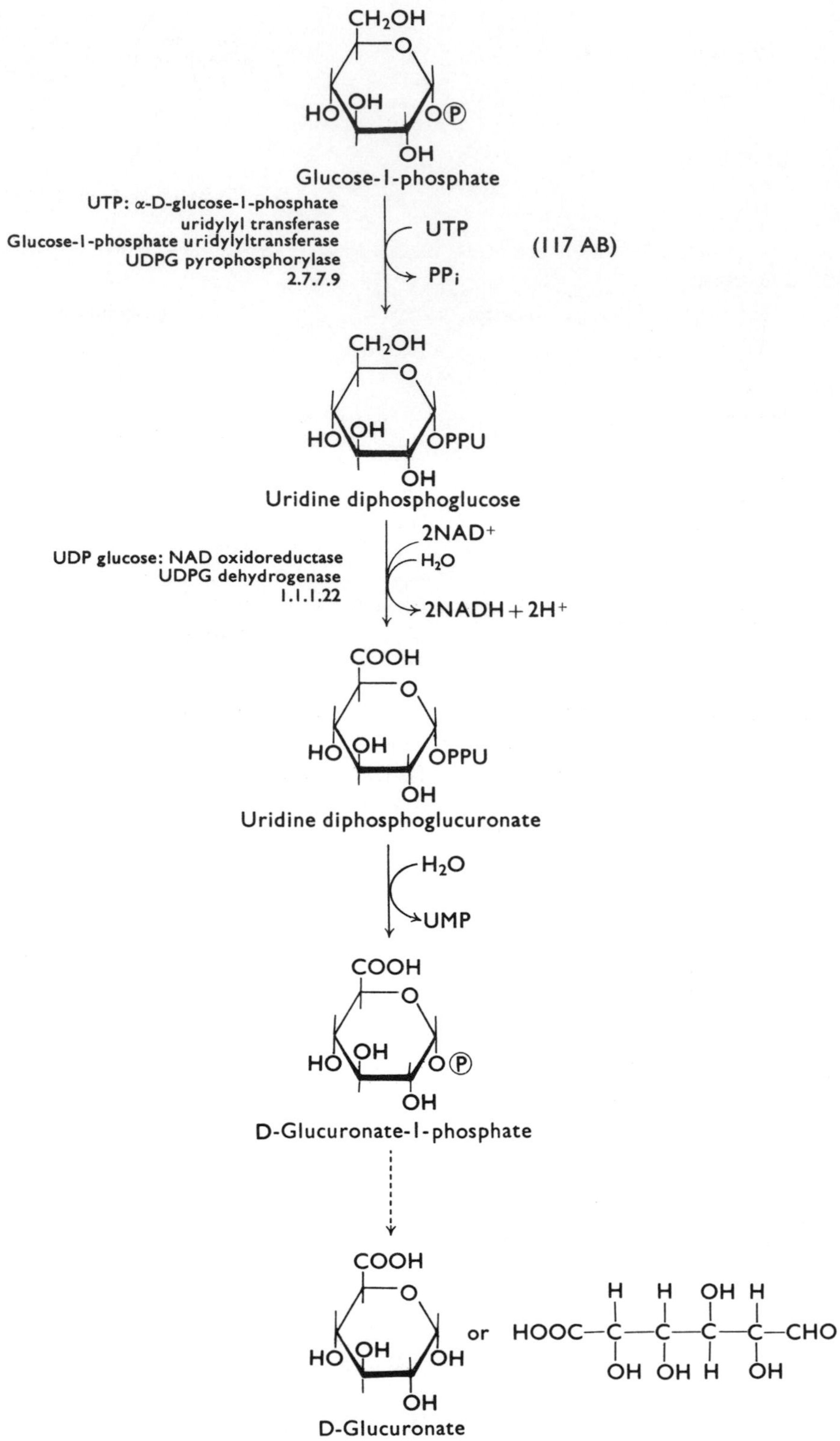

Glucose-1-phosphate (A)

UTP: α-D-glucose-1-phosphate uridylyl transferase
Glucose-1-phosphate uridylyltransferase
UDPG pyrophosphorylase
2.7.7.9

UTP → PP_i (117 AB)

Uridine diphosphoglucose (B)

UDP glucose: NAD oxidoreductase
UDPG dehydrogenase
1.1.1.22

$2NAD^+$, H_2O → $2NADH + 2H^+$

Uridine diphosphoglucuronate (C)

H_2O → UMP

D-Glucuronate-1-phosphate (D)

D-Glucuronate or HOOC—CH(OH)—CH(OH)—CH(H)(OH)... (E)

HOOC—C(H)(OH)—C(H)(OH)—C(OH)(H)—C(H)(OH)—CHO

2. From Inositol

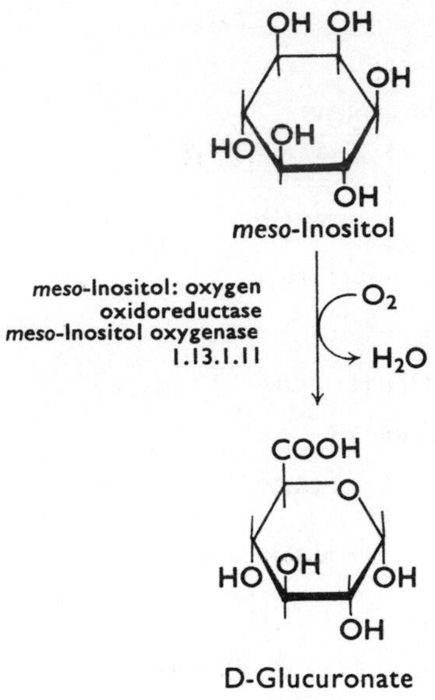

OH OH
OH
HO OH
OH
meso-Inositol **A**

meso-Inositol: oxygen oxidoreductase
meso-Inositol oxygenase
1.13.1.11

O_2 → H_2O

COOH
O
HO OH OH
OH
D-Glucuronate **B**

Breakdown of Glucuronate to Xylulose

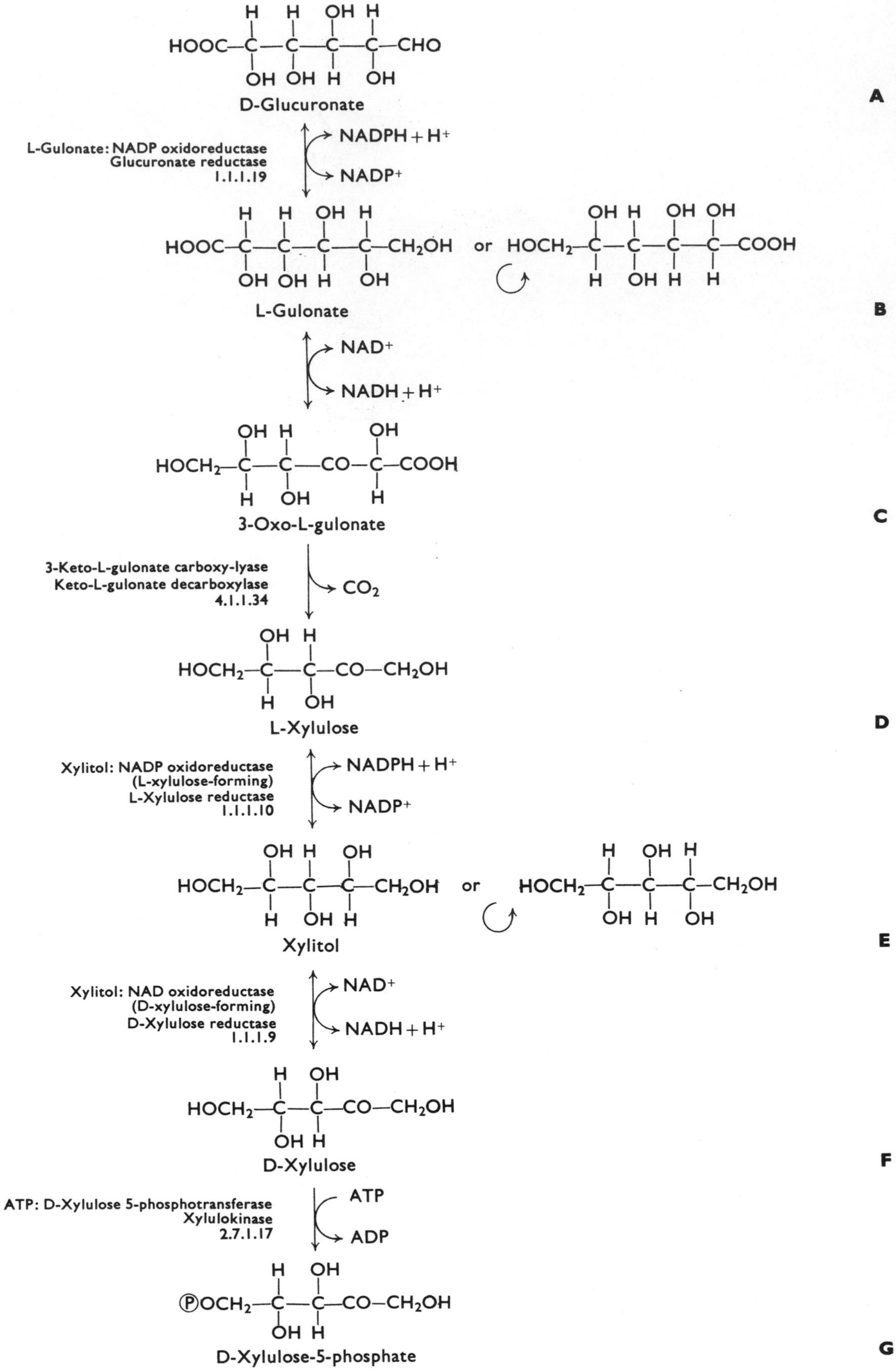

```
        H   H   OH  H
        |   |   |   |
HOOC—C—C—C—C—CHO
        |   |   |   |
        OH  OH  H   OH
```

D-Glucuronate

A

L-Gulonate: NADP oxidoreductase
Glucuronate reductase
1.1.1.19

$NADPH + H^+$ → $NADP^+$

```
        H   H   OH  H                                   OH  H   OH  OH
        |   |   |   |                                   |   |   |   |
HOOC—C—C—C—C—CH2OH     or     HOCH2—C—C—C—C—COOH
        |   |   |   |                                   |   |   |   |
        OH  OH  H   OH                 (rotated)         H   OH  H   H
```

L-Gulonate

B

NAD^+ → $NADH + H^+$

```
         OH  H        OH
         |   |        |
HOCH2—C—C—CO—C—COOH
         |   |        |
         H   OH       H
```

3-Oxo-L-gulonate

C

3-Keto-L-gulonate carboxy-lyase
Keto-L-gulonate decarboxylase
4.1.1.34

→ CO_2

```
         OH  H
         |   |
HOCH2—C—C—CO—CH2OH
         |   |
         H   OH
```

L-Xylulose

D

Xylitol: NADP oxidoreductase
(L-xylulose-forming)
L-Xylulose reductase
1.1.1.10

$NADPH + H^+$ → $NADP^+$

```
         OH  H   OH                               H   OH  H
         |   |   |                                |   |   |
HOCH2—C—C—C—CH2OH     or     HOCH2—C—C—C—CH2OH
         |   |   |                                |   |   |
         H   OH  H             (rotated)          OH  H   OH
```

Xylitol

E

Xylitol: NAD oxidoreductase
(D-xylulose-forming)
D-Xylulose reductase
1.1.1.9

NAD^+ → $NADH + H^+$

```
         H   OH
         |   |
HOCH2—C—C—CO—CH2OH
         |   |
         OH  H
```

D-Xylulose

F

ATP: D-Xylulose 5-phosphotransferase
Xylulokinase
2.7.1.17

ATP → ADP

```
         H   OH
         |   |
ⓅOCH2—C—C—CO—CH2OH
         |   |
         OH  H
```

D-Xylulose-5-phosphate

G

The symbol ↻ implies that the molecule is rotated through 180°.

Alternative Breakdown of Glucuronate (Bacteria)

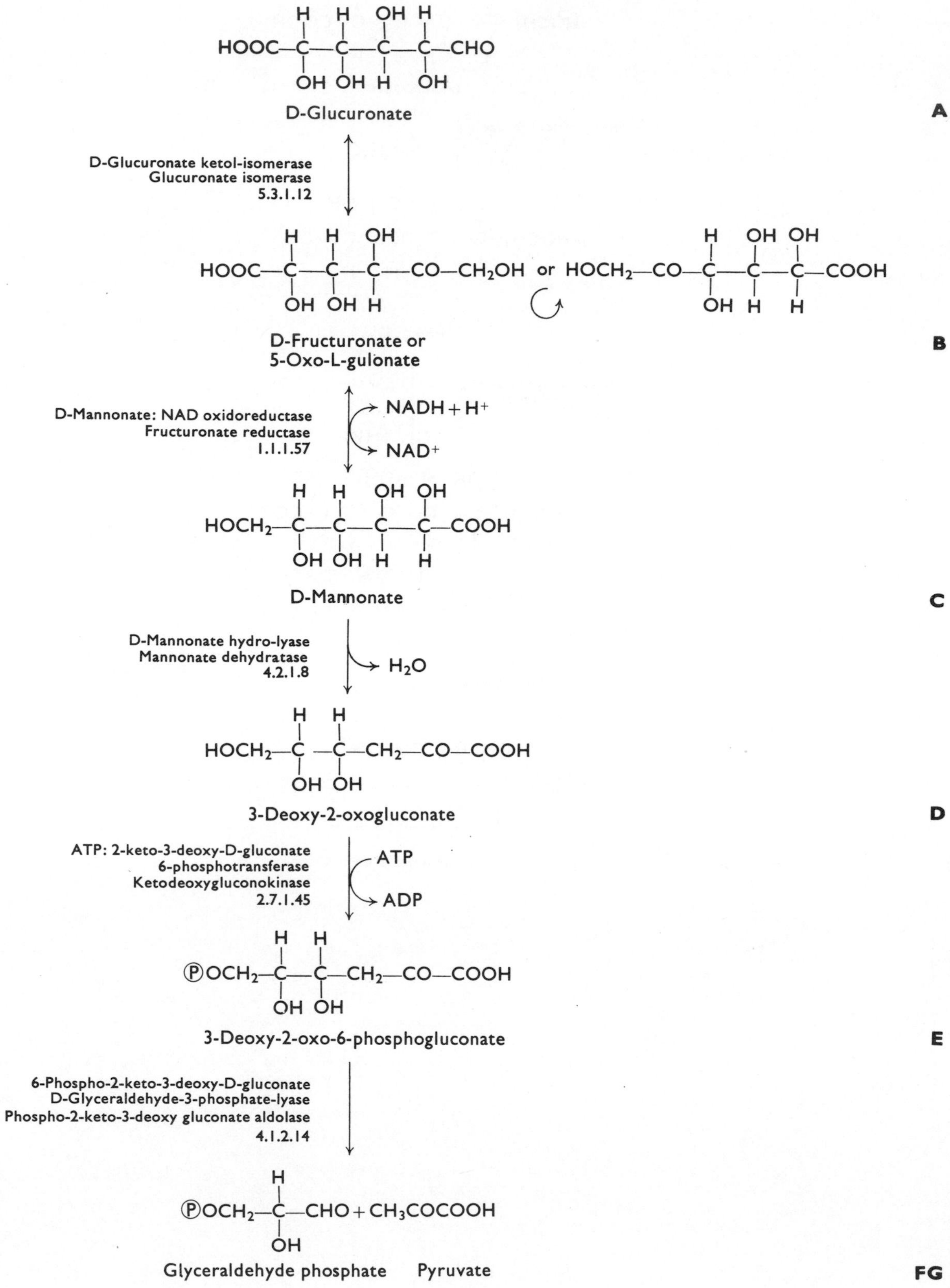

$$HOOC-CH(OH)-CH(OH)-CH(OH)-CH(OH)-CHO$$

(H above, OH below; H above, OH below; OH above, H below; H above, OH below)

D-Glucuronate

A

↕ D-Glucuronate ketol-isomerase
Glucuronate isomerase
5.3.1.12

$$HOOC-CH(OH)-CH(OH)-CH(OH)-CO-CH_2OH \text{ or } HOCH_2-CO-CH(OH)-CH(OH)-CH(OH)-COOH$$

D-Fructuronate or
5-Oxo-L-gulonate

B

↕ D-Mannonate: NAD oxidoreductase
Fructuronate reductase
1.1.1.57
$NADH + H^+$ / NAD^+

$$HOCH_2-CH(OH)-CH(OH)-CH(OH)-CH(OH)-COOH$$

D-Mannonate

C

↓ D-Mannonate hydro-lyase
Mannonate dehydratase
4.2.1.8
→ H_2O

$$HOCH_2-CH(OH)-CH(OH)-CH_2-CO-COOH$$

3-Deoxy-2-oxogluconate

D

↓ ATP: 2-keto-3-deoxy-D-gluconate
6-phosphotransferase
Ketodeoxygluconokinase
2.7.1.45
ATP → ADP

$$ⓅOCH_2-CH(OH)-CH(OH)-CH_2-CO-COOH$$

3-Deoxy-2-oxo-6-phosphogluconate

E

↓ 6-Phospho-2-keto-3-deoxy-D-gluconate
D-Glyceraldehyde-3-phosphate-lyase
Phospho-2-keto-3-deoxy gluconate aldolase
4.1.2.14

$$ⓅOCH_2-CH(OH)-CHO + CH_3COCOOH$$

Glyceraldehyde phosphate Pyruvate

FG

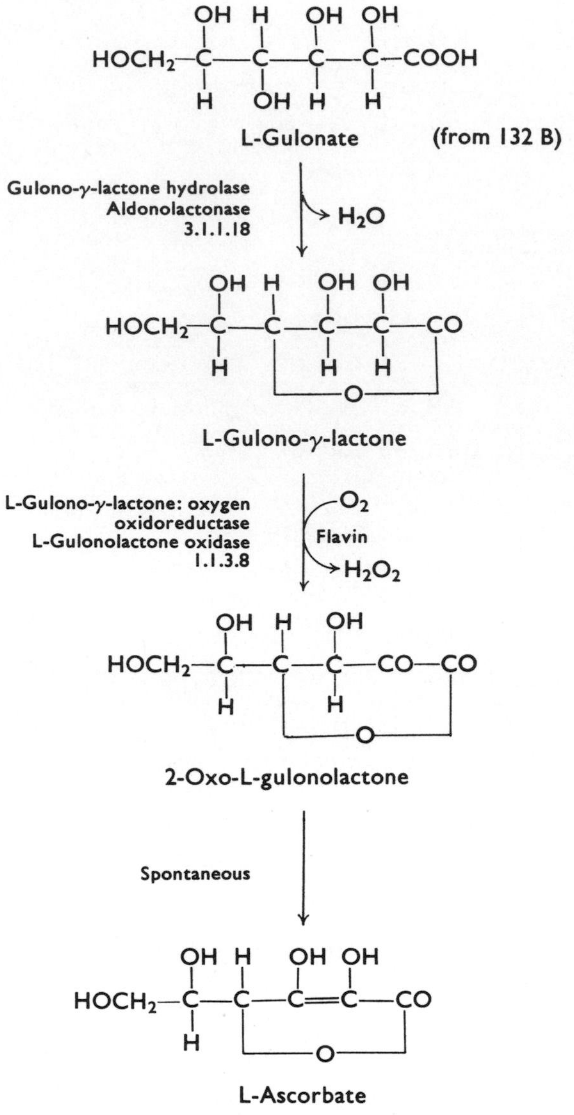

$HOCH_2-CH(OH)-CH(OH)-CH(OH)-CH(OH)-COOH$ (C2: OH, H; C3: H, OH; C4: OH, H; C5: OH, H)

L-Gulonate (from 132 B) **A**

Gulono-γ-lactone hydrolase
Aldonolactonase
3.1.1.18
→ H_2O

$HOCH_2-C-C-C-C-CO$ (lactone ring via O)

L-Gulono-γ-lactone **B**

L-Gulono-γ-lactone: oxygen oxidoreductase
L-Gulonolactone oxidase
1.1.3.8
O_2, Flavin → H_2O_2

$HOCH_2-C-C-C-CO-CO$ (lactone ring via O)

2-Oxo-L-gulonolactone **C**

Spontaneous

$HOCH_2-C-C-C{=}C-CO$ (lactone ring via O)

L-Ascorbate **D**

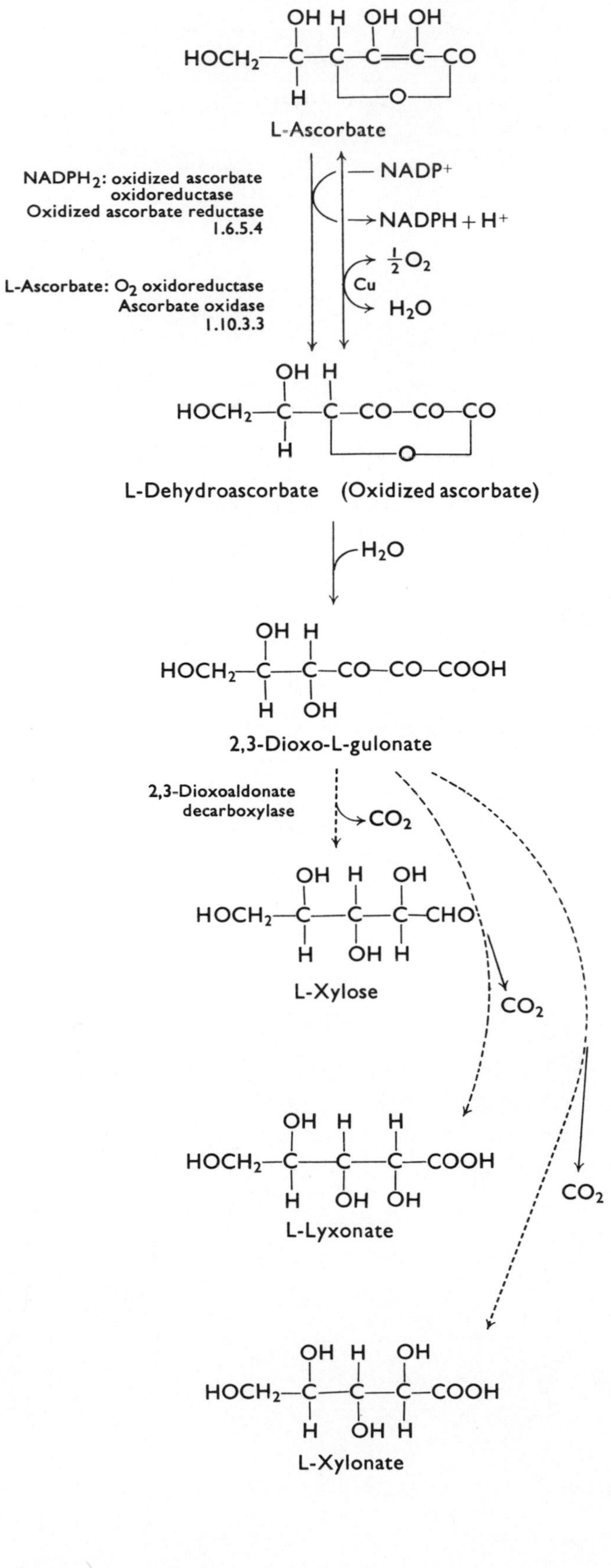

$HOCH_2-C(OH)(H)-C(H)-C(OH)=C(OH)-CO$ (ring: –O–)

L-Ascorbate **A**

NADPH2: oxidized ascorbate oxidoreductase
Oxidized ascorbate reductase
1.6.5.4

$NADP^+$ → $NADPH + H^+$

L-Ascorbate: O_2 oxidoreductase
Ascorbate oxidase
1.10.3.3

$\frac{1}{2}O_2$ → H_2O (Cu)

$HOCH_2-C(OH)(H)-C(H)-CO-CO-CO$ (ring: –O–)

L-Dehydroascorbate (Oxidized ascorbate) **B**

H_2O

$HOCH_2-C(OH)(H)-C(H)(OH)-CO-CO-COOH$

2,3-Dioxo-L-gulonate **C**

2,3-Dioxoaldonate decarboxylase

CO_2

$HOCH_2-C(OH)(H)-C(H)(OH)-C(OH)(H)-CHO$

L-Xylose **D**

CO_2

$HOCH_2-C(OH)(H)-C(H)(OH)-C(H)(OH)-COOH$

L-Lyxonate **E**

CO_2

$HOCH_2-C(OH)(H)-C(H)(OH)-C(OH)(H)-COOH$

L-Xylonate **F**

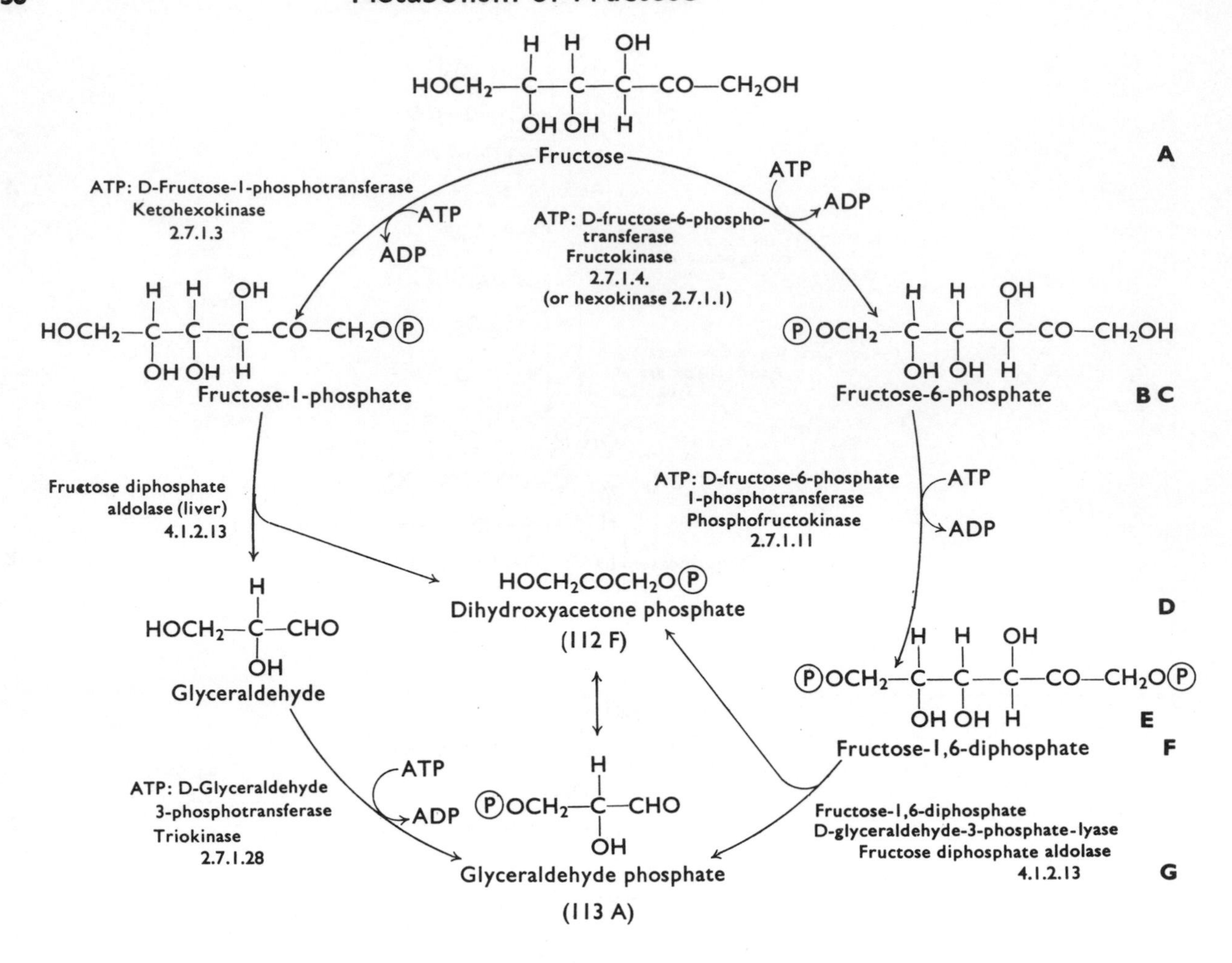

Fructose
A
ATP: D-Fructose-1-phosphotransferase
Ketohexokinase
2.7.1.3
ATP
ADP
ATP: D-fructose-6-phospho-transferase
Fructokinase
2.7.1.4.
(or hexokinase 2.7.1.1)
Fructose-1-phosphate
Fructose-6-phosphate
B C
Fructose diphosphate
aldolase (liver)
4.1.2.13
ATP: D-fructose-6-phosphate
1-phosphotransferase
Phosphofructokinase
2.7.1.11
$HOCH_2COCH_2O$Ⓟ
Dihydroxyacetone phosphate
(112 F)
D
Glyceraldehyde
E
Fructose-1,6-diphosphate
F
ATP: D-Glyceraldehyde
3-phosphotransferase
Triokinase
2.7.1.28
Glyceraldehyde phosphate
(113 A)
Fructose-1,6-diphosphate
D-glyceraldehyde-3-phosphate-lyase
Fructose diphosphate aldolase
4.1.2.13
G

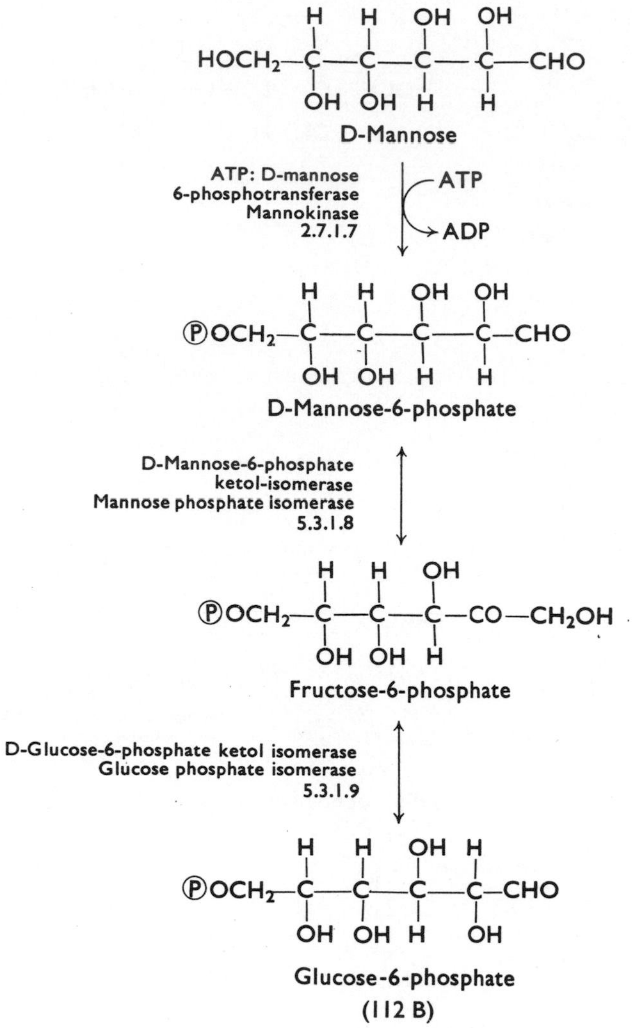

```
        H    H    OH   OH
        |    |    |    |
HOCH2—C——C——C——C—CHO
        |    |    |    |
        OH   OH   H    H
```

D-Mannose **A**

ATP: D-mannose 6-phosphotransferase
Mannokinase
2.7.1.7

ATP → ADP

```
        H    H    OH   OH
        |    |    |    |
ⓅOCH2—C——C——C——C—CHO
        |    |    |    |
        OH   OH   H    H
```

D-Mannose-6-phosphate **B**

D-Mannose-6-phosphate ketol-isomerase
Mannose phosphate isomerase
5.3.1.8

⇅

```
        H    H    OH
        |    |    |
ⓅOCH2—C——C——C—CO—CH2OH
        |    |    |
        OH   OH   H
```

Fructose-6-phosphate **C**

D-Glucose-6-phosphate ketol isomerase
Glucose phosphate isomerase
5.3.1.9

⇅

```
        H    H    OH   H
        |    |    |    |
ⓅOCH2—C——C——C——C—CHO
        |    |    |    |
        OH   OH   H    OH
```

Glucose-6-phosphate
(112 B) **D**

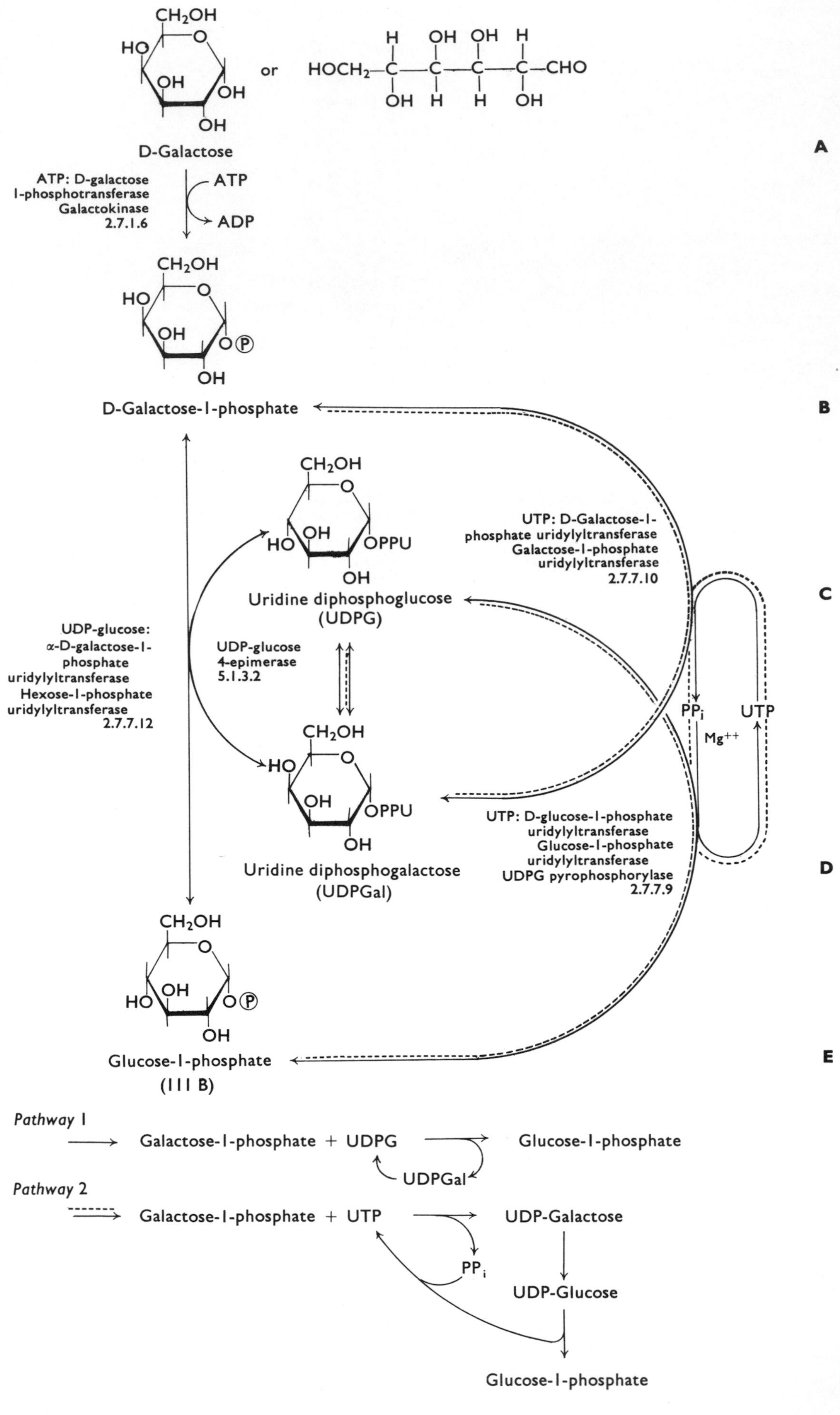

CH_2OH HO O OH OH OH or $HOCH_2$—C(H)(OH)—C(OH)(H)—C(OH)(H)—C(H)(OH)—CHO

D-Galactose

A

ATP: D-galactose 1-phosphotransferase
Galactokinase
2.7.1.6

ATP → ADP

CH_2OH HO O OH O(P) OH

D-Galactose-1-phosphate

B

CH_2OH O HO OH OPPU OH

Uridine diphosphoglucose (UDPG)

UTP: D-Galactose-1-phosphate uridylyltransferase
Galactose-1-phosphate uridylyltransferase
2.7.7.10

C

UDP-glucose: α-D-galactose-1-phosphate uridylyltransferase
Hexose-1-phosphate uridylyltransferase
2.7.7.12

UDP-glucose 4-epimerase
5.1.3.2

PP_i UTP Mg^{++}

CH_2OH HO O OH OPPU OH

Uridine diphosphogalactose (UDPGal)

UTP: D-glucose-1-phosphate uridylyltransferase
Glucose-1-phosphate uridylyltransferase
UDPG pyrophosphorylase
2.7.7.9

D

CH_2OH O HO OH O(P) OH

Glucose-1-phosphate (III B)

E

Pathway 1

→ Galactose-1-phosphate + UDPG → Glucose-1-phosphate
UDPGal

Pathway 2

⇢ Galactose-1-phosphate + UTP → UDP-Galactose
PP_i
UDP-Glucose
Glucose-1-phosphate

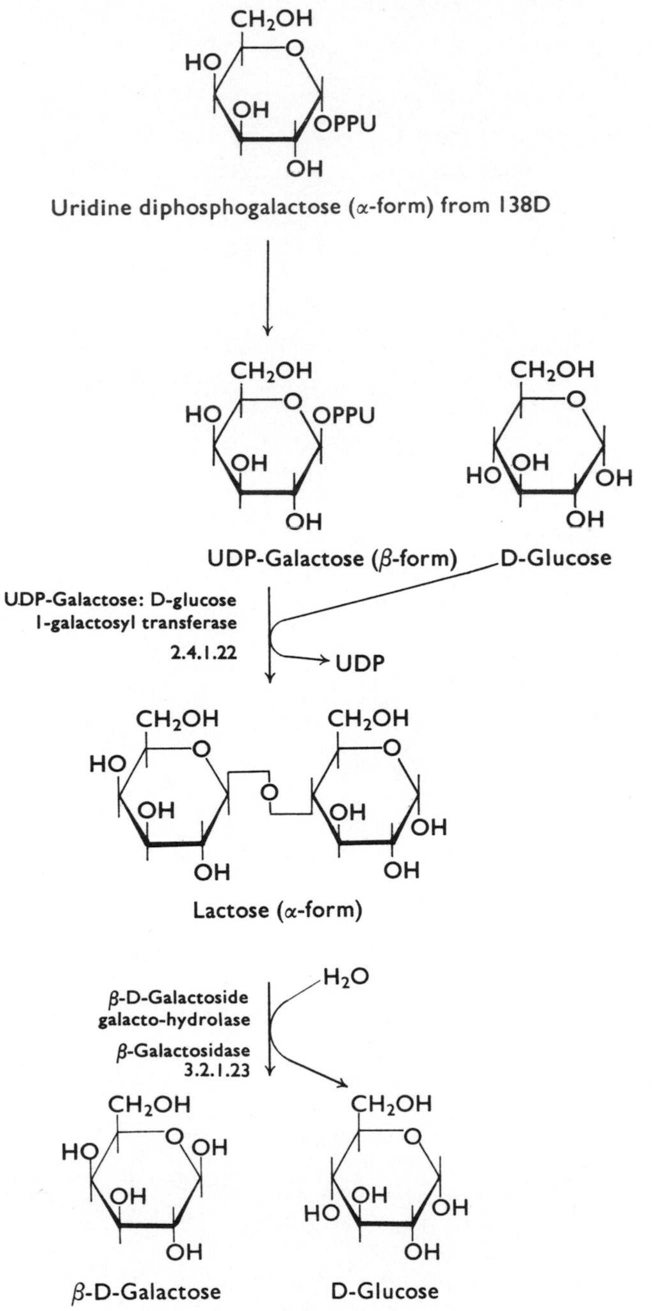

CH_2OH HO OH OPPU OH

Uridine diphosphogalactose (α-form) from 138D

A

CH_2OH HO OPPU OH OH

UDP-Galactose (β-form)

CH_2OH HO OH OH OH

D-Glucose

BC

UDP-Galactose: D-glucose 1-galactosyl transferase 2.4.1.22

→ UDP

CH_2OH HO OH OH O CH_2OH OH OH OH

Lactose (α-form)

CD

β-D-Galactoside galacto-hydrolase

β-Galactosidase 3.2.1.23

H_2O

CH_2OH HO OH OH OH

β-D-Galactose

CH_2OH HO OH OH OH

D-Glucose

EF

Formation of GDP- L-Fucose

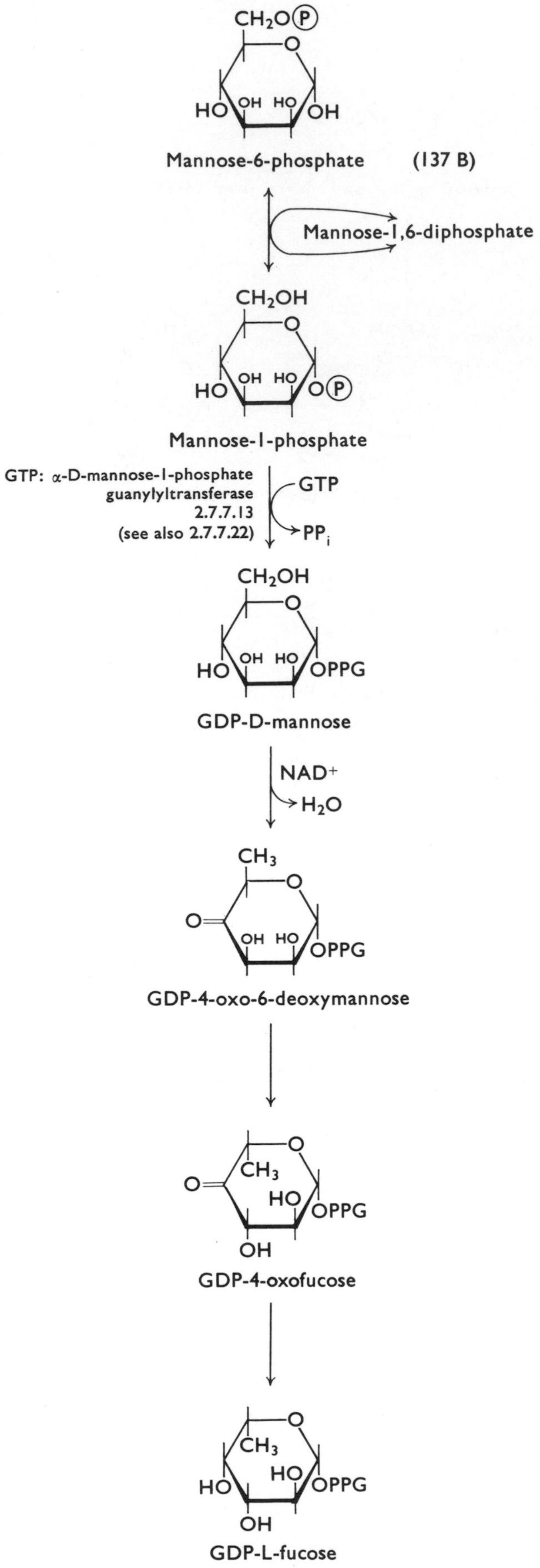

CH_2OⓅ

HO OH HO OH

Mannose-6-phosphate (137 B) **A**

Mannose-1,6-diphosphate

CH_2OH

HO OH HO OⓅ

Mannose-1-phosphate **B**

GTP: α-D-mannose-1-phosphate guanylyltransferase 2.7.7.13 (see also 2.7.7.22)

GTP → PP_i

CH_2OH

HO OH HO OPPG

GDP-D-mannose **C**

NAD^+ → H_2O

CH_3

O= OH HO OPPG

GDP-4-oxo-6-deoxymannose **D**

O= CH_3 HO OPPG OH

GDP-4-oxofucose **E**

CH_3 HO HO OPPG OH

GDP-L-fucose **F**

Metabolism of Pentoses and Pentitols

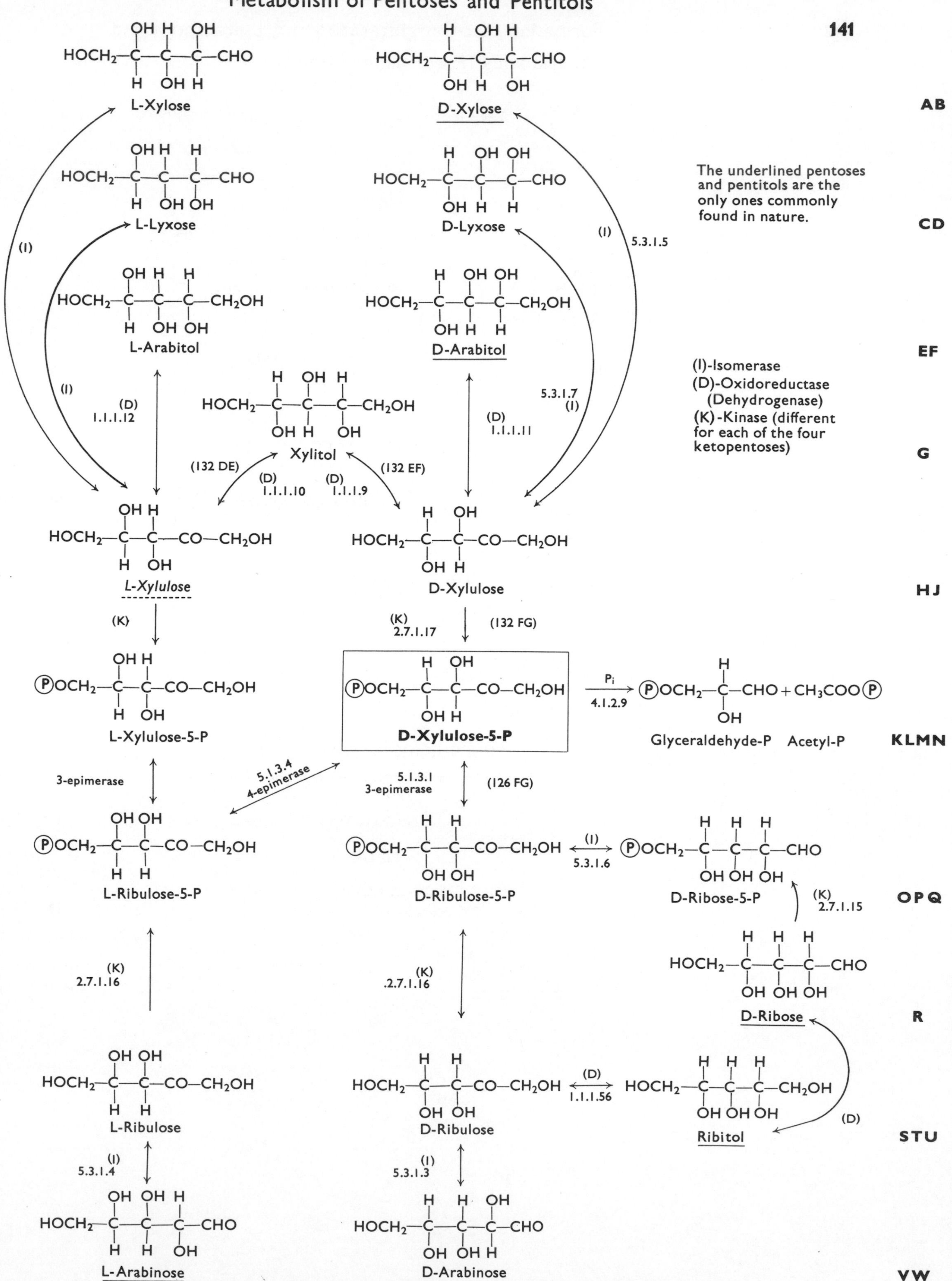

D-Xylulose-5-phosphate is the focal point of pentose and pentitol metabolism.

Formation of 2-oxoglutarate from L-arabinose and hexaric acids (*Pseudomonadaceae*)

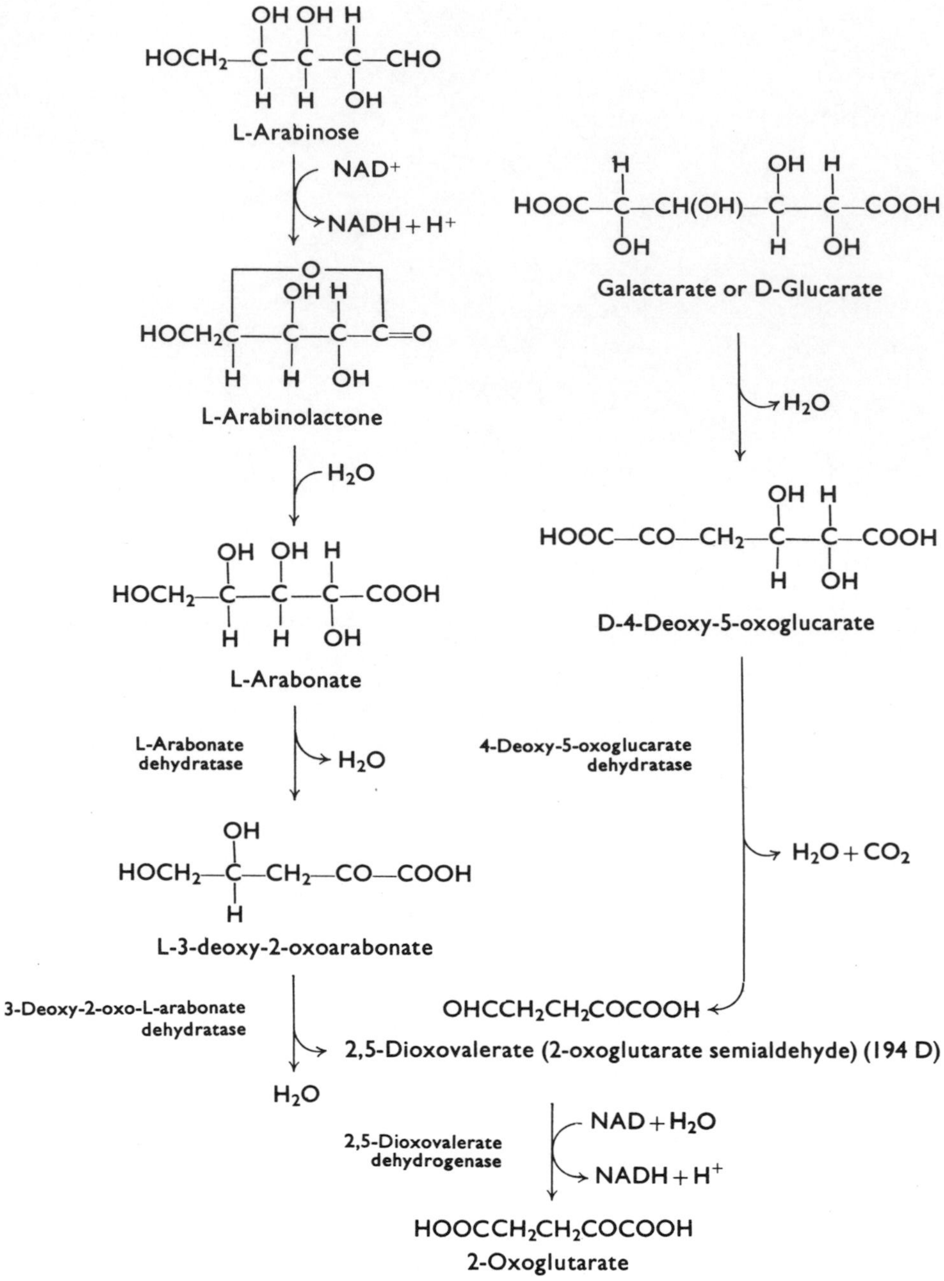

$HOCH_2$—CH(OH)—CH(OH)—CH(OH)—CHO

L-Arabinose

↓ NAD^+ → $NADH + H^+$

L-Arabinolactone

↓ H_2O

$HOCH_2$—CH(OH)—CH(OH)—CH(OH)—COOH

L-Arabonate

L-Arabonate dehydratase ↓ → H_2O

$HOCH_2$—CH(OH)—CH_2—CO—COOH

L-3-deoxy-2-oxoarabonate

3-Deoxy-2-oxo-L-arabonate dehydratase ↓ H_2O

HOOC—CH(OH)—CH(OH)—CH(OH)—CH(OH)—COOH

Galactarate or D-Glucarate

↓ → H_2O

HOOC—CO—CH_2—CH(OH)—CH(OH)—COOH

D-4-Deoxy-5-oxoglucarate

4-Deoxy-5-oxoglucarate dehydratase ↓ → $H_2O + CO_2$

$OHCCH_2CH_2COCOOH$

2,5-Dioxovalerate (2-oxoglutarate semialdehyde) (194 D)

2,5-Dioxovalerate dehydrogenase ↓ $NAD + H_2O$ → $NADH + H^+$

$HOOCCH_2CH_2COCOOH$

2-Oxoglutarate

E F G H J K L M

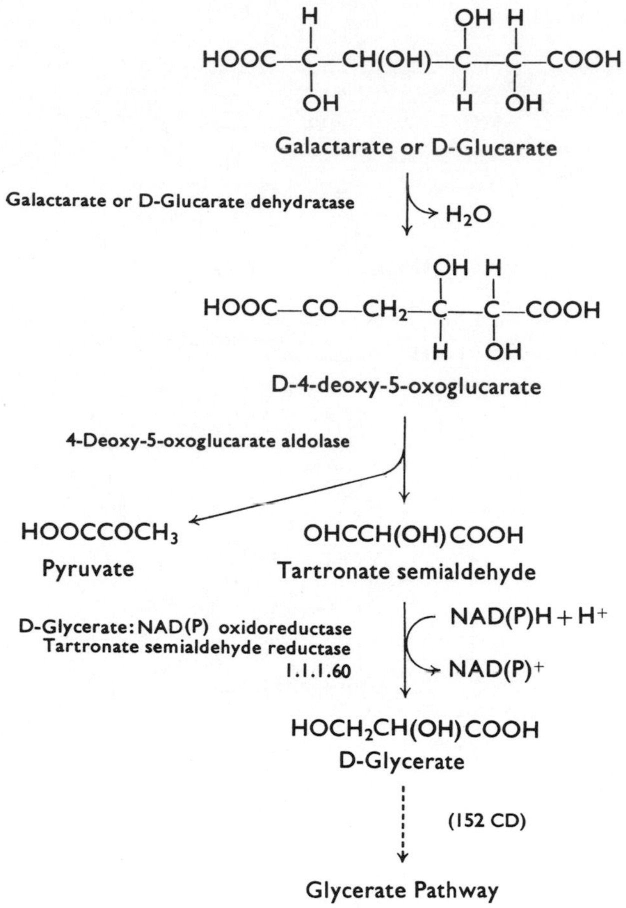

$$HOOC-\underset{OH}{\overset{H}{C}}-CH(OH)-\underset{H}{\overset{OH}{C}}-\underset{OH}{\overset{H}{C}}-COOH$$

Galactarate or D-Glucarate **A**

Galactarate or D-Glucarate dehydratase → H_2O

$$HOOC-CO-CH_2-\underset{H}{\overset{OH}{C}}-\underset{OH}{\overset{H}{C}}-COOH$$

D-4-deoxy-5-oxoglucarate **B**

4-Deoxy-5-oxoglucarate aldolase

$HOOCCOCH_3$ Pyruvate

OHCCH(OH)COOH Tartronate semialdehyde **C**

D-Glycerate: NAD(P) oxidoreductase
Tartronate semialdehyde reductase
1.1.1.60

$NAD(P)H + H^+$ → $NAD(P)^+$

$HOCH_2CH(OH)COOH$
D-Glycerate **D**

(152 CD)

Glycerate Pathway

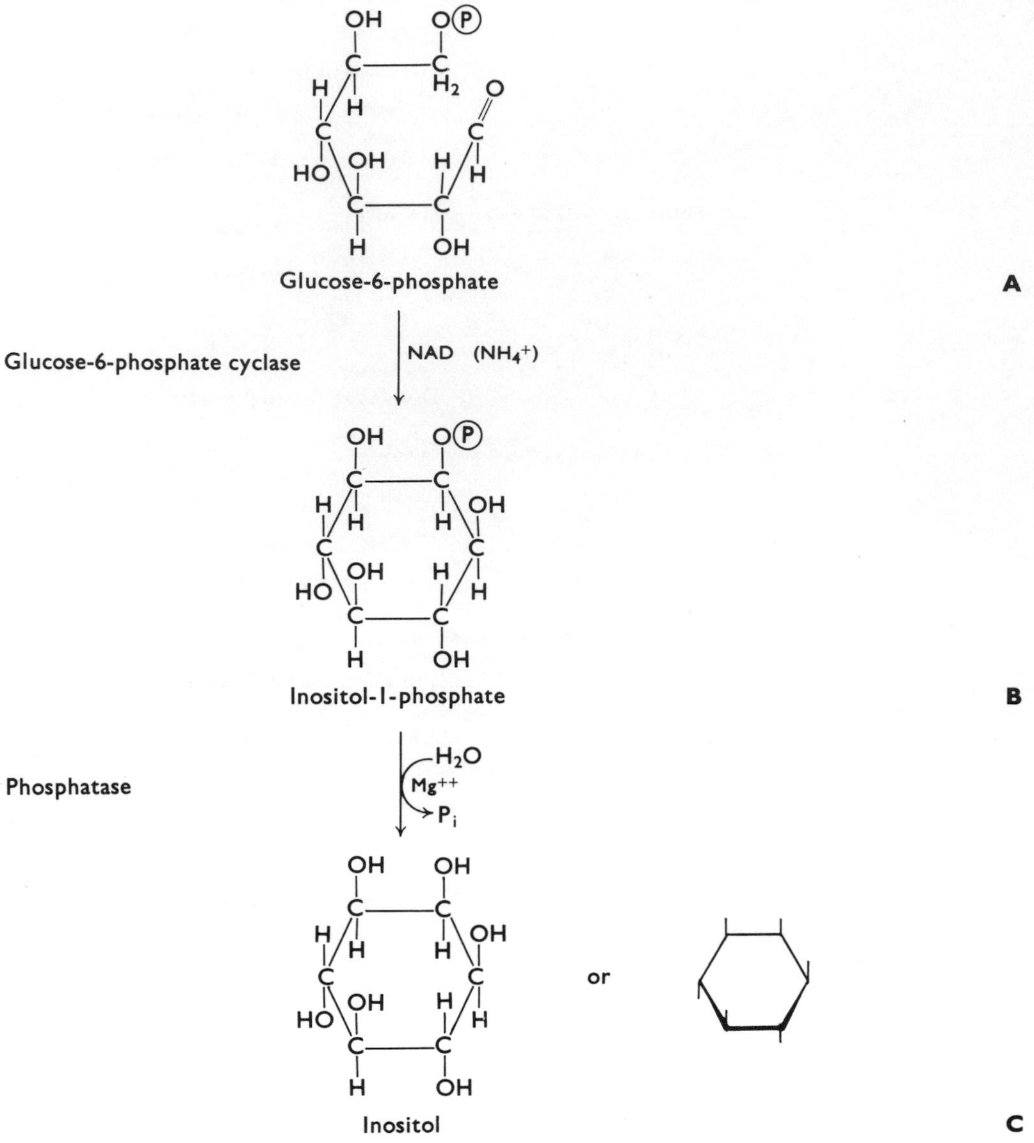

Glucose-6-phosphate

A

Glucose-6-phosphate cyclase — NAD (NH_4^+)

Inositol-1-phosphate

B

Phosphatase — H_2O, Mg^{++}, P_i

Inositol

or

C

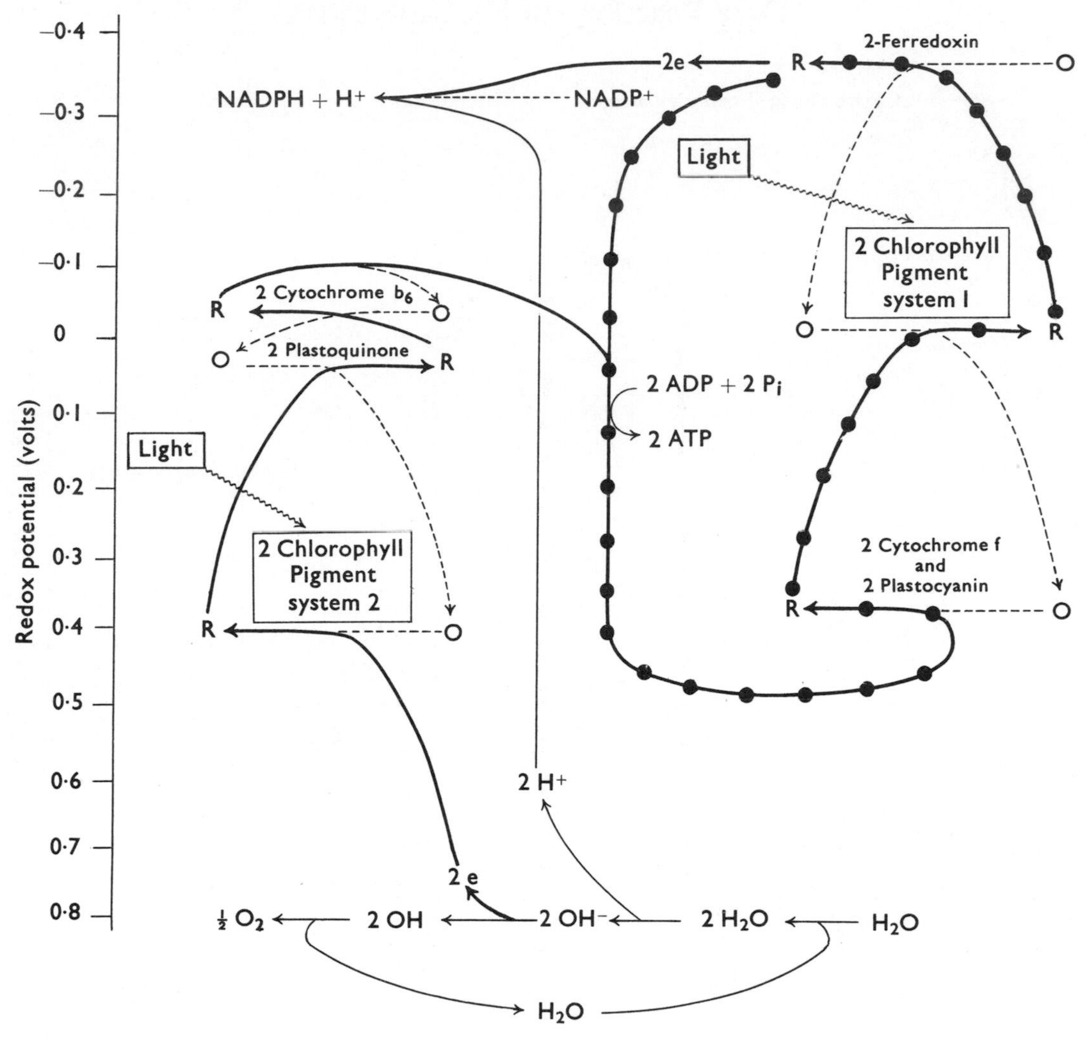

Pathway of electron transport.

Pathway of cyclic photophosphorylation.

Net result $H_2O + NADP^+ + 2\,ADP + 2\,P_i \rightarrow NADPH + H^+ + 2\,ATP + \frac{1}{2}\,O_2 + 2\,H_2O$

The redox potentials of some of the above systems are still uncertain.

R and O signify the reduced and oxidized forms of the co-factors.

Interrelations of Light and Dark Reactions of Photosynthesis

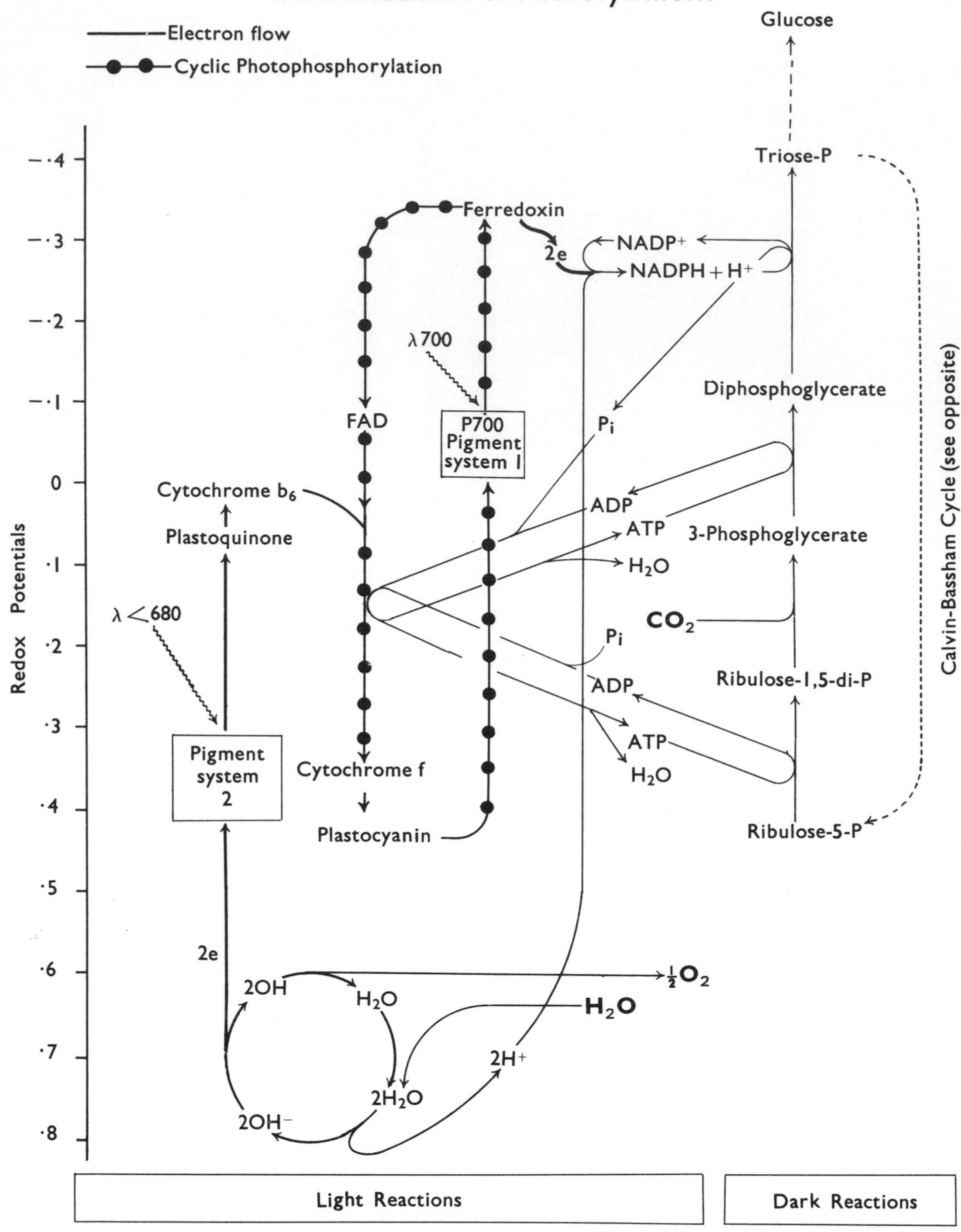

Net result of Light Reactions:

$$H_2O + NADP^+ + 2ADP + 2P_i \rightarrow NADPH + H^+ + 2ATP + \tfrac{1}{2}O_2 + 2H_2O$$

Additional ATP is obtained by cyclic photophosphorylation.

$$n\ ADP + n\ P_i \rightarrow n\ ATP + n\ H_2O$$

The Reduced NADP reduces diphosphoglycerate to triose phosphate. (In effect reduces carbon dioxide to hexose.)

The ATP phosphorylates 3-phosphoglycerate and ribulose-6-phosphate to the diphosphorylated compounds.

Pigment System 1 includes chlorophyll **a** and carotenoids.

Pigment System 2 (accessory pigment system) includes chlorophyll **a** and **b**.

The requirements of the dark reaction for ATP and reduced NADP can be seen from the equation

$$3\ CO_2 + 9\ ATP + 6\ NADPH + 6\ H^+ + 5\ H_2O \rightarrow \text{Triose phosphate} + 9\ ADP + 8\ P_i + 6\ NADP^+$$

Carbon Dioxide Utilization:
Calvin Bassham Cycle

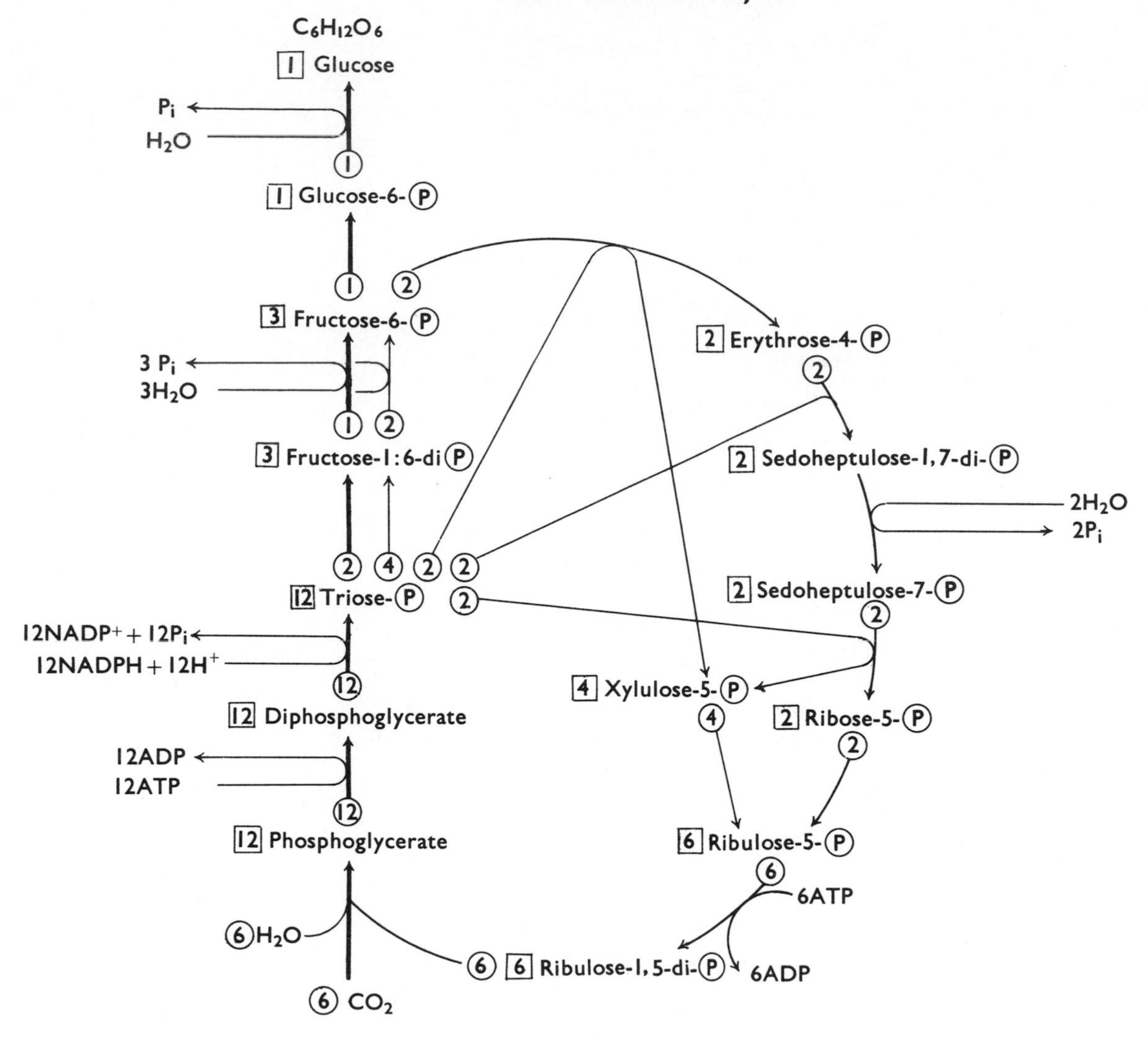

Net result

$6CO_2 + 12NADPH + 12H^+ + 18ATP + 12H_2O \rightarrow C_6H_{12}O_6 + 12NADP^+ + 18ADP + 18P_i$

Eliminating $18ATP + 18H_2O \rightarrow 18ADP + 18P_i$

leaves $6CO_2 + 12NADPH + 12H^+ \rightarrow C_6H_{12}O_6 + 12NADP^+ + 6H_2O$

Carbon dioxide Glucose

The cycle regenerates 6 molecules of ribulose diphosphate from 10 molecules (4+2+2+2) of triose phosphate (i.e. 10 $C_3 \rightarrow 6\ C_5$).

The figures in squares denote the number of molecules resulting from previous reaction(s). The figures in circles denote the number of these molecules utilized in different pathways.

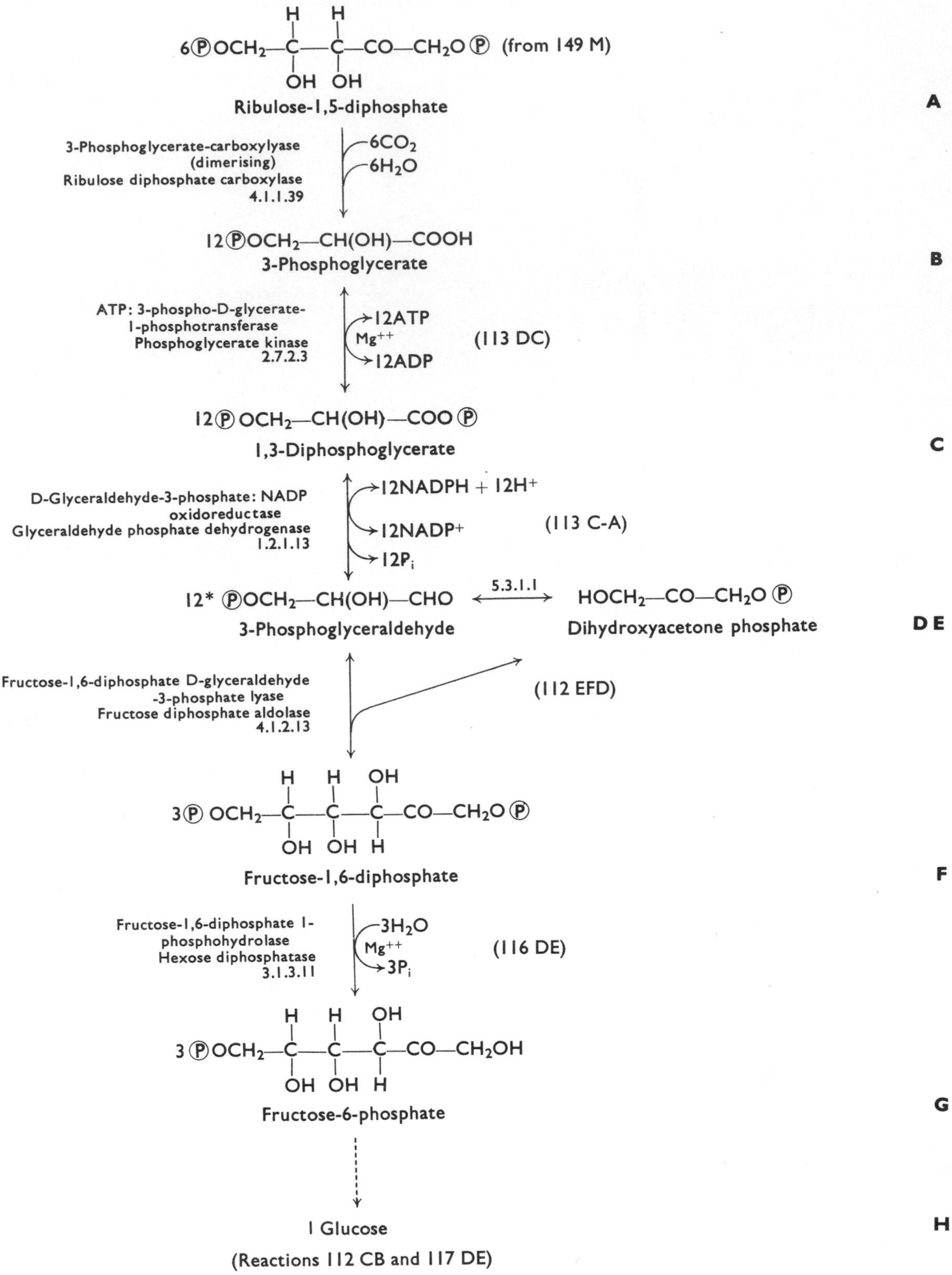

$$6\text{Ⓟ}OCH_2\text{—}\underset{OH}{\overset{H}{C}}\text{—}\underset{OH}{\overset{H}{C}}\text{—}CO\text{—}CH_2O\text{Ⓟ}$$ (from 149 M)

Ribulose-1,5-diphosphate — A

3-Phosphoglycerate-carboxylyase (dimerising)
Ribulose diphosphate carboxylase
4.1.1.39
$6CO_2$, $6H_2O$ ↓

$12\text{Ⓟ}OCH_2\text{—}CH(OH)\text{—}COOH$

3-Phosphoglycerate — B

ATP: 3-phospho-D-glycerate-1-phosphotransferase
Phosphoglycerate kinase
2.7.2.3
→12ATP, Mg^{++}, →12ADP ↕ (113 DC)

$12\text{Ⓟ}OCH_2\text{—}CH(OH)\text{—}COO\text{Ⓟ}$

1,3-Diphosphoglycerate — C

D-Glyceraldehyde-3-phosphate: NADP oxidoreductase
Glyceraldehyde phosphate dehydrogenase
1.2.1.13
→12NADPH + $12H^+$, →$12NADP^+$, →$12P_i$ ↕ (113 C-A)

$12^*\ \text{Ⓟ}OCH_2\text{—}CH(OH)\text{—}CHO$ ⟷ (5.3.1.1) $HOCH_2\text{—}CO\text{—}CH_2O\text{Ⓟ}$

3-Phosphoglyceraldehyde — Dihydroxyacetone phosphate — DE

Fructose-1,6-diphosphate D-glyceraldehyde-3-phosphate lyase
Fructose diphosphate aldolase
4.1.2.13
↕ (112 EFD)

$$3\text{Ⓟ}OCH_2\text{—}\underset{OH}{\overset{H}{C}}\text{—}\underset{OH}{\overset{H}{C}}\text{—}\underset{H}{\overset{OH}{C}}\text{—}CO\text{—}CH_2O\text{Ⓟ}$$

Fructose-1,6-diphosphate — F

Fructose-1,6-diphosphate 1-phosphohydrolase
Hexose diphosphatase
3.1.3.11
$3H_2O$, Mg^{++}, →$3P_i$ ↓ (116 DE)

$$3\text{Ⓟ}OCH_2\text{—}\underset{OH}{\overset{H}{C}}\text{—}\underset{OH}{\overset{H}{C}}\text{—}\underset{H}{\overset{OH}{C}}\text{—}CO\text{—}CH_2OH$$

Fructose-6-phosphate — G

⋮ ↓

1 Glucose — H

(Reactions 112 CB and 117 DE)

* Six molecules of triose phosphate form three molecules of fructose-6-phosphate, one of which is converted to glucose. The other two molecules of fructose-6-phosphate and the remaining six molecules of triose phosphate are utilized in the reactions shown opposite.

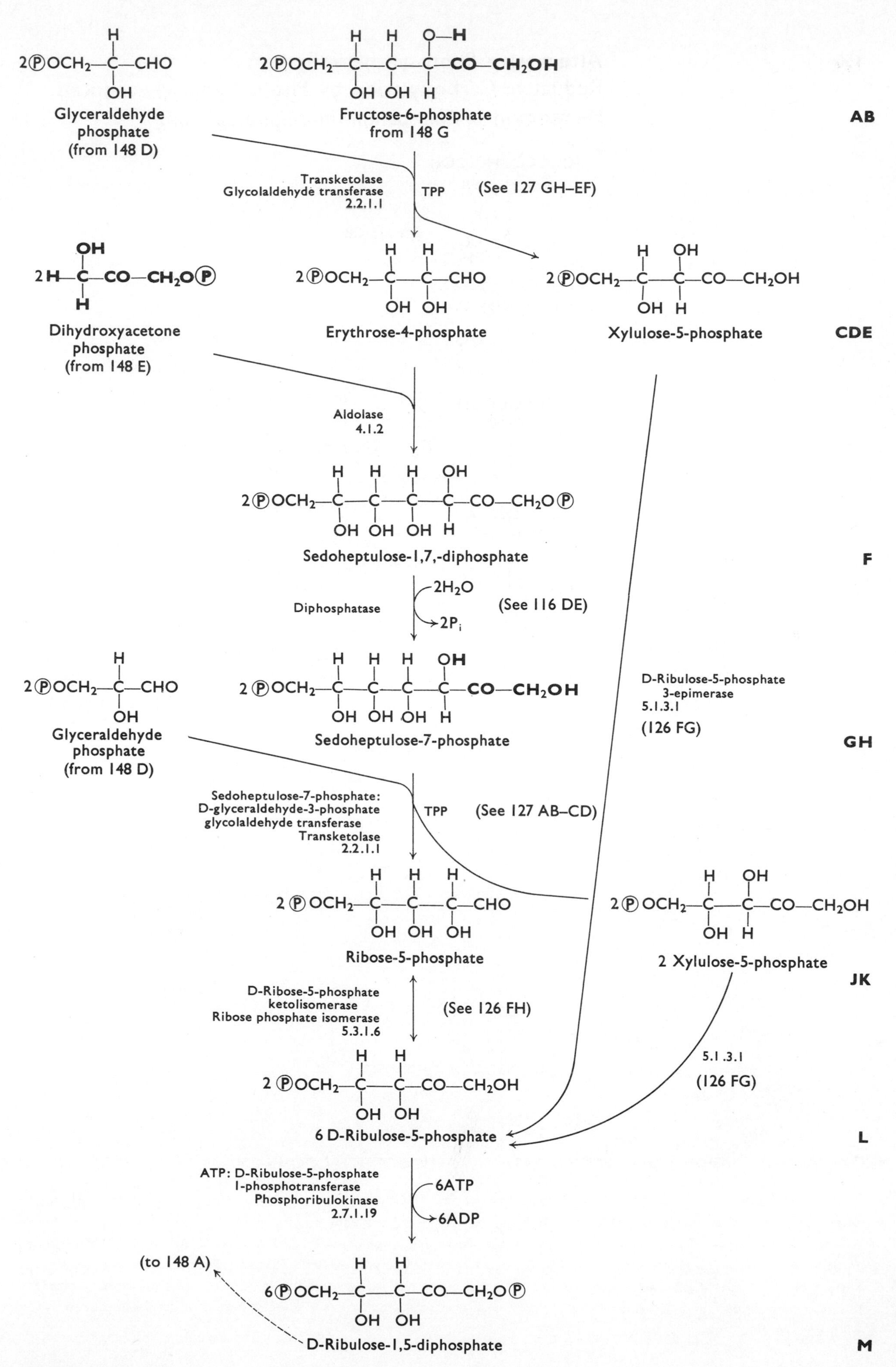

See note on page 127.

Alternative Photosynthetic System
Reductive Carboxylation by Photochemically-reduced Ferredoxin in *Chlorobium thiosulphatophilum*

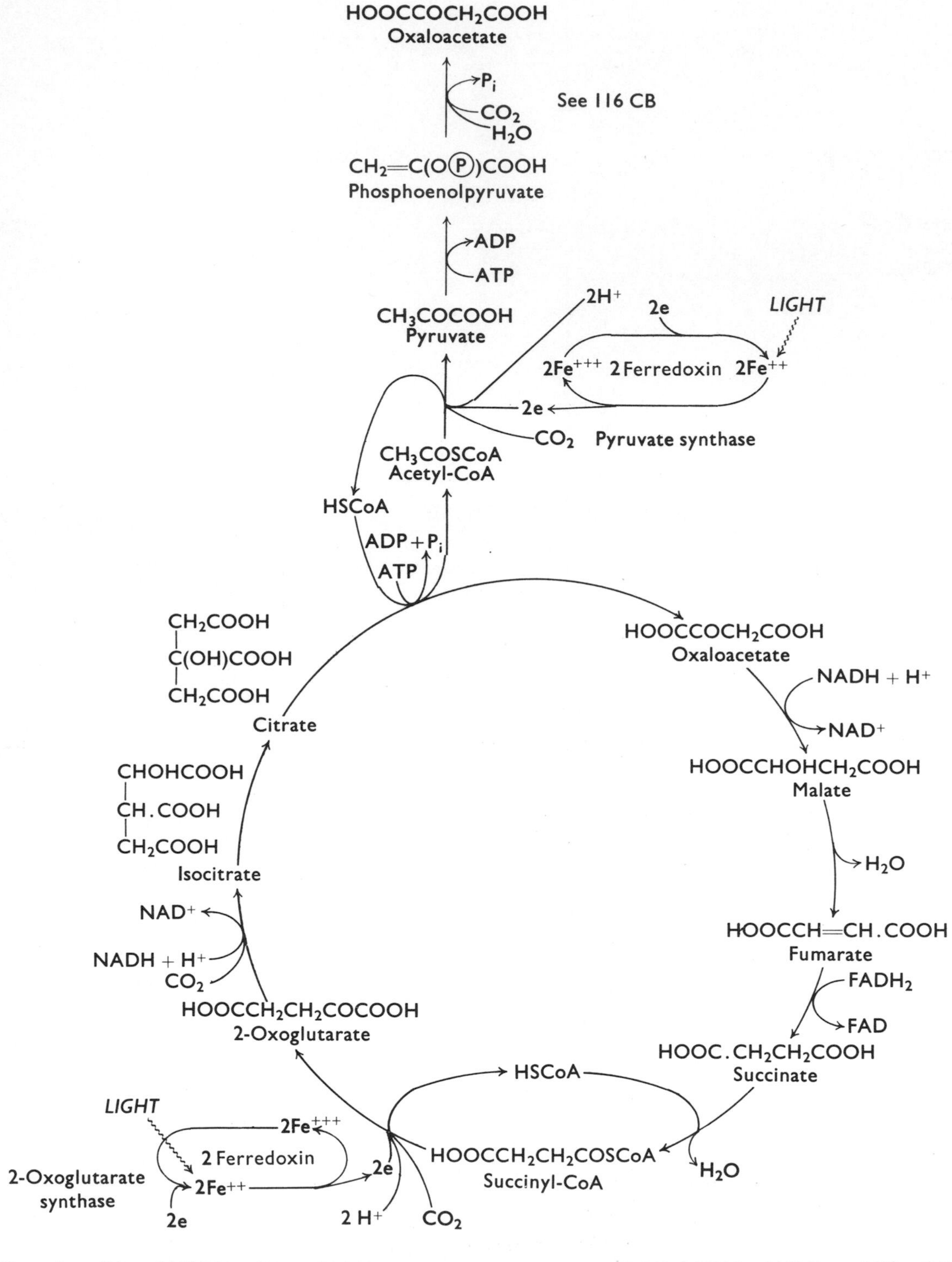

$$4CO_2 + 4e + 4H^+ + 2NADH + 2H^+ + FADH_2 + 2ATP \longrightarrow HOOCCOCH_2COOH + 2NAD^+ + FAD + 2ADP + 2P_i + H_2O$$

The lower cycle is the tricarboxylic acid cycle in reverse. The thermodynamically-unlikely reductive carboxylation of acetyl-CoA and of succinyl-CoA are made possible because of the very strong reducing activity of reduced ferredoxin which is obtained photochemically. This is also responsible for the ATP and reduced NAD and NADP required in other reactions. The electrons involved in the reduction of the oxidized ferredoxin, and the protons involved in the reduction of succinyl-CoA and of acetyl-CoA are products of the rather complex oxidations of thiosulphate or other similar compounds. The electrons fill the 'holes' in the ferredoxin molecules produced by the photochemical removal of high-energy electrons.

Anaplerotic route for formation from acetyl-CoA of oxaloacetate necessary to perpetuate TCA cycle.

For discussion see pages 40–41.

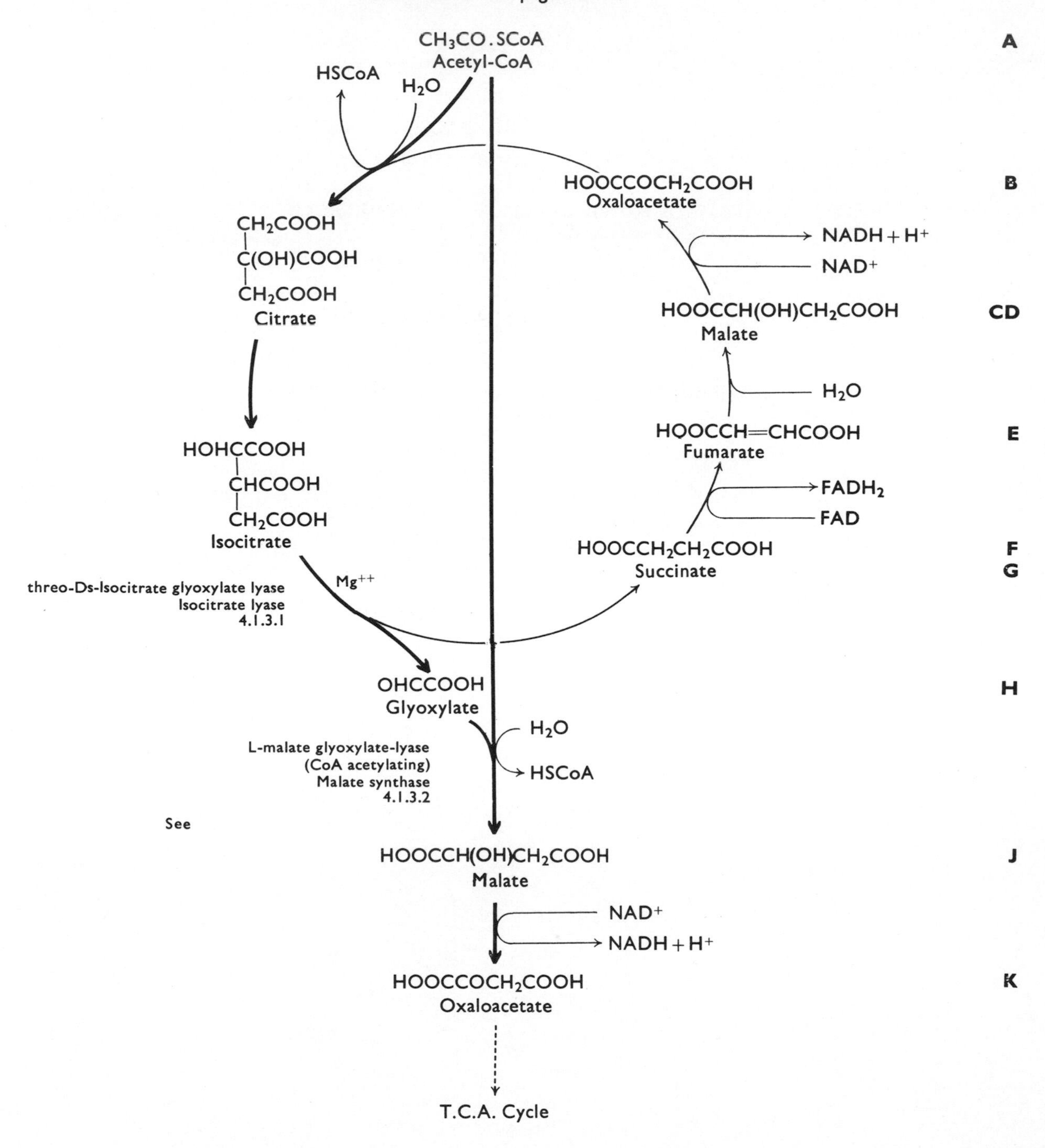

Net result

$$2CH_3CO.SCoA + 3H_2O + 2NAD^+ + FAD \longrightarrow \begin{matrix} COCOOH \\ | \\ CH_2COOH \end{matrix} + 2HSCoA + 2NADH + 2H^+ + FADH_2$$

2 Acetyl-CoA → Oxaloacetate

Glyoxylate Utilization (Glycerate Pathway) in microorganisms and plants

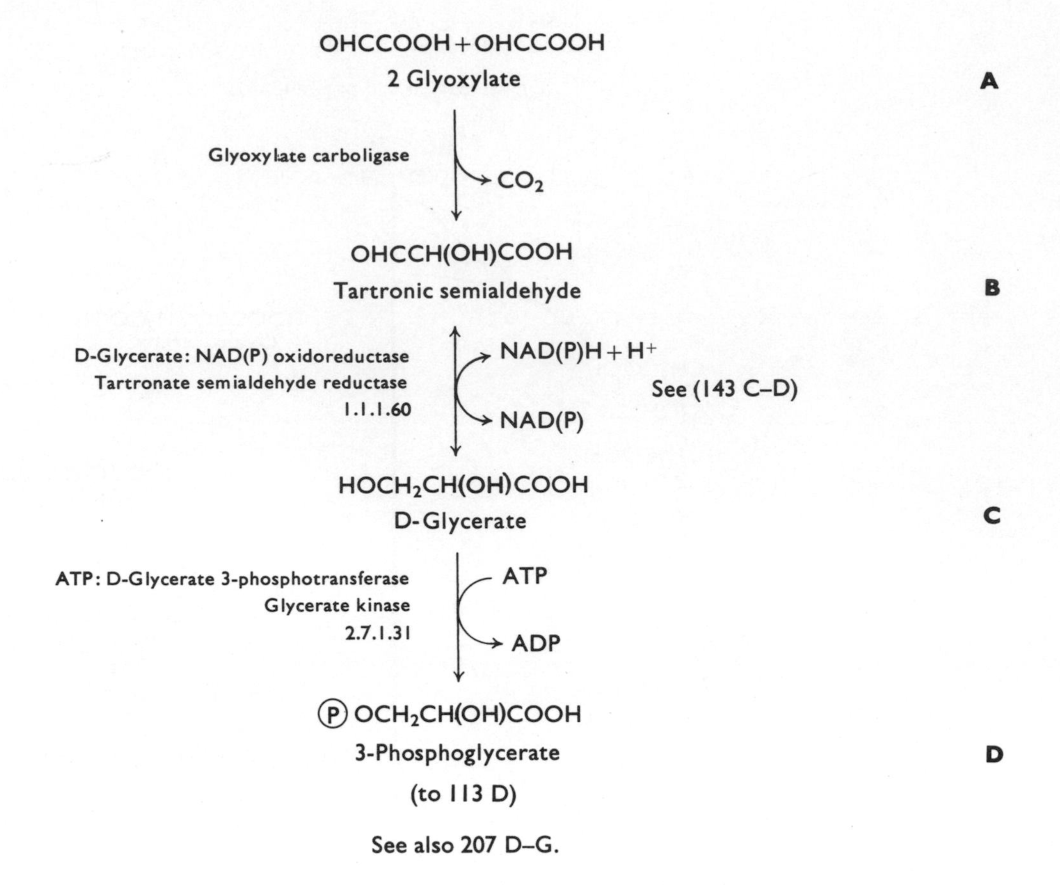

See also 207 D–G.

Oxalate is so highly oxidized that it must be reduced before becoming available for biosynthetic purposes. Energy can be obtained by oxidation to CO_2 and hydrogen (carried by NAD or NADP). The reduced NADP can also be utilized to reduce another molecule of oxalate to glyoxylate which can then be metabolized through the glycerate pathway (page 152). For further discussion see page 42.

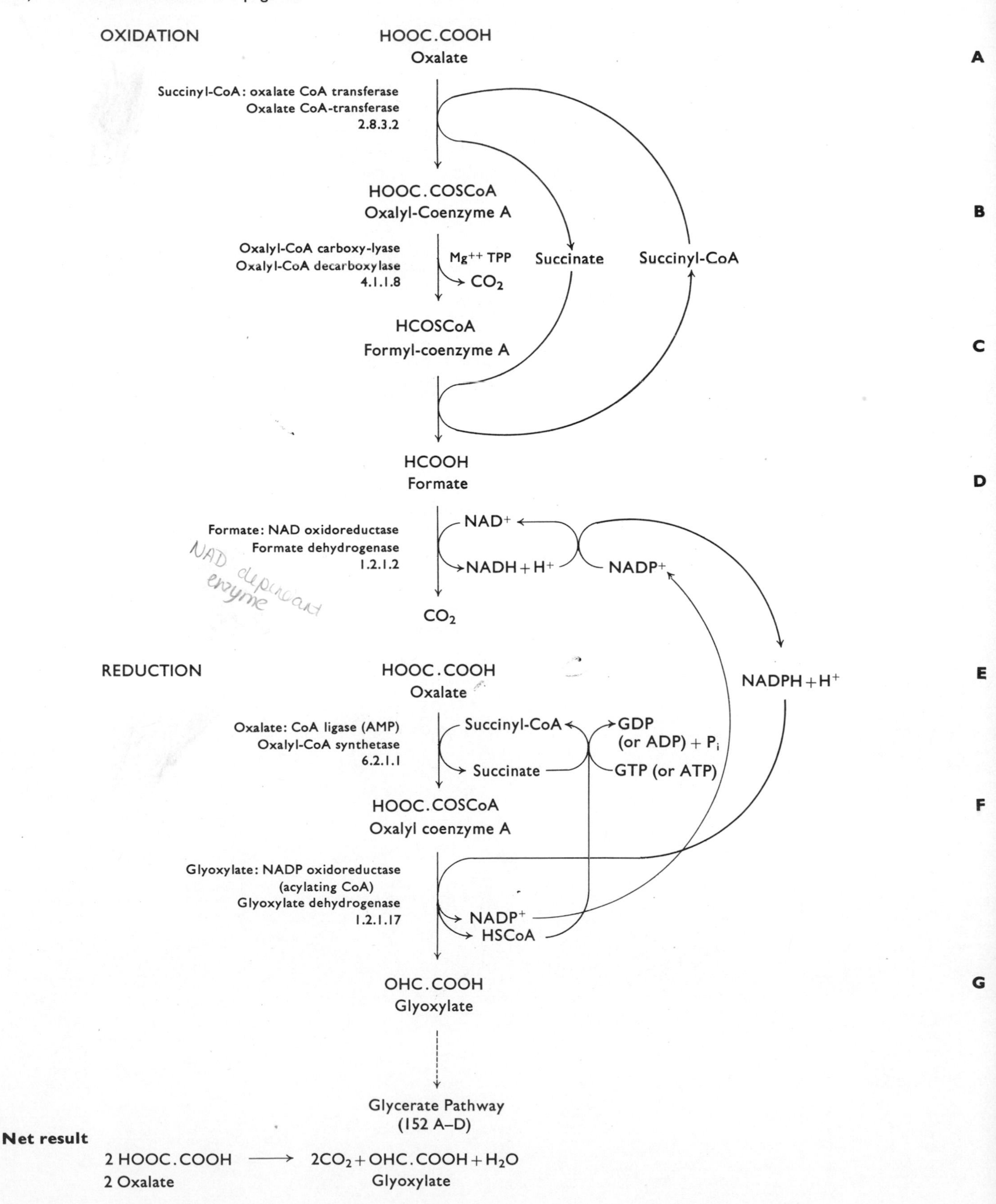

Net result

$$2\,HOOC.COOH \longrightarrow 2CO_2 + OHC.COOH + H_2O$$

2 Oxalate → Glyoxylate

Fluorescent species of *Pseudomonas*

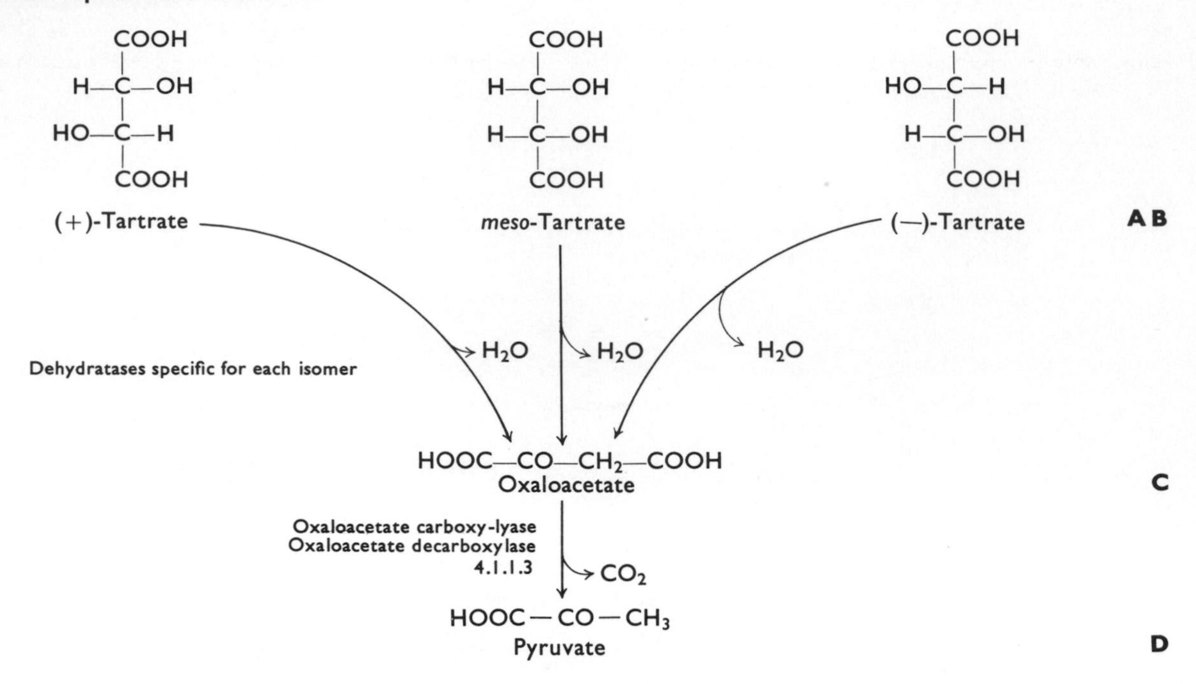

Pseudomonas acidovorans

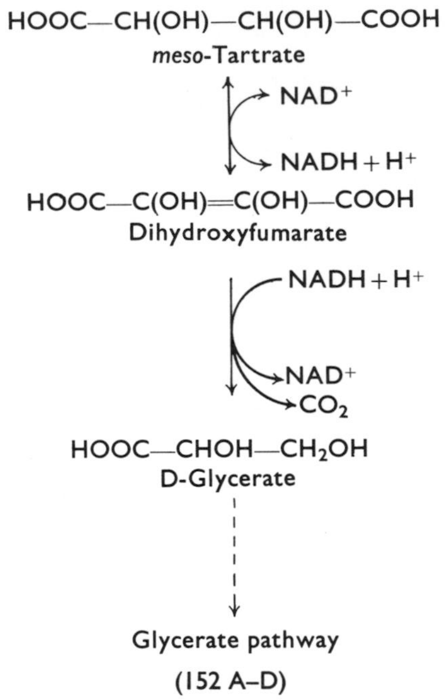

HOOC—CH(OH)—CH(OH)—COOH
meso-Tartrate **E**

↕ NAD^+ / $NADH + H^+$

HOOC—C(OH)=C(OH)—COOH
Dihydroxyfumarate **F**

↓ $NADH + H^+$ → NAD^+, CO_2

HOOC—CHOH—CH_2OH
D-Glycerate

⇣

Glycerate pathway **G**
(152 A–D)

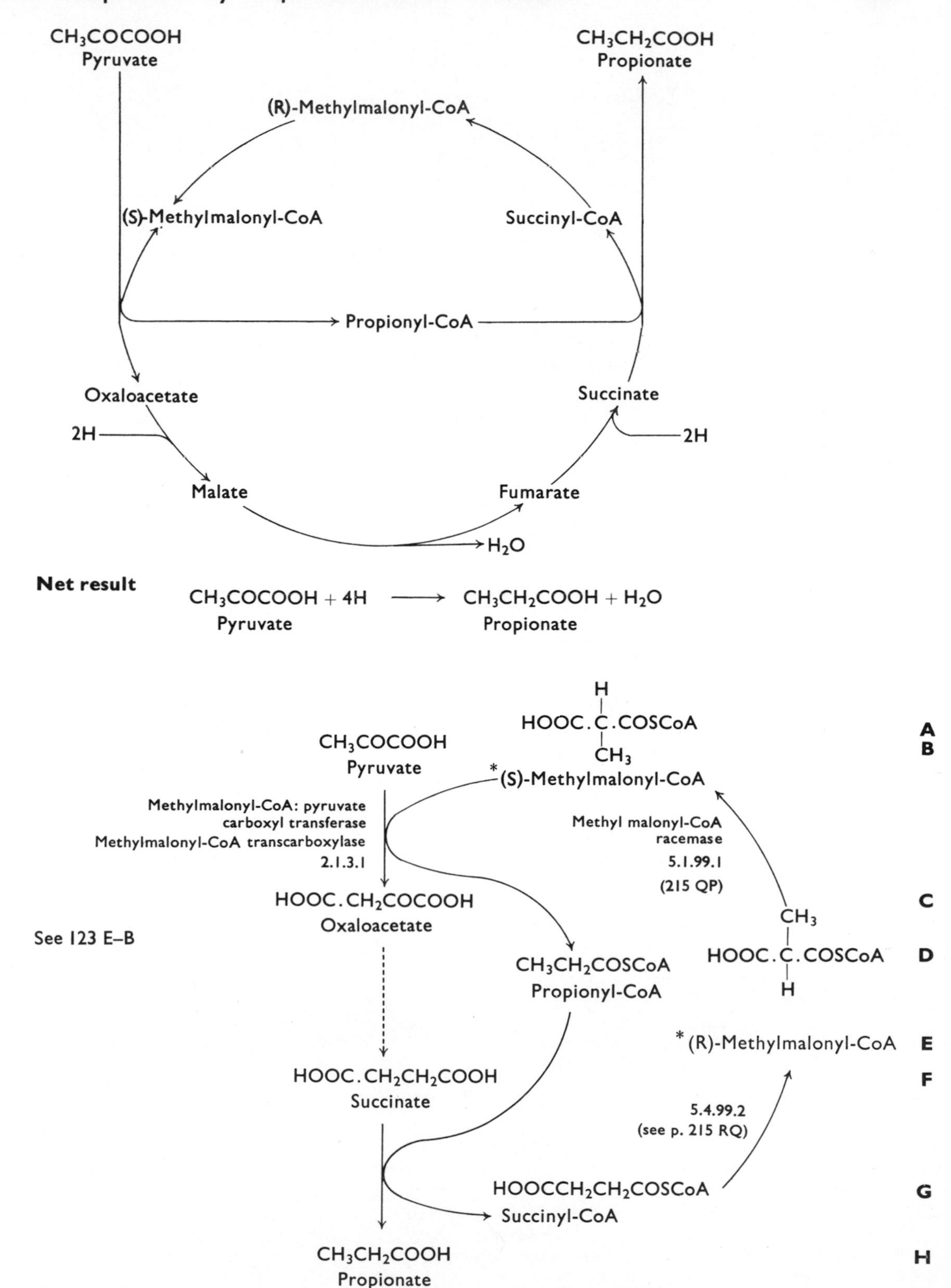

For conversion of propionyl-CoA into succinyl-CoA, see 215 O–R.

* For significance of prefixes S (from the Latin 'sinister' meaning 'left') and R (from 'rectus' meaning 'right') see Sequence Rule proposed by R. S. Cahn, C.K. Ingold and V. Prelog in *Experientia* (1956) **12**, 81. R.S. Cahn has written an introduction to the Sequence Rule in *Journal of Chemical Education* (1964) **41**, 116.

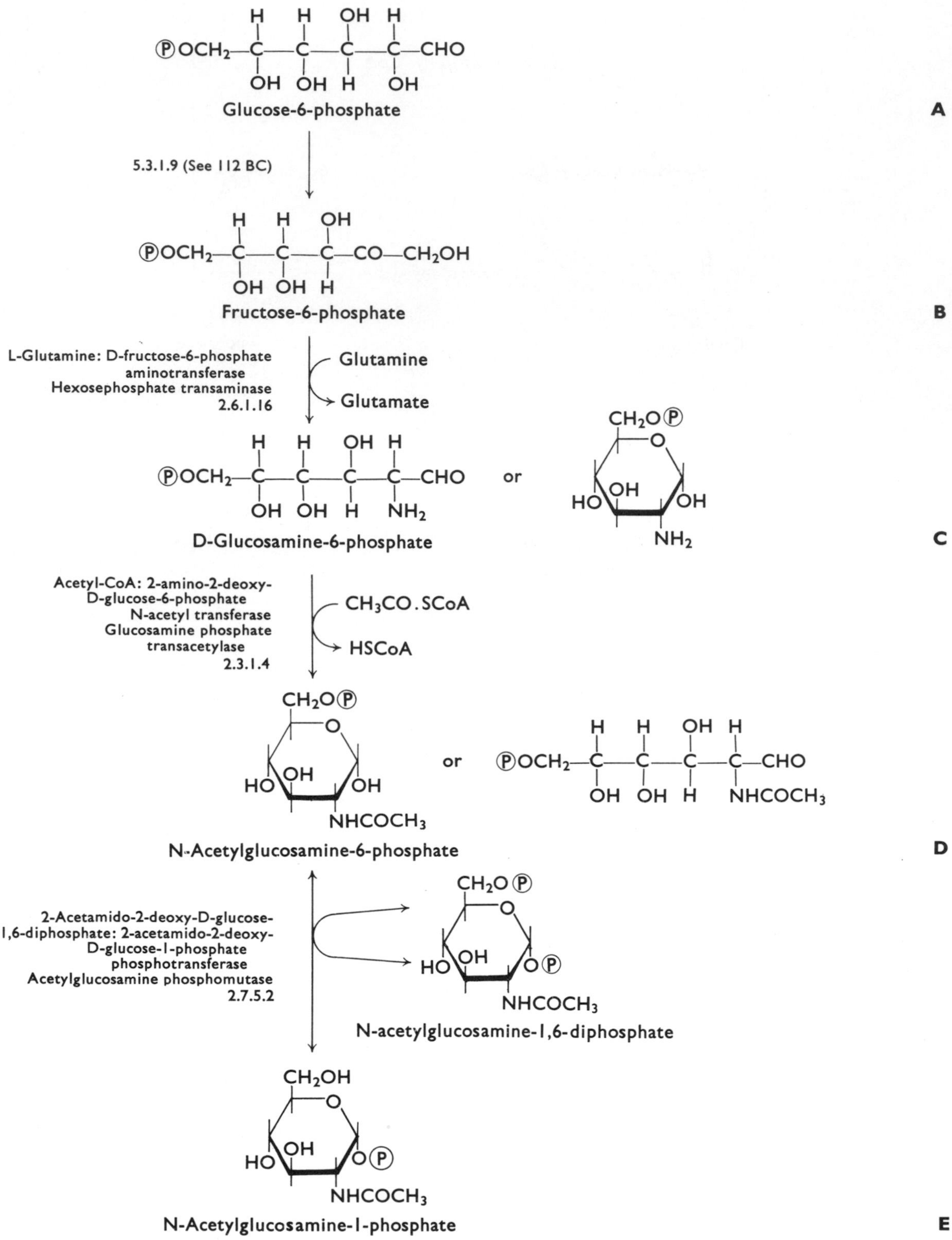

ⓅOCH_2—CH(OH)—CH(OH)—CH(OH)—CH(OH)—CHO
Glucose-6-phosphate **A**

5.3.1.9 (See 112 BC)

ⓅOCH_2—CH(OH)—CH(OH)—CH(OH)—CO—CH_2OH
Fructose-6-phosphate **B**

L-Glutamine: D-fructose-6-phosphate aminotransferase
Hexosephosphate transaminase
2.6.1.16

Glutamine → Glutamate

ⓅOCH_2—CH(OH)—CH(OH)—CH(OH)—CH(NH_2)—CHO or (ring form: CH_2OⓅ, HO, OH, OH, NH_2)
D-Glucosamine-6-phosphate **C**

Acetyl-CoA: 2-amino-2-deoxy-D-glucose-6-phosphate N-acetyl transferase
Glucosamine phosphate transacetylase
2.3.1.4

$CH_3CO.SCoA$ → HSCoA

(ring form: CH_2OⓅ, HO, OH, OH, $NHCOCH_3$) or ⓅOCH_2—CH(OH)—CH(OH)—CH(OH)—CH($NHCOCH_3$)—CHO
N-Acetylglucosamine-6-phosphate **D**

2-Acetamido-2-deoxy-D-glucose-1,6-diphosphate: 2-acetamido-2-deoxy-D-glucose-1-phosphate phosphotransferase
Acetylglucosamine phosphomutase
2.7.5.2

(ring form: CH_2OⓅ, HO, OH, OⓅ, $NHCOCH_3$)
N-acetylglucosamine-1,6-diphosphate

(ring form: CH_2OH, HO, OH, OⓅ, $NHCOCH_3$)
N-Acetylglucosamine-1-phosphate **E**

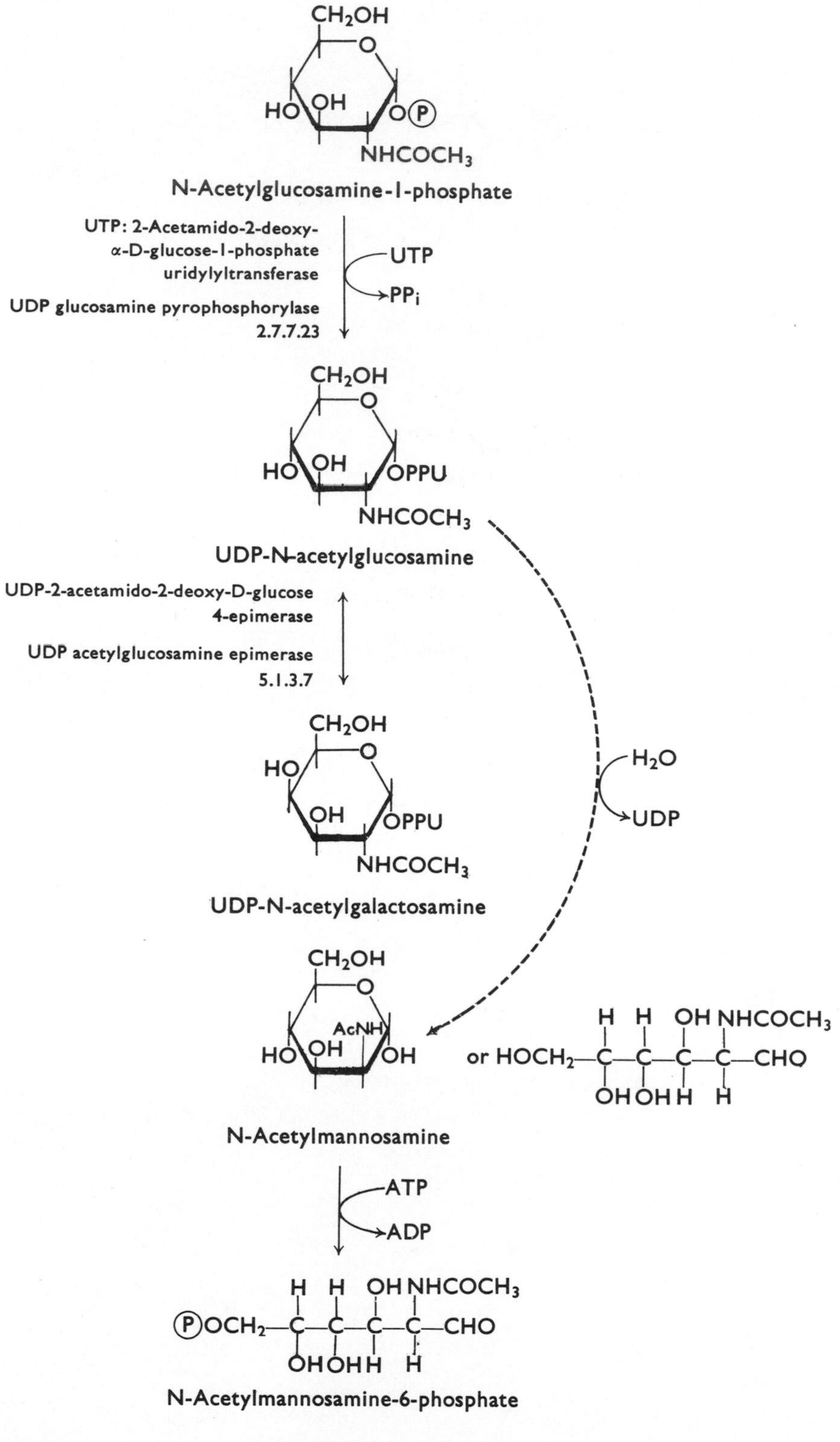

CH_2OH ... HO, OH, O(P), $NHCOCH_3$

N-Acetylglucosamine-1-phosphate

A

UTP: 2-Acetamido-2-deoxy-α-D-glucose-1-phosphate uridylyltransferase

UDP glucosamine pyrophosphorylase 2.7.7.23

UTP → PP_i

CH_2OH ... HO, OH, OPPU, $NHCOCH_3$

UDP-N-acetylglucosamine

B

UDP-2-acetamido-2-deoxy-D-glucose 4-epimerase

UDP acetylglucosamine epimerase 5.1.3.7

H_2O → UDP

CH_2OH ... HO, OH, OPPU, $NHCOCH_3$

UDP-N-acetylgalactosamine

C

CH_2OH ... AcNH, HO, OH, OH

or $HOCH_2$—C(OH)(H)—C(OH)(H)—C(H)(OH)—C(H)($NHCOCH_3$)—CHO

N-Acetylmannosamine

D

ATP → ADP

(P)OCH_2—C(OH)(H)—C(OH)(H)—C(H)(OH)—C(H)($NHCOCH_3$)—CHO

N-Acetylmannosamine-6-phosphate

E

Synthesis of UDP-muramate

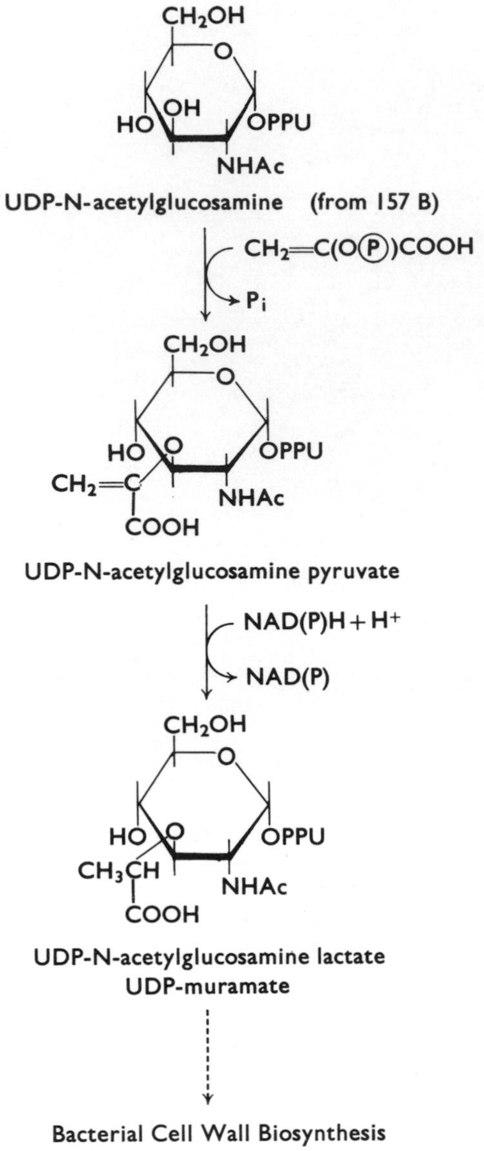

CH_2OH, HO, OH, OPPU, NHAc

UDP-N-acetylglucosamine (from 157 B) **A**

$CH_2{=}C(O(P))COOH$ → P_i

CH_2OH, HO, O, OPPU, $CH_2{=}C$, NHAc, COOH

UDP-N-acetylglucosamine pyruvate **B**

$NAD(P)H + H^+$ → NAD(P)

CH_2OH, HO, O, OPPU, CH_3CH, NHAc, COOH

UDP-N-acetylglucosamine lactate
UDP-muramate **C**

Bacterial Cell Wall Biosynthesis

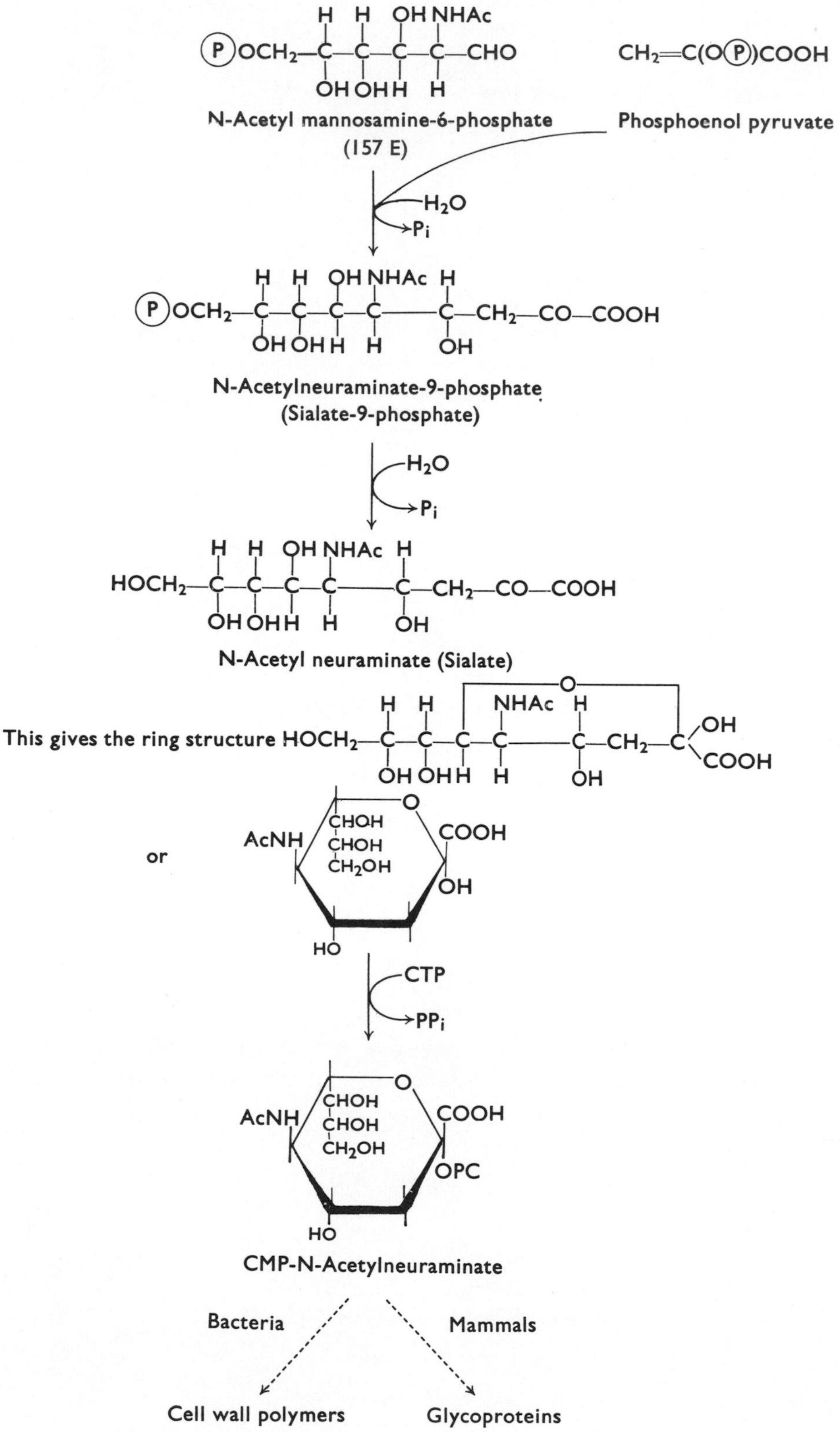

$(P)OCH_2-CH(OH)-CH(OH)-CH(OH)-CH(NHAc)-CHO$ (substituents: H H OH NHAc above; OH OH H H below)

N-Acetyl mannosamine-6-phosphate (157 E)

$CH_2{=}C(O(P))COOH$

Phosphoenol pyruvate

AB

$\downarrow$ $-H_2O$, $\rightarrow P_i$

$(P)OCH_2-C-C-C-C-C-CH_2-CO-COOH$ (substituents: H H OH NHAc H above; OH OH H H OH below)

N-Acetylneuraminate-9-phosphate
(Sialate-9-phosphate)

C

$\downarrow$ $-H_2O$, $\rightarrow P_i$

$HOCH_2-C-C-C-C-C-CH_2-CO-COOH$ (substituents: H H OH NHAc H above; OH OH H H OH below)

N-Acetyl neuraminate (Sialate)

D

This gives the ring structure $HOCH_2-C-C-C-C-C-CH_2-C(OH)(COOH)$ (substituents: H H, –O– ring bridge, NHAc H above; OH OH H H OH below)

or

Ring form: AcNH, CHOH, CHOH, CH_2OH, O, COOH, OH, HO

$\downarrow$ $-CTP$, $\rightarrow PP_i$

Ring form: AcNH, CHOH, CHOH, CH_2OH, O, COOH, OPC, HO

CMP-N-Acetylneuraminate

E

Bacteria → Cell wall polymers

Mammals → Glycoproteins

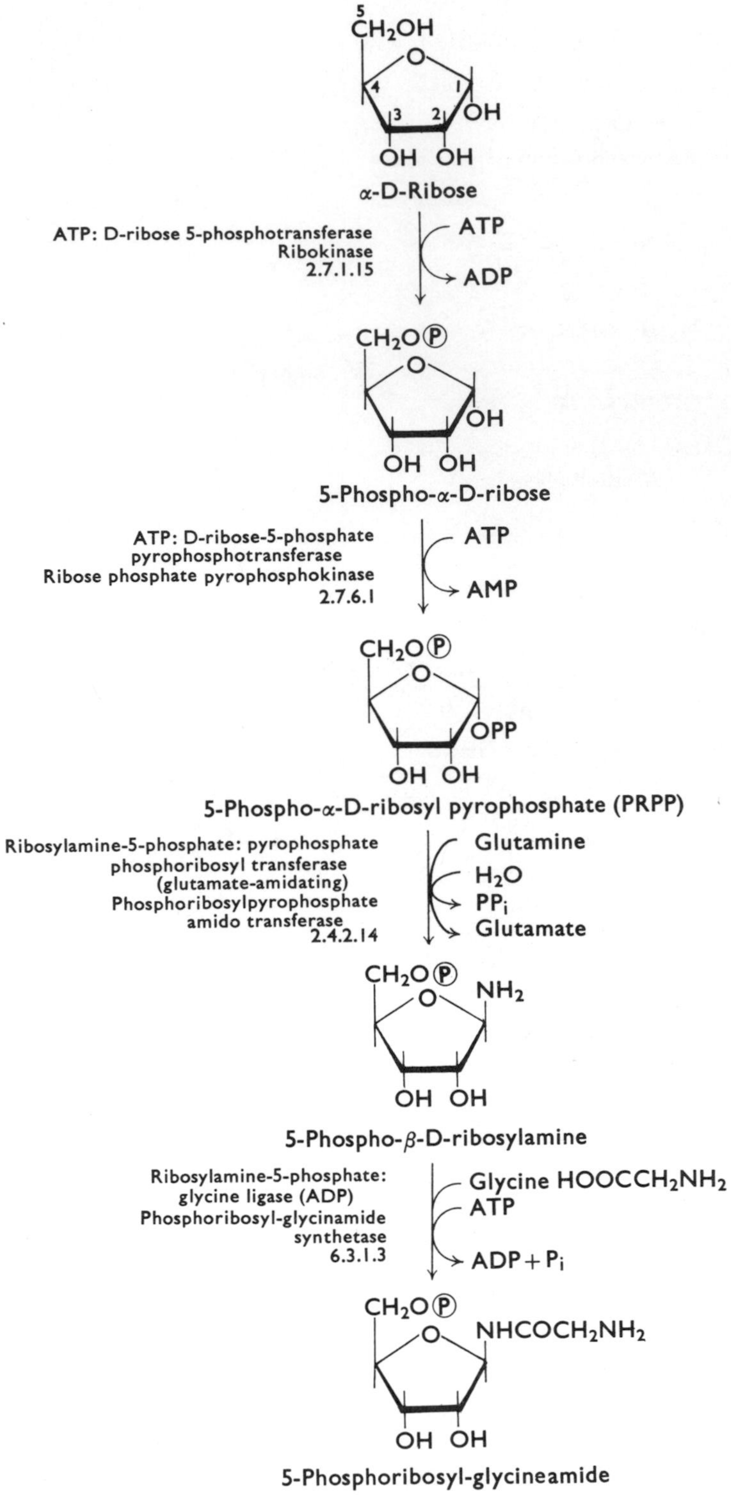

α-D-Ribose — A

ATP: D-ribose 5-phosphotransferase
Ribokinase
2.7.1.15
ATP → ADP

5-Phospho-α-D-ribose — B

ATP: D-ribose-5-phosphate pyrophosphotransferase
Ribose phosphate pyrophosphokinase
2.7.6.1
ATP → AMP

5-Phospho-α-D-ribosyl pyrophosphate (PRPP) — C

Ribosylamine-5-phosphate: pyrophosphate phosphoribosyl transferase (glutamate-amidating)
Phosphoribosylpyrophosphate amido transferase
2.4.2.14
Glutamine, H_2O → PP_i, Glutamate

5-Phospho-β-D-ribosylamine — D

Ribosylamine-5-phosphate: glycine ligase (ADP)
Phosphoribosyl-glycinamide synthetase
6.3.1.3
Glycine $HOOCCH_2NH_2$, ATP → ADP + P_i

5-Phosphoribosyl-glycineamide — E

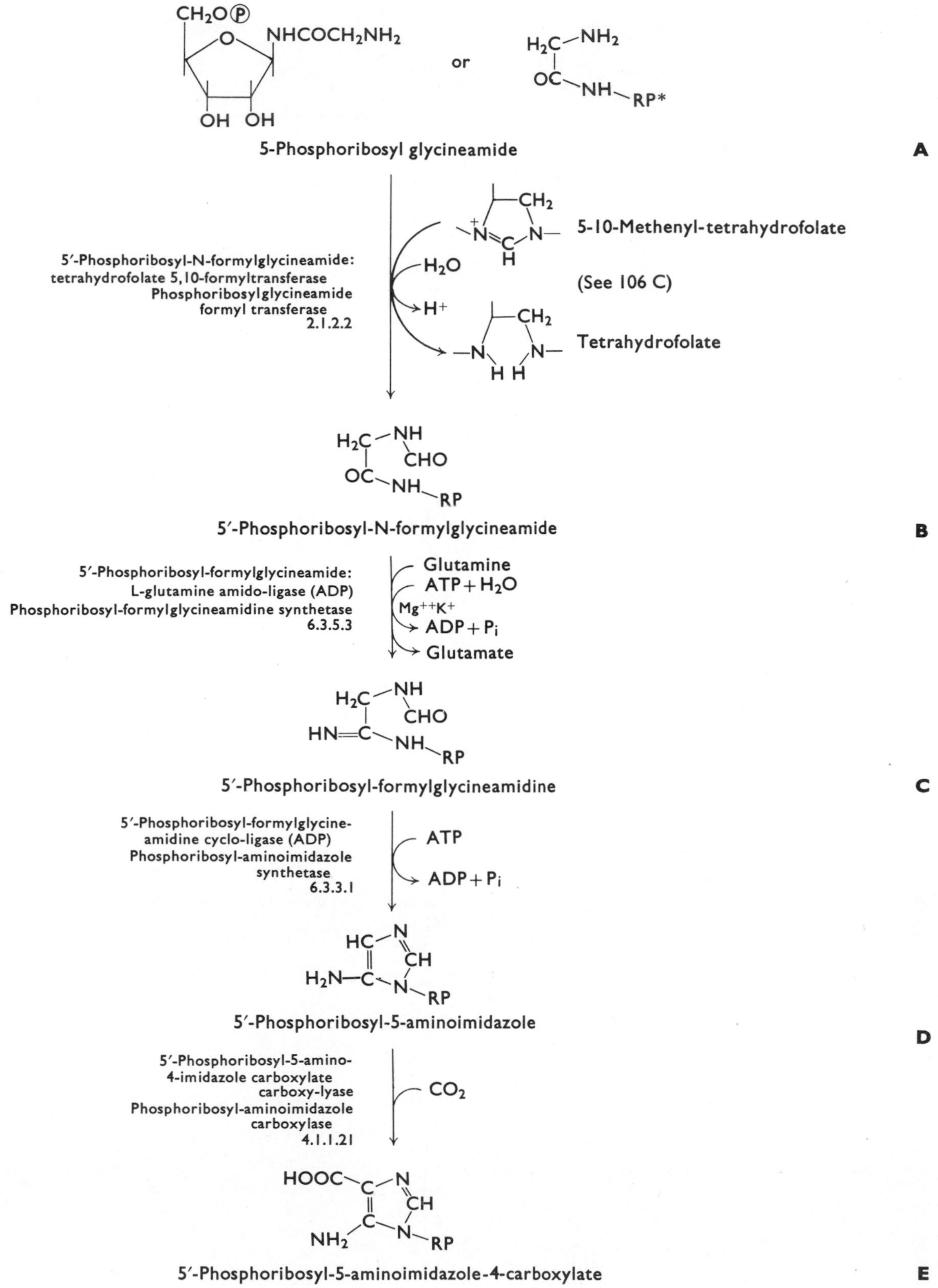

CH_2OⓅ ... $NHCOCH_2NH_2$ ... OH OH

or

H_2C—NH_2 / OC—NH—RP*

5-Phosphoribosyl glycineamide **A**

5′-Phosphoribosyl-N-formylglycineamide:
tetrahydrofolate 5,10-formyltransferase
Phosphoribosylglycineamide
formyl transferase
2.1.2.2

5-10-Methenyl-tetrahydrofolate

H_2O

(See 106 C)

H^+

Tetrahydrofolate

H_2C—NH—CHO / OC—NH—RP

5′-Phosphoribosyl-N-formylglycineamide **B**

5′-Phosphoribosyl-formylglycineamide:
L-glutamine amido-ligase (ADP)
Phosphoribosyl-formylglycineamidine synthetase
6.3.5.3

Glutamine
ATP + H_2O
$Mg^{++}K^+$
ADP + P_i
Glutamate

H_2C—NH—CHO / HN=C—NH—RP

5′-Phosphoribosyl-formylglycineamidine **C**

5′-Phosphoribosyl-formylglycine-
amidine cyclo-ligase (ADP)
Phosphoribosyl-aminoimidazole
synthetase
6.3.3.1

ATP
ADP + P_i

HC—N=CH / H_2N—C—N—RP

5′-Phosphoribosyl-5-aminoimidazole **D**

5′-Phosphoribosyl-5-amino-
4-imidazole carboxylate
carboxy-lyase
Phosphoribosyl-aminoimidazole
carboxylase
4.1.1.21

CO_2

HOOC—C—N=CH / NH_2—C—N—RP

5′-Phosphoribosyl-5-aminoimidazole-4-carboxylate **E**

* RP represents the phosphoribosyl group.

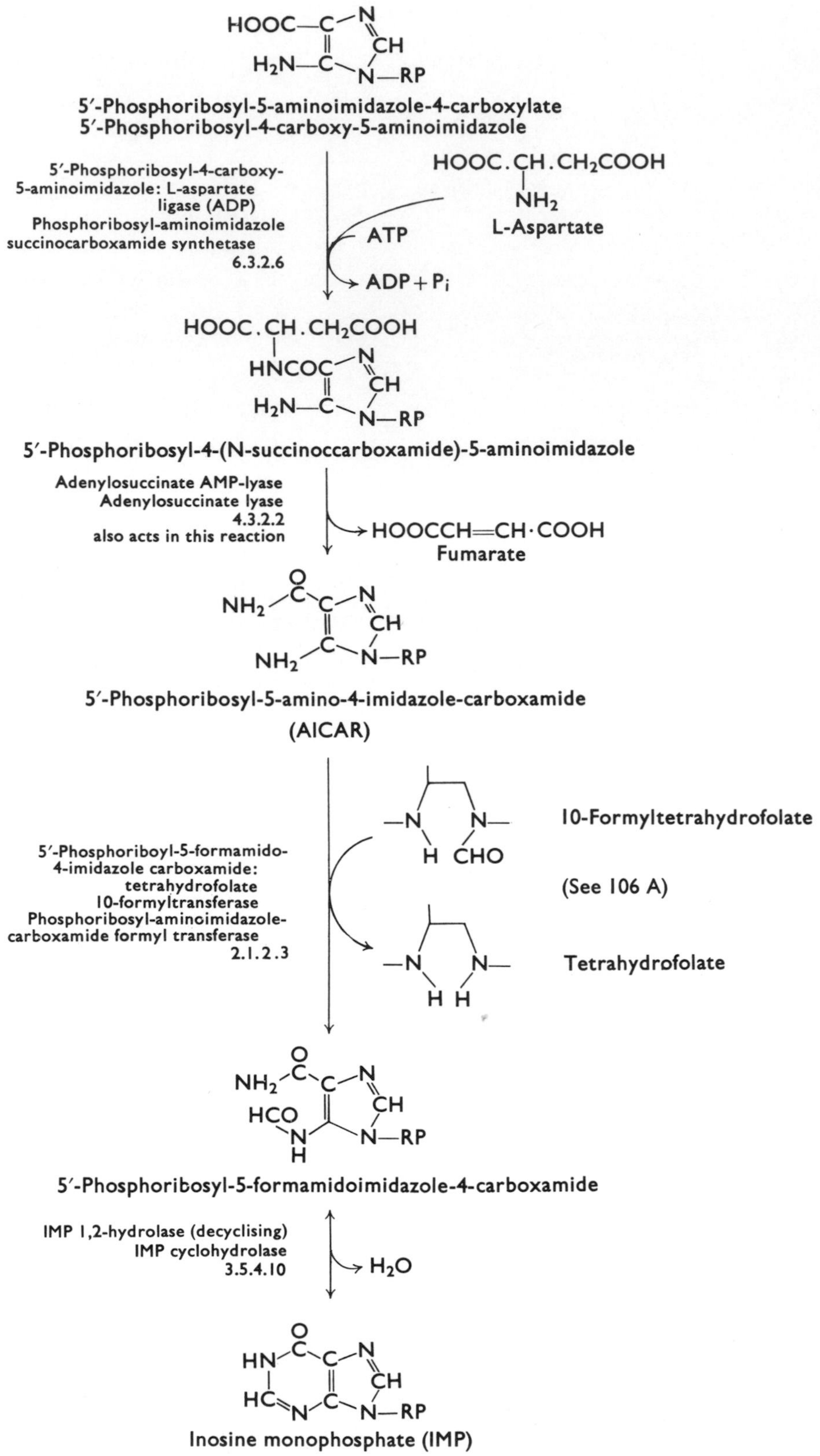

A

5′-Phosphoribosyl-5-aminoimidazole-4-carboxylate
5′-Phosphoribosyl-4-carboxy-5-aminoimidazole

5′-Phosphoribosyl-4-carboxy-5-aminoimidazole: L-aspartate ligase (ADP)
Phosphoribosyl-aminoimidazole succinocarboxamide synthetase
6.3.2.6

$HOOC.CH(NH_2).CH_2COOH$
L-Aspartate

ATP → ADP + P_i

B

5′-Phosphoribosyl-4-(N-succinoccarboxamide)-5-aminoimidazole

Adenylosuccinate AMP-lyase
Adenylosuccinate lyase
4.3.2.2
also acts in this reaction

→ HOOCCH=CH·COOH
Fumarate

C

5′-Phosphoribosyl-5-amino-4-imidazole-carboxamide
(AICAR)

5′-Phosphoriboyl-5-formamido-4-imidazole carboxamide: tetrahydrofolate 10-formyltransferase
Phosphoribosyl-aminoimidazole-carboxamide formyl transferase
2.1.2.3

10-Formyltetrahydrofolate

(See 106 A)

Tetrahydrofolate

D

5′-Phosphoribosyl-5-formamidoimidazole-4-carboxamide

IMP 1,2-hydrolase (decyclising)
IMP cyclohydrolase
3.5.4.10

→ H_2O

E

Inosine monophosphate (IMP)

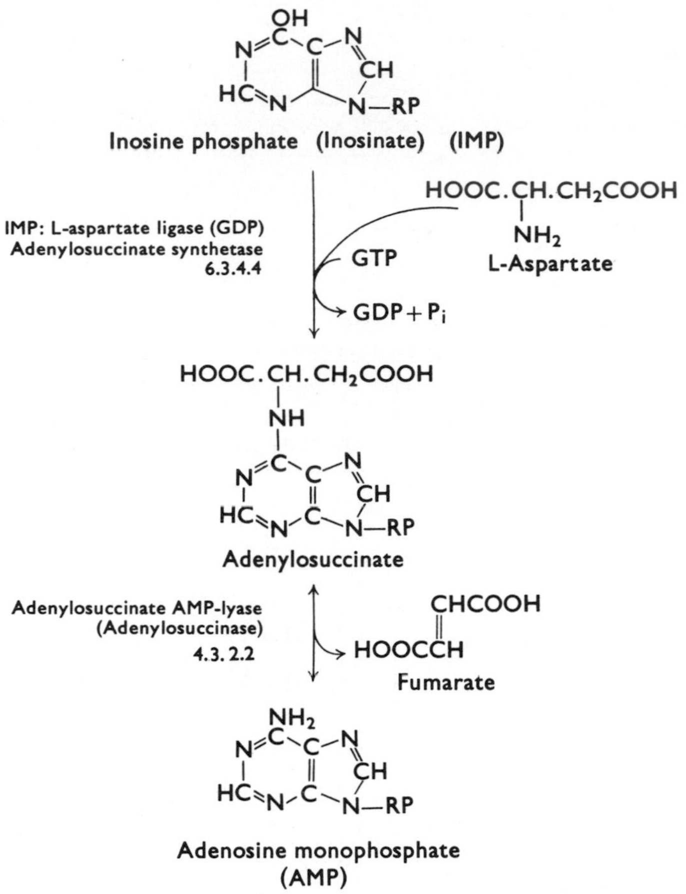

Inosine phosphate (Inosinate) (IMP) **A**

IMP: L-aspartate ligase (GDP)
Adenylosuccinate synthetase
6.3.4.4

HOOC.CH.CH_2COOH, NH_2 — L-Aspartate

GTP → GDP + P_i

Adenylosuccinate **B**

Adenylosuccinate AMP-lyase
(Adenylosuccinase)
4.3.2.2

→ HOOCCH=CHCOOH
Fumarate

Adenosine monophosphate
(AMP) **C**

Formation of Xanthosine and Guanosine Phosphates

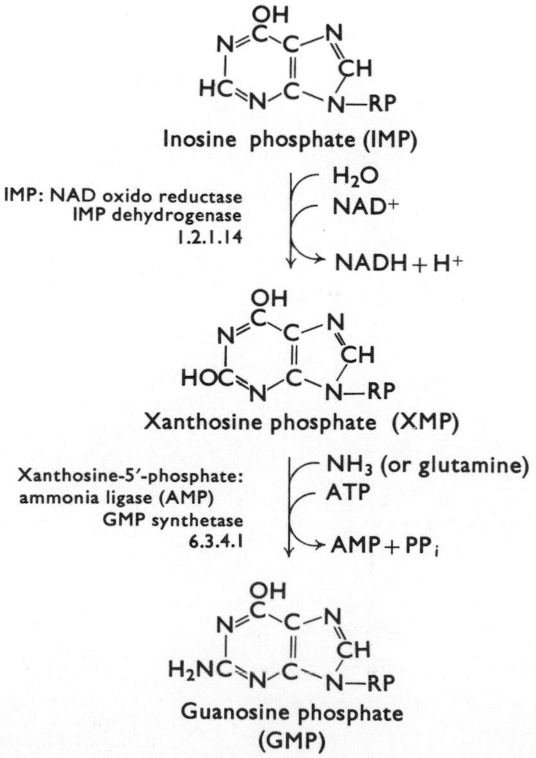

Inosine phosphate (IMP) **D**

IMP: NAD oxido reductase
IMP dehydrogenase
1.2.1.14

H_2O, NAD^+ → NADH + H^+

Xanthosine phosphate (XMP) **E**

Xanthosine-5′-phosphate:
ammonia ligase (AMP)
GMP synthetase
6.3.4.1

NH_3 (or glutamine), ATP → AMP + PP_i

Guanosine phosphate
(GMP) **F**

Catabolism of Purines

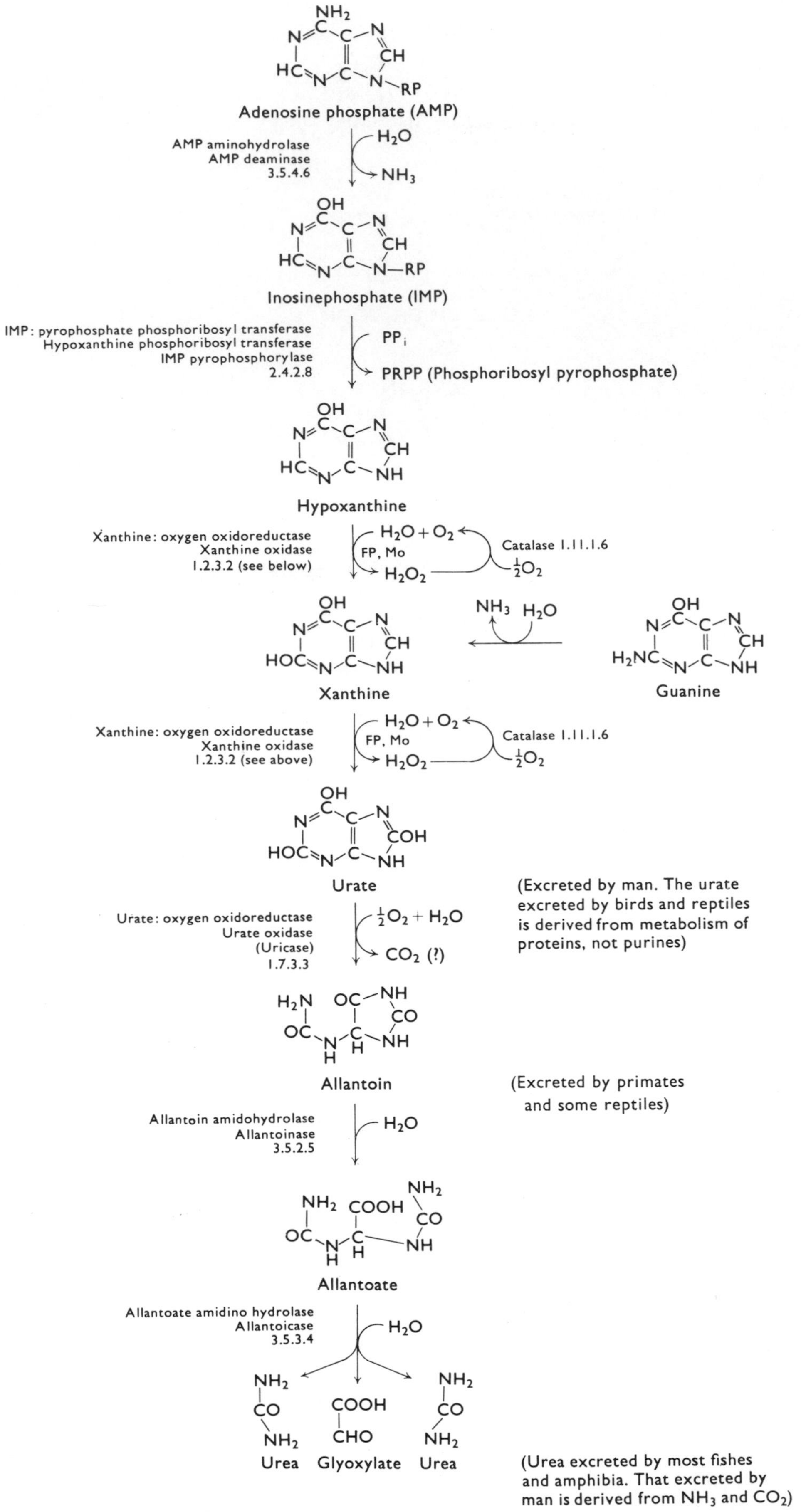

Adenosine phosphate (AMP)

AMP aminohydrolase
AMP deaminase
3.5.4.6

H_2O → NH_3

A

Inosinephosphate (IMP)

IMP: pyrophosphate phosphoribosyl transferase
Hypoxanthine phosphoribosyl transferase
IMP pyrophosphorylase
2.4.2.8

PP_i → PRPP (Phosphoribosyl pyrophosphate)

B

Hypoxanthine

Xanthine: oxygen oxidoreductase
Xanthine oxidase
1.2.3.2 (see below)

$H_2O + O_2$; FP, Mo; → H_2O_2 — Catalase 1.11.1.6 — $\frac{1}{2}O_2$

C

Xanthine ← Guanine (H_2O, NH_3)

D E

Xanthine: oxygen oxidoreductase
Xanthine oxidase
1.2.3.2 (see above)

$H_2O + O_2$; FP, Mo; → H_2O_2 — Catalase 1.11.1.6 — $\frac{1}{2}O_2$

Urate

(Excreted by man. The urate excreted by birds and reptiles is derived from metabolism of proteins, not purines)

F

Urate: oxygen oxidoreductase
Urate oxidase
(Uricase)
1.7.3.3

$\frac{1}{2}O_2 + H_2O$ → CO_2 (?)

Allantoin

(Excreted by primates and some reptiles)

G

Allantoin amidohydrolase
Allantoinase
3.5.2.5

H_2O

Allantoate

H

Allantoate amidino hydrolase
Allantoicase
3.5.3.4

H_2O

Urea Glyoxylate Urea

(Urea excreted by most fishes and amphibia. That excreted by man is derived from NH_3 and CO_2)

J K

Biosynthesis of Pyrimidines and their Nucleotides

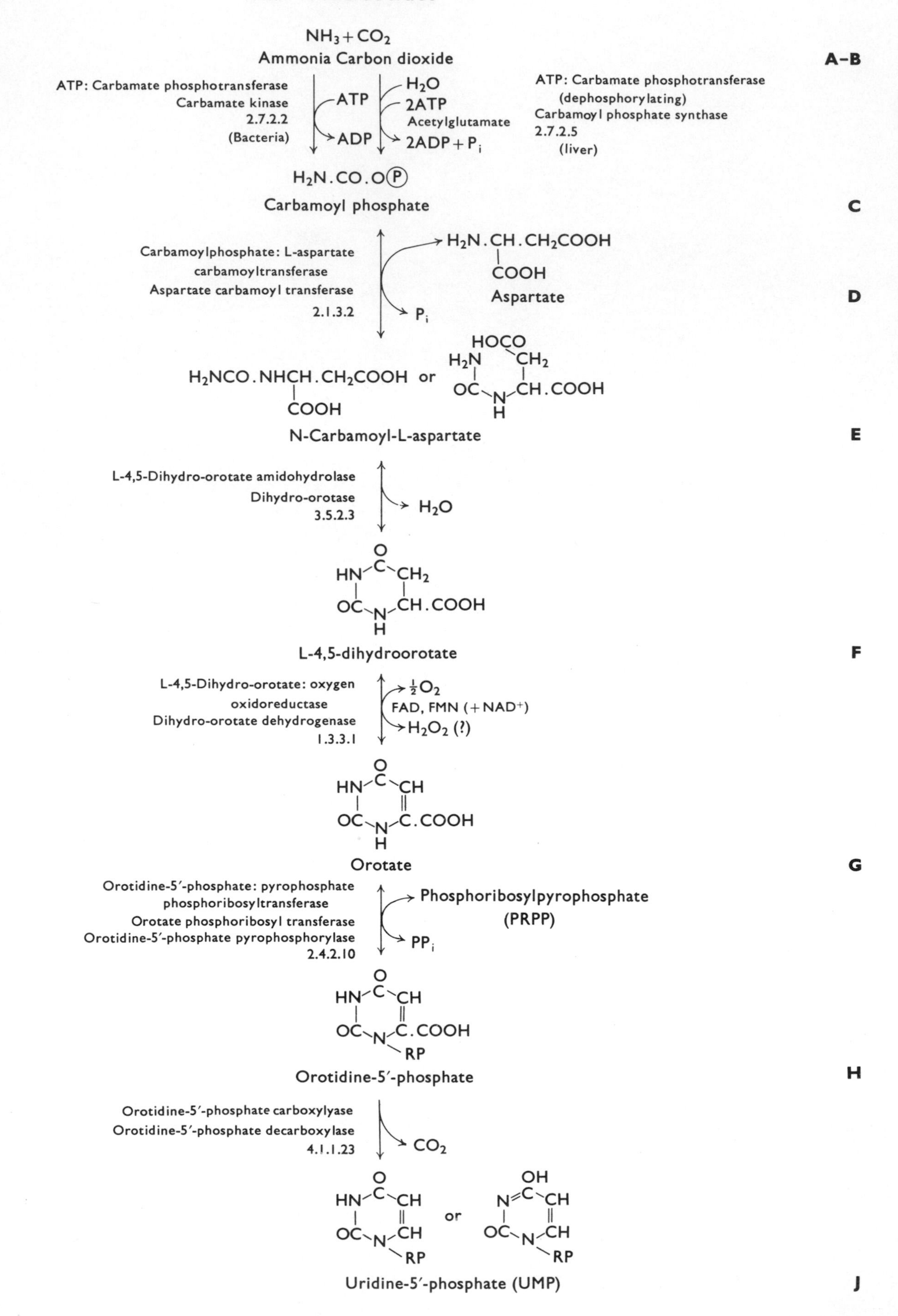

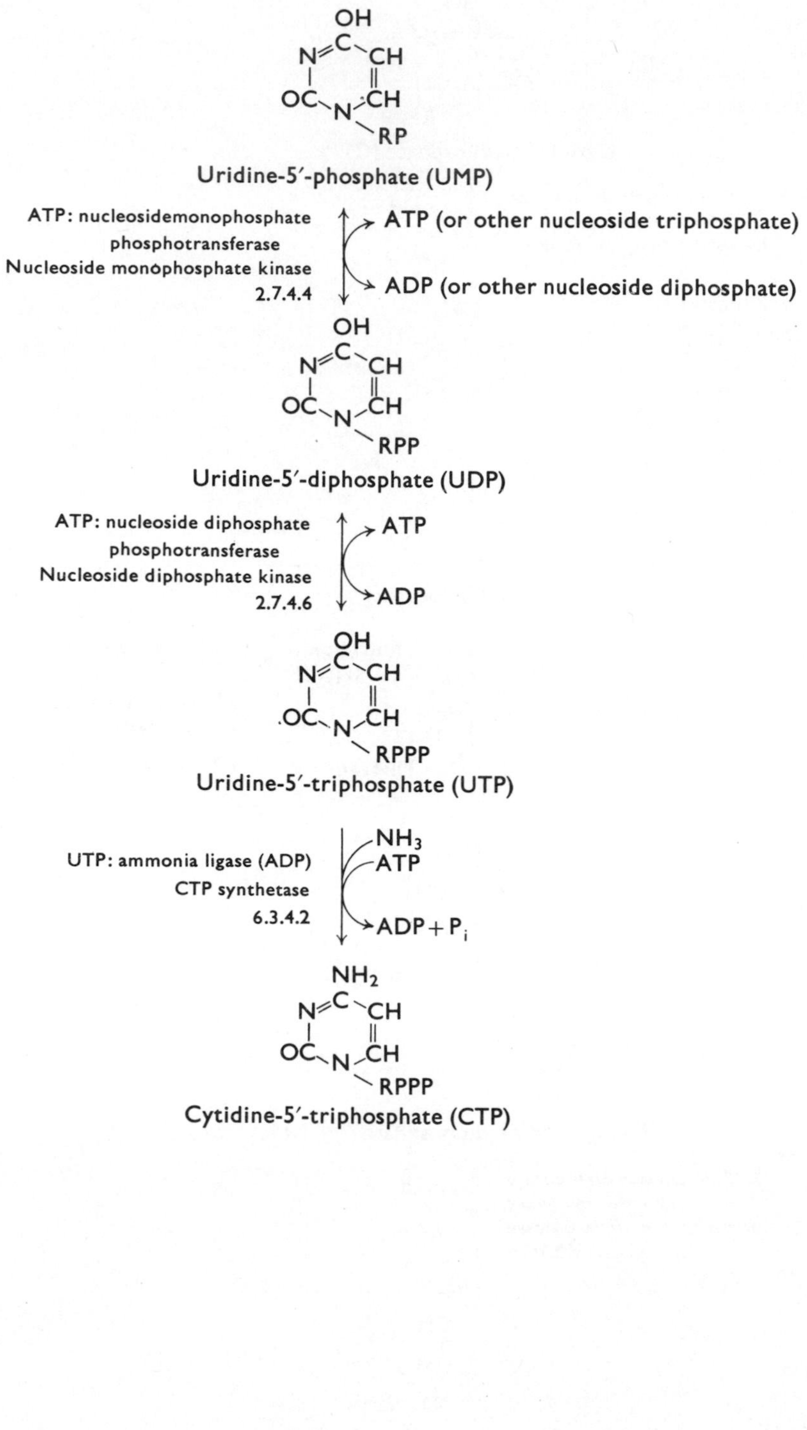

Uridine-5′-phosphate (UMP) **A**

ATP: nucleosidemonophosphate phosphotransferase
Nucleoside monophosphate kinase
2.7.4.4

ATP (or other nucleoside triphosphate)
ADP (or other nucleoside diphosphate)

Uridine-5′-diphosphate (UDP) **B**

ATP: nucleoside diphosphate phosphotransferase
Nucleoside diphosphate kinase
2.7.4.6

ATP
ADP

Uridine-5′-triphosphate (UTP) **C**

UTP: ammonia ligase (ADP)
CTP synthetase
6.3.4.2

NH_3
ATP
ADP + P_i

Cytidine-5′-triphosphate (CTP) **D**

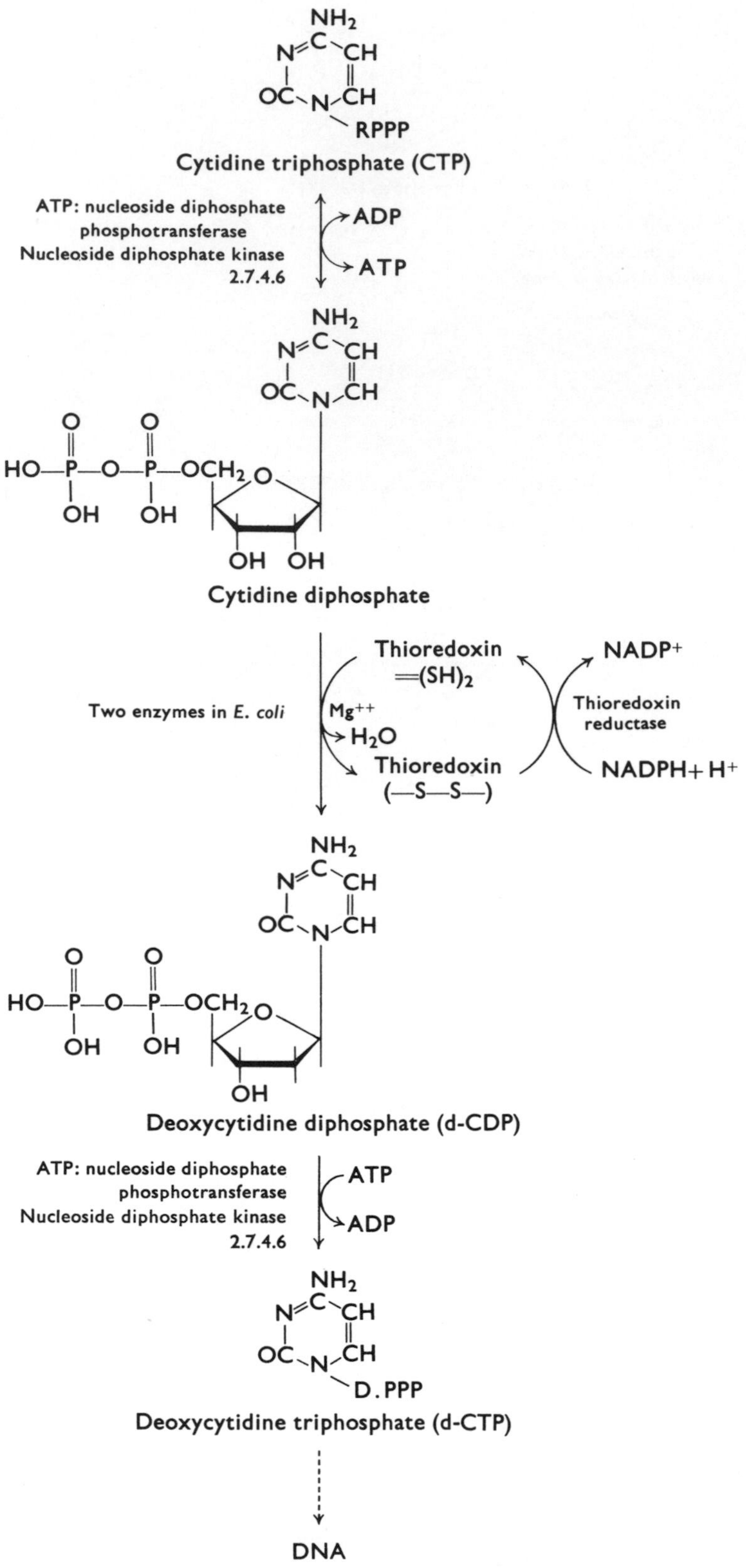

NH$_2$ / N=C–CH / OC–N–CH / RPPP

Cytidine triphosphate (CTP) **A**

ATP: nucleoside diphosphate phosphotransferase
Nucleoside diphosphate kinase
2.7.4.6

ADP / ATP

Cytidine diphosphate **B**

Two enzymes in *E. coli*

Thioredoxin =(SH)$_2$

Mg^{++}

H_2O

Thioredoxin (—S—S—)

$NADP^+$

Thioredoxin reductase

$NADPH + H^+$

Deoxycytidine diphosphate (d-CDP) **C**

ATP: nucleoside diphosphate phosphotransferase
Nucleoside diphosphate kinase
2.7.4.6

ATP / ADP

NH$_2$ / N=C–CH / OC–N–CH / D.PPP

Deoxycytidine triphosphate (d-CTP) **D**

DNA **E**

(D=Deoxyribose)

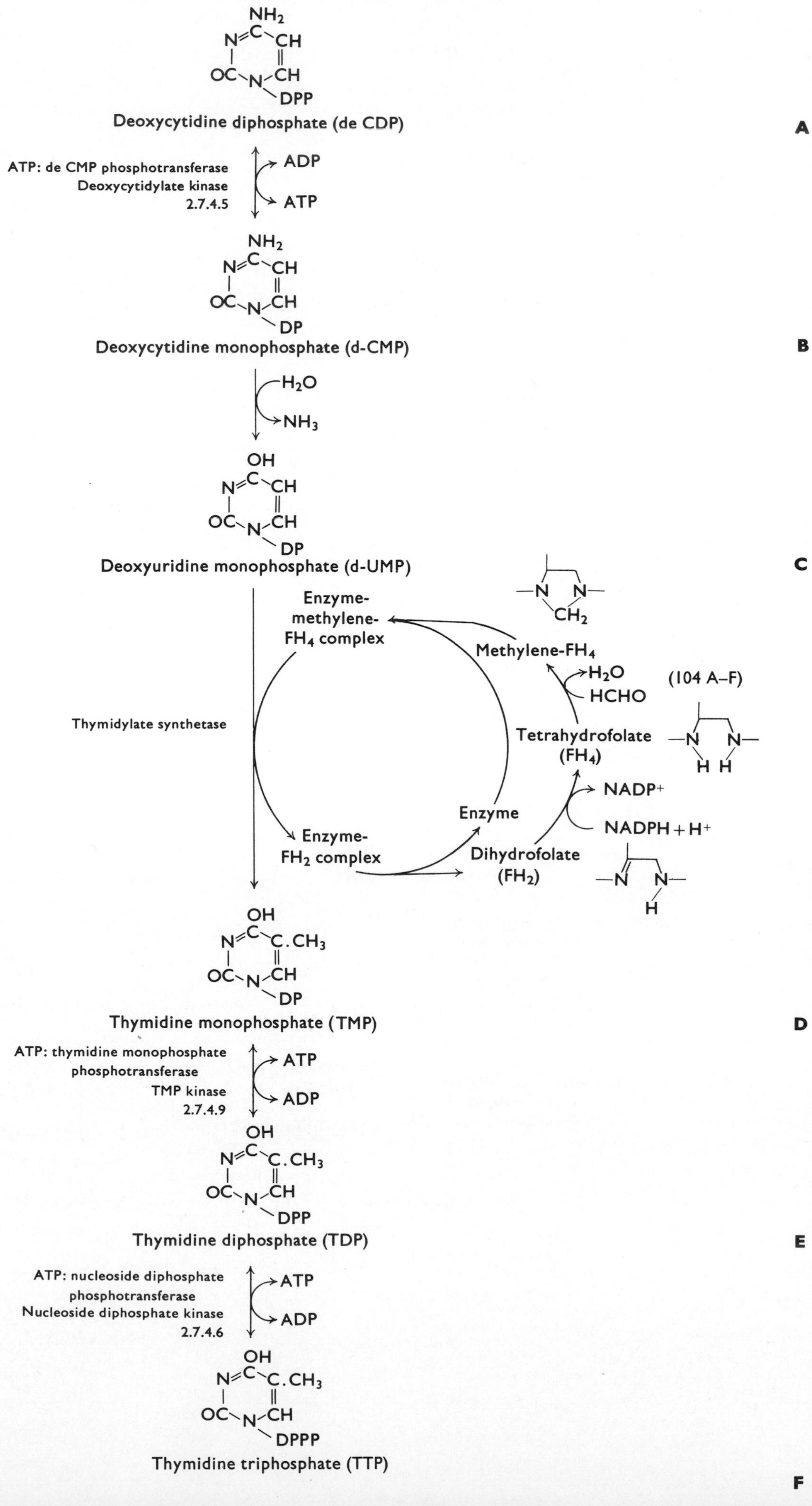

NH_2
N=C–CH
OC–N–CH
DPP
Deoxycytidine diphosphate (de CDP) **A**

ATP: de CMP phosphotransferase
Deoxycytidylate kinase
2.7.4.5
ADP
ATP

NH_2
N=C–CH
OC–N–CH
DP
Deoxycytidine monophosphate (d-CMP) **B**

H_2O
NH_3

OH
N=C–CH
OC–N–CH
DP
Deoxyuridine monophosphate (d-UMP) **C**

Thymidylate synthetase

Enzyme-methylene-FH_4 complex

Methylene-FH_4 (—N N— CH_2)

H_2O
HCHO

(104 A–F)

Tetrahydrofolate (FH_4) (—N N— H H)

$NADP^+$
$NADPH + H^+$

Enzyme

Enzyme-FH_2 complex

Dihydrofolate (FH_2) (—N N— H)

OH
N=C–C.CH_3
OC–N–CH
DP
Thymidine monophosphate (TMP) **D**

ATP: thymidine monophosphate phosphotransferase
TMP kinase
2.7.4.9
ATP
ADP

OH
N=C–C.CH_3
OC–N–CH
DPP
Thymidine diphosphate (TDP) **E**

ATP: nucleoside diphosphate phosphotransferase
Nucleoside diphosphate kinase
2.7.4.6
ATP
ADP

OH
N=C–C.CH_3
OC–N–CH
DPPP
Thymidine triphosphate (TTP)

F

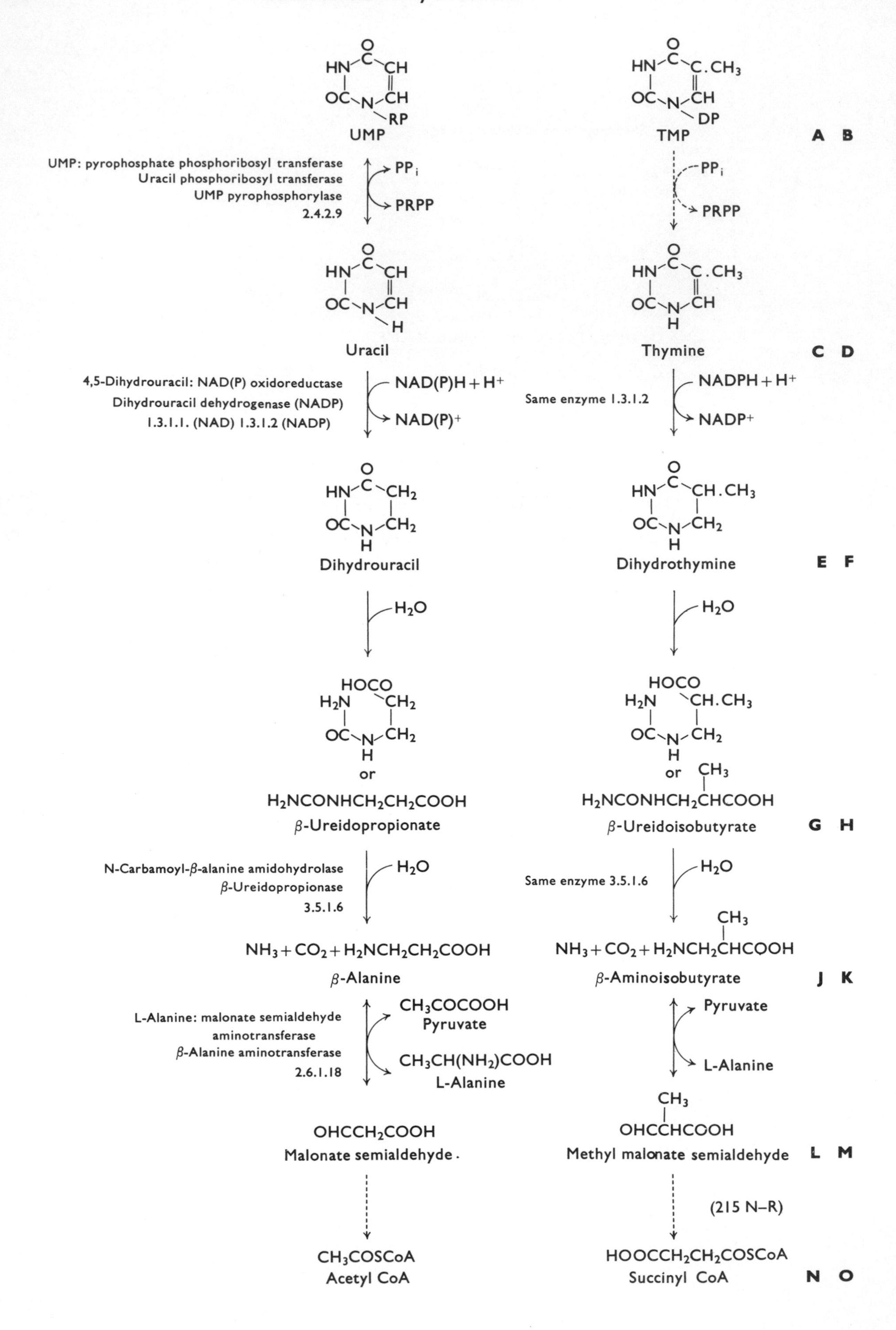
UMP
TMP
A B
UMP: pyrophosphate phosphoribosyl transferase
Uracil phosphoribosyl transferase
UMP pyrophosphorylase
2.4.2.9
PP_i
PRPP
Uracil
Thymine
C D
4,5-Dihydrouracil: NAD(P) oxidoreductase
Dihydrouracil dehydrogenase (NADP)
1.3.1.1. (NAD) 1.3.1.2 (NADP)
$NAD(P)H + H^+$
$NAD(P)^+$
Same enzyme 1.3.1.2
$NADPH + H^+$
$NADP^+$
Dihydrouracil
Dihydrothymine
E F
H_2O
or
$H_2NCONHCH_2CH_2COOH$
β-Ureidopropionate
$H_2NCONHCH_2CH(CH_3)COOH$
β-Ureidoisobutyrate
G H
N-Carbamoyl-β-alanine amidohydrolase
β-Ureidopropionase
3.5.1.6
Same enzyme 3.5.1.6
$NH_3 + CO_2 + H_2NCH_2CH_2COOH$
β-Alanine
$NH_3 + CO_2 + H_2NCH_2CH(CH_3)COOH$
β-Aminoisobutyrate
J K
L-Alanine: malonate semialdehyde
aminotransferase
β-Alanine aminotransferase
2.6.1.18
$CH_3COCOOH$
Pyruvate
$CH_3CH(NH_2)COOH$
L-Alanine
$OHCCH_2COOH$
Malonate semialdehyde
$OHCCH(CH_3)COOH$
Methyl malonate semialdehyde
L M
(215 N–R)
$CH_3COSCoA$
Acetyl CoA
$HOOCCH_2CH_2COSCoA$
Succinyl CoA
N O

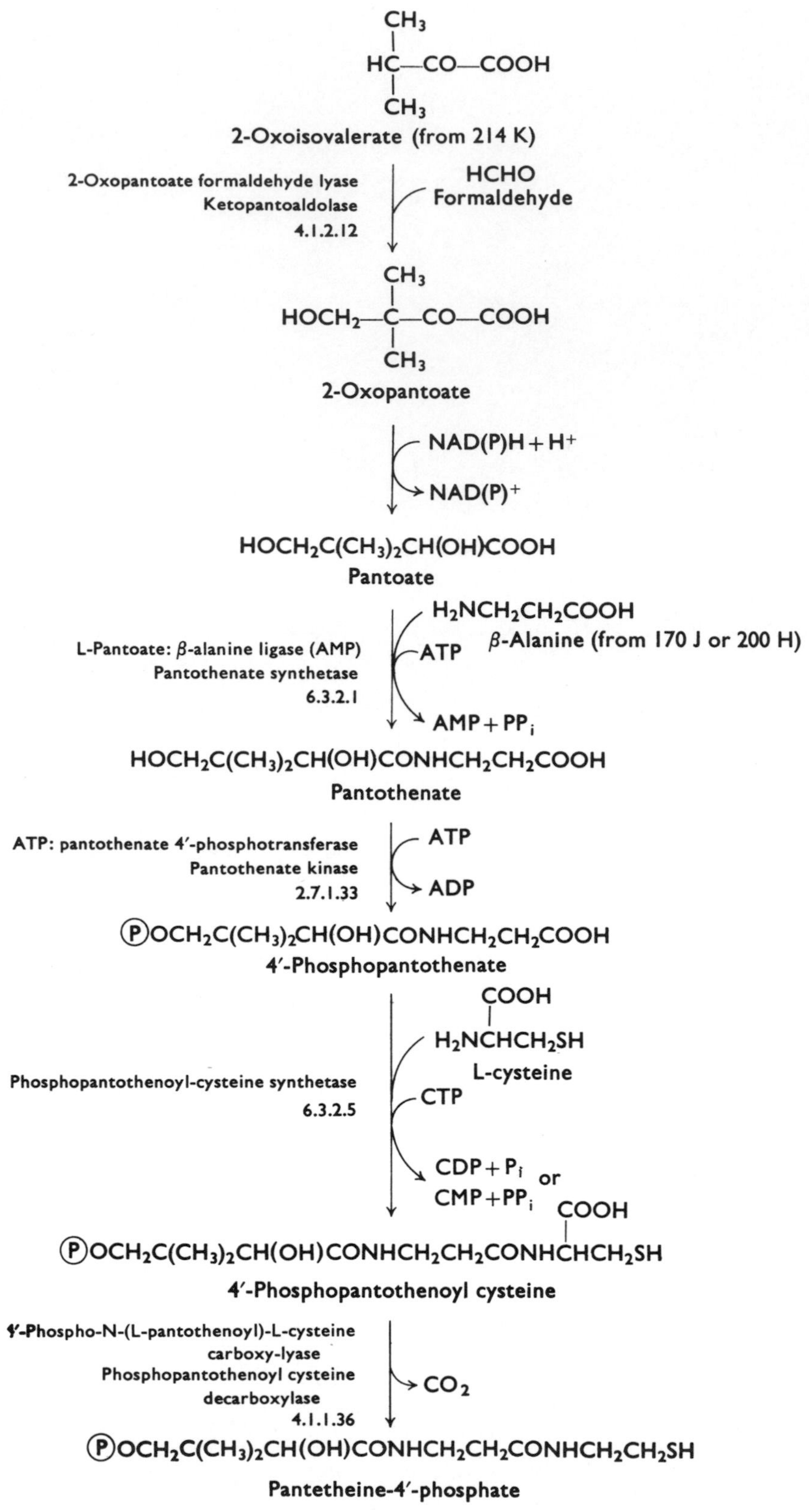

CH_3
|
$HC—CO—COOH$
|
CH_3

2-Oxoisovalerate (from 214 K) **A**

2-Oxopantoate formaldehyde lyase
Ketopantoaldolase
4.1.2.12

↓ ← HCHO Formaldehyde

CH_3
|
$HOCH_2—C—CO—COOH$
|
CH_3

2-Oxopantoate **B**

↓ $NAD(P)H + H^+$ → $NAD(P)^+$

$HOCH_2C(CH_3)_2CH(OH)COOH$

Pantoate **C**

L-Pantoate: β-alanine ligase (AMP)
Pantothenate synthetase
6.3.2.1

↓ ← $H_2NCH_2CH_2COOH$ β-Alanine (from 170 J or 200 H)
← ATP
→ $AMP + PP_i$

$HOCH_2C(CH_3)_2CH(OH)CONHCH_2CH_2COOH$

Pantothenate **D**

ATP: pantothenate 4′-phosphotransferase
Pantothenate kinase
2.7.1.33

↓ ATP → ADP

Ⓟ$OCH_2C(CH_3)_2CH(OH)CONHCH_2CH_2COOH$

4′-Phosphopantothenate **E**

Phosphopantothenoyl-cysteine synthetase
6.3.2.5

↓ ← $H_2NCH(COOH)CH_2SH$ L-cysteine
← CTP
→ $CDP + P_i$ or $CMP + PP_i$

Ⓟ$OCH_2C(CH_3)_2CH(OH)CONHCH_2CH_2CONHCH(COOH)CH_2SH$

4′-Phosphopantothenoyl cysteine **F**

4′-Phospho-N-(L-pantothenoyl)-L-cysteine carboxy-lyase
Phosphopantothenoyl cysteine decarboxylase
4.1.1.36

↓ → CO_2

Ⓟ$OCH_2C(CH_3)_2CH(OH)CONHCH_2CH_2CONHCH_2CH_2SH$

Pantetheine-4′-phosphate **G**

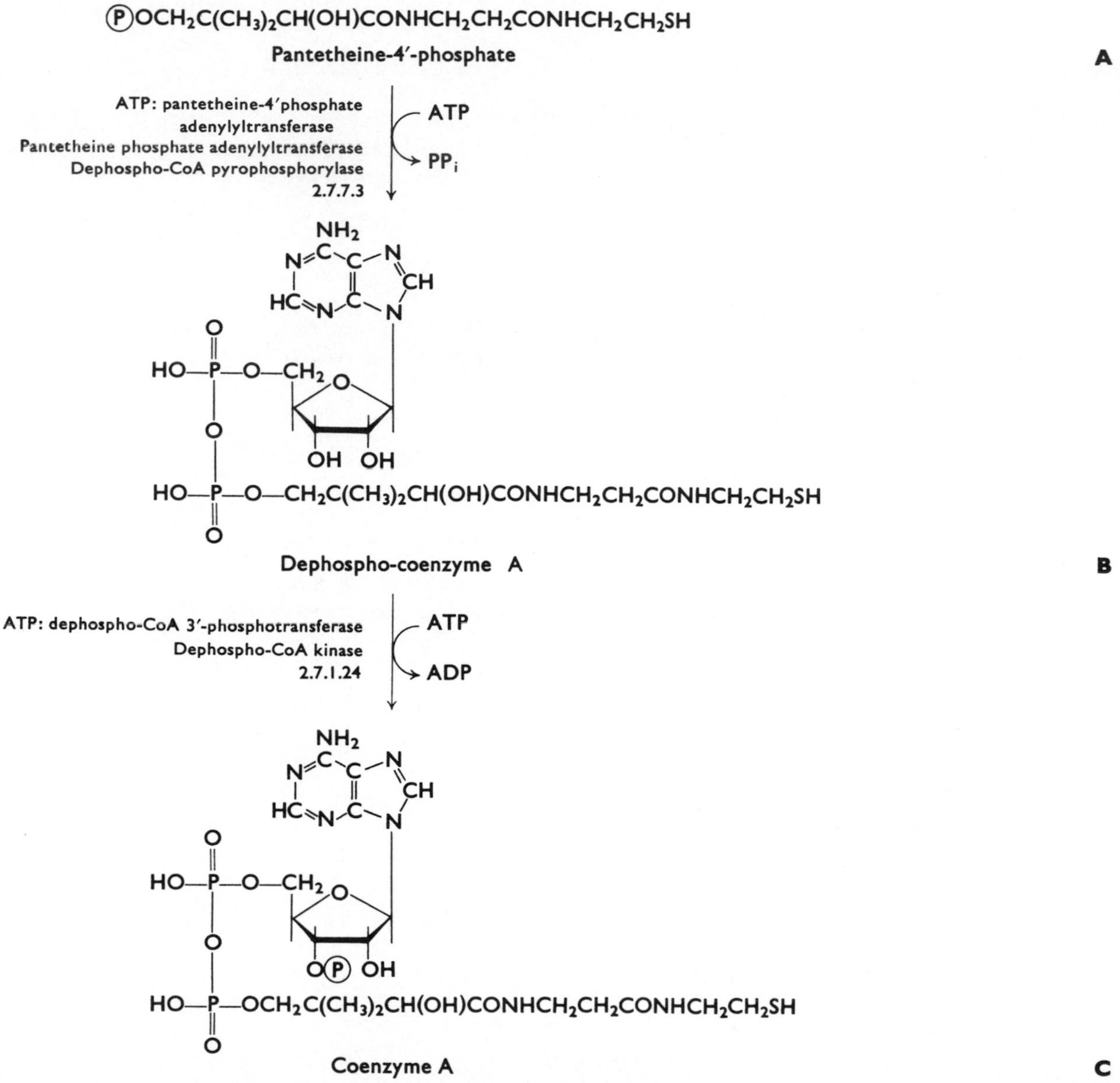

Ⓟ$OCH_2C(CH_3)_2CH(OH)CONHCH_2CH_2CONHCH_2CH_2SH$

Pantetheine-4′-phosphate **A**

ATP: pantetheine-4′phosphate adenylyltransferase
Pantetheine phosphate adenylyltransferase
Dephospho-CoA pyrophosphorylase
2.7.7.3

ATP → PP_i

Dephospho-coenzyme A **B**

ATP: dephospho-CoA 3′-phosphotransferase
Dephospho-CoA kinase
2.7.1.24

ATP → ADP

Coenzyme A **C**

Source of constituents of Coenzyme A molecule

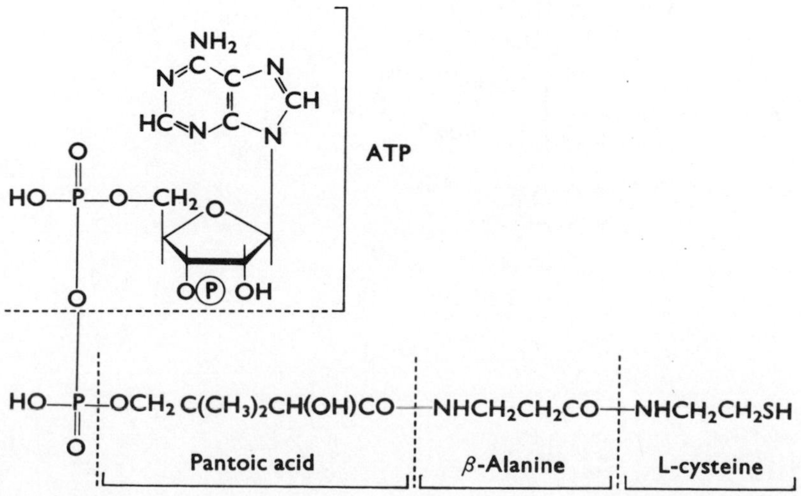

ATP

$HO-P-OCH_2C(CH_3)_2CH(OH)CO$ | $NHCH_2CH_2CO$ | $NHCH_2CH_2SH$

Pantoic acid | β-Alanine | L-cysteine

Biosynthesis of Fatty Acids
(in *E. coli*, yeast, and extra-mitochondrially in mammals)

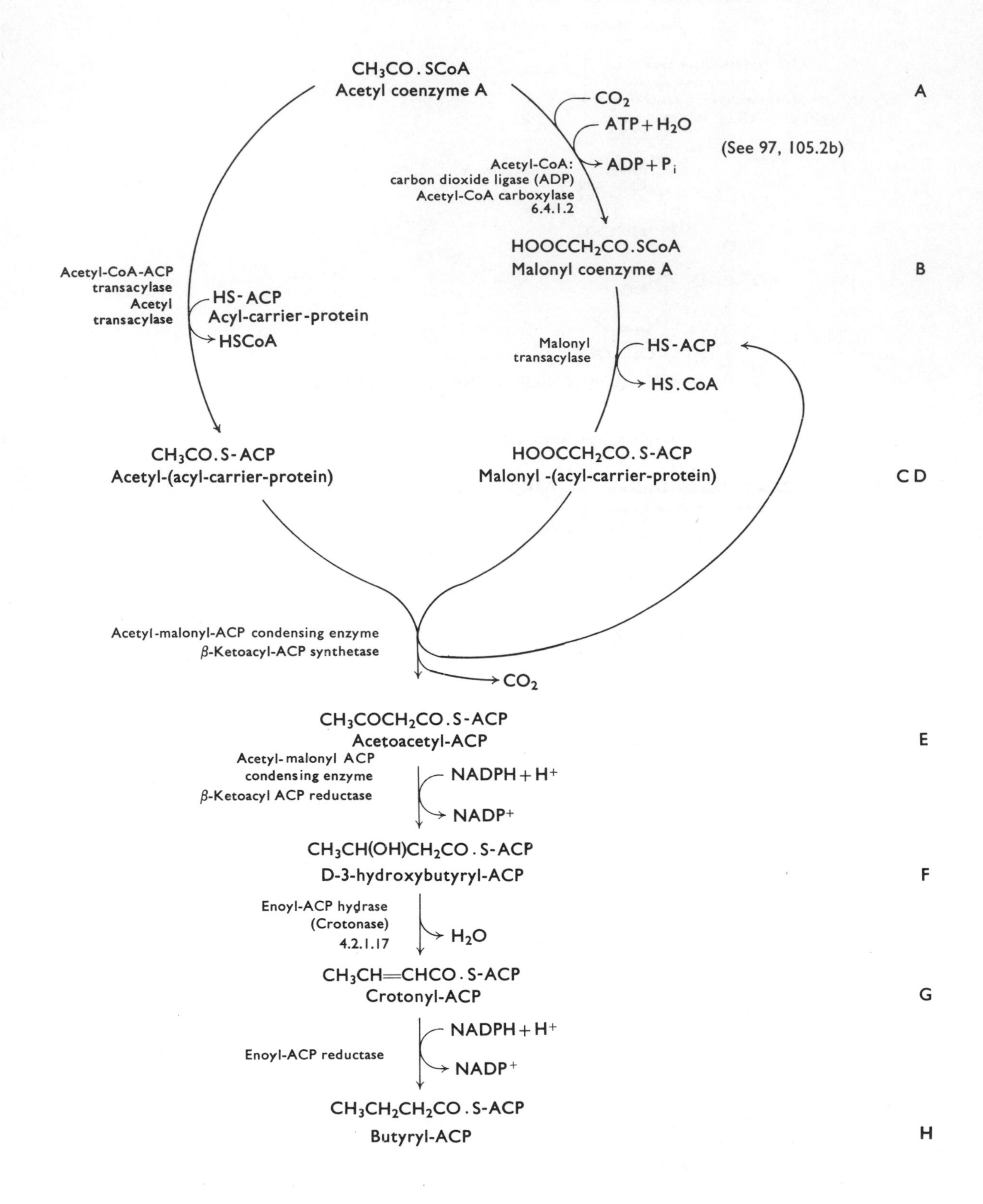

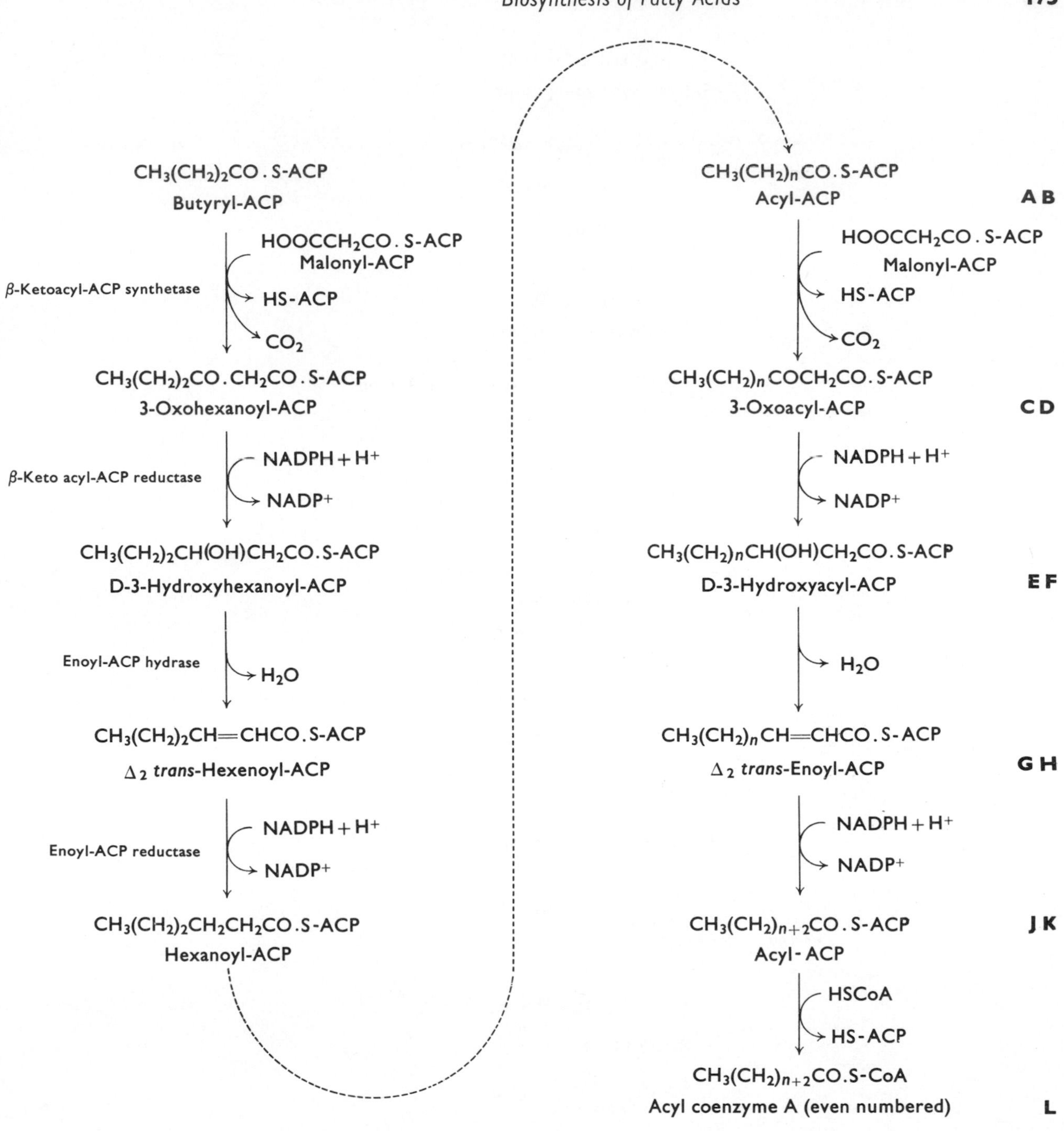

Fatty acids with an odd number of carbon atoms are formed by initial condensation of *propionyl*-ACP with malonyl-ACP

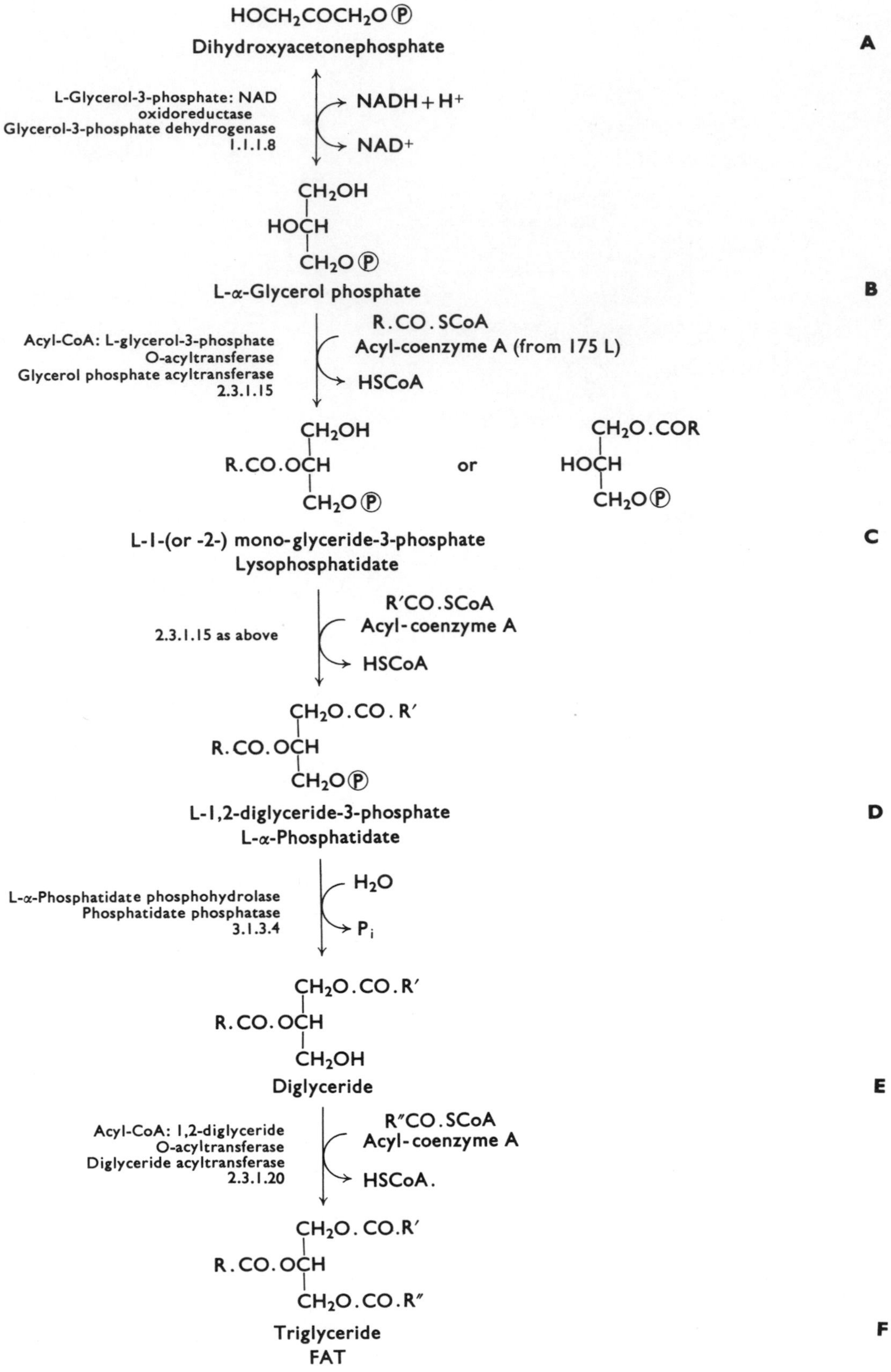

Formation of Di- and Tri-glycerides

$HOCH_2COCH_2O$Ⓟ
Dihydroxyacetonephosphate — A

L-Glycerol-3-phosphate: NAD oxidoreductase
Glycerol-3-phosphate dehydrogenase
1.1.1.8

$\updownarrow$ $NADH + H^+$ / NAD^+

CH_2OH
$HOCH$
CH_2OⓅ
L-α-Glycerol phosphate — B

Acyl-CoA: L-glycerol-3-phosphate O-acyltransferase
Glycerol phosphate acyltransferase
2.3.1.15

$\downarrow$ R.CO.SCoA Acyl-coenzyme A (from 175 L) → HSCoA

CH_2OH
$R.CO.OCH$
CH_2OⓅ

or

$CH_2O.COR$
$HOCH$
CH_2OⓅ

L-1-(or -2-) mono-glyceride-3-phosphate
Lysophosphatidate — C

2.3.1.15 as above

$\downarrow$ R′CO.SCoA Acyl-coenzyme A → HSCoA

$CH_2O.CO.R'$
$R.CO.OCH$
CH_2OⓅ
L-1,2-diglyceride-3-phosphate
L-α-Phosphatidate — D

L-α-Phosphatidate phosphohydrolase
Phosphatidate phosphatase
3.1.3.4

$\downarrow$ H_2O → P_i

$CH_2O.CO.R'$
$R.CO.OCH$
CH_2OH
Diglyceride — E

Acyl-CoA: 1,2-diglyceride O-acyltransferase
Diglyceride acyltransferase
2.3.1.20

$\downarrow$ R″CO.SCoA Acyl-coenzyme A → HSCoA.

$CH_2O.CO.R'$
$R.CO.OCH$
$CH_2O.CO.R''$
Triglyceride
FAT — F

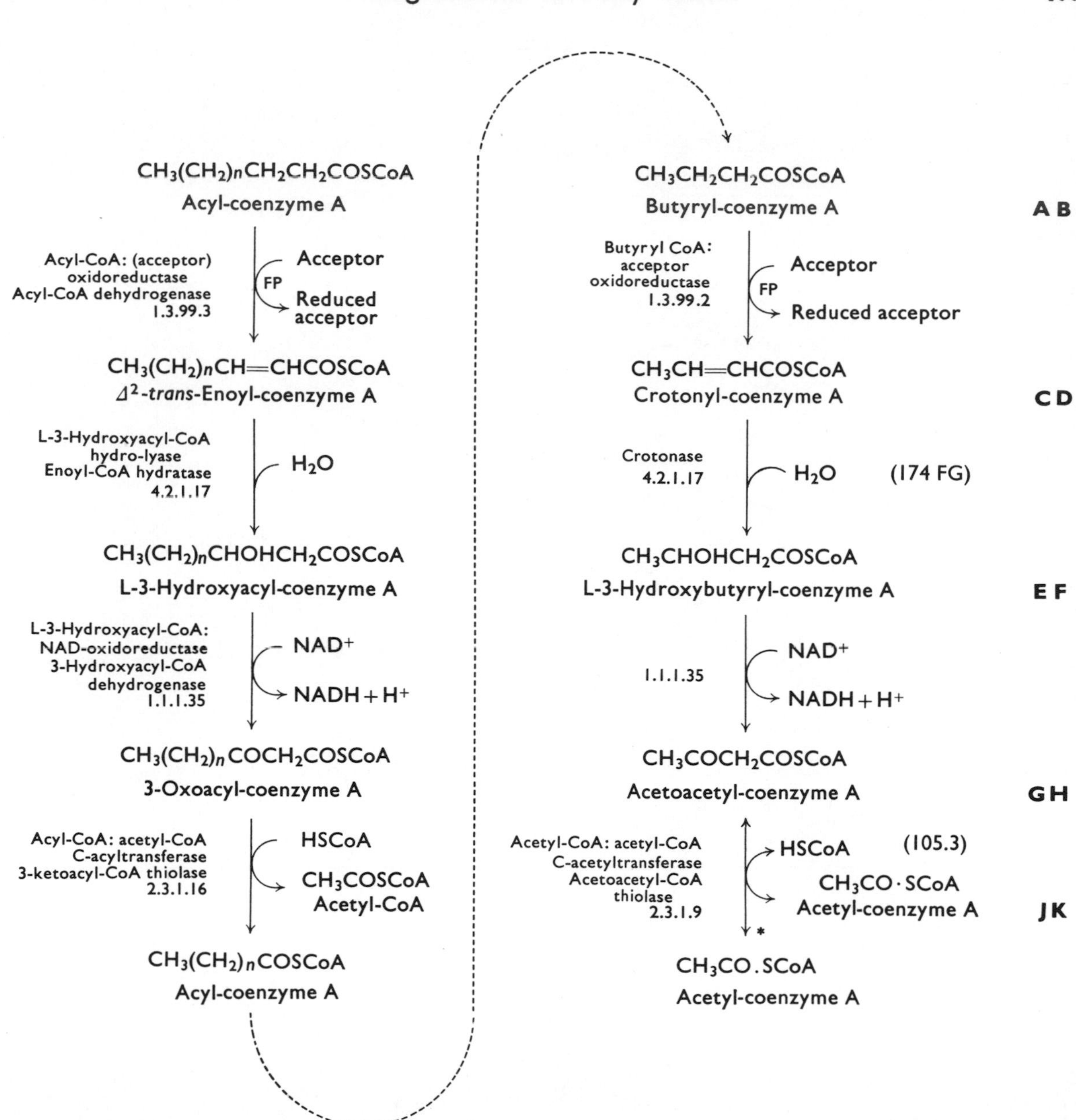

* Fatty acids with an odd number of carbon atoms yield ultimately one molecule of propionyl-CoA which may be metabolized by the pathway 215 N–R.

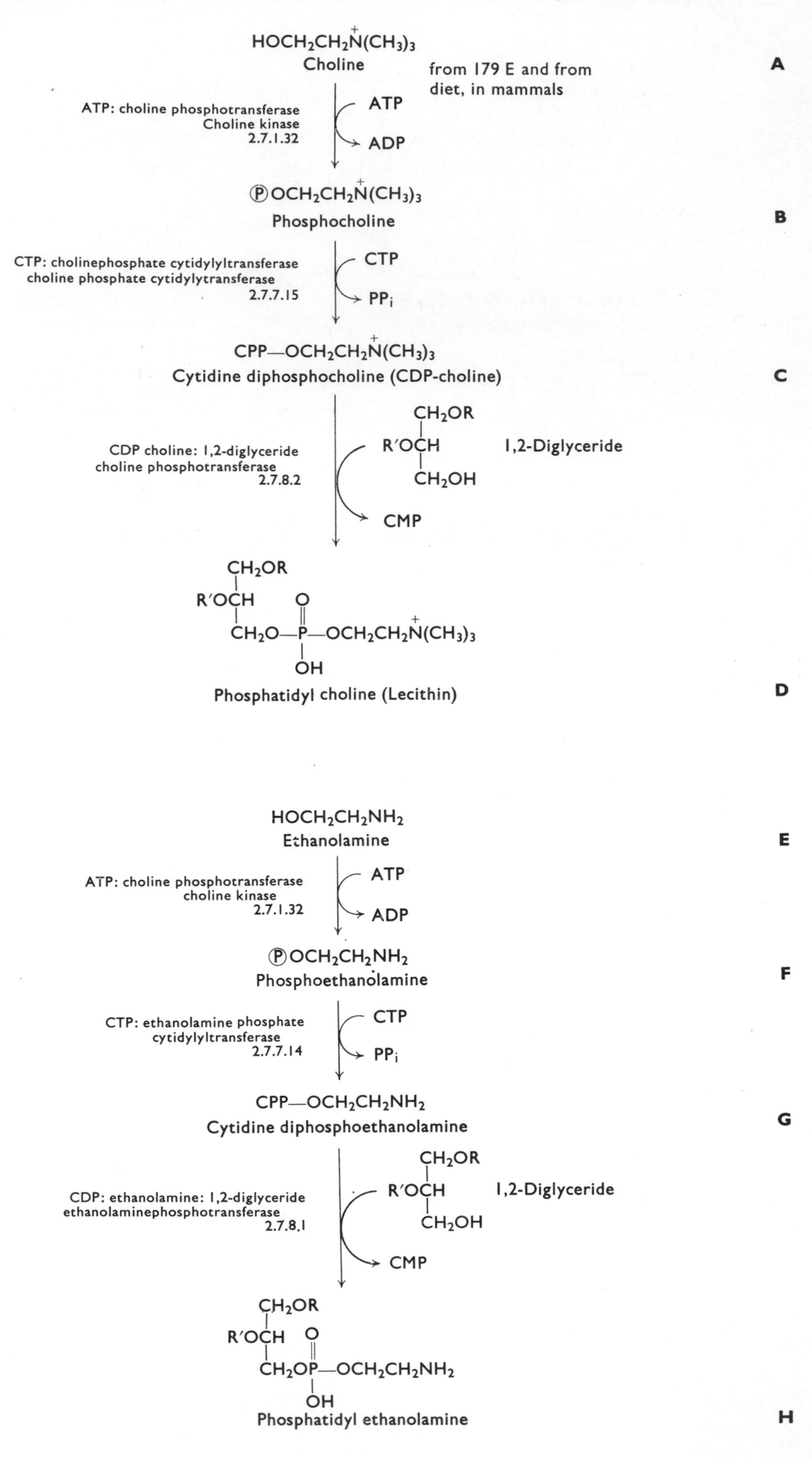
HOCH₂CH₂N⁺(CH₃)₃
Choline
from 179 E and from diet, in mammals
A
ATP: choline phosphotransferase
Choline kinase
2.7.1.32
ATP
ADP
Ⓟ OCH₂CH₂N⁺(CH₃)₃
Phosphocholine
B
CTP: cholinephosphate cytidylyltransferase
choline phosphate cytidylytransferase
2.7.7.15
CTP
PPi
CPP—OCH₂CH₂N⁺(CH₃)₃
Cytidine diphosphocholine (CDP-choline)
C
CDP choline: 1,2-diglyceride
choline phosphotransferase
2.7.8.2
CH₂OR
R′OCH
CH₂OH
1,2-Diglyceride
CMP
CH₂OR
R′OCH
O
CH₂O—P—OCH₂CH₂N⁺(CH₃)₃
OH
Phosphatidyl choline (Lecithin)
D
HOCH₂CH₂NH₂
Ethanolamine
E
ATP: choline phosphotransferase
choline kinase
2.7.1.32
ATP
ADP
Ⓟ OCH₂CH₂NH₂
Phosphoethanolamine
F
CTP: ethanolamine phosphate
cytidylyltransferase
2.7.7.14
CTP
PPi
CPP—OCH₂CH₂NH₂
Cytidine diphosphoethanolamine
G
CDP: ethanolamine: 1,2-diglyceride
ethanolaminephosphotransferase
2.7.8.1
CH₂OR
R′OCH
CH₂OH
1,2-Diglyceride
CMP
CH₂OR
R′OCH
O
CH₂OP—OCH₂CH₂NH₂
OH
Phosphatidyl ethanolamine
H

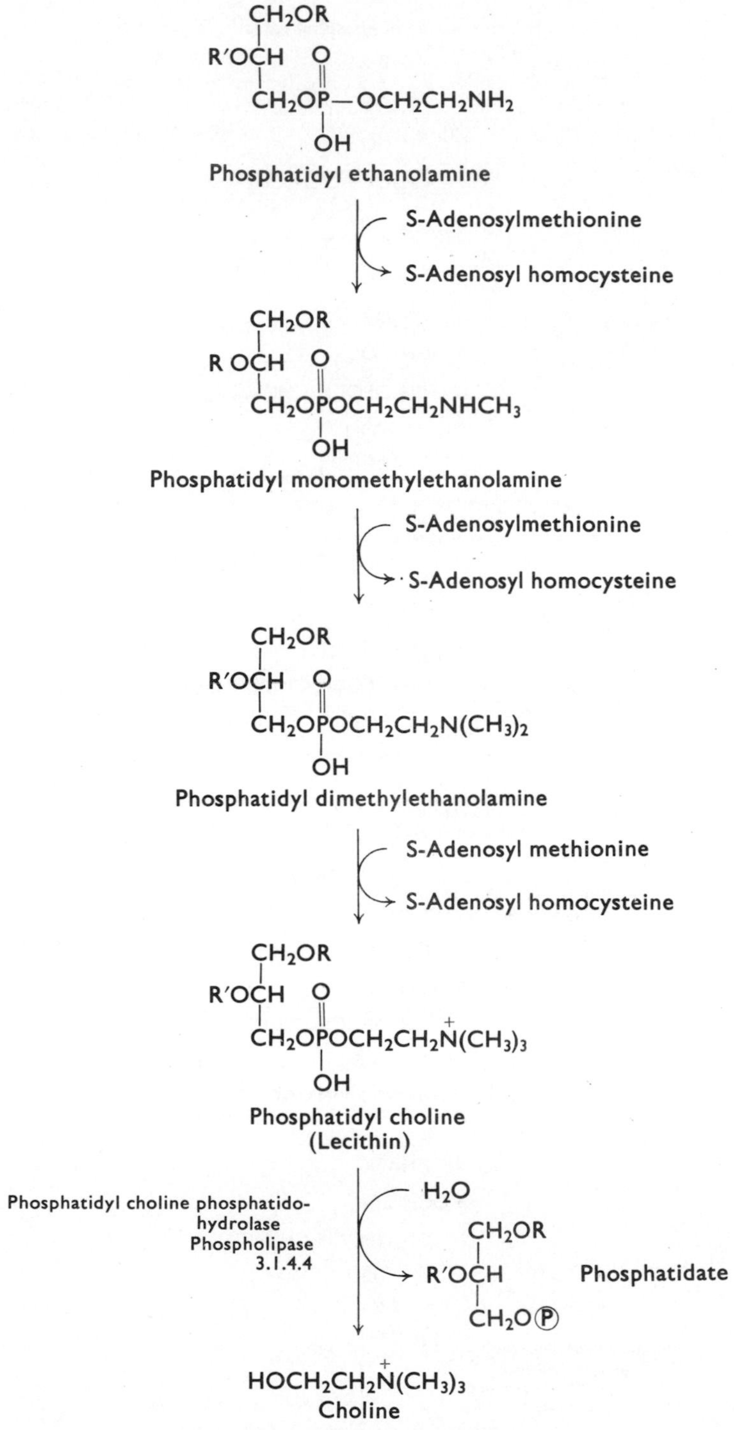

$$\begin{array}{l} CH_2OR \\ | \\ R'OCH \quad O \\ | \qquad\quad \| \\ CH_2OP{-}OCH_2CH_2NH_2 \\ \qquad\quad | \\ \qquad\quad OH \end{array}$$

Phosphatidyl ethanolamine **A**

↓ S-Adenosylmethionine → S-Adenosyl homocysteine

$$\begin{array}{l} CH_2OR \\ | \\ R\,OCH \quad O \\ | \qquad\quad \| \\ CH_2OPOCH_2CH_2NHCH_3 \\ \qquad\quad | \\ \qquad\quad OH \end{array}$$

Phosphatidyl monomethylethanolamine **B**

↓ S-Adenosylmethionine → S-Adenosyl homocysteine

$$\begin{array}{l} CH_2OR \\ | \\ R'OCH \quad O \\ | \qquad\quad \| \\ CH_2OPOCH_2CH_2N(CH_3)_2 \\ \qquad\quad | \\ \qquad\quad OH \end{array}$$

Phosphatidyl dimethylethanolamine **C**

↓ S-Adenosyl methionine → S-Adenosyl homocysteine

$$\begin{array}{l} CH_2OR \\ | \\ R'OCH \quad O \\ | \qquad\quad \| \\ CH_2OPOCH_2CH_2\overset{+}{N}(CH_3)_3 \\ \qquad\quad | \\ \qquad\quad OH \end{array}$$

Phosphatidyl choline
(Lecithin) **D**

Phosphatidyl choline phosphatidohydrolase
Phospholipase
3.1.4.4

↓ H_2O → Phosphatidate

$$\begin{array}{l} CH_2OR \\ | \\ R'OCH \\ | \\ CH_2O\text{Ⓟ} \end{array}$$

$$HOCH_2CH_2\overset{+}{N}(CH_3)_3$$

Choline **E**

(See 178 A, 210 K)

Hydrolysis of Lecithin by Phospholipases

Points of action of phospholipases A, B, C and D

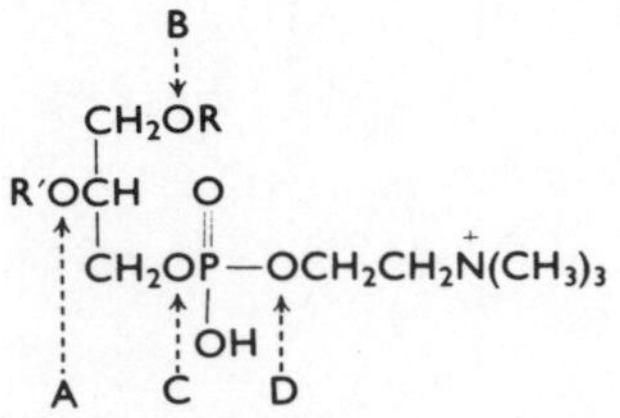

$$\begin{array}{l} \quad\;\; \overset{B}{\downarrow} \\ \;\; CH_2OR \\ R'OCH \quad O \\ \uparrow A \quad\;\; \| \\ \;\; CH_2OP{-}OCH_2CH_2\overset{+}{N}(CH_3)_3 \\ \qquad\; \uparrow C \;\; | \;\; \uparrow D \\ \qquad\quad\; OH \\ A \quad\; C \quad\; D \end{array}$$

A

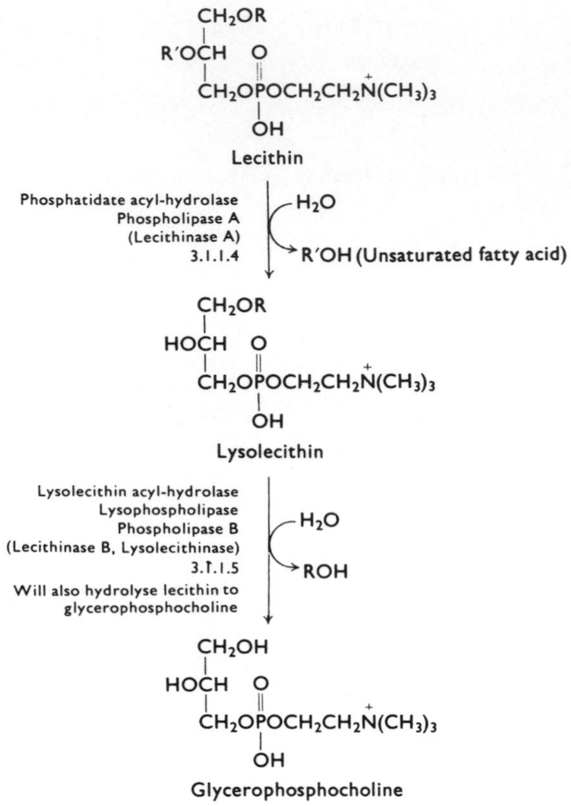

$$\begin{array}{l} CH_2OR \\ R'OCH \quad O \\ \qquad\quad\; \| \\ CH_2OPOCH_2CH_2\overset{+}{N}(CH_3)_3 \\ \qquad\quad\; | \\ \qquad\quad OH \end{array}$$

Lecithin

B

Phosphatidate acyl-hydrolase
Phospholipase A
(Lecithinase A)
3.1.1.4

H_2O → $R'OH$ (Unsaturated fatty acid)

$$\begin{array}{l} CH_2OR \\ HOCH \quad O \\ \qquad\quad\; \| \\ CH_2OPOCH_2CH_2\overset{+}{N}(CH_3)_3 \\ \qquad\quad\; | \\ \qquad\quad OH \end{array}$$

Lysolecithin

C

Lysolecithin acyl-hydrolase
Lysophospholipase
Phospholipase B
(Lecithinase B, Lysolecithinase)
3.1.1.5
Will also hydrolyse lecithin to glycerophosphocholine

H_2O → ROH

$$\begin{array}{l} CH_2OH \\ HOCH \quad O \\ \qquad\quad\; \| \\ CH_2OPOCH_2CH_2\overset{+}{N}(CH_3)_3 \\ \qquad\quad\; | \\ \qquad\quad OH \end{array}$$

Glycerophosphocholine

D

$$\begin{array}{l} CH_2OR \\ R'OCH \quad O \\ \qquad\quad\; \| \\ CH_2OPOCH_2CH_2\overset{+}{N}(CH_3)_3 \\ \qquad\quad\; | \\ \qquad\quad OH \end{array}$$

Lecithin

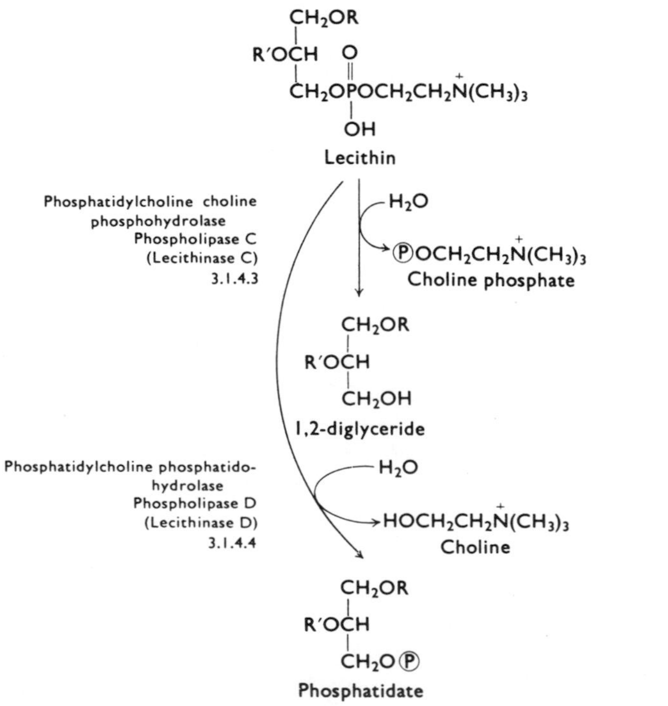

E

Phosphatidylcholine choline phosphohydrolase
Phospholipase C
(Lecithinase C)
3.1.4.3

H_2O → Ⓟ$OCH_2CH_2\overset{+}{N}(CH_3)_3$
Choline phosphate

$$\begin{array}{l} CH_2OR \\ R'OCH \\ CH_2OH \end{array}$$

1,2-diglyceride

F

Phosphatidylcholine phosphatido-hydrolase
Phospholipase D
(Lecithinase D)
3.1.4.4

H_2O → $HOCH_2CH_2\overset{+}{N}(CH_3)_3$
Choline

$$\begin{array}{l} CH_2OR \\ R'OCH \\ CH_2O\text{Ⓟ} \end{array}$$

Phosphatidate

G

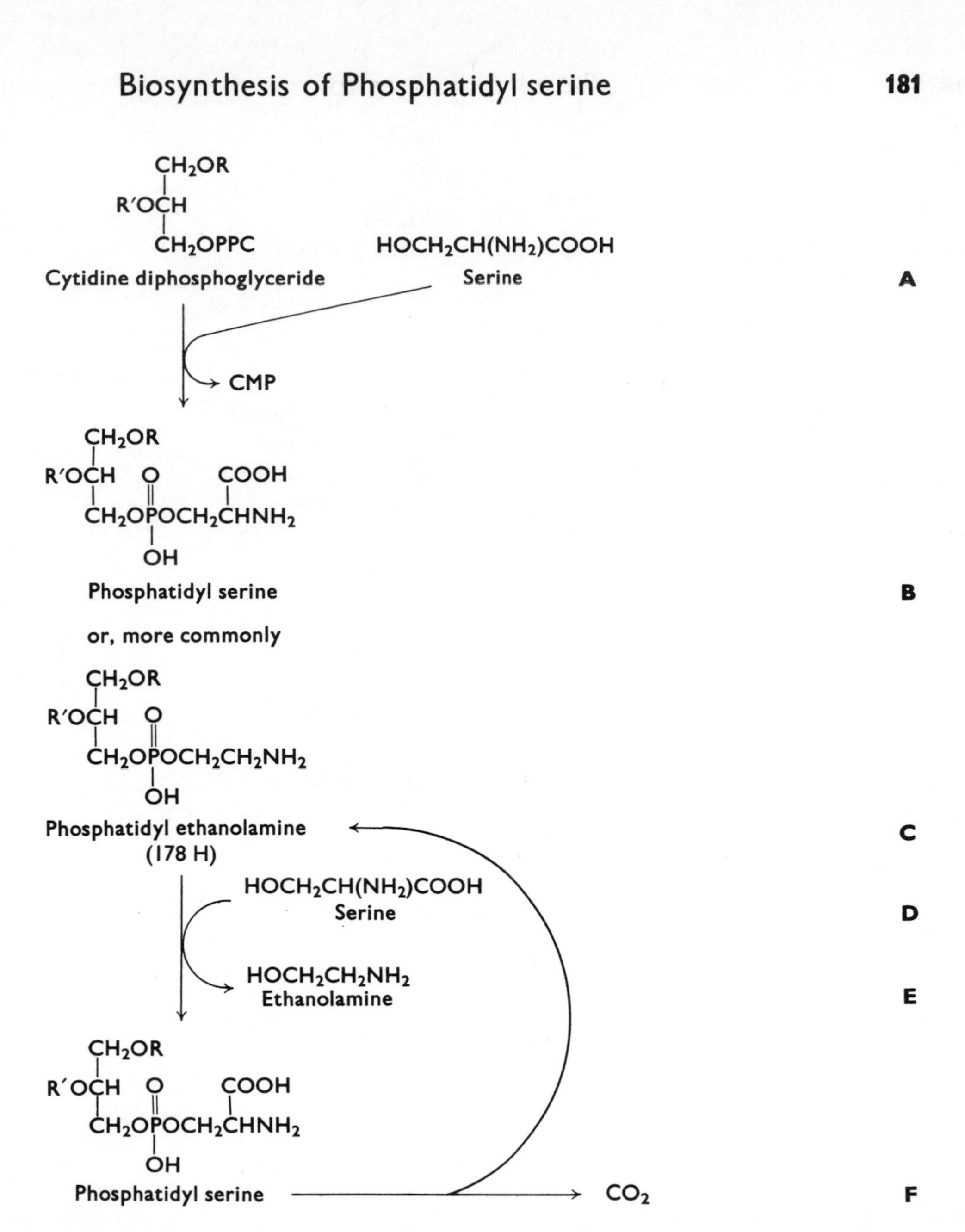
CH2OR
R′OCH
CH2OPPC
Cytidine diphosphoglyceride
HOCH2CH(NH2)COOH
Serine
A
CMP
CH2OR
R′OCH O COOH
CH2OPOCH2CHNH2
OH
Phosphatidyl serine
B
or, more commonly
CH2OR
R′OCH O
CH2OPOCH2CH2NH2
OH
Phosphatidyl ethanolamine
(178 H)
C
HOCH2CH(NH2)COOH
Serine
D
HOCH2CH2NH2
Ethanolamine
E
CH2OR
R′OCH O COOH
CH2OPOCH2CHNH2
OH
Phosphatidyl serine
CO2
F

Phosphatidyl Inositol

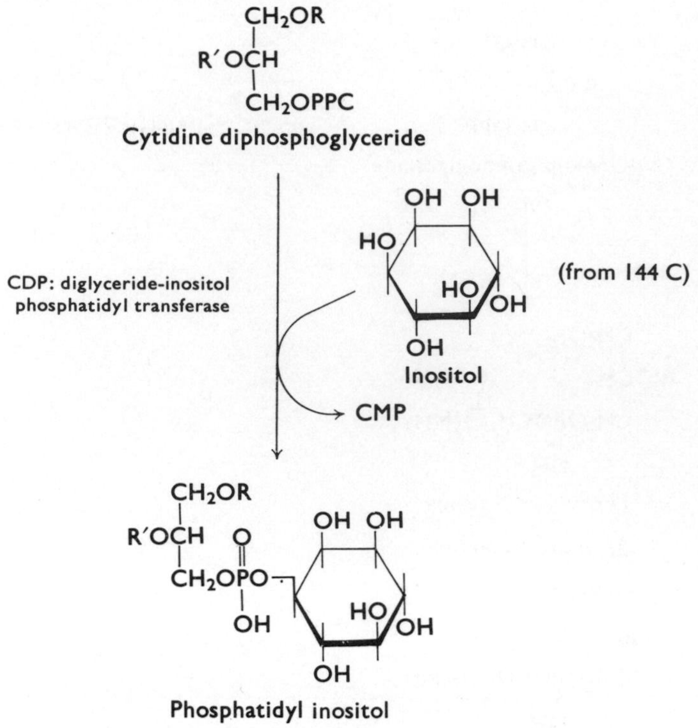

CH_2OR
$R'OCH$
CH_2OPPC

Cytidine diphosphoglyceride

A

CDP: diglyceride-inositol phosphatidyl transferase

Inositol (from 144 C)

→ CMP

CH_2OR
$R'OCH$
$CH_2OP(=O)(OH)O$–inositol

Phosphatidyl inositol

B

$CH_3(CH_2)_{14}COSCoA$
Palmitoyl coenzyme A **A**

$\downarrow$ $NADPH + H^+$ → $HSCoA$, $NADP^+$

$CH_3(CH_2)_{14}CHO$
Palmitaldehyde **B**

$\downarrow$ $HOOC{-}CH(NH_2)CH_2OH$ (L-Serine); Mn^{++} Pyr. → CO_2

$CH_3(CH_2)_{14}CH(OH)CH(NH_2)CH_2OH$
Dihydrosphingosine **C**

$\downarrow$ Flavoprotein → Flavoprotein 2H

$CH_3(CH_2)_{12}CH{=}CH.CH(OH)CH(NH_2)CH_2OH$
Sphingosine **D**

$\downarrow$ Acyl-CoA → HSCoA

$CH_3(CH_2)_{12}CH{=}CHCH(OH)CH(NHAcyl)CH_2OH$
N-Acylsphingosine
Ceramide **E**

$\downarrow$ $CPP{-}OCH_2CH_2\overset{+}{N}(CH_3)_3$ (CDP-choline) → CMP

$CH_3(CH_2)_{12}CH{=}CHCH(OH)CH(NHAcyl)CH_2OP(=O)(OH)OCH_2CH_2\overset{+}{N}(CH_3)_3$
Sphingomyelin **F**

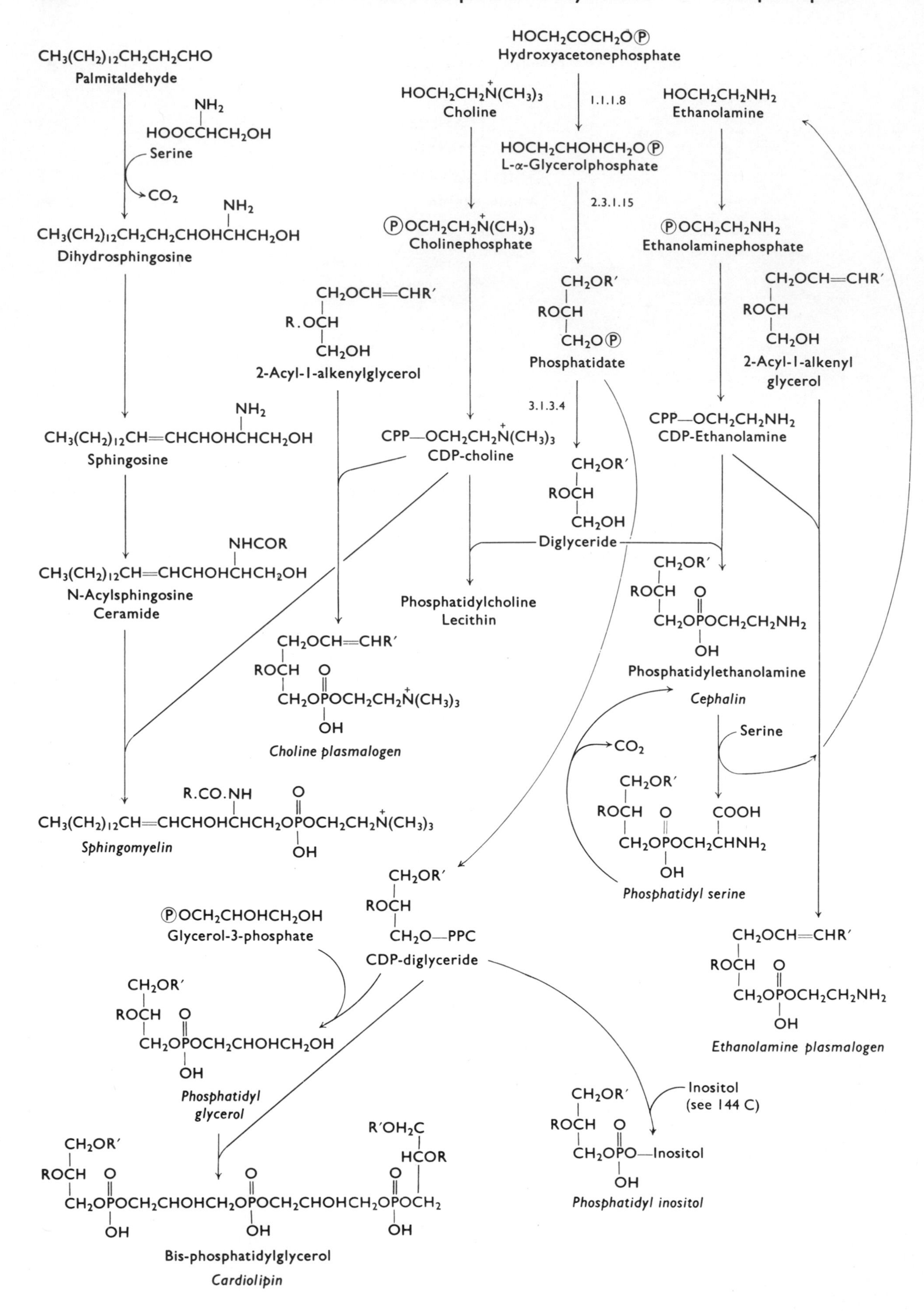

HOCH2COCH2OⓅ
Hydroxyacetonephosphate
CH3(CH2)12CH2CH2CHO
Palmitaldehyde
HOCH2CH2N(CH3)3
Choline
1.1.1.8
HOCH2CH2NH2
Ethanolamine
HOOCCHCH2OH
NH2
Serine
CO2
HOCH2CHOHCH2OⓅ
L-α-Glycerolphosphate
2.3.1.15
CH3(CH2)12CH2CH2CHOHCHCH2OH
Dihydrosphingosine
ⓅOCH2CH2N(CH3)3
Cholinephosphate
ⓅOCH2CH2NH2
Ethanolaminephosphate
CH2OCH=CHR'
R.OCH
CH2OH
2-Acyl-1-alkenylglycerol
CH2OR'
ROCH
CH2OⓅ
Phosphatidate
2-Acyl-1-alkenyl glycerol
3.1.3.4
CH3(CH2)12CH=CHCHOHCHCH2OH
Sphingosine
CPP—OCH2CH2N(CH3)3
CDP-choline
CPP—OCH2CH2NH2
CDP-Ethanolamine
CH2OH
Diglyceride
NHCOR
CH3(CH2)12CH=CHCHOHCHCH2OH
N-Acylsphingosine
Ceramide
Phosphatidylcholine
Lecithin
CH2OPOCH2CH2NH2
OH
Phosphatidylethanolamine
Cephalin
CH2OPOCH2CH2N(CH3)3
Choline plasmalogen
Serine
COOH
CH2OPOCH2CHNH2
Phosphatidyl serine
R.CO.NH
CH3(CH2)12CH=CHCHOHCHCH2OPOCH2CH2N(CH3)3
Sphingomyelin
ⓅOCH2CHOHCH2OH
Glycerol-3-phosphate
CH2O—PPC
CDP-diglyceride
Ethanolamine plasmalogen
CH2OPOCH2CHOHCH2OH
Phosphatidyl glycerol
Inositol
(see 144 C)
CH2OPO—Inositol
Phosphatidyl inositol
R'OH2C
HCOR
CH2OPOCH2CHOHCH2OPOCH2CHOHCH2OPOCH2
Bis-phosphatidylglycerol
Cardiolipin

Breakdown of Phospholipids

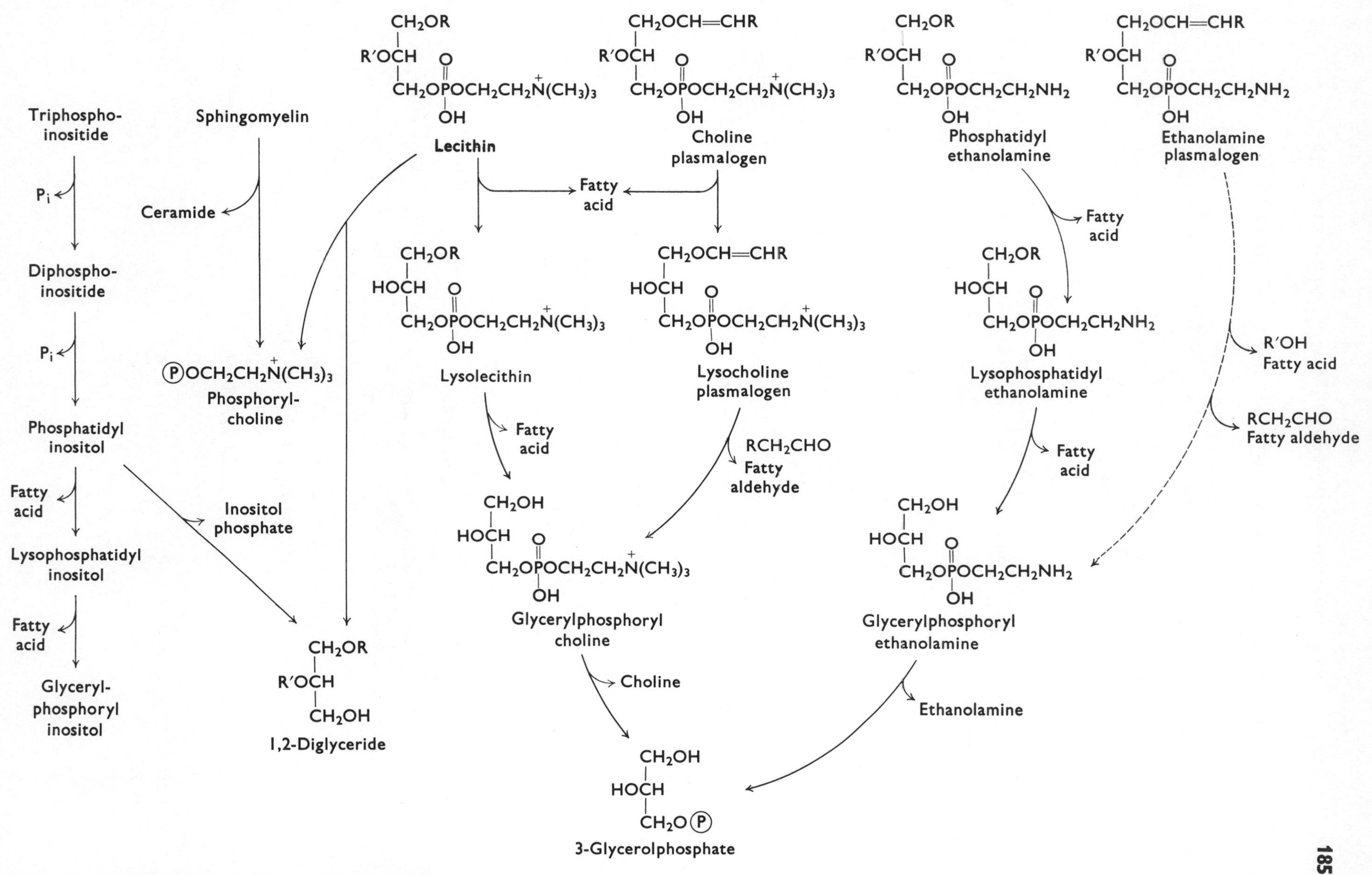

All these reactions are hydrolytic fissions: the participating H_2O has been omitted.

Several steps in the sequence are uncertain. One possible pathway is presented

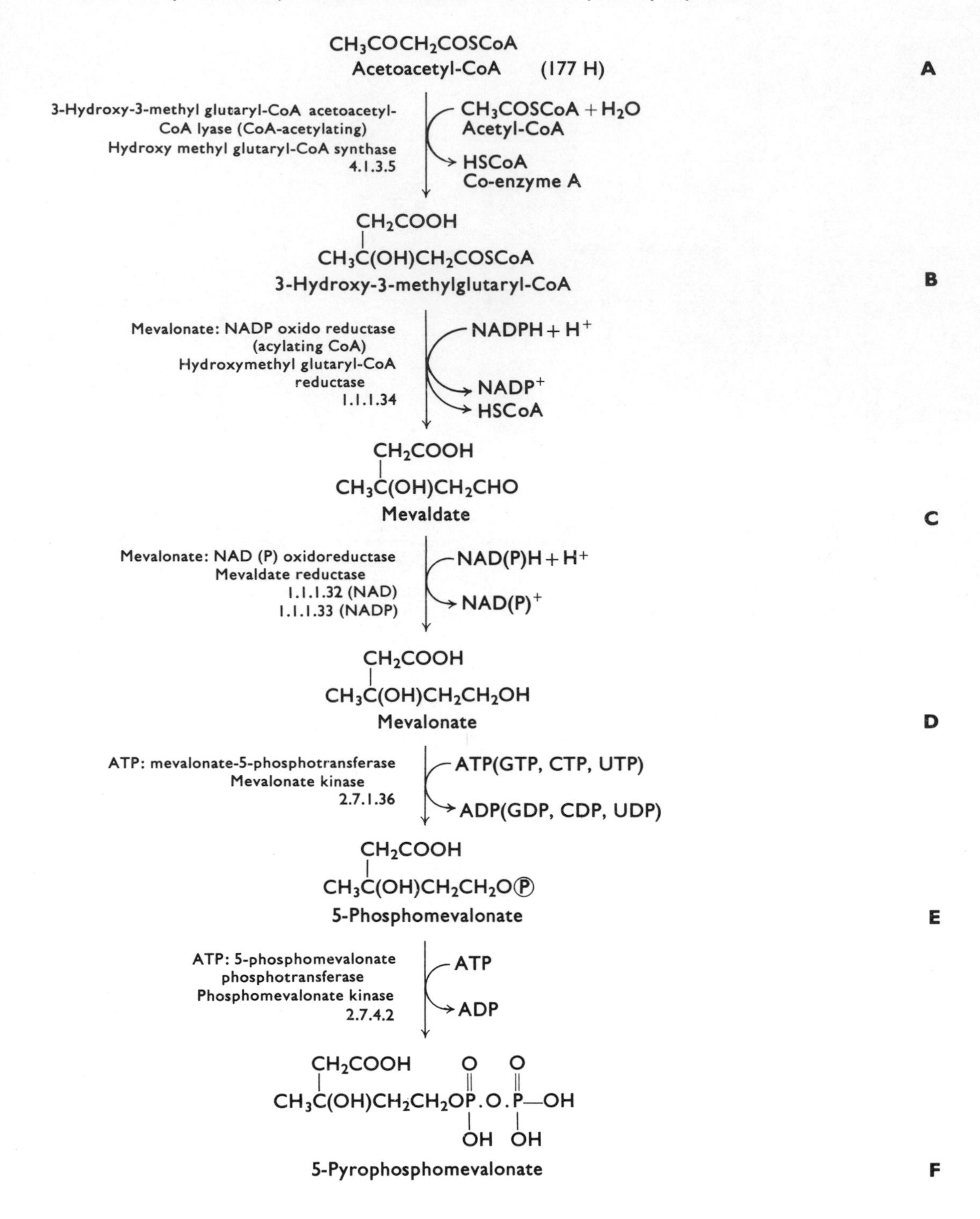

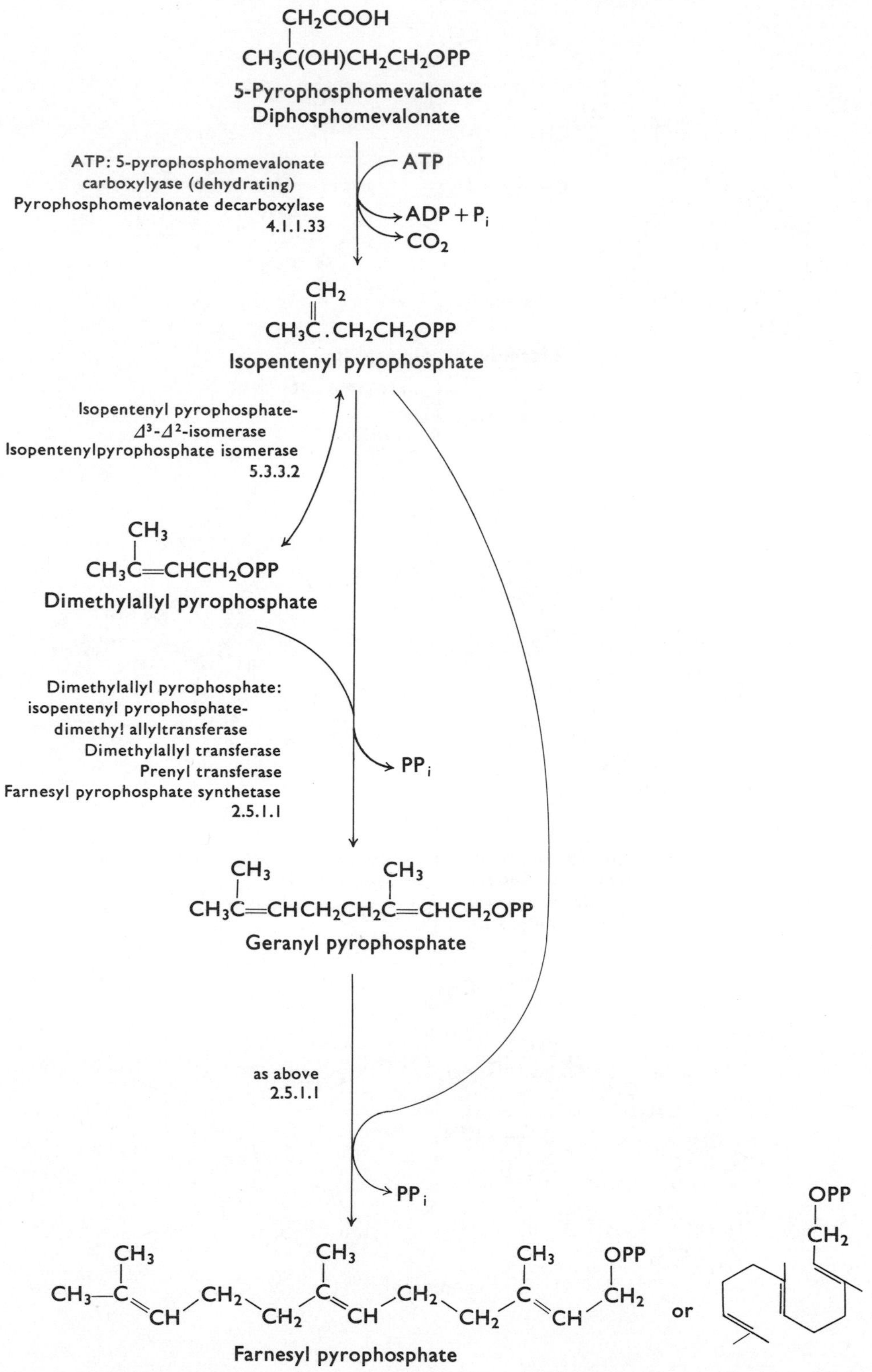

CH_2COOH
$CH_3C(OH)CH_2CH_2OPP$

5-Pyrophosphomevalonate
Diphosphomevalonate

A

ATP: 5-pyrophosphomevalonate carboxylyase (dehydrating)
Pyrophosphomevalonate decarboxylase
4.1.1.33

ATP → ADP + P_i, CO_2

CH_2
$CH_3C.CH_2CH_2OPP$

Isopentenyl pyrophosphate

B

Isopentenyl pyrophosphate-Δ^3-Δ^2-isomerase
Isopentenylpyrophosphate isomerase
5.3.3.2

CH_3
$CH_3C{=}CHCH_2OPP$

Dimethylallyl pyrophosphate

C

Dimethylallyl pyrophosphate: isopentenyl pyrophosphate-dimethyl allyltransferase
Dimethylallyl transferase
Prenyl transferase
Farnesyl pyrophosphate synthetase
2.5.1.1

PP_i

CH_3 CH_3
$CH_3C{=}CHCH_2CH_2C{=}CHCH_2OPP$

Geranyl pyrophosphate

D

as above
2.5.1.1

PP_i

$CH_3{-}C(CH_3){=}CH{-}CH_2{-}CH_2{-}C(CH_3){=}CH{-}CH_2{-}CH_2{-}C(CH_3){=}CH{-}CH_2{-}OPP$ or [skeletal structure with CH_2 and OPP]

Farnesyl pyrophosphate

E

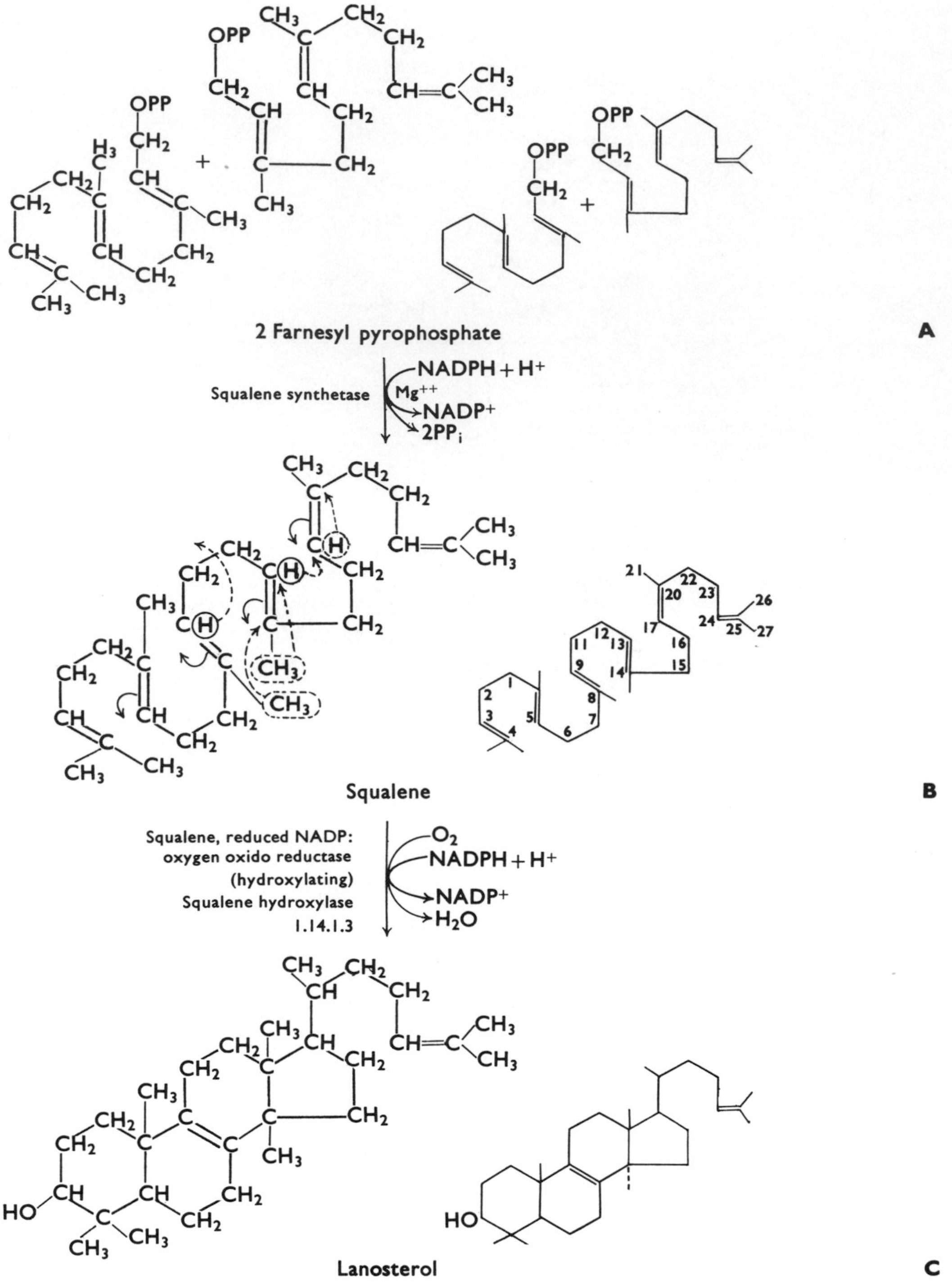

2 Farnesyl pyrophosphate

A

Squalene synthetase; $NADPH + H^+$; Mg^{++}; $NADP^+$; $2PP_i$

Squalene

B

Squalene, reduced NADP: oxygen oxido reductase (hydroxylating) Squalene hydroxylase 1.14.1.3; O_2; $NADPH + H^+$; $NADP^+$; H_2O

Lanosterol

C

The conversion of lanosterol into cholesterol is still uncertain but may proceed by the following sequence. The reactions involve the removal of three methyl groups (C-4, C-4 and C-14), the shift of the Δ^8 double bond to the Δ^5 position, and the reduction of the side chain double bond.

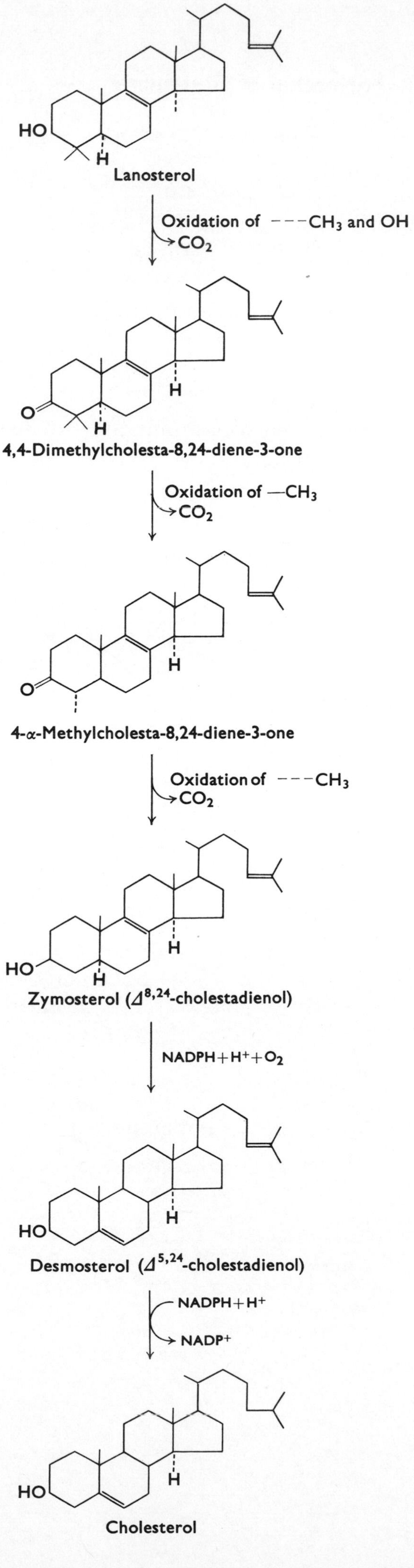

HO
H
Lanosterol

A

Oxidation of ---CH_3 and OH
$\rightarrow CO_2$

H
O
H
4,4-Dimethylcholesta-8,24-diene-3-one

B

Oxidation of —CH_3
$\rightarrow CO_2$

H
O
4-α-Methylcholesta-8,24-diene-3-one

C

Oxidation of ---CH_3
$\rightarrow CO_2$

H
HO
H
Zymosterol ($\Delta^{8,24}$-cholestadienol)

D

NADPH+H^++O_2

H
HO
Desmosterol ($\Delta^{5,24}$-cholestadienol)

E

NADPH+H^+
$\rightarrow$ $NADP^+$

H
HO
Cholesterol

F

Formation of Glutamate

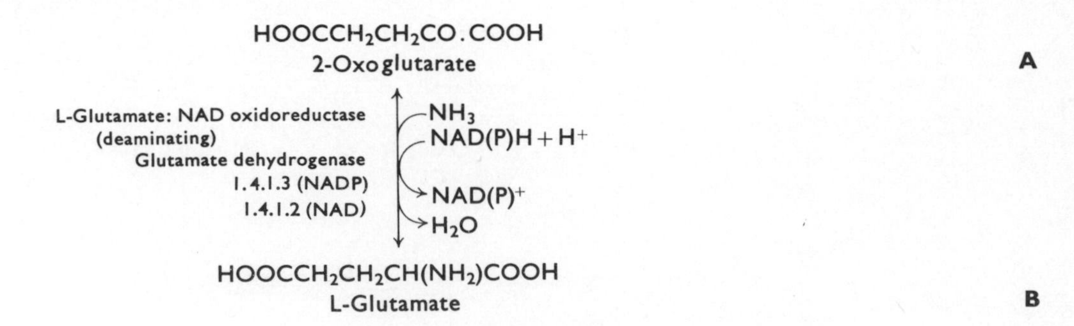

A

B

Formation of Glutamate from Aspartate

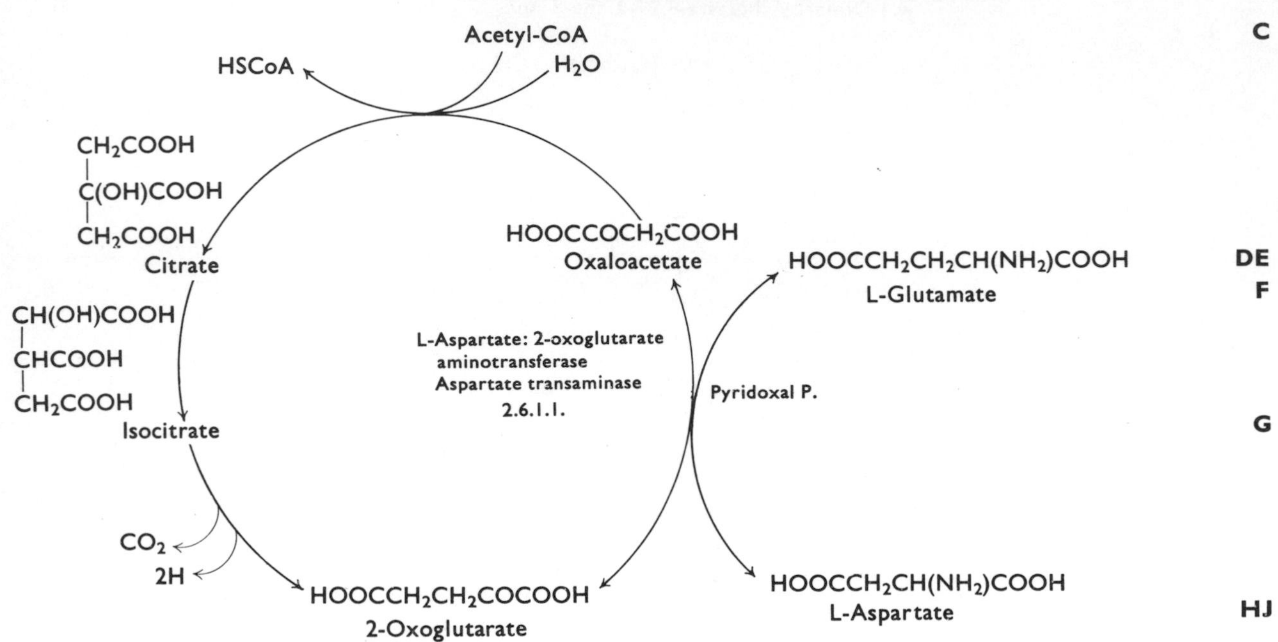

C

DE
F

G

HJ

Net result

$$HOOCCH_2CH(NH_2)COOH + CH_3COSCoA + H_2O \longrightarrow HOOCCH_2CH_2CH(NH_2)COOH + CO_2 + 2H + HSCoA$$

Aspartate — Glutamate

Synthesis of Glutamine

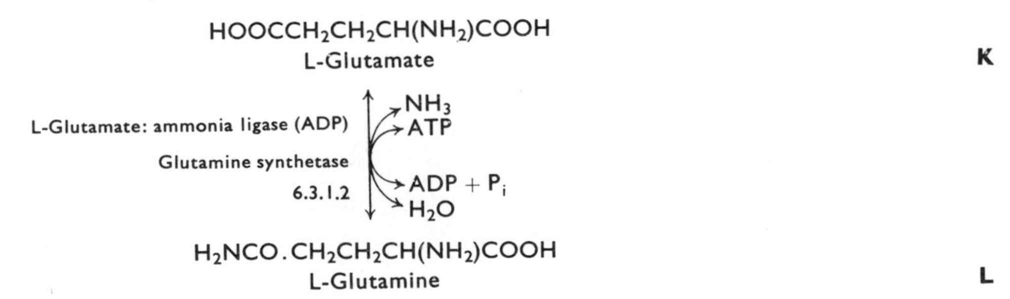

K

L

Reactions of Glutamate
(in *C. tetanomorphum*)

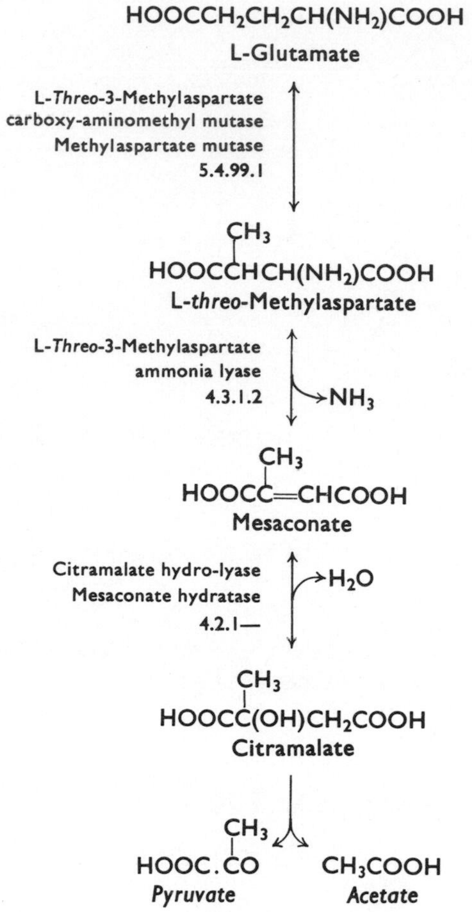

$HOOCCH_2CH_2CH(NH_2)COOH$
L-Glutamate — **A**

↕ L-*Threo*-3-Methylaspartate carboxy-aminomethyl mutase
Methylaspartate mutase
5.4.99.1

$HOOCCH(CH_3)CH(NH_2)COOH$
L-*threo*-Methylaspartate — **B**

↕ L-*Threo*-3-Methylaspartate ammonia lyase
4.3.1.2
→ NH_3

$HOOCC(CH_3){=}CHCOOH$
Mesaconate — **C**

↕ Citramalate hydro-lyase
Mesaconate hydratase
4.2.1—
→ H_2O

$HOOCC(CH_3)(OH)CH_2COOH$
Citramalate — **D**

↓

$HOOC.CO.CH_3$ — *Pyruvate*
CH_3COOH — *Acetate* — **E F**

(in *E. coli*)

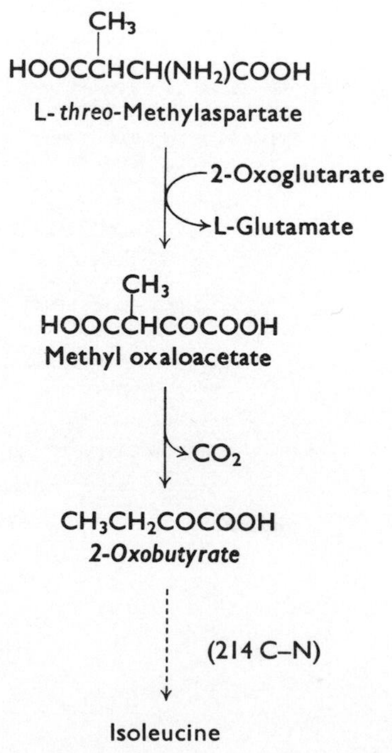

$HOOCCH(CH_3)CH(NH_2)COOH$
L-*threo*-Methylaspartate — **G**

↓ 2-Oxoglutarate → L-Glutamate

$HOOCCH(CH_3)COCOOH$
Methyl oxaloacetate — **H**

↓ → CO_2

$CH_3CH_2COCOOH$
2-Oxobutyrate — **J**

⇣ (214 C–N)

Isoleucine — **K**

Oxidation of Glutamate to Aspartate

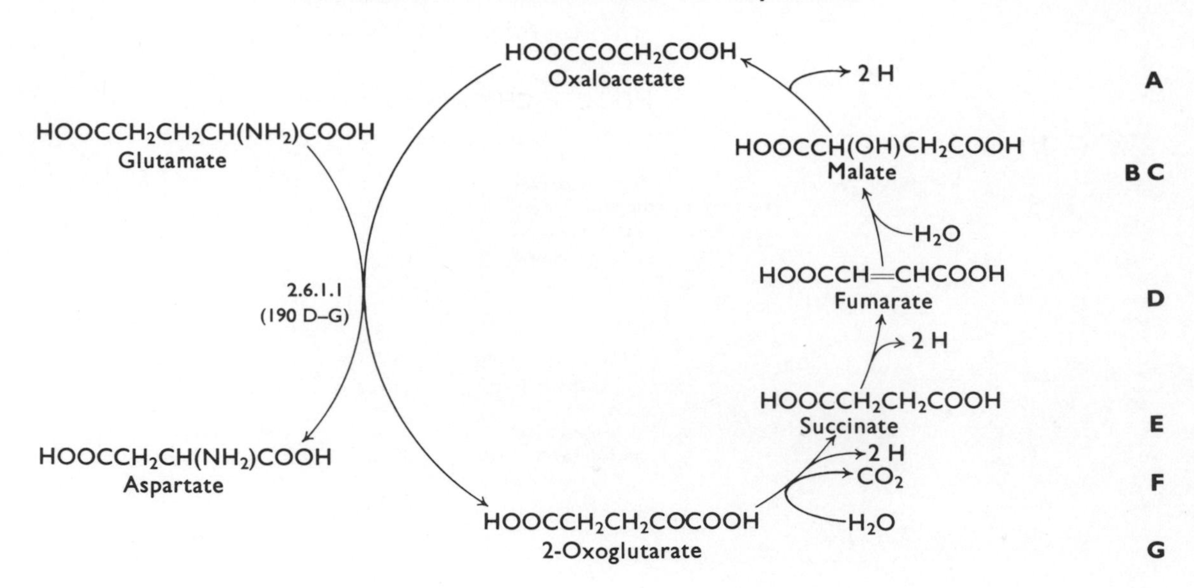

Net result

$$\text{HOOCCH}_2\text{CH}_2\text{CH(NH}_2\text{)COOH} + 2\ \text{H}_2\text{O} \longrightarrow \text{HOOCCH}_2\text{CH(NH}_2\text{)COOH} + \text{CO}_2 + 6\ \text{H}$$

Glutamate → Aspartate

$$\text{or HOOCCH}_2\text{CH}_2\text{CH(NH}_2\text{)COOH} + 1\tfrac{1}{2}\ \text{O}_2 \longrightarrow \text{HOOCCH}_2\text{CH(NH}_2\text{)COOH} + \text{CO}_2 + \text{H}_2\text{O}$$

Glutamate oxidized to Aspartate

Oxidation of Glutamate to Succinate (in Brain)

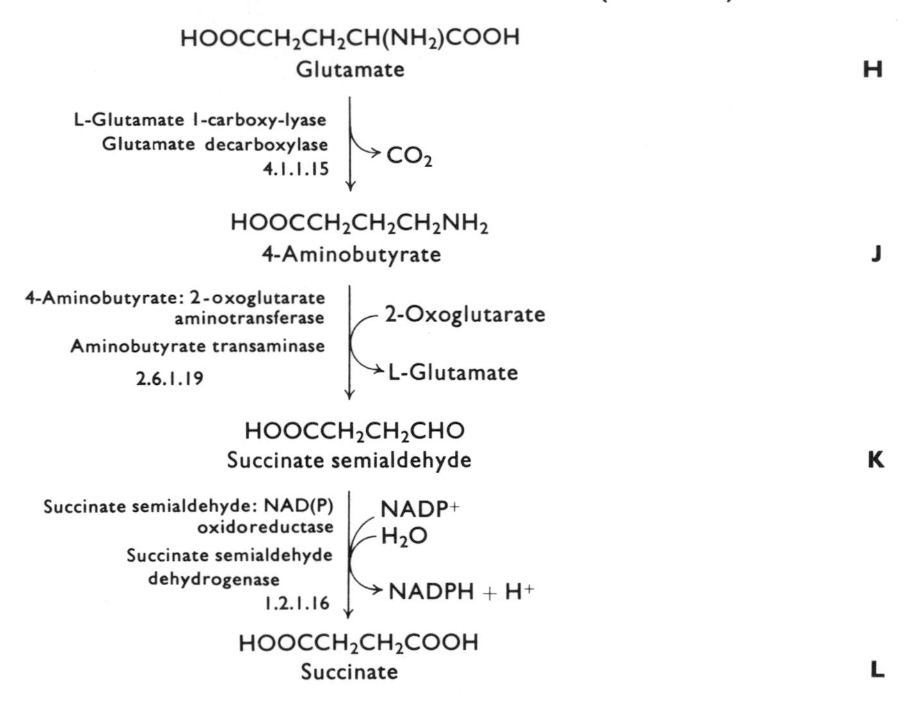

Formation and Degradation of Proline and 4-Hydroxyproline

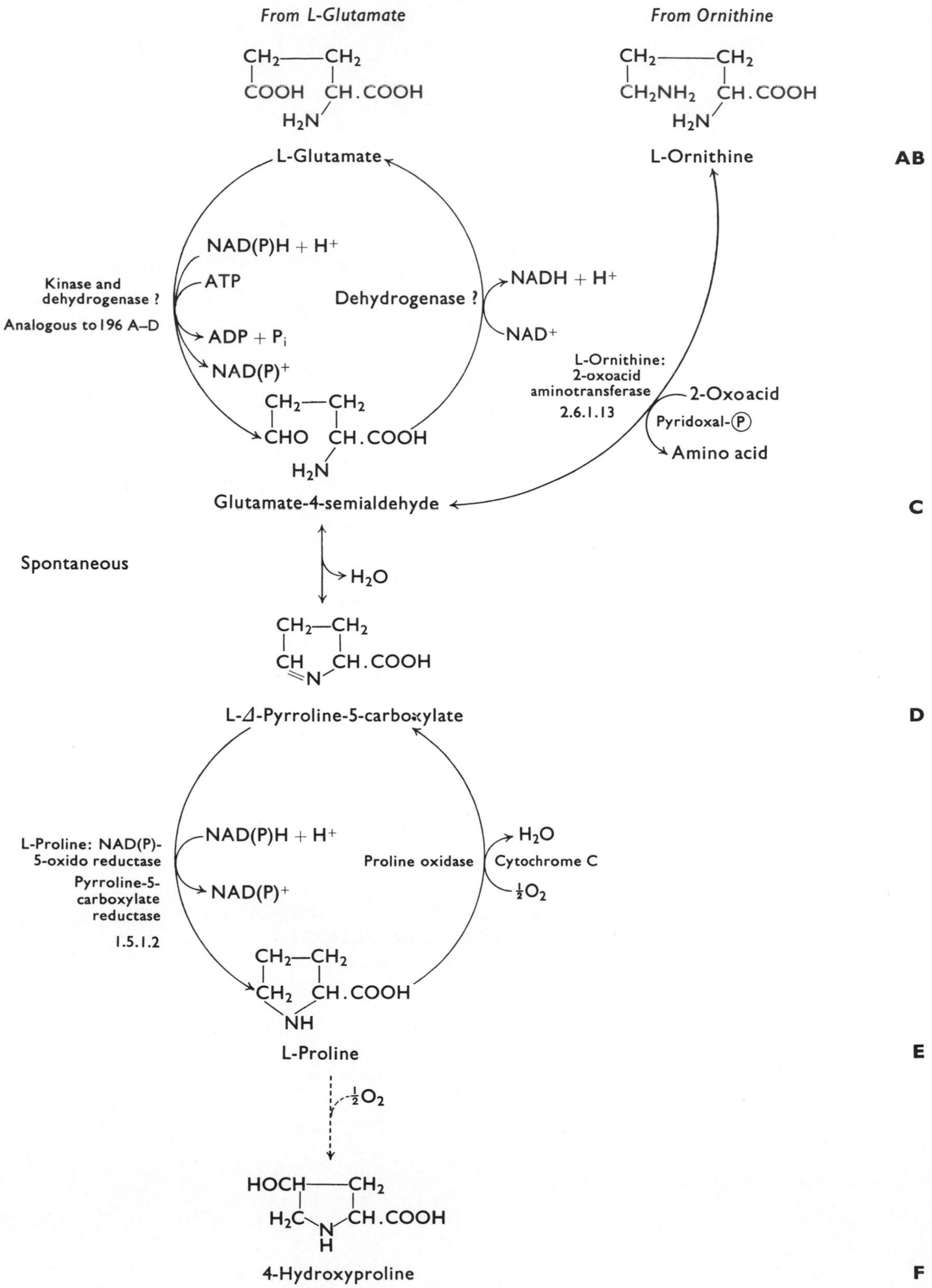

From L-Glutamate

CH_2—CH_2 | COOH CH.COOH | H_2N

L-Glutamate

From Ornithine

CH_2—CH_2 | CH_2NH_2 CH.COOH | H_2N

L-Ornithine

AB

Kinase and dehydrogenase ?
Analogous to 196 A–D

NAD(P)H + H^+
ATP
ADP + P_i
$NAD(P)^+$

Dehydrogenase ?
NADH + H^+
NAD^+

L-Ornithine: 2-oxoacid aminotransferase
2.6.1.13

2-Oxoacid
Pyridoxal-(P)
Amino acid

CH_2—CH_2 | CHO CH.COOH | H_2N

Glutamate-4-semialdehyde

C

Spontaneous

H_2O

CH_2—CH_2 | CH CH.COOH | =N—

L-Δ-Pyrroline-5-carboxylate

D

L-Proline: NAD(P)-5-oxido reductase
Pyrroline-5-carboxylate reductase
1.5.1.2

NAD(P)H + H^+
$NAD(P)^+$

Proline oxidase
H_2O
Cytochrome C
$\frac{1}{2}O_2$

CH_2—CH_2 | CH_2 CH.COOH | NH

L-Proline

E

$\frac{1}{2}O_2$

HOCH—CH_2 | H_2C CH.COOH | N H

4-Hydroxyproline

F

The oxidation of proline to hydroxyproline appears to take place after incorporation of the proline into a peptide.

Breakdown of Hydroxyproline (Bacteria)

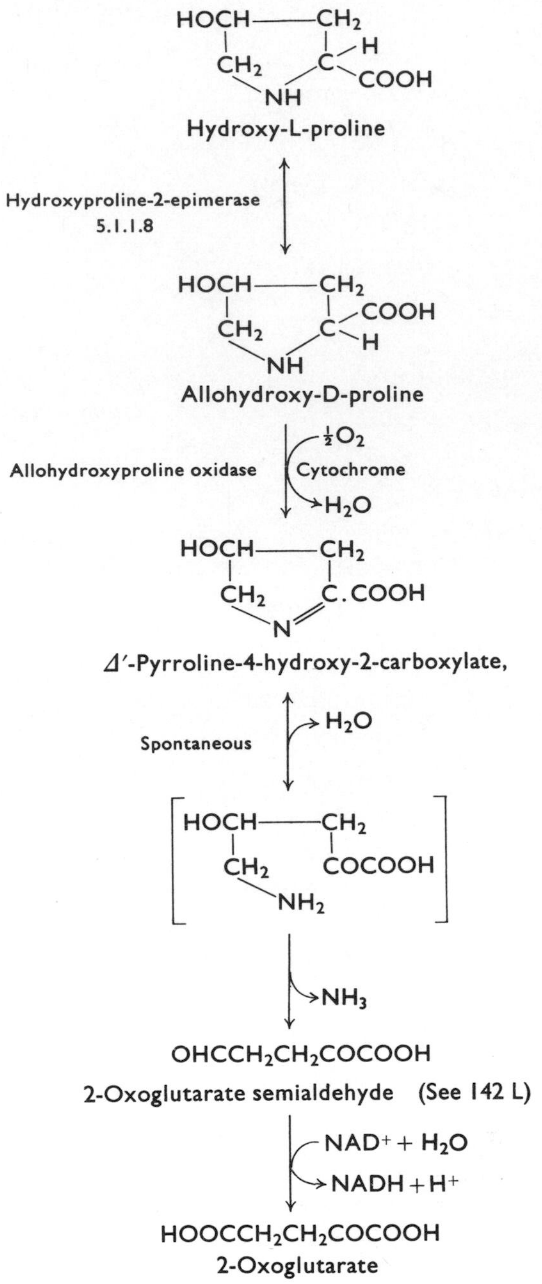

HOCH——CH$_2$ / CH$_2$ / C(H)(COOH) / NH

Hydroxy-L-proline — **A**

↕ Hydroxyproline-2-epimerase 5.1.1.8

HOCH——CH$_2$ / CH$_2$ / C(COOH)(H) / NH

Allohydroxy-D-proline — **B**

↓ Allohydroxyproline oxidase ($\frac{1}{2}O_2$ → Cytochrome → H_2O)

HOCH——CH$_2$ / CH$_2$ / C.COOH / N

Δ′-Pyrroline-4-hydroxy-2-carboxylate, — **C**

↕ Spontaneous (→ H_2O)

[HOCH——CH$_2$ / CH$_2$ / COCOOH / NH$_2$]

↓ (→ NH_3)

$OHCCH_2CH_2COCOOH$

2-Oxoglutarate semialdehyde (See 142 L) — **D**

↓ ($NAD^+ + H_2O$ → $NADH + H^+$)

$HOOCCH_2CH_2COCOOH$

2-Oxoglutarate — **E**

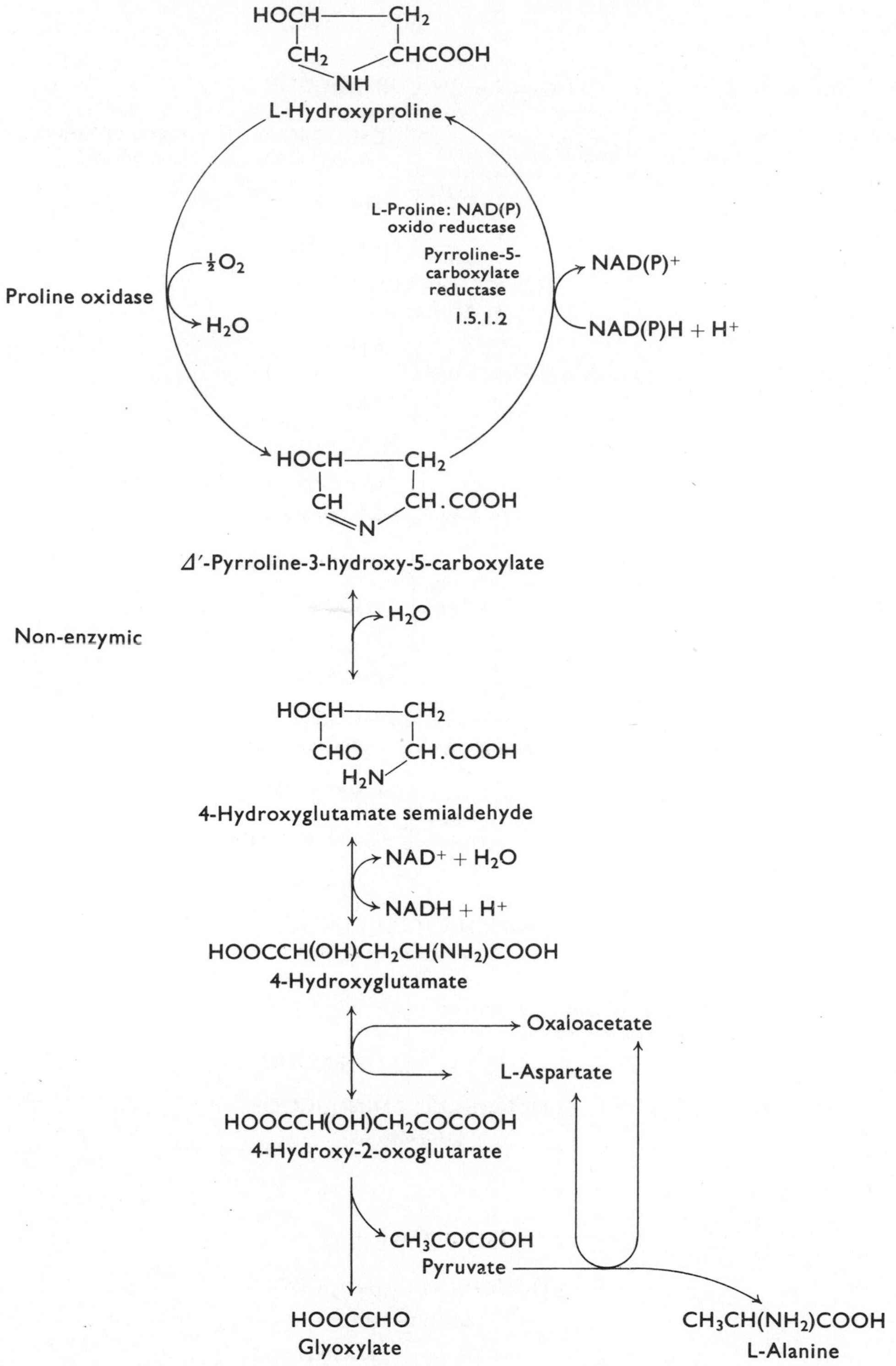

HOCH—CH$_2$ / CH$_2$ CHCOOH / NH

L-Hydroxyproline — A

Proline oxidase: $\frac{1}{2}O_2$ → H_2O

L-Proline: NAD(P) oxido reductase

Pyrroline-5-carboxylate reductase

1.5.1.2

$NAD(P)^+$ ← $NAD(P)H + H^+$

HOCH—CH$_2$ / CH CH.COOH / N

Δ′-Pyrroline-3-hydroxy-5-carboxylate — B

Non-enzymic

H_2O

HOCH—CH$_2$ / CHO CH.COOH / H_2N

4-Hydroxyglutamate semialdehyde — C

$NAD^+ + H_2O$

$NADH + H^+$

$HOOCCH(OH)CH_2CH(NH_2)COOH$

4-Hydroxyglutamate — D

Oxaloacetate

L-Aspartate

$HOOCCH(OH)CH_2COCOOH$

4-Hydroxy-2-oxoglutarate — E

$CH_3COCOOH$

Pyruvate — F

HOOCCHO

Glyoxylate

$CH_3CH(NH_2)COOH$

L-Alanine — GH

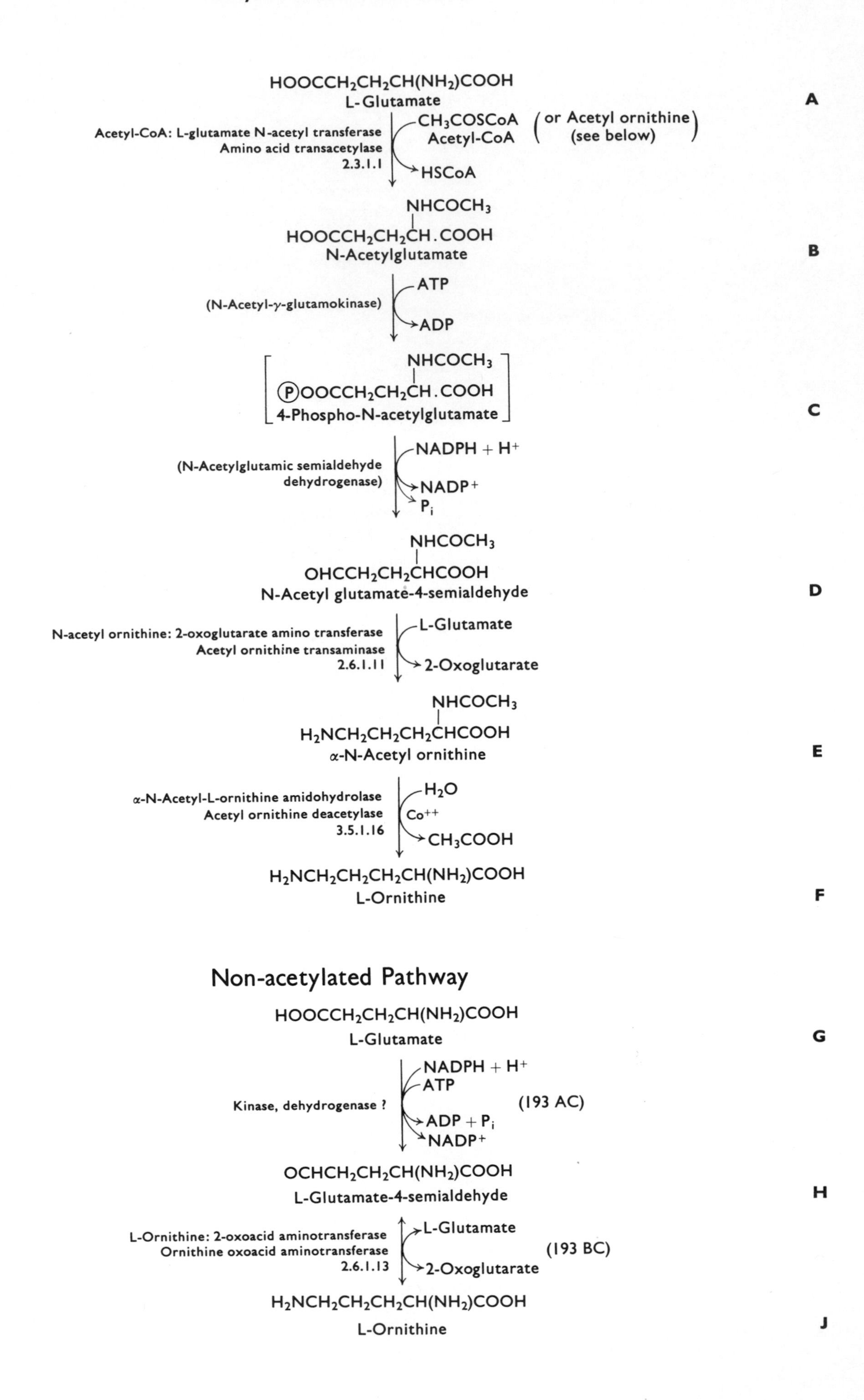
HOOCCH2CH2CH(NH2)COOH
L-Glutamate
A
Acetyl-CoA: L-glutamate N-acetyl transferase
Amino acid transacetylase
2.3.1.1
CH3COSCoA
Acetyl-CoA
(or Acetyl ornithine (see below))
HSCoA
NHCOCH3
HOOCCH2CH2CH.COOH
N-Acetylglutamate
B
(N-Acetyl-γ-glutamokinase)
ATP
ADP
NHCOCH3
(P)OOCCH2CH2CH.COOH
4-Phospho-N-acetylglutamate
C
(N-Acetylglutamic semialdehyde dehydrogenase)
NADPH + H+
NADP+
Pi
NHCOCH3
OHCCH2CH2CHCOOH
N-Acetyl glutamate-4-semialdehyde
D
N-acetyl ornithine: 2-oxoglutarate amino transferase
Acetyl ornithine transaminase
2.6.1.11
L-Glutamate
2-Oxoglutarate
NHCOCH3
H2NCH2CH2CH2CHCOOH
α-N-Acetyl ornithine
E
α-N-Acetyl-L-ornithine amidohydrolase
Acetyl ornithine deacetylase
3.5.1.16
H2O
Co++
CH3COOH
H2NCH2CH2CH2CH(NH2)COOH
L-Ornithine
F

Non-acetylated Pathway

HOOCCH2CH2CH(NH2)COOH
L-Glutamate
G
Kinase, dehydrogenase ?
NADPH + H+
ATP
ADP + Pi
NADP+
(193 AC)
OCHCH2CH2CH(NH2)COOH
L-Glutamate-4-semialdehyde
H
L-Ornithine: 2-oxoacid aminotransferase
Ornithine oxoacid aminotransferase
2.6.1.13
L-Glutamate
2-Oxoglutarate
(193 BC)
H2NCH2CH2CH2CH(NH2)COOH
L-Ornithine
J

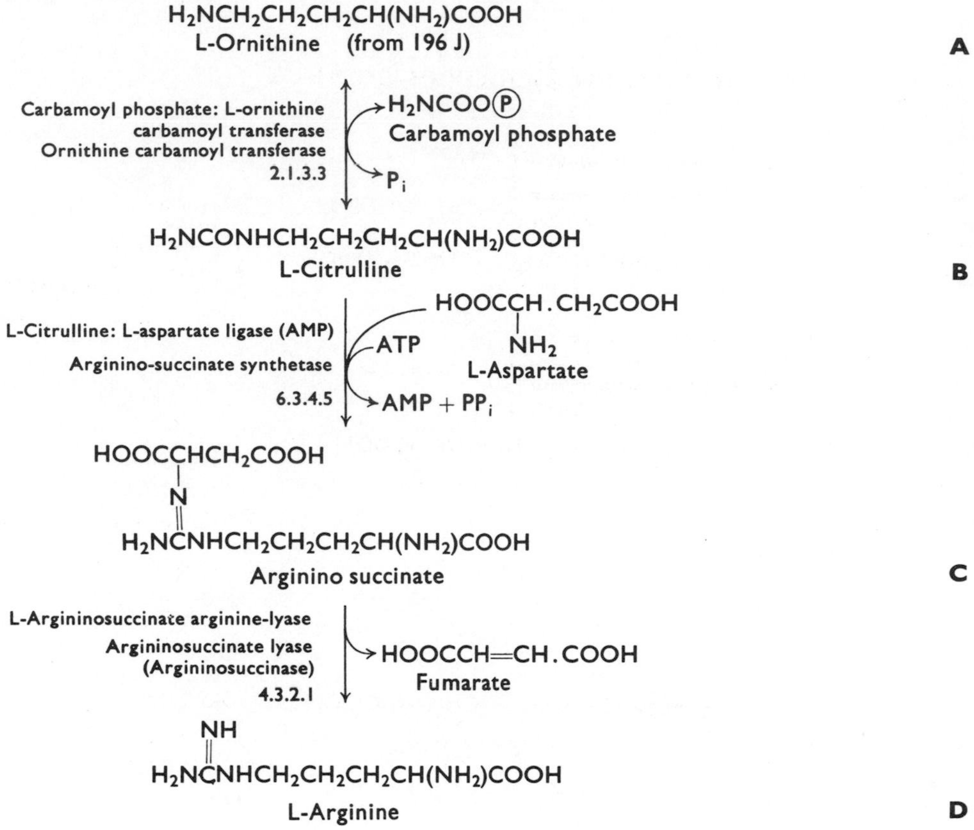

$H_2NCH_2CH_2CH_2CH(NH_2)COOH$
L-Ornithine (from 196 J) — **A**

Carbamoyl phosphate: L-ornithine carbamoyl transferase
Ornithine carbamoyl transferase
2.1.3.3

→ H_2NCOOⓅ Carbamoyl phosphate
→ P_i

$H_2NCONHCH_2CH_2CH_2CH(NH_2)COOH$
L-Citrulline — **B**

L-Citrulline: L-aspartate ligase (AMP)
Arginino-succinate synthetase
6.3.4.5

$HOOCCH.CH_2COOH$ (with NH_2 on the CH) L-Aspartate
ATP
→ AMP + PP_i

$HOOCCHCH_2COOH$
|
N
‖
$H_2NCNHCH_2CH_2CH_2CH(NH_2)COOH$
Arginino succinate — **C**

L-Argininosuccinate arginine-lyase
Argininosuccinate lyase
(Argininosuccinase)
4.3.2.1

→ $HOOCCH{=}CH.COOH$ Fumarate

NH
‖
$H_2NCNHCH_2CH_2CH_2CH(NH_2)COOH$
L-Arginine — **D**

Breakdown of Arginine

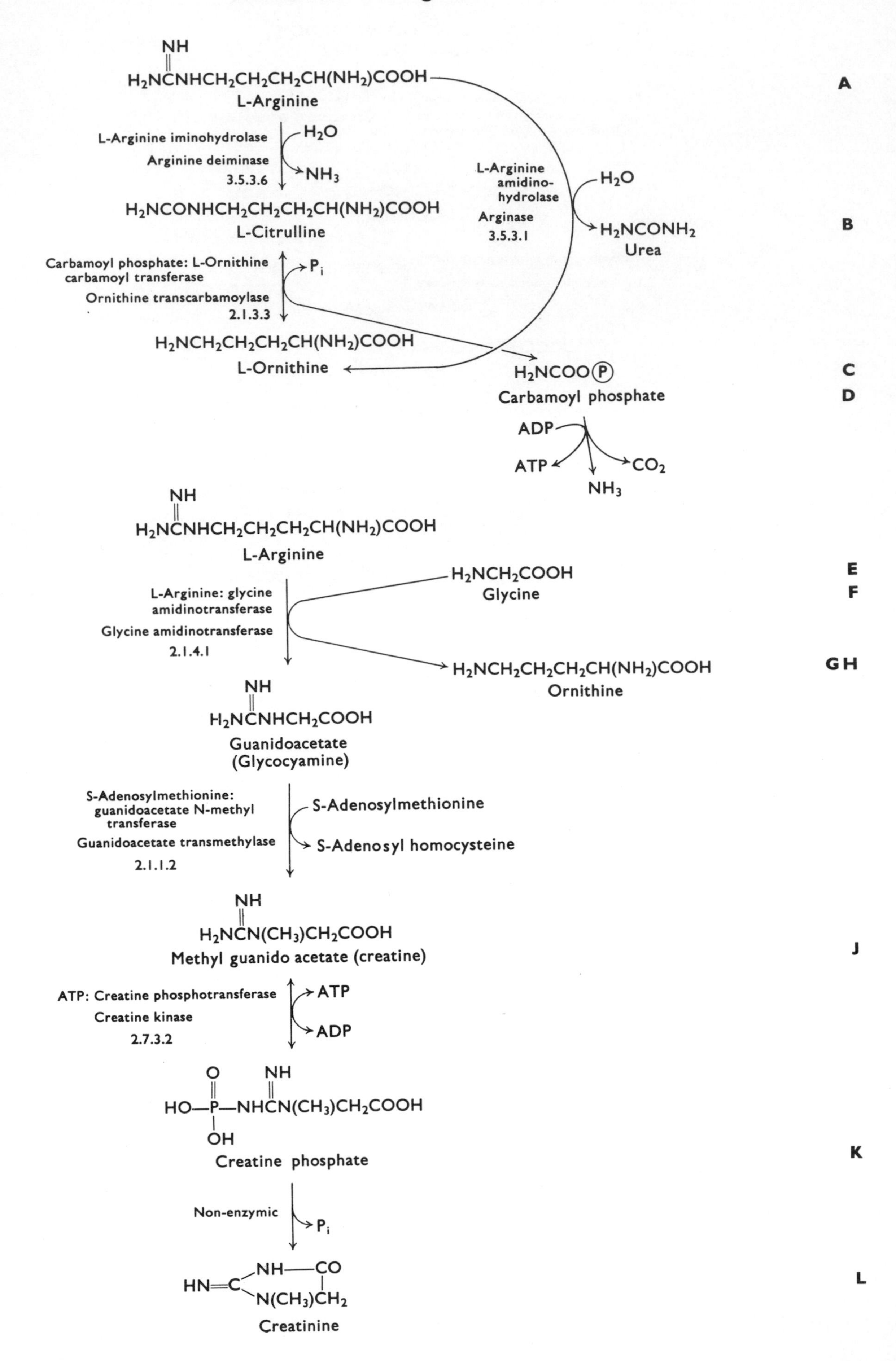
NH
H2NCNHCH2CH2CH2CH(NH2)COOH
L-Arginine
L-Arginine iminohydrolase
Arginine deiminase
3.5.3.6
H2O
NH3
H2NCONHCH2CH2CH2CH(NH2)COOH
L-Citrulline
L-Arginine amidino-hydrolase
Arginase
3.5.3.1
H2O
H2NCONH2
Urea
Carbamoyl phosphate: L-Ornithine carbamoyl transferase
Ornithine transcarbamoylase
2.1.3.3
Pi
H2NCH2CH2CH2CH(NH2)COOH
L-Ornithine
H2NCOO(P)
Carbamoyl phosphate
ADP
ATP
CO2
NH3
A
B
C
D
NH
H2NCNHCH2CH2CH2CH(NH2)COOH
L-Arginine
L-Arginine: glycine amidinotransferase
Glycine amidinotransferase
2.1.4.1
H2NCH2COOH
Glycine
H2NCH2CH2CH2CH(NH2)COOH
Ornithine
E
F
GH
NH
H2NCNHCH2COOH
Guanidoacetate
(Glycocyamine)
S-Adenosylmethionine: guanidoacetate N-methyl transferase
Guanidoacetate transmethylase
2.1.1.2
S-Adenosylmethionine
S-Adenosyl homocysteine
NH
H2NCN(CH3)CH2COOH
Methyl guanido acetate (creatine)
J
ATP: Creatine phosphotransferase
Creatine kinase
2.7.3.2
ATP
ADP
O NH
HO—P—NHCN(CH3)CH2COOH
OH
Creatine phosphate
K
Non-enzymic
Pi
NH—CO
HN=C
N(CH3)CH2
Creatinine
L

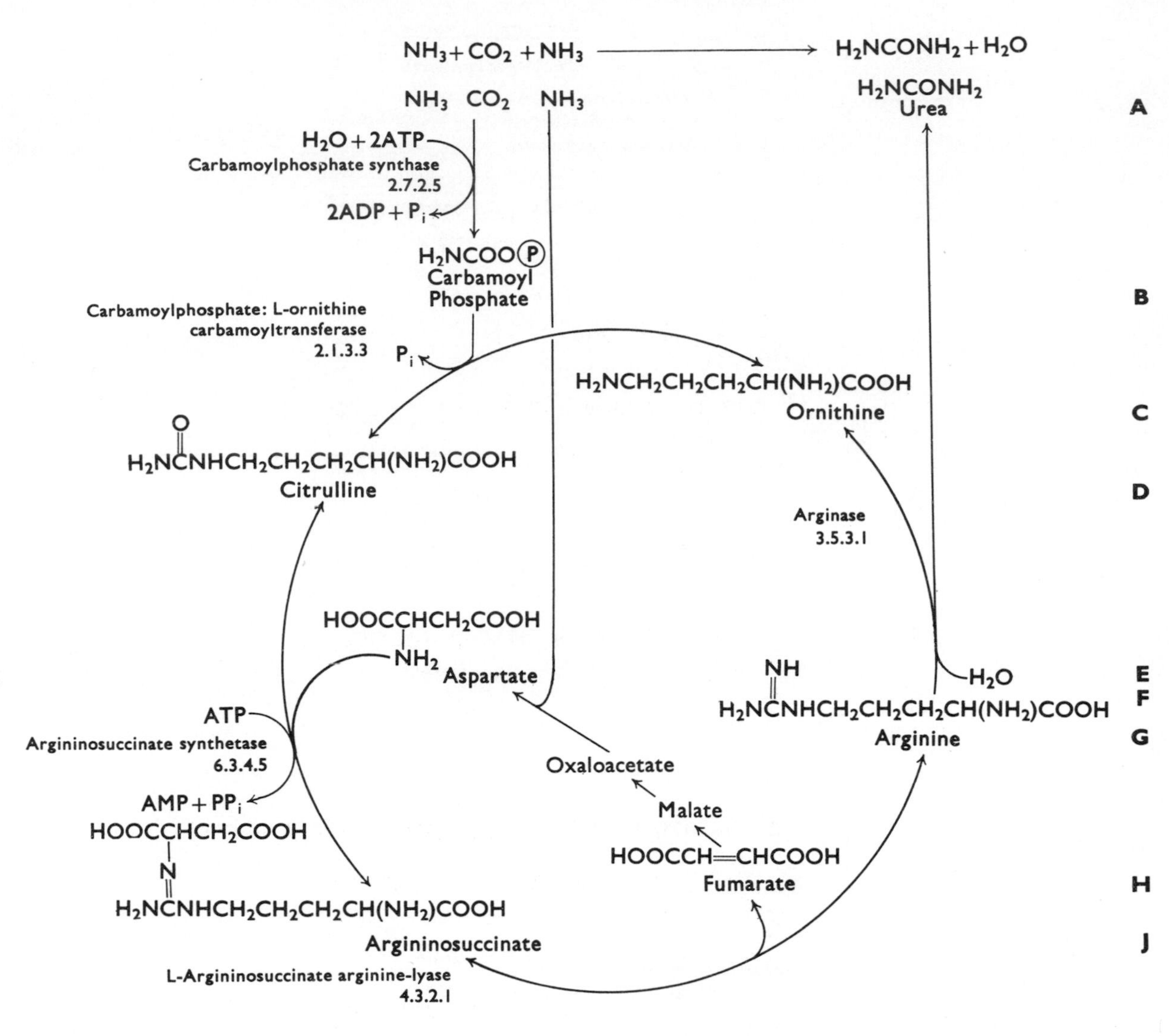

Net result

$$3ATP + 2H_2O + NH_3 + CO_2 + NH_3 \longrightarrow 2ADP + 2P_i + AMP + PP_i + H_2NCONH_2$$

Biosynthesis Reactions of Aspartate

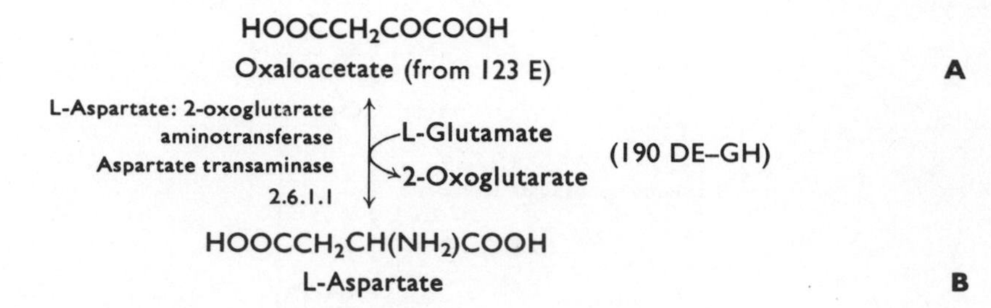

From Fumarate (in microorganisms)

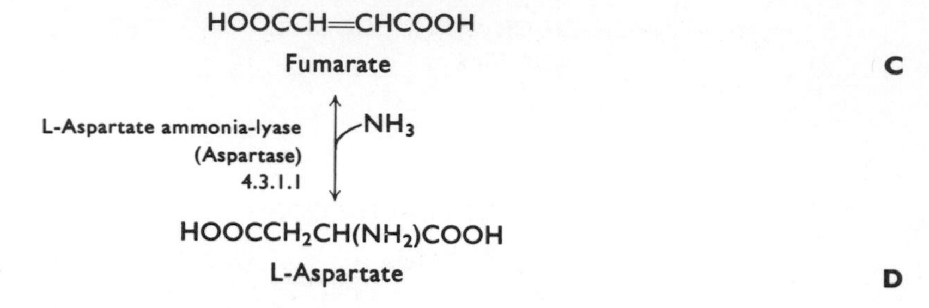

Formation of Asparagine (microorganisms)

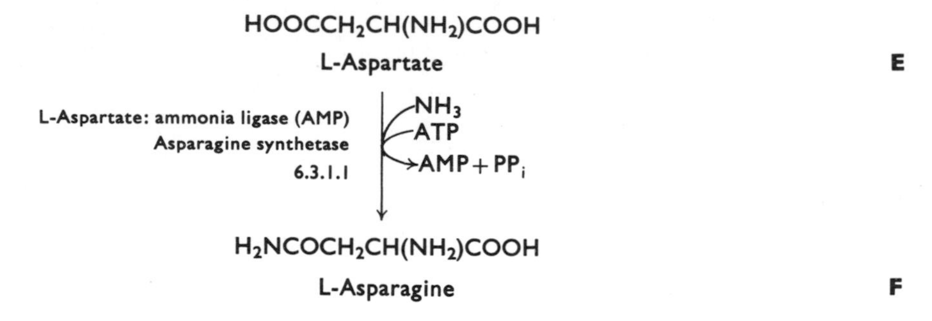

Formation of β-Alanine

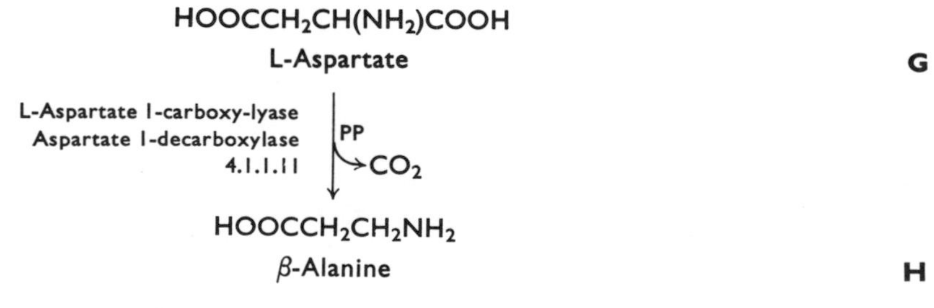

For formation of L-Alanine see 213 HJ

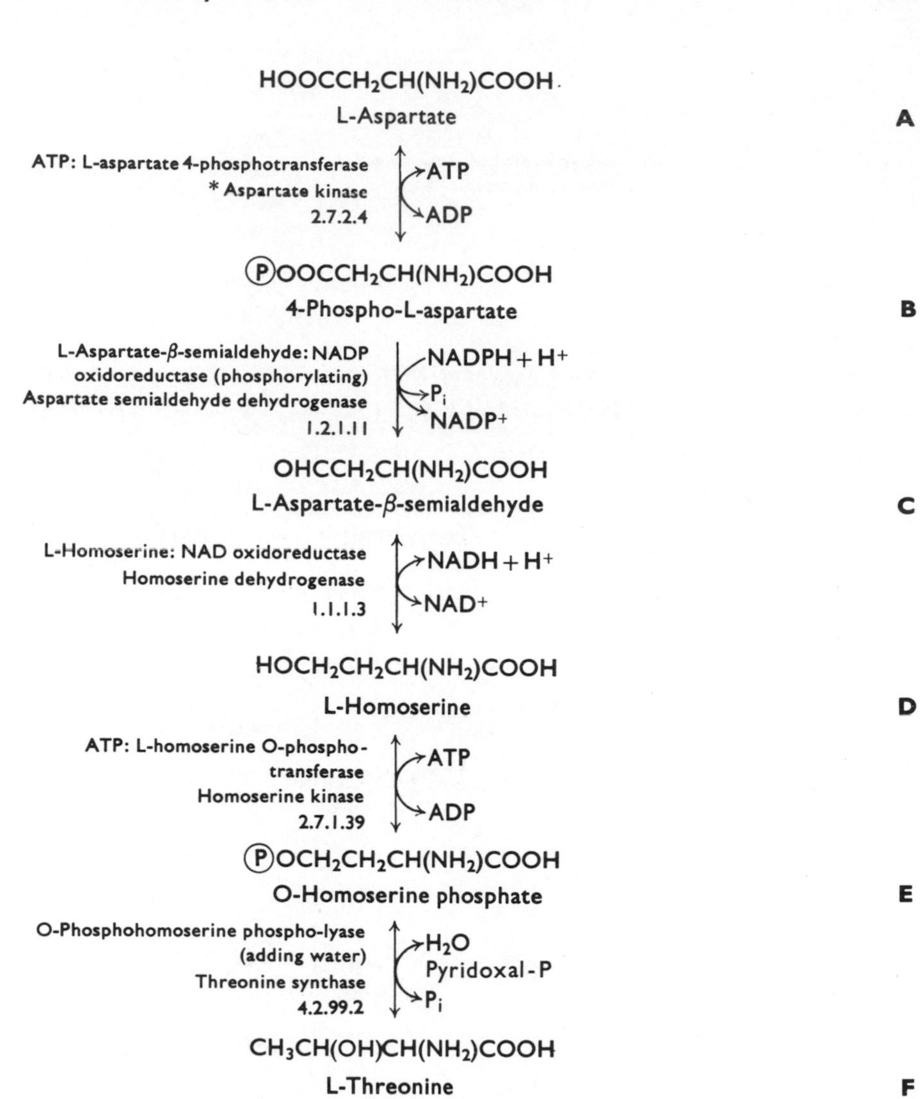

* At least three aspartate kinases are known which are repressed I by threonine, II by methionine and III by lysine.

Alternative route

$CH_3CHO + HCH(NH_2)COOH$

Acetaldehyde Glycine **GH**

L-Threonine acetaldehyde-lyase
Threonine aldolase
4.1.2.5

↕ Pyridoxal-P

$CH_3CH(OH)CH(NH_2)COOH$

L-Threonine **J**

Degradation of Threonine

$CH_3CH(OH)CH(NH_2)COOH$
L-Threonine — **A**

L-Threonine hydrolyase (deaminating)
Threonine dehydratase
4.2.1.16

↓ H_2O, Pyridoxal P. → H_2O, NH_3

$CH_3CH_2COCOOH$
2-Oxobutyrate — **B**

An aldolase in liver splits threonine into acetaldehyde and glycine

$$CH_3CH(OH)CH(NH_2)COOH \rightleftharpoons CH_3CHO + CH_2(NH_2)COOH$$

Conversion to Lactate

$CH_3CH(OH)CH(NH_2)COOH$
L-Threonine — **C**

L-Threonine dehydrogenase

↓ NAD^+ → $NADH + H^+$

[$CH_3COCH(NH_2)COOH$
2-Amino-3-oxobutyrate] — **D**

↓ → CO_2

$CH_3COCH_2NH_2$
Aminoacetone — **E**

↓ $\frac{1}{2}O_2$ → NH_3

CH_3COCHO
Methyl glyoxal — **F**

Glyoxylase ?

↓ H_2O, Glutathione

$CH_3CH(OH)COOH$
D-Lactate — **G**

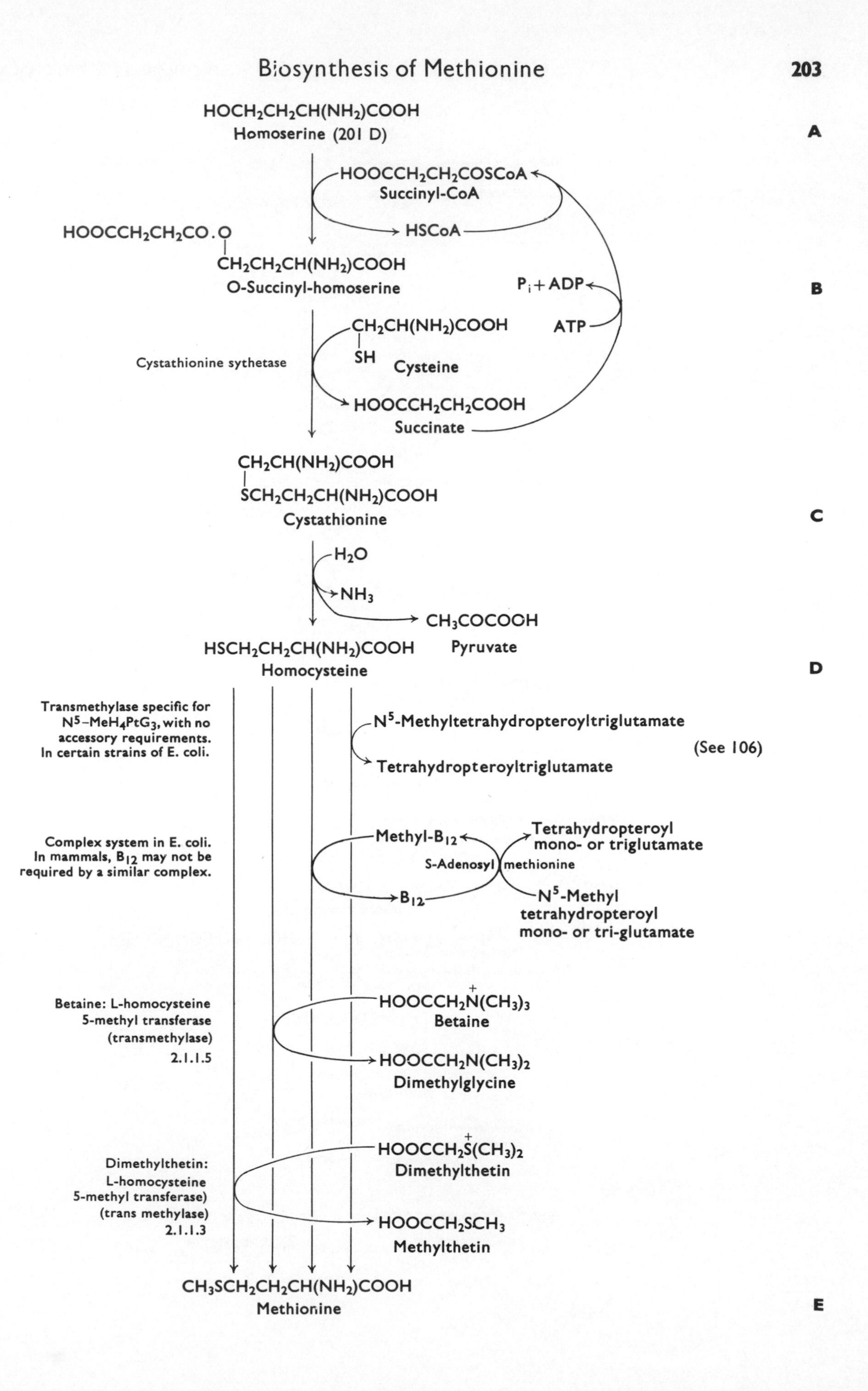

HOCH₂CH₂CH(NH₂)COOH
Homoserine (201 D)
A
HOOCCH₂CH₂COSCoA
Succinyl-CoA
HSCoA
HOOCCH₂CH₂CO.O
CH₂CH₂CH(NH₂)COOH
O-Succinyl-homoserine
Pi + ADP
ATP
B
Cystathionine sythetase
CH₂CH(NH₂)COOH
SH
Cysteine
HOOCCH₂CH₂COOH
Succinate
CH₂CH(NH₂)COOH
SCH₂CH₂CH(NH₂)COOH
Cystathionine
C
H₂O
NH₃
CH₃COCOOH
Pyruvate
HSCH₂CH₂CH(NH₂)COOH
Homocysteine
D
Transmethylase specific for N5-MeH4PtG3, with no accessory requirements. In certain strains of E. coli.
N5-Methyltetrahydropteroyltriglutamate
(See 106)
Tetrahydropteroyltriglutamate
Complex system in E. coli. In mammals, B12 may not be required by a similar complex.
Methyl-B12
Tetrahydropteroyl mono- or triglutamate
S-Adenosyl methionine
B12
N5-Methyl tetrahydropteroyl mono- or tri-glutamate
Betaine: L-homocysteine 5-methyl transferase (transmethylase) 2.1.1.5
HOOCCH₂N⁺(CH₃)₃
Betaine
HOOCCH₂N(CH₃)₂
Dimethylglycine
Dimethylthetin: L-homocysteine 5-methyl transferase) (trans methylase) 2.1.1.3
HOOCCH₂S⁺(CH₃)₂
Dimethylthetin
HOOCCH₂SCH₃
Methylthetin
CH₃SCH₂CH₂CH(NH₂)COOH
Methionine
E

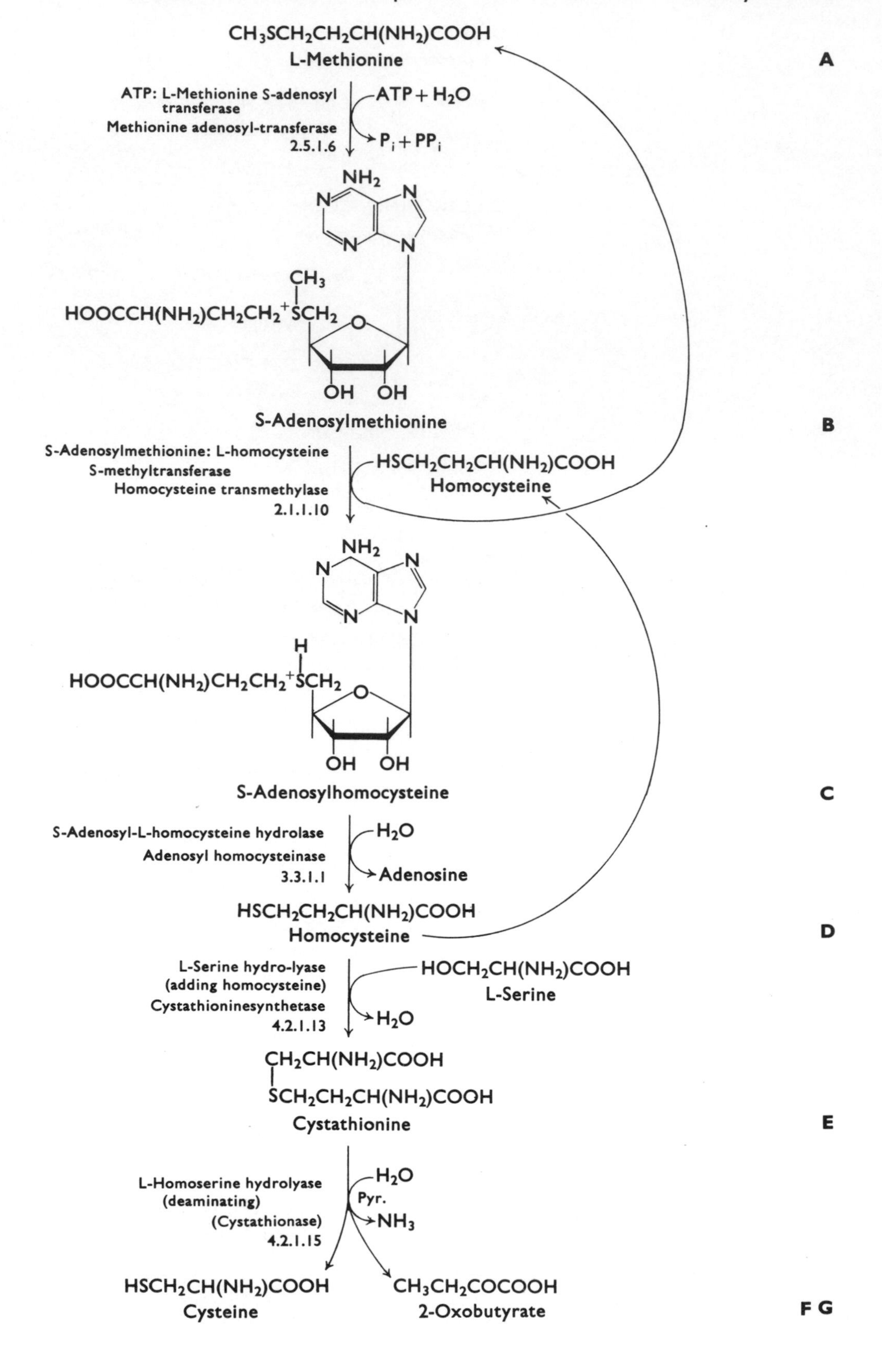

CH3SCH2CH2CH(NH2)COOH
L-Methionine
A
ATP: L-Methionine S-adenosyl transferase
Methionine adenosyl-transferase
2.5.1.6
ATP + H2O
Pi + PPi
NH2
N
N
N
N
CH3
HOOCCH(NH2)CH2CH2+SCH2
O
OH OH
S-Adenosylmethionine
B
S-Adenosylmethionine: L-homocysteine S-methyltransferase
Homocysteine transmethylase
2.1.1.10
HSCH2CH2CH(NH2)COOH
Homocysteine
NH2
N
N
N
N
H
HOOCCH(NH2)CH2CH2+SCH2
O
OH OH
S-Adenosylhomocysteine
C
S-Adenosyl-L-homocysteine hydrolase
Adenosyl homocysteinase
3.3.1.1
H2O
Adenosine
HSCH2CH2CH(NH2)COOH
Homocysteine
D
L-Serine hydro-lyase (adding homocysteine)
Cystathioninesynthetase
4.2.1.13
HOCH2CH(NH2)COOH
L-Serine
H2O
CH2CH(NH2)COOH
SCH2CH2CH(NH2)COOH
Cystathionine
E
L-Homoserine hydrolyase (deaminating)
(Cystathionase)
4.2.1.15
H2O
Pyr.
NH3
HSCH2CH(NH2)COOH
Cysteine
CH3CH2COCOOH
2-Oxobutyrate
F G

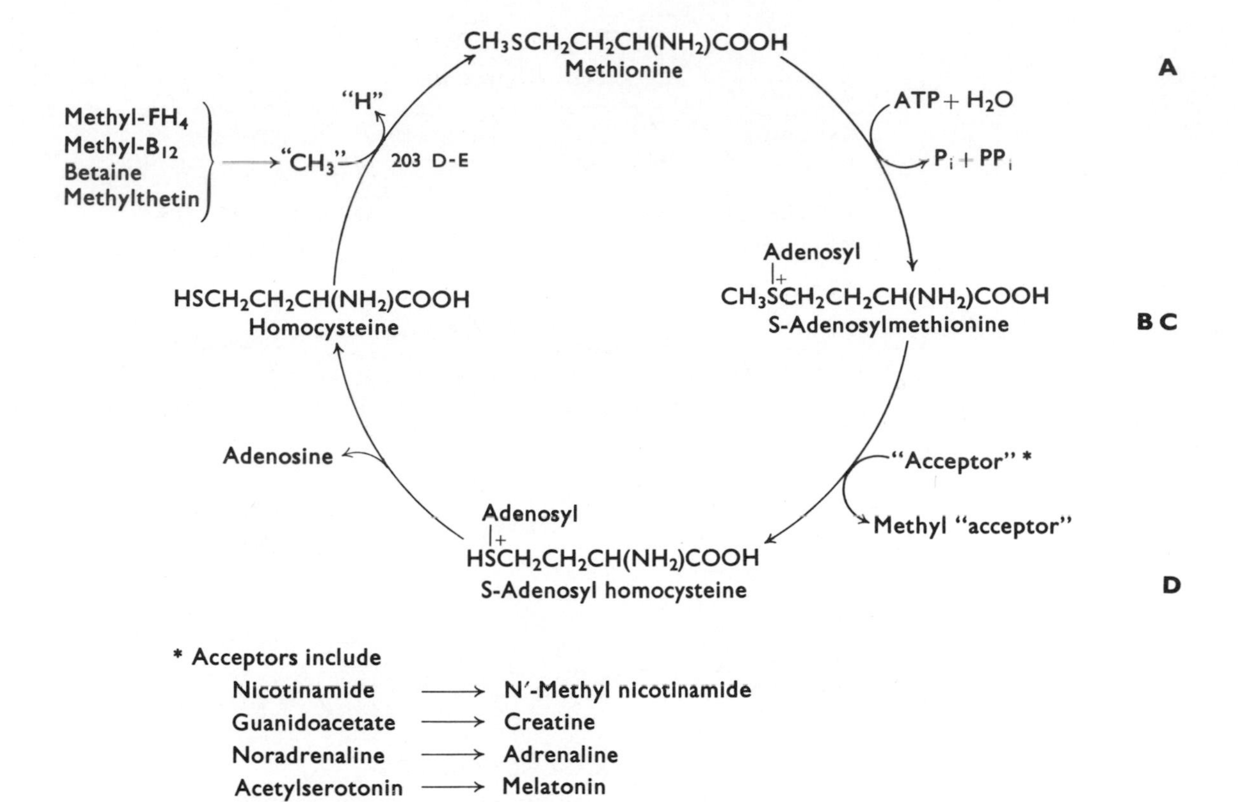

* Acceptors include

Nicotinamide	⟶	N′-Methyl nicotinamide
Guanidoacetate	⟶	Creatine
Noradrenaline	⟶	Adrenaline
Acetylserotonin	⟶	Melatonin
Phosphatidyl ethanolamine	⟶	Phosphatidyl choline
Polynucleotides	⟶	Methylated polynucleotides
Homocysteine	⟶	Methionine

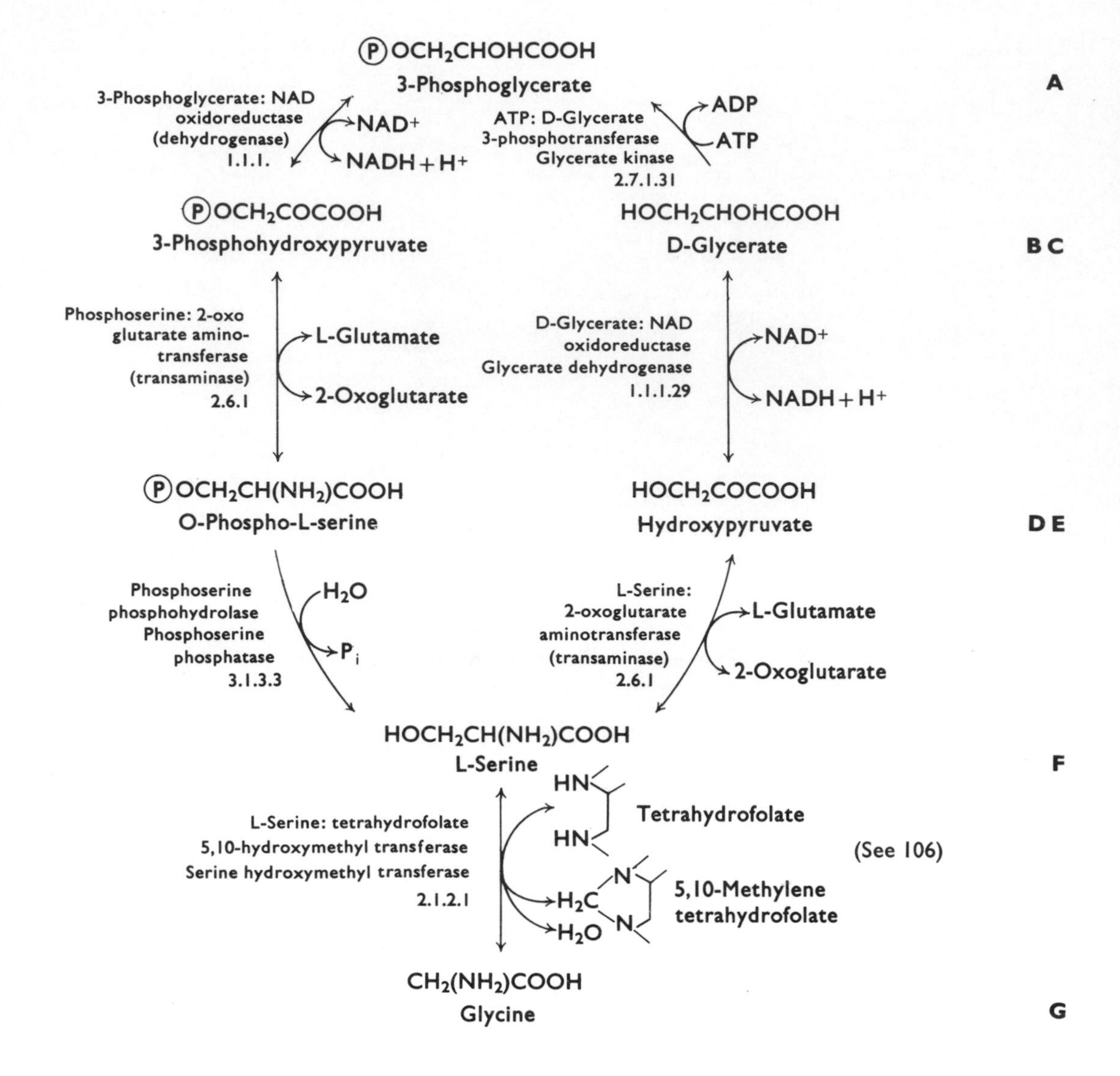

Ⓟ$OCH_2CHOHCOOH$
3-Phosphoglycerate
A
3-Phosphoglycerate: NAD oxidoreductase (dehydrogenase) 1.1.1.
NAD^+
$NADH + H^+$
ATP: D-Glycerate 3-phosphotransferase Glycerate kinase 2.7.1.31
ADP
ATP
Ⓟ$OCH_2COCOOH$
3-Phosphohydroxypyruvate
$HOCH_2CHOHCOOH$
D-Glycerate
B C
Phosphoserine: 2-oxo glutarate amino-transferase (transaminase) 2.6.1
L-Glutamate
2-Oxoglutarate
D-Glycerate: NAD oxidoreductase Glycerate dehydrogenase 1.1.1.29
NAD^+
$NADH + H^+$
Ⓟ$OCH_2CH(NH_2)COOH$
O-Phospho-L-serine
$HOCH_2COCOOH$
Hydroxypyruvate
D E
Phosphoserine phosphohydrolase Phosphoserine phosphatase 3.1.3.3
H_2O
P_i
L-Serine: 2-oxoglutarate aminotransferase (transaminase) 2.6.1
L-Glutamate
2-Oxoglutarate
$HOCH_2CH(NH_2)COOH$
L-Serine
F
L-Serine: tetrahydrofolate 5,10-hydroxymethyl transferase Serine hydroxymethyl transferase 2.1.2.1
HN
HN
Tetrahydrofolate
(See 106)
H_2C
N
N
H_2O
5,10-Methylene tetrahydrofolate
$CH_2(NH_2)COOH$
Glycine
G

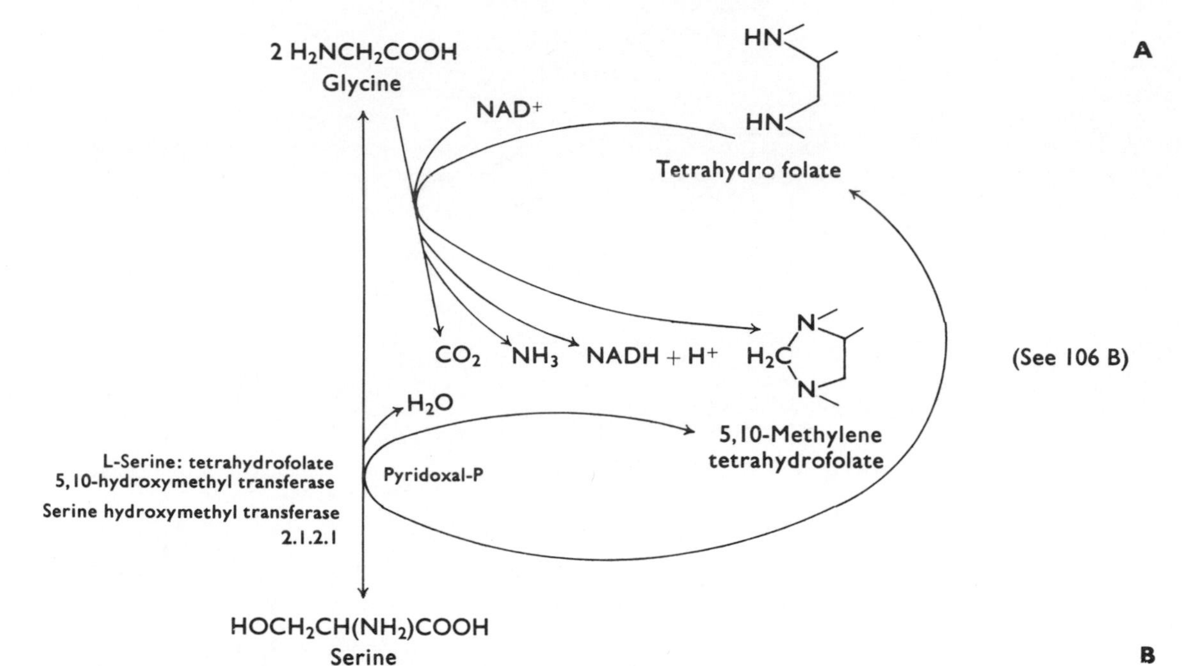

L-Serine hydrolyase (deaminating)
L-Serine dehydratase
Serine deaminase
4.2.1.13

H_2O, Pyridoxal-P, H_2O, NH_3

$CH_3COCOOH$
Pyruvate **C**

Net result

$$2\ CH_2(NH_2)COOH + NAD^+ + H_2O \longrightarrow CH_3COCOOH + CO_2 + 2\ NH_3 + NADH + H^+$$

Glycine → Pyruvate

3-Hydroxyaspartate Pathway
(in *Micrococcus denitrificans*)

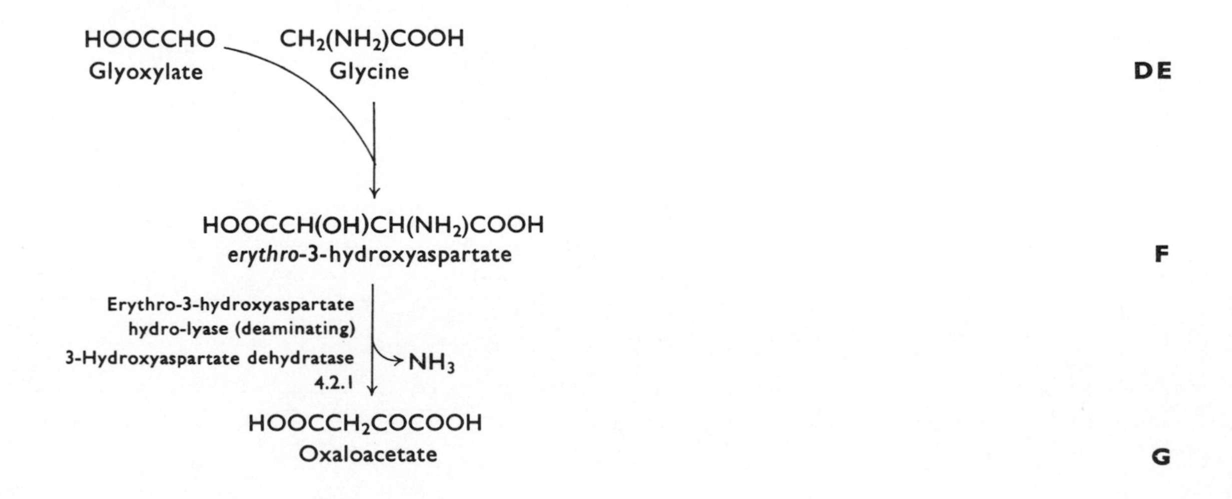

Synthesis of Glutathione

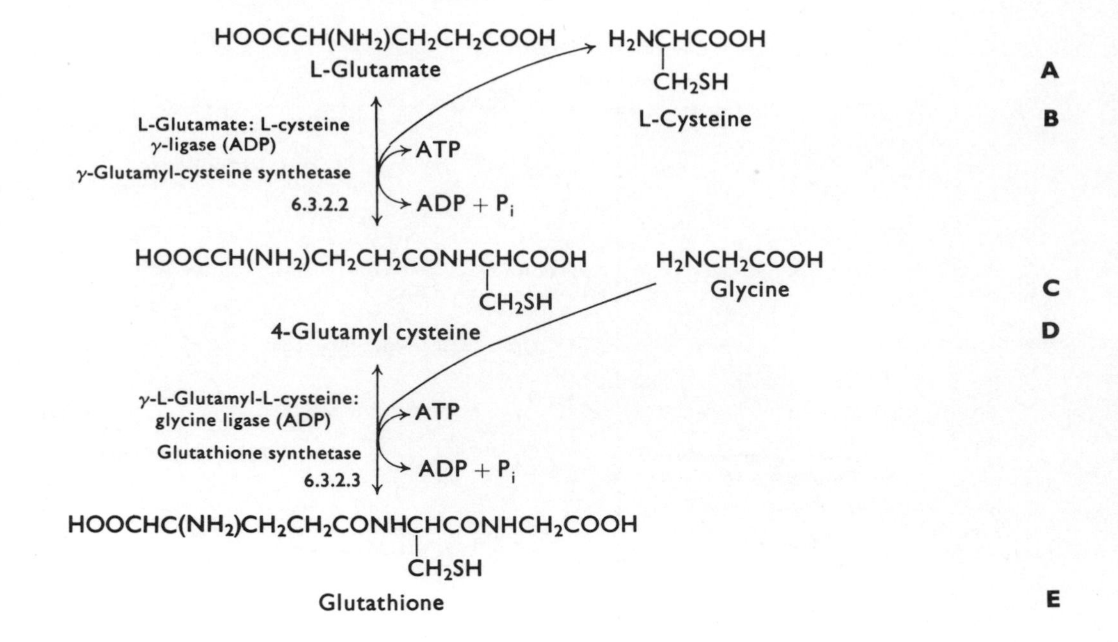

HOOCCH(NH2)CH2CH2COOH
L-Glutamate
H2NCHCOOH
CH2SH
L-Cysteine
A
B
L-Glutamate: L-cysteine γ-ligase (ADP)
γ-Glutamyl-cysteine synthetase
6.3.2.2
ATP
ADP + Pi
HOOCCH(NH2)CH2CH2CONHCHCOOH
CH2SH
4-Glutamyl cysteine
H2NCH2COOH
Glycine
C
D
γ-L-Glutamyl-L-cysteine: glycine ligase (ADP)
Glutathione synthetase
6.3.2.3
ATP
ADP + Pi
HOOCHC(NH2)CH2CH2CONHCHCONHCH2COOH
CH2SH
Glutathione
E

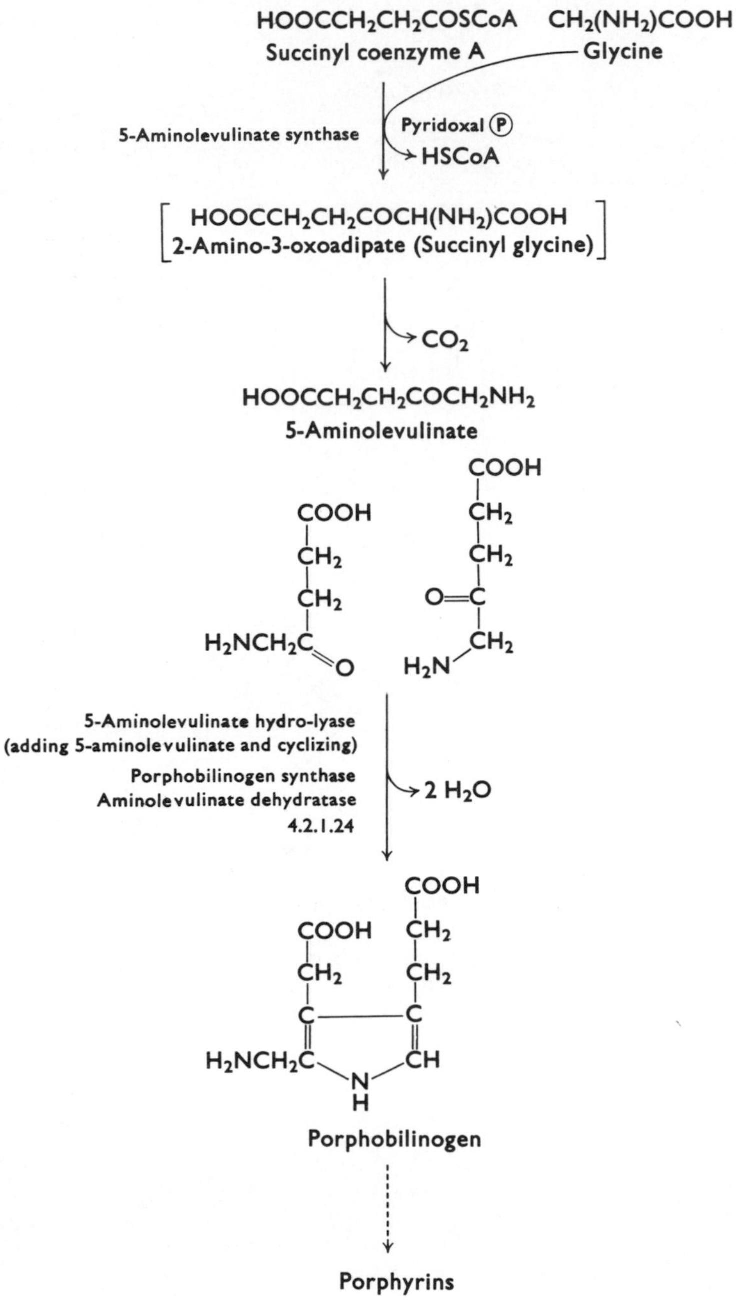

$HOOCCH_2CH_2COSCoA$ — Succinyl coenzyme A (A)

$CH_2(NH_2)COOH$ — Glycine (B)

5-Aminolevulinate synthase; Pyridoxal Ⓟ; → HSCoA

[$HOOCCH_2CH_2COCH(NH_2)COOH$ — 2-Amino-3-oxoadipate (Succinyl glycine)] (C)

→ CO_2

$HOOCCH_2CH_2COCH_2NH_2$

5-Aminolevulinate (D)

2 × 5-Aminolevulinate: $H_2NCH_2C(=O)CH_2CH_2COOH$ + $H_2NCH_2C(=O)CH_2CH_2COOH$ (E)

5-Aminolevulinate hydro-lyase
(adding 5-aminolevulinate and cyclizing)
Porphobilinogen synthase
Aminolevulinate dehydratase
4.2.1.24

→ 2 H_2O

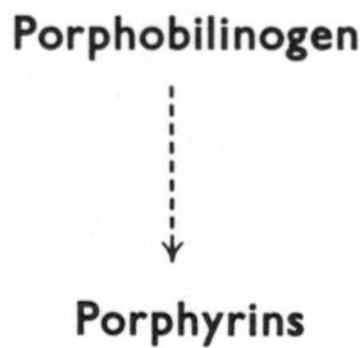

Porphobilinogen (F)

Porphyrins (G)

Interrelationships of Methylated Glycine and Choline Derivatives

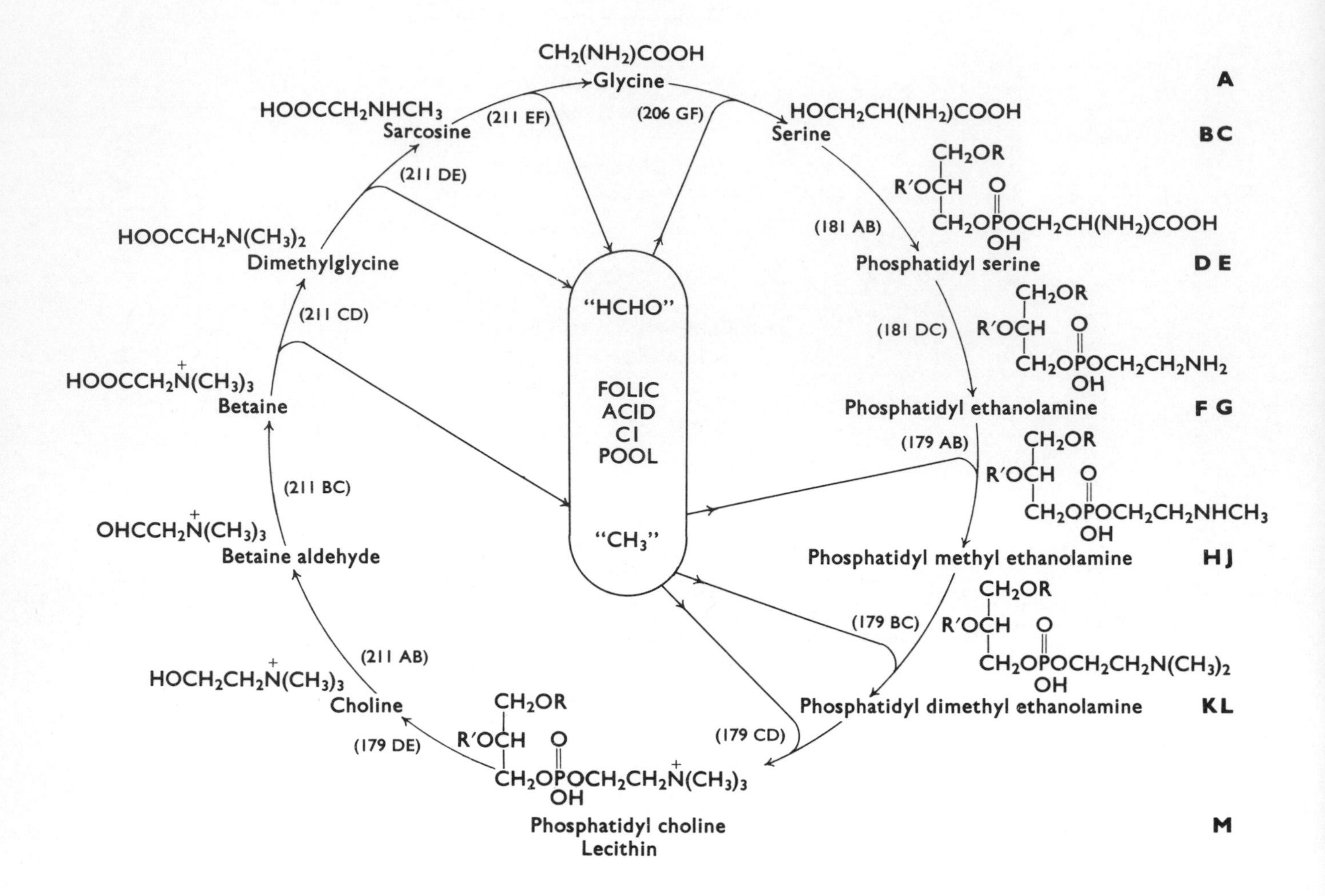

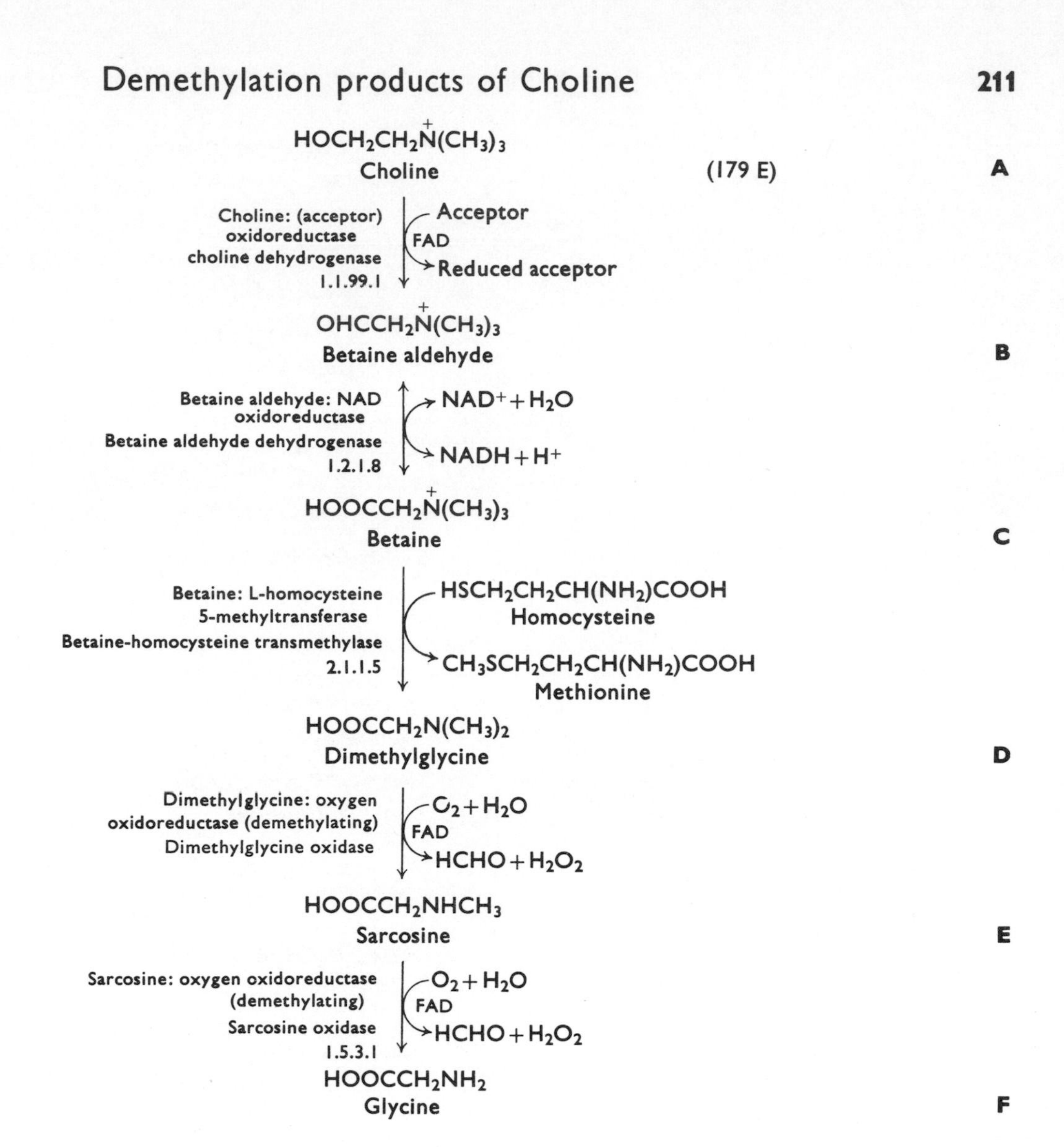
$HOCH_2CH_2\overset{+}{N}(CH_3)_3$
Choline
(179 E)
A
Choline: (acceptor) oxidoreductase
choline dehydrogenase
1.1.99.1
Acceptor
FAD
Reduced acceptor
$OHCCH_2\overset{+}{N}(CH_3)_3$
Betaine aldehyde
B
Betaine aldehyde: NAD oxidoreductase
Betaine aldehyde dehydrogenase
1.2.1.8
$NAD^+ + H_2O$
$NADH + H^+$
$HOOCCH_2\overset{+}{N}(CH_3)_3$
Betaine
C
Betaine: L-homocysteine 5-methyltransferase
Betaine-homocysteine transmethylase
2.1.1.5
$HSCH_2CH_2CH(NH_2)COOH$
Homocysteine
$CH_3SCH_2CH_2CH(NH_2)COOH$
Methionine
$HOOCCH_2N(CH_3)_2$
Dimethylglycine
D
Dimethylglycine: oxygen oxidoreductase (demethylating)
Dimethylglycine oxidase
$O_2 + H_2O$
FAD
$HCHO + H_2O_2$
$HOOCCH_2NHCH_3$
Sarcosine
E
Sarcosine: oxygen oxidoreductase (demethylating)
Sarcosine oxidase
1.5.3.1
$O_2 + H_2O$
FAD
$HCHO + H_2O_2$
$HOOCCH_2NH_2$
Glycine
F

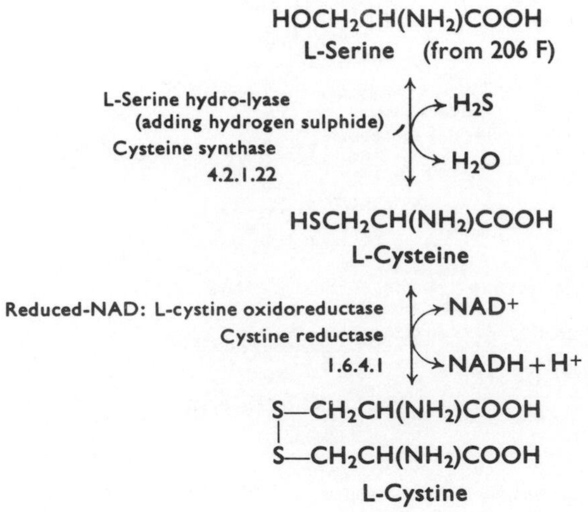

Biosynthesis of Cysteine and Cystine

$HOCH_2CH(NH_2)COOH$
L-Serine (from 206 F) **A**

L-Serine hydro-lyase (adding hydrogen sulphide)
Cysteine synthase
4.2.1.22
↕ → H_2S, → H_2O

$HSCH_2CH(NH_2)COOH$
L-Cysteine **B**

Reduced-NAD: L-cystine oxidoreductase
Cystine reductase
1.6.4.1
↕ → NAD^+, → $NADH + H^+$

$S—CH_2CH(NH_2)COOH$
$|$
$S—CH_2CH(NH_2)COOH$
L-Cystine **C**

For biosynthesis from methionine and serine, see page 204.

Breakdown of L-Cysteine

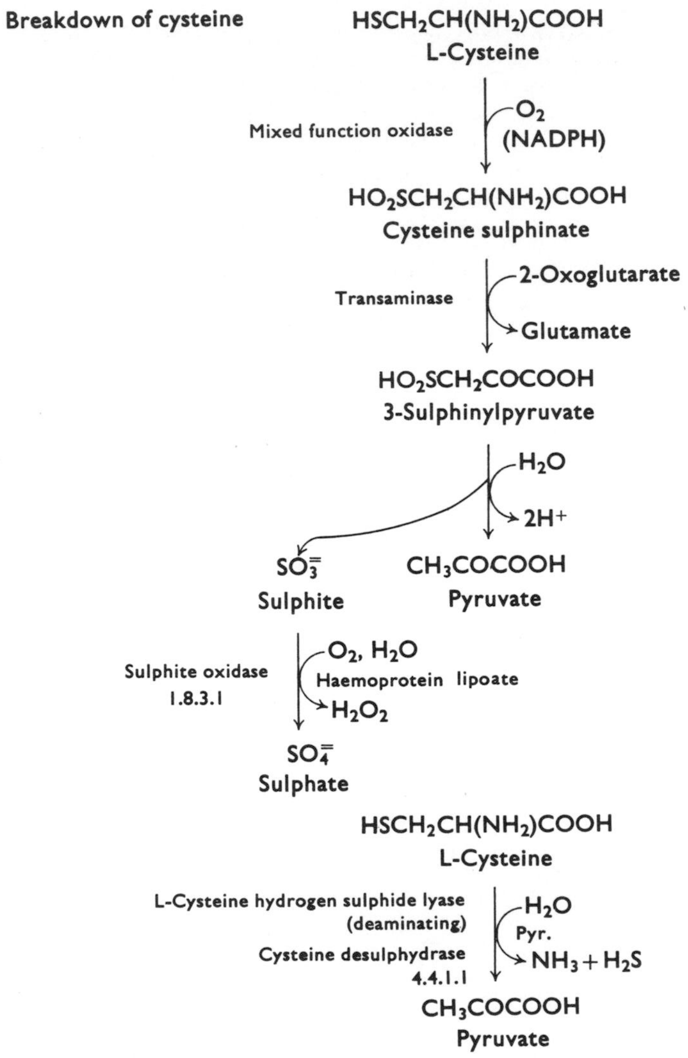

Breakdown of cysteine

$HSCH_2CH(NH_2)COOH$
L-Cysteine **D**

Mixed function oxidase
↓ O_2 (NADPH)

$HO_2SCH_2CH(NH_2)COOH$
Cysteine sulphinate **E**

Transaminase
↓ 2-Oxoglutarate → Glutamate

$HO_2SCH_2COCOOH$
3-Sulphinylpyruvate **F**

↓ H_2O → $2H^+$

$SO_3^=$ Sulphite; $CH_3COCOOH$ Pyruvate **G**

H

Sulphite oxidase
1.8.3.1
↓ O_2, H_2O; Haemoprotein lipoate → H_2O_2

$SO_4^=$
Sulphate **J**

$HSCH_2CH(NH_2)COOH$
L-Cysteine **K**

L-Cysteine hydrogen sulphide lyase (deaminating)
Cysteine desulphydrase
4.4.1.1
↓ H_2O, Pyr. → $NH_3 + H_2S$

$CH_3COCOOH$
Pyruvate **L**

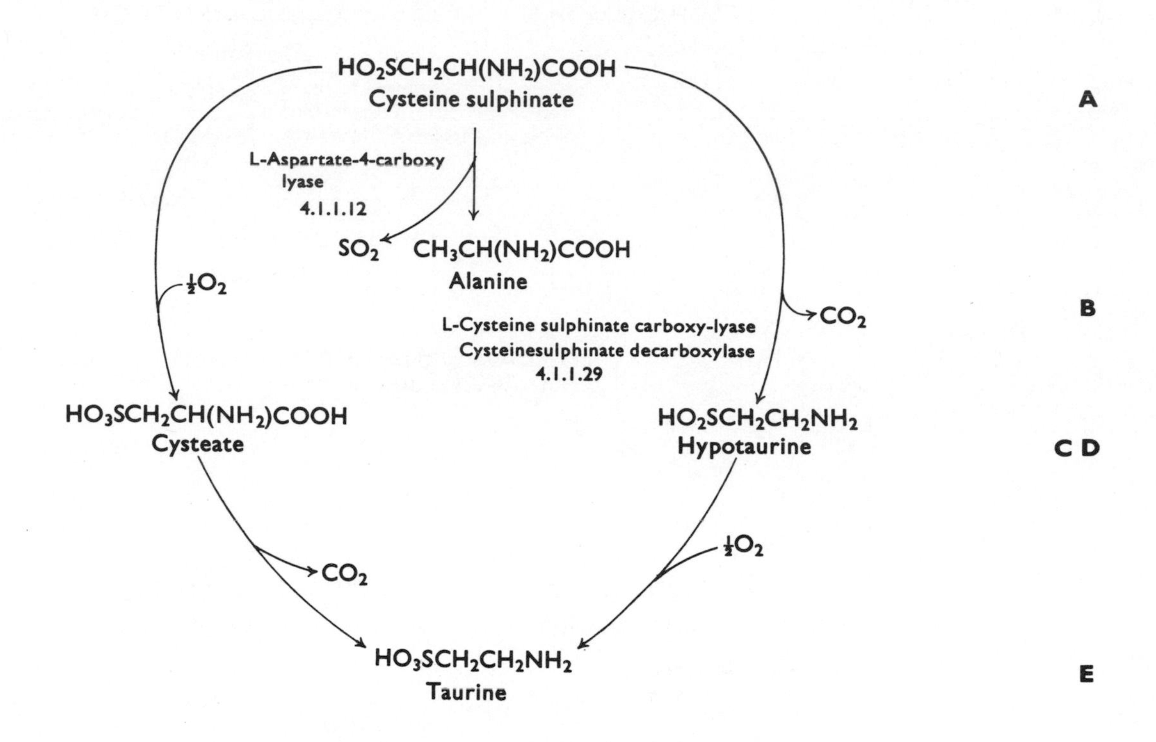

Biosynthesis of L-Alanine

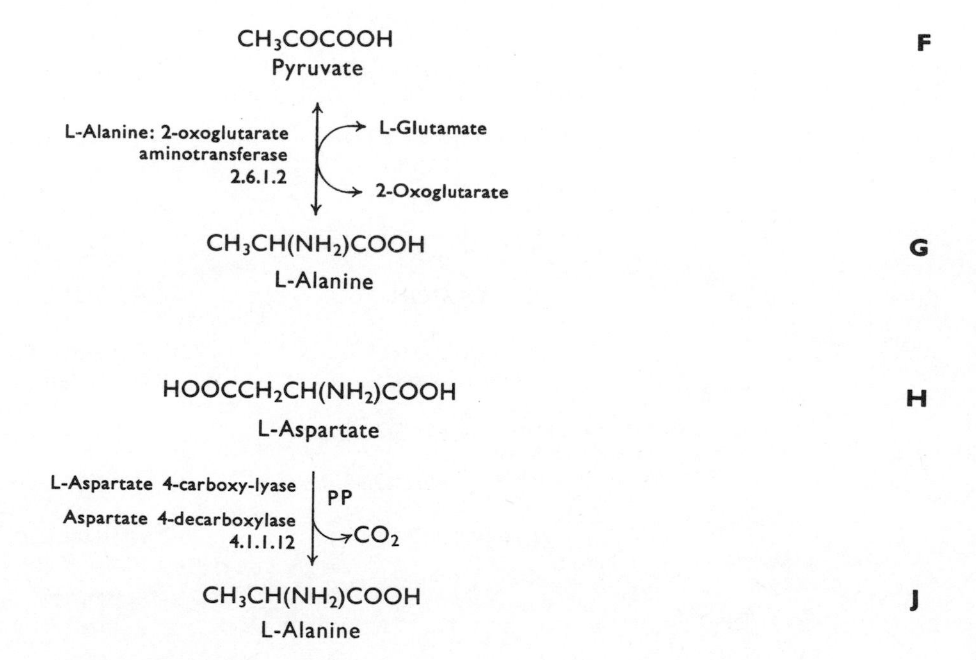

Biosynthesis of Valine and Isoleucine

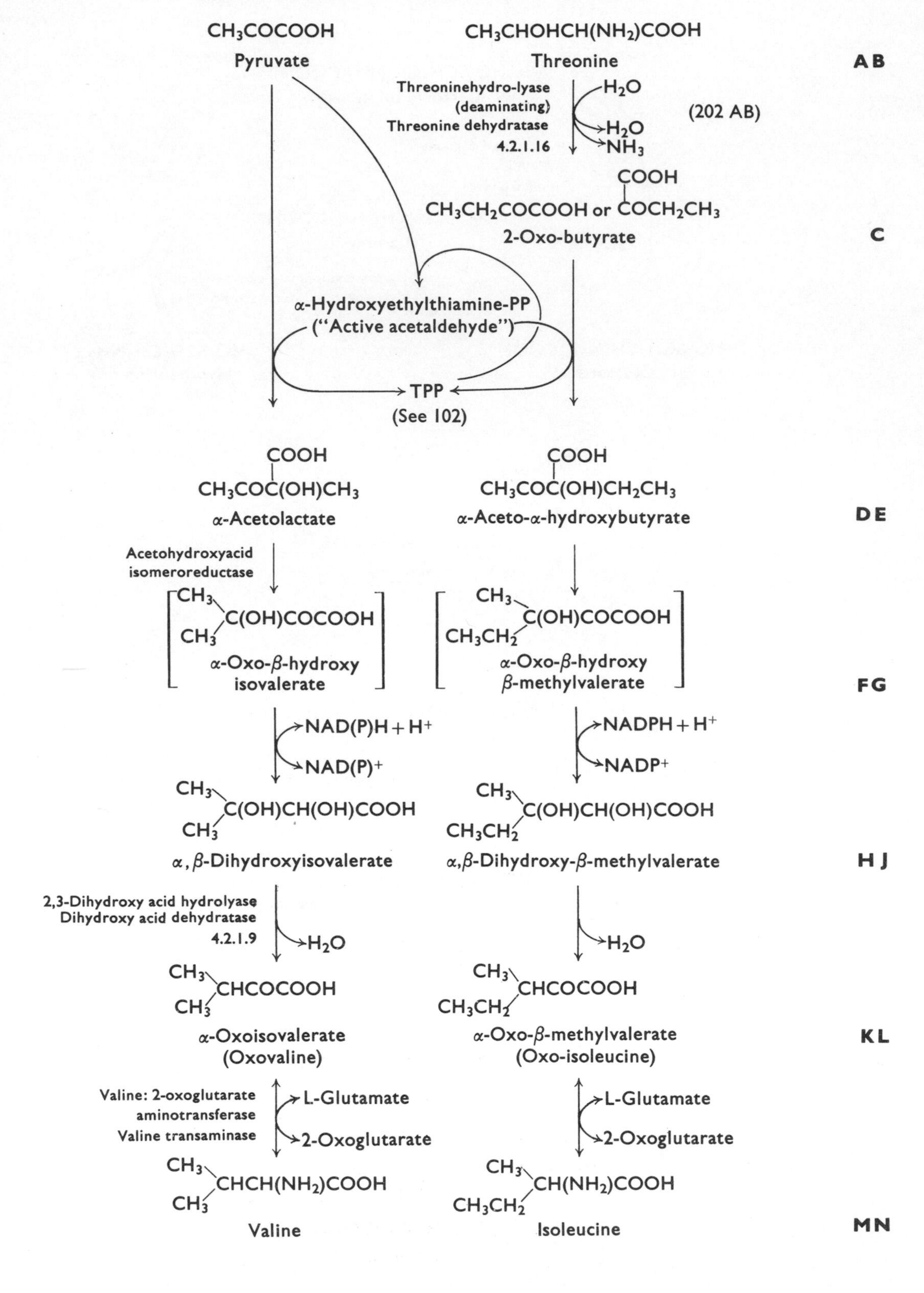

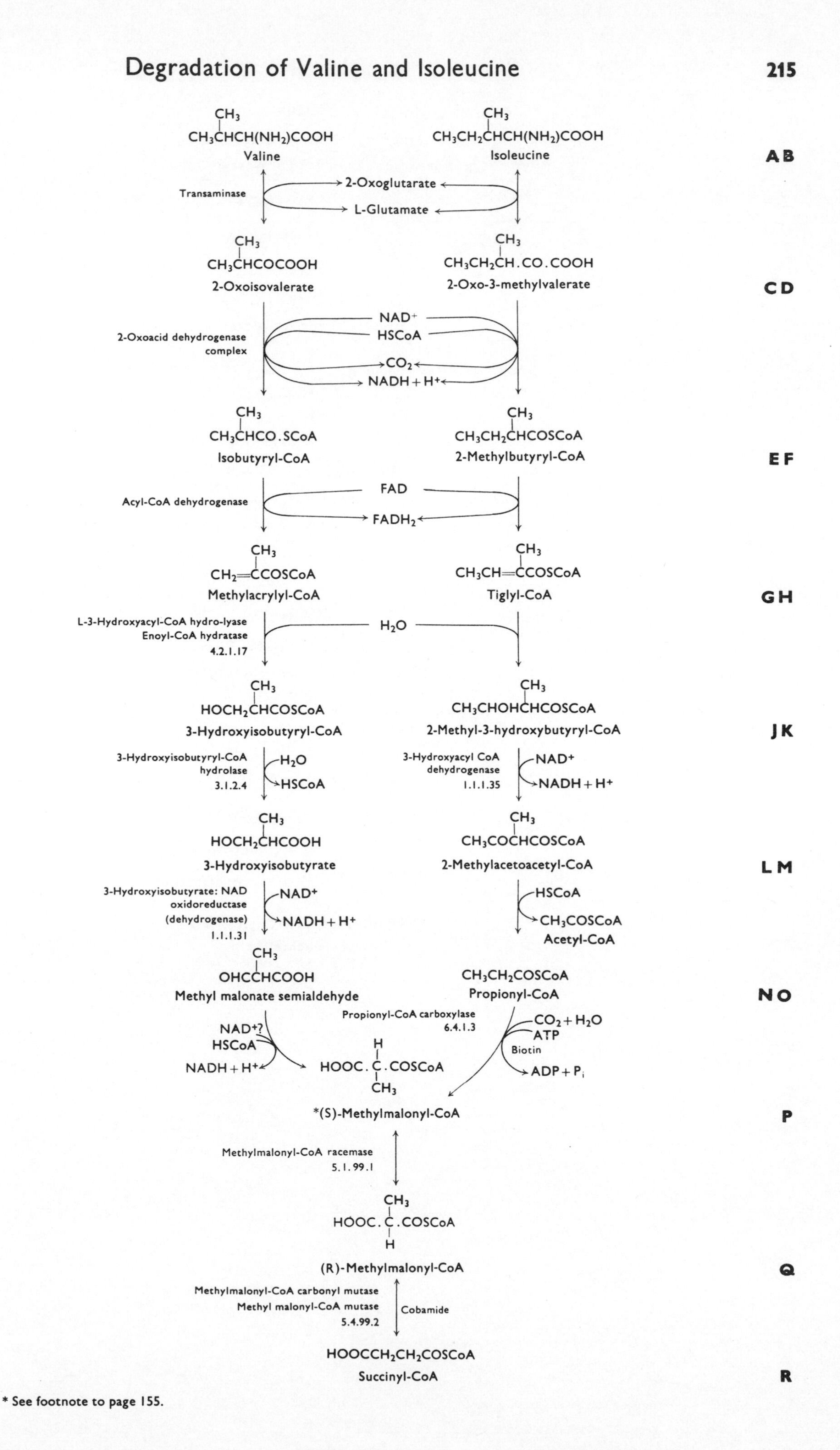

* See footnote to page 155.

Biosynthesis of Leucine

COOH
$(CH_3)_2CHCO$ $CH_3COSCoA$
2-Oxoisovalerate (from 215 C) Acetyl-CoA — **A B**

↓ ← H_2O → HSCoA

COOH
$(CH_3)_2CHC(OH)CH_2COOH$
3-Carboxy-3-hydroxyisocaproate — **C**

2-Hydroxy-3-carboxyisocaproate isomerase ⇅ → H_2O

[COOH
$(CH_3)_2CHC{=}CHCOOH$
Dimethyl citraconate] — **D**

Ditto ⇅ ← H_2O

COOH
$(CH_3)_2CHCHCH(OH)COOH$
3-Carboxy-2-hydroxyisocaproate — **E**

2-Hydroxy-3-carboxyisocaproate dehydrogenase ↓ ← NAD^+ → $NADH + H^+$, CO_2

$(CH_3)_2CHCH_2COCOOH$
2-Oxoisocaproate (Oxoleucine) — **F**

L-Leucine: 2-oxoglutarate aminotransferase
Leucine transaminase
2.6.1.6 ⇅ ← L-Glutamate → 2-Oxoglutarate

$(CH_3)_2CHCH_2CH(NH_2)COOH$
L-Leucine — **G**

Degradation of Leucine

$(CH_3)_2CHCH_2CH(NH_2)COOH$
L-Leucine **A**

L-Leucine: 2-oxoglutarate aminotransferase
Leucine transaminase
2.6.1.6
(↕) → 2-Oxoglutarate; → L-Glutamate

$(CH_3)_2CHCH_2COCOOH$
2-Oxoisocaproate **B**

2-Oxoacid dehydrogenase complex
↓ NAD^+, HSCoA → CO_2, $NADH + H^+$

$(CH_3)_2CHCH_2COSCoA$
Isovaleryl coenzyme A **C**

Acyl-CoA dehydrogenase
1.3.99.3
↓ FAD → $FADH_2$

$CH_3C(CH_3){=}CHCOSCoA$
3-Methylcrotonyl coenzyme A **D**

3-Methyl crotonyl-CoA: carbon dioxide ligase (ADP)
Methyl crotonyl-CoA carboxylase
6.4.1.4
↓ $CO_2 + H_2O$, ATP, Biotin → $ADP + P_i$

$HOOCCH_2C(CH_3){=}CHCOSCoA$
3-Methylglutaconyl coenzyme A **E**

3-Hydroxy-3-methyl glutanyl-CoA hydrolyase
Methylglutaconyl-CoA hydratase
4.2.1.18
↓ H_2O

$HOOCCH_2C(CH_3)(OH)CH_2COSCoA$
3-Hydroxy-3-methylglutaryl coenzyme A **F**

3-Hydroxy-3-methyl glutaryl-CoA acetoacetate-lyase
Hydroxymethylglutaryl CoA lyase
4.1.3.4

$HOOCCH_2COCH_3$
Acetoacetate

$CH_3COSCoA$
Acetyl coenzyme A **G**

CH_2CH_2COOH
|
$COCOOH$
2-Oxoglutarate — $CH_3COSCoA + H_2O$ Acetyl coenzyme A — **A**

↓ → HSCoA

CH_2CH_2COOH
|
$C(OH)COOH$
|
CH_2COOH
Homocitrate — **B**

Homocitrate hydrolyase ↕ → H_2O

CH_2CH_2COOH
|
$C.COOH$
||
$CH.COOH$
cis-Homoaconitate — **C**

↕ ← H_2O

CH_2CH_2COOH
|
$CH.COOH$
|
$CH(OH)COOH$
Homoisocitrate — **D**

↕ $NAD(P)^+$ → $NAD(P)H + H^+$

CH_2CH_2COOH
|
$CHCOOH$
|
$COCOOH$
Oxaloglutarate — **E**

↓ → CO_2

CH_2CH_2COOH
|
CH_2
|
$COCOOH$
2-Oxoadipate — **F**

$HOOCCH_2CH_2CH_2COCOOH$
2-Oxoadipate **A**

↕ → L-Glutamate
→ 2-Oxoglutarate

$HOOCCH_2CH_2CH_2CH(NH_2)COOH$
2-Aminoadipate **B**

2-Aminoadipate reductase
↓ $NADH + H^+$, ATP, Mg^{++} → $AMP + PP_i$, NAD^+

$OHCCH_2CH_2CH_2CH(NH_2)COOH$
2-Aminoadipate semialdehyde **C**

↓ $H_2NCH(COOH)CH_2CH_2COOH$ (L-Glutamate) → H_2O

$$\left[\begin{array}{l} N{-}CH(COOH)CH_2CH_2COOH \\ \| \\ HC{-}CH_2CH_2CH_2CH(NH_2)COOH \end{array}\right]$$ **D**

"Saccharopine reductase"
↓ $NAD(P) + H^+$ → $NAD(P)^+$

$$\begin{array}{l} NH{-}CH(COOH)CH_2CH_2COOH \\ | \\ CH_2CH_2CH_2CH_2CH(NH_2)COOH \end{array}$$
Saccharopine **E**

Saccharopine: NAD oxidoreductase
↓ NAD^+, H_2O → $NADH + H^+$, 2-Oxoglutarate

$H_2NCH_2CH_2CH_2CH_2CH(NH_2)COOH$
L-Lysine **F**

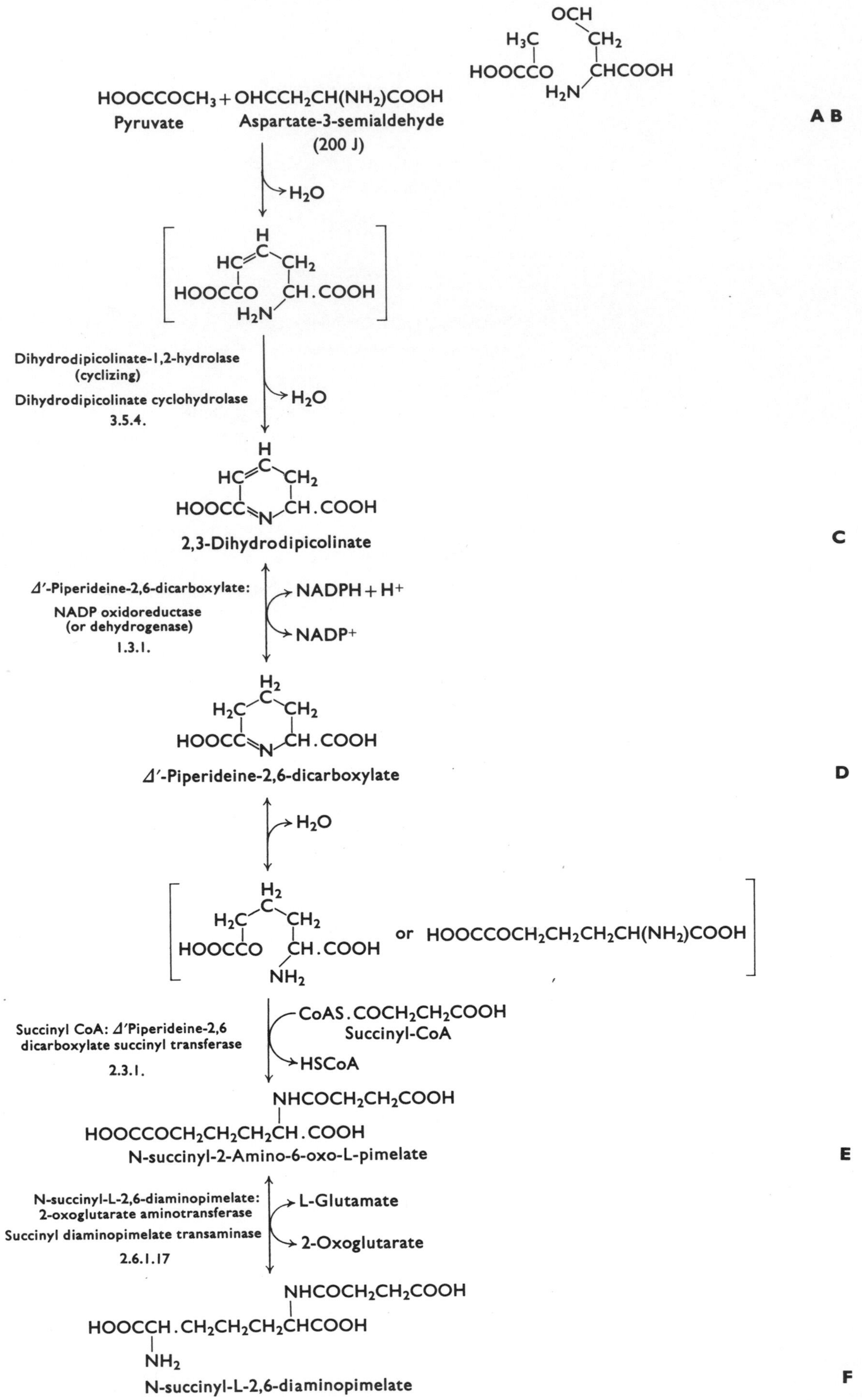

A B

$HOOCCOCH_3 + OHCCH_2CH(NH_2)COOH$

Pyruvate — Aspartate-3-semialdehyde (200 J)

[cyclic intermediate: OCH, H_3C, CH_2, HOOCCO, CHCOOH, H_2N]

↓ $\rightarrow H_2O$

[H; HC=C–CH_2; HOOCCO CH.COOH; H_2N]

Dihydrodipicolinate-1,2-hydrolase (cyclizing)

Dihydrodipicolinate cyclohydrolase

3.5.4.

↓ $\rightarrow H_2O$

C

H; HC=C–CH_2; HOOCC=N–CH.COOH

2,3-Dihydrodipicolinate

Δ′-Piperideine-2,6-dicarboxylate: NADP oxidoreductase (or dehydrogenase)

1.3.1.

↕ $NADPH + H^+$ / $NADP^+$

D

H_2; H_2C–C–CH_2; HOOCC=N–CH.COOH

Δ′-Piperideine-2,6-dicarboxylate

↕ H_2O

[H_2; H_2C–C–CH_2; HOOCCO CH.COOH; NH_2 or $HOOCCOCH_2CH_2CH_2CH(NH_2)COOH$]

Succinyl CoA: Δ′Piperideine-2,6 dicarboxylate succinyl transferase

2.3.1.

↓ $CoAS.COCH_2CH_2COOH$ (Succinyl-CoA) / HSCoA

E

$NHCOCH_2CH_2COOH$

$HOOCCOCH_2CH_2CH_2CH.COOH$

N-succinyl-2-Amino-6-oxo-L-pimelate

N-succinyl-L-2,6-diaminopimelate: 2-oxoglutarate aminotransferase

Succinyl diaminopimelate transaminase

2.6.1.17

↕ L-Glutamate / 2-Oxoglutarate

F

$NHCOCH_2CH_2COOH$

$HOOCCH.CH_2CH_2CH_2CHCOOH$

NH_2

N-succinyl-L-2,6-diaminopimelate

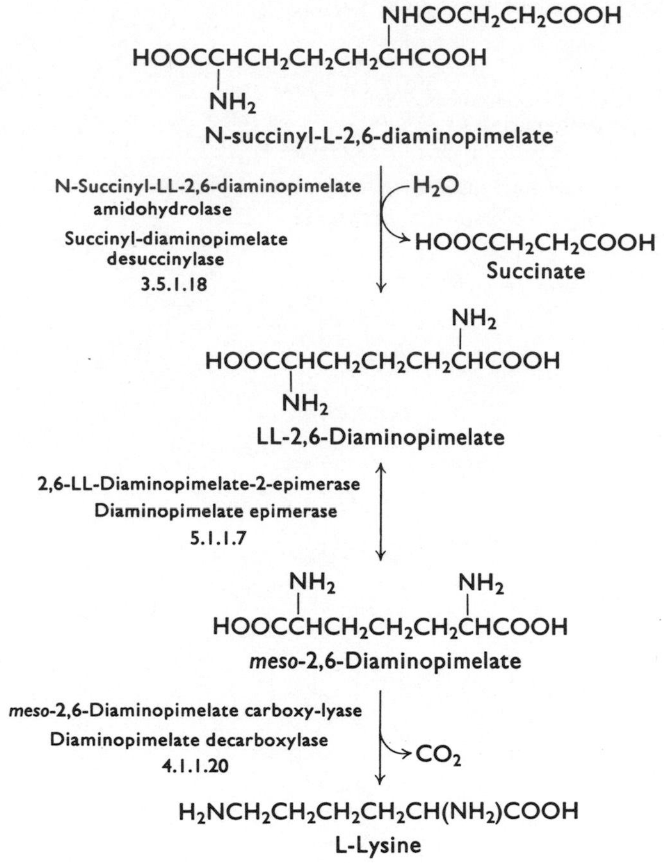

$$\begin{array}{c} \text{NHCOCH}_2\text{CH}_2\text{COOH} \\ | \\ \text{HOOCCHCH}_2\text{CH}_2\text{CH}_2\text{CHCOOH} \\ | \\ \text{NH}_2 \end{array}$$

N-succinyl-L-2,6-diaminopimelate **A**

N-Succinyl-LL-2,6-diaminopimelate amidohydrolase
Succinyl-diaminopimelate desuccinylase
3.5.1.18

$\downarrow$ H_2O → $HOOCCH_2CH_2COOH$ Succinate

$$\begin{array}{c} \text{NH}_2 \\ | \\ \text{HOOCCHCH}_2\text{CH}_2\text{CH}_2\text{CHCOOH} \\ | \\ \text{NH}_2 \end{array}$$

LL-2,6-Diaminopimelate **B**

2,6-LL-Diaminopimelate-2-epimerase
Diaminopimelate epimerase
5.1.1.7

$\updownarrow$

$$\begin{array}{c} \text{NH}_2 \qquad\qquad \text{NH}_2 \\ | \qquad\qquad\quad | \\ \text{HOOCCHCH}_2\text{CH}_2\text{CH}_2\text{CHCOOH} \end{array}$$

meso-2,6-Diaminopimelate **C**

meso-2,6-Diaminopimelate carboxy-lyase
Diaminopimelate decarboxylase
4.1.1.20

$\downarrow$ → CO_2

$H_2NCH_2CH_2CH_2CH_2CH(NH_2)COOH$

L-Lysine **D**

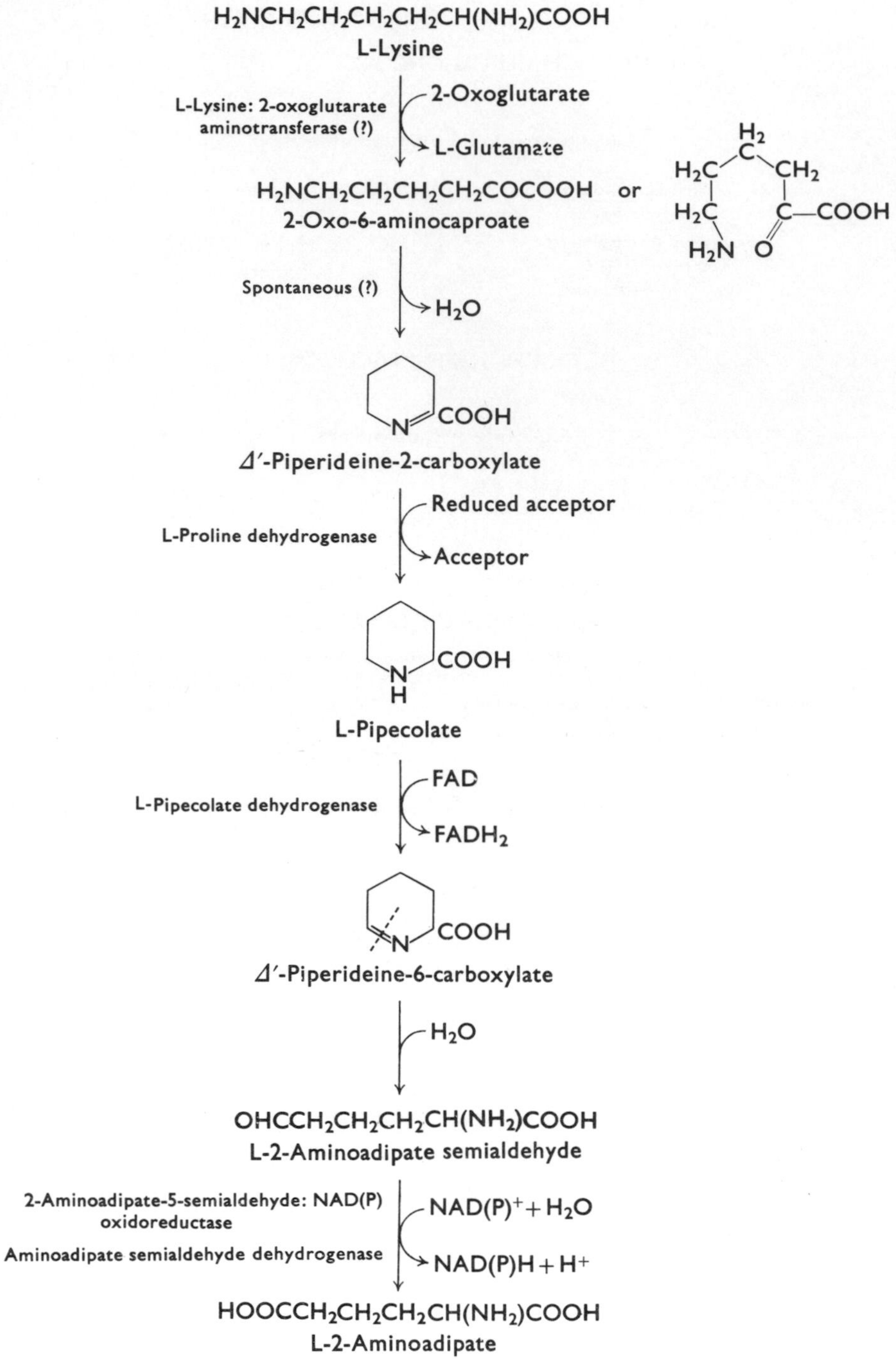

Catabolism of Lysine

$H_2NCH_2CH_2CH_2CH_2CH(NH_2)COOH$
L-Lysine **A**

L-Lysine: 2-oxoglutarate aminotransferase (?)
2-Oxoglutarate → L-Glutamate

$H_2NCH_2CH_2CH_2CH_2COCOOH$ or (cyclic form: H_2C, CH_2, H_2C, C—COOH, H_2N, O)
2-Oxo-6-aminocaproate **B**

Spontaneous (?)
→ H_2O

COOH
Δ′-Piperideine-2-carboxylate **C**

L-Proline dehydrogenase
Reduced acceptor → Acceptor

COOH
L-Pipecolate **D**

L-Pipecolate dehydrogenase
FAD → $FADH_2$

COOH
Δ′-Piperideine-6-carboxylate **E**

H_2O

$OHCCH_2CH_2CH_2CH(NH_2)COOH$
L-2-Aminoadipate semialdehyde **F**

2-Aminoadipate-5-semialdehyde: NAD(P) oxidoreductase
Aminoadipate semialdehyde dehydrogenase
$NAD(P)^+ + H_2O$ → $NAD(P)H + H^+$

$HOOCCH_2CH_2CH_2CH(NH_2)COOH$
L-2-Aminoadipate **G**

$HOOCCH_2CH_2CH_2CH(NH_2)COOH$
2-Aminoadipate — **A**

Aminoadipate: 2-oxoglutarate aminotransferase
↓ 2-Oxoglutarate → L-Glutamate

$HOOCCH_2CH_2CH_2COCOOH$
2-Oxoadipate — **B**

2-Oxoacid dehydrogenase complex
↓ NAD^+, HSCoA, TPP, lipoate etc. → CO_2, $NADH + H^+$

$HOOCCH_2CH_2CH_2COSCoA$
Glutarylcoenzyme A — **C**

↓ FAD → $FADH_2$

[$HOOCCH_2CH{=}CHCOSCoA$
Glutaconyl coenzyme A] — **D**

↓ → CO_2

$CH_3CH{=}CHCOSCoA$
Crotonyl coenzyme A — **E**

L-3-Hydroxyacyl-CoA hydro-lyase
Enoyl-CoA hydratase
(crotonase)
4.2.1.17
↓ H_2O

$CH_3CH(OH)CH_2COSCoA$
3-Hydroxybutyryl coenzyme A — **F**

L-3-Hydroxyacyl-CoA: NAD oxidoreductase
3-Hydroxyacyl-CoA dehydrogenase
1.1.1.35
⇅ NAD^+ → $NADH + H^+$

$CH_3COCH_2COSCoA$
Acetoacetyl coenzyme A — **G**

Acetyl-CoA: acetyl-CoA C-acetyl transferase
Acetyl-CoA transacetylase
Acetoacetyl-CoA thiolase
2.3.1.9
⇅ → HSCoA

$2CH_3COSCoA$
Acetyl coenzyme A — **H**

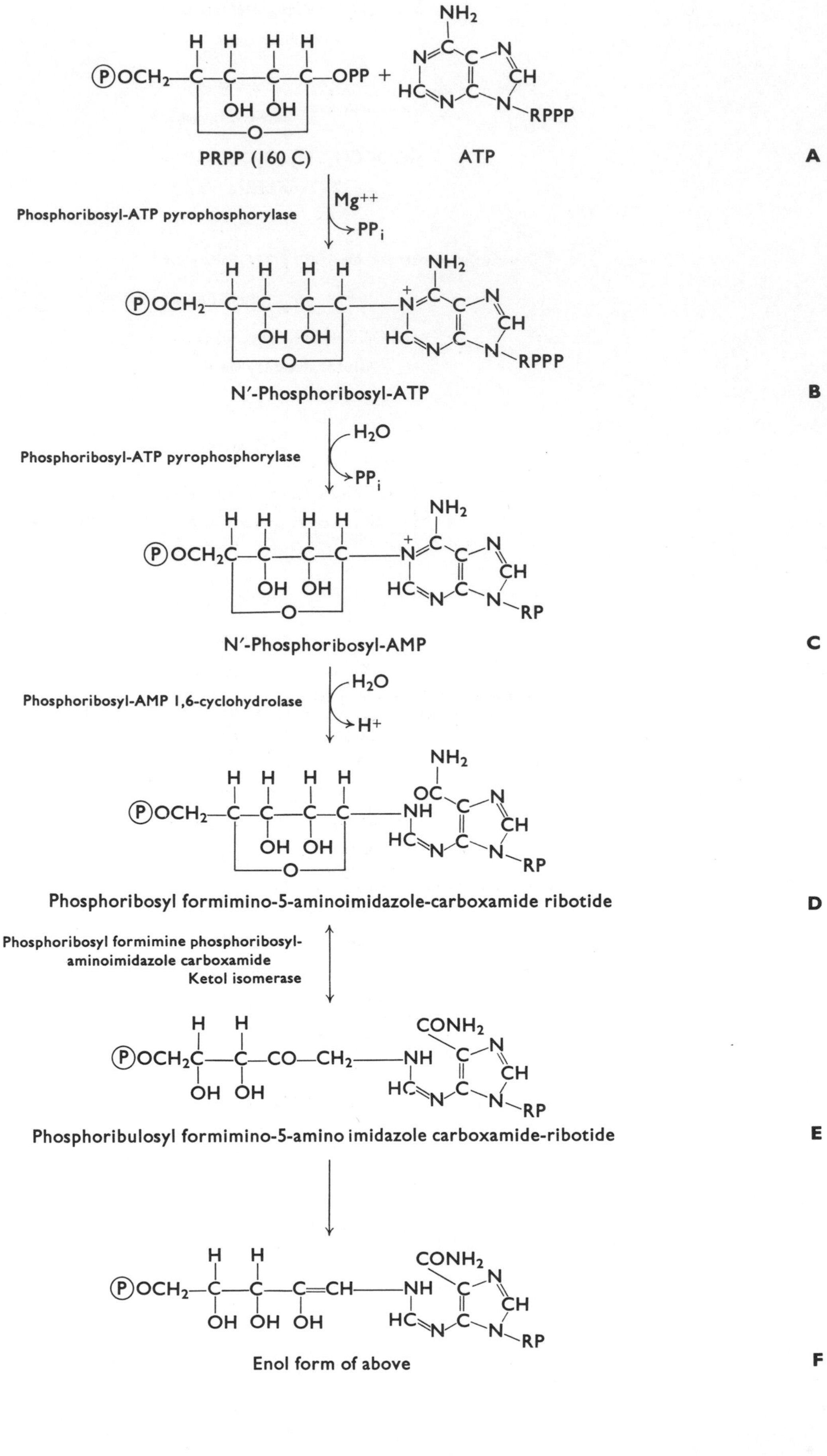

PRPP (160 C) + ATP — A

Phosphoribosyl-ATP pyrophosphorylase (Mg^{++}; → PP_i)

N′-Phosphoribosyl-ATP — B

Phosphoribosyl-ATP pyrophosphorylase (H_2O; → PP_i)

N′-Phosphoribosyl-AMP — C

Phosphoribosyl-AMP 1,6-cyclohydrolase (H_2O; → H^+)

Phosphoribosyl formimino-5-aminoimidazole-carboxamide ribotide — D

Phosphoribosyl formimine phosphoribosyl-aminoimidazole carboxamide Ketol isomerase

Phosphoribulosyl formimino-5-amino imidazole carboxamide-ribotide — E

Enol form of above — F

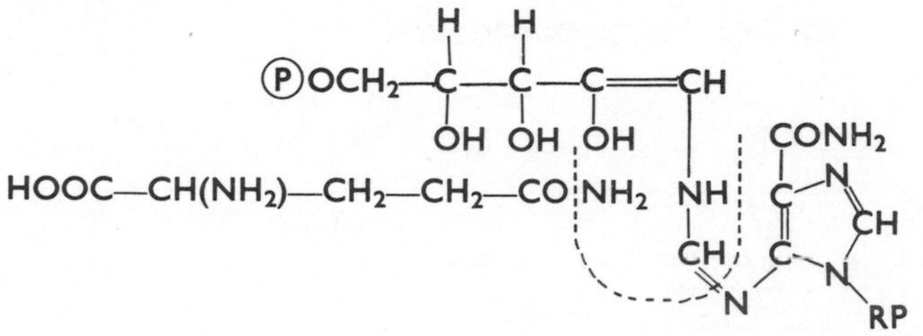

ⓅOCH_2—CH(OH)—CH(OH)—C(OH)=CH— ... $HOOC—CH(NH_2)—CH_2—CH_2—CO$ NH_2 NH ... CH=N ... $CONH_2$ ... RP

L-Glutamine amidotransferase

Phosphoribulosyl-AICAR (enol form) (162 C) **A B**

$HOOCCH(NH_2)CH_2CH_2COOH$

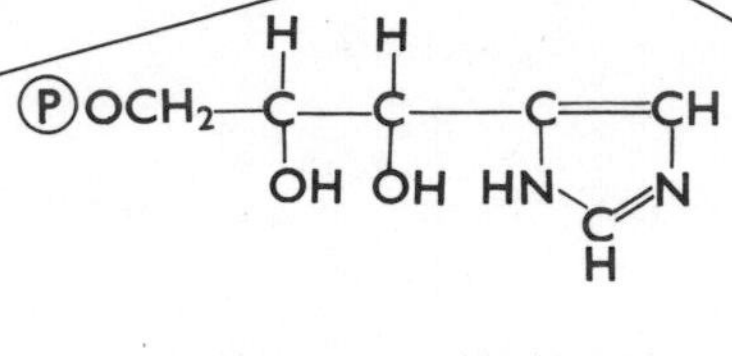

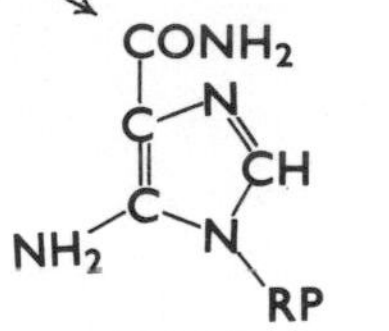

ⓅOCH_2—CH(OH)—CH(OH)—C=CH (imidazole)

$CONH_2$ ... NH_2 ... RP

Glutamate Imidazole glycerol phosphate AICAR **C D E**

D-*erythro*-imidazoleglycerolphosphate hydro-lyase
Imidazole glycerolphosphate dehydratase
4.2.1.19
(Same enzyme as 3.1.3.15)

Mn^{++} → H_2O

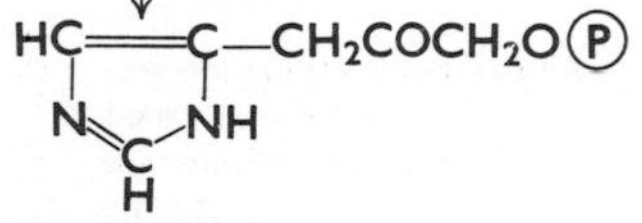

HC=C—CH_2COCH_2OⓅ (N—CH—NH)

Imidazole acetol phosphate **F**

L-Histidinol phosphate: 2-oxoglutarate aminotransferase
Histidinol phosphate transaminase
2.6.1.9

L-Glutamate → Pyr. → 2-Oxoglutarate

HC=C—$CH_2CH(NH_2)CH_2O$Ⓟ (N—CH—NH)

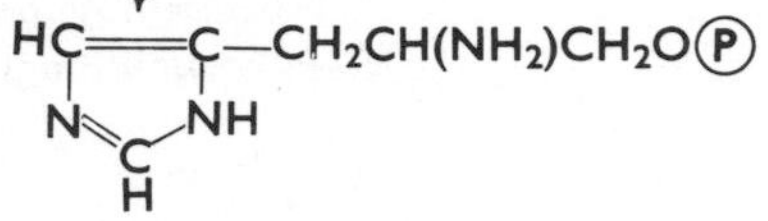

L-Histidinol phosphate **G**

L-Histidinol phosphate phosphohydrolase
Histidinol phosphatase
3.1.3.15
(Same enzyme as 4.2.1.19)

H_2O → Pyr. → P_i

HC=C—$CH_2CH(NH_2)CH_2OH$ (N—CH—NH)

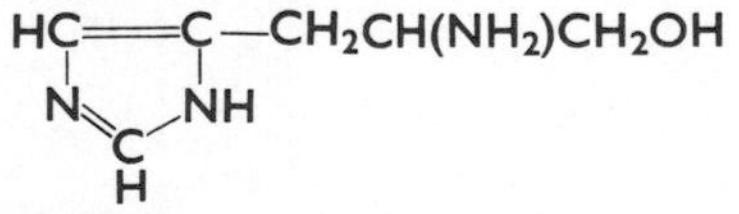

L-Histidinol **H**

L-Histidinol: NAD oxidoreductase
Histidinol dehydrogenase
1.1.1.23

NAD^+ → $NADH+H^+$

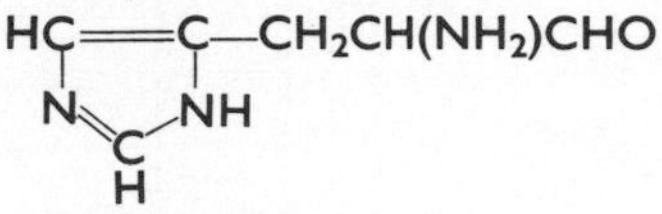

HC=C—$CH_2CH(NH_2)CHO$ (N—CH—NH)

L-Histidinal **J**

1.1.1.23 as above

$NAD^+ + H_2O$ → $NADH+H^+$

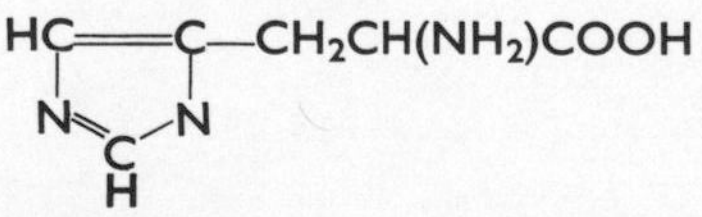

HC=C—$CH_2CH(NH_2)COOH$ (N—CH—N)

L-Histidine **K**

Catabolism of Histidine

CH=C—CH₂CH(NH₂)COOH (imidazole ring: N, NH, CH)

L-Histidine — **A**

L-Histidine ammonia lyase (Histidase) 4.3.1.3 → NH_3

CH=C—CH=CHCOOH (imidazole ring: N, NH, CH)

Urocanate — **B**

Urocanase ⇅ H_2O

CO—CHCH₂CH₂COOH (ring: N, NH, CH)

4-Imidazolone-5-propionate

$\frac{1}{2}O_2$ (FAD) →

CO—CHCH₂CH₂COOH (ring: HN, NH, C=O)

Hydantoin-5-propionate — **CD**

(Excreted by rat, man and monkey)

4-Imidazolone-5-propionate amidohydrolase
Imidazolone propionase
3.5.2.7 ← H_2O

HOOC—CHCH₂CH₂COOH (HN=CH—NH)

Formiminoglutamate
(α-Formamidinoglutarate) — **E**

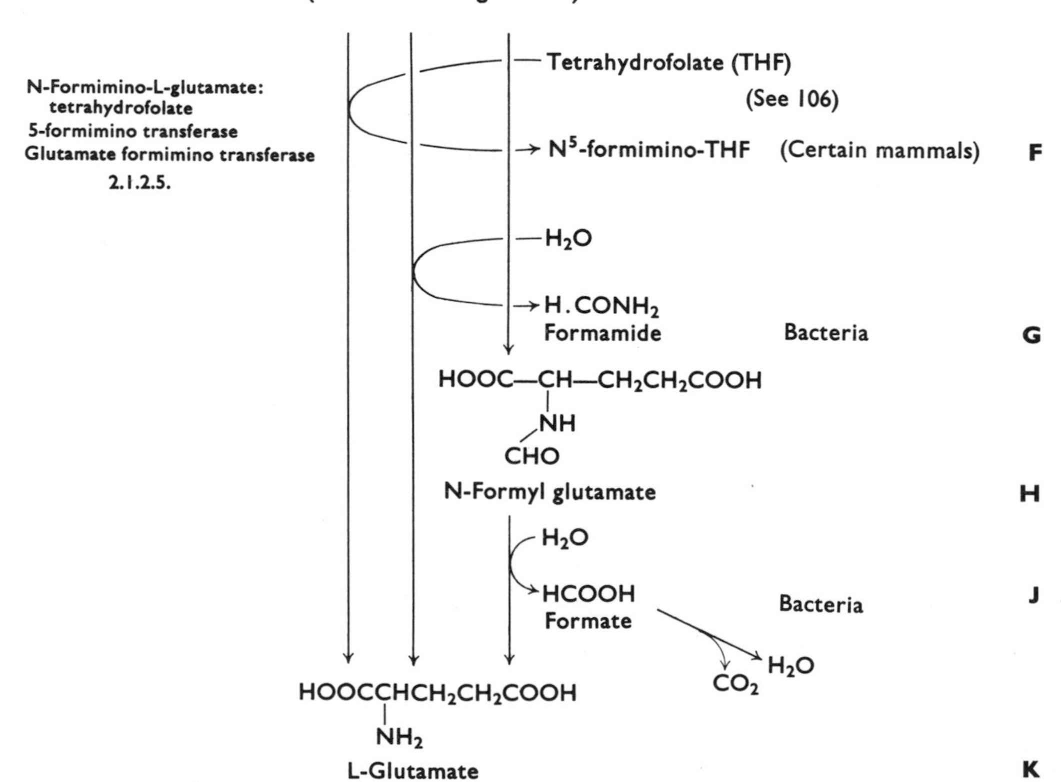

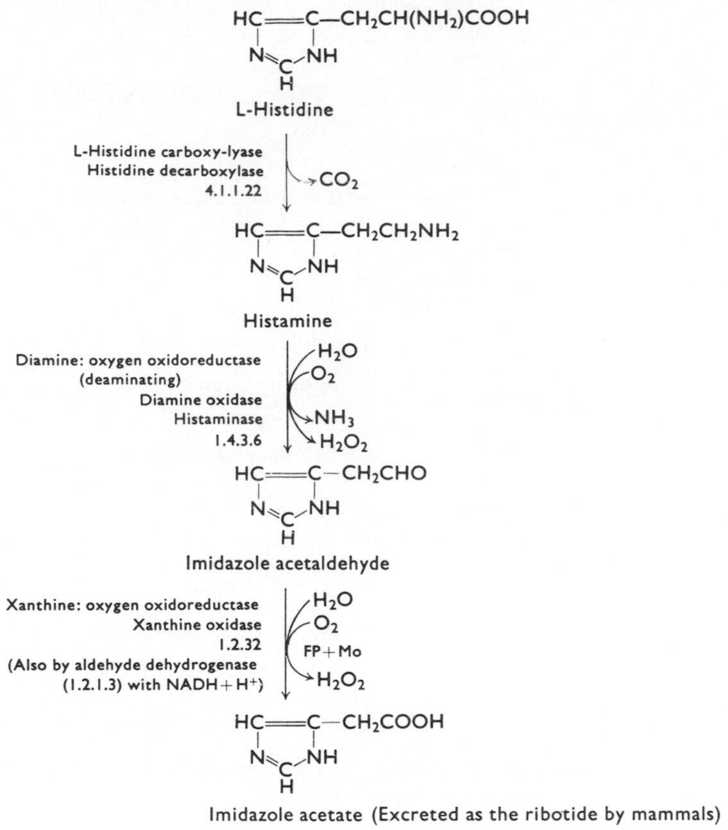

$HC{=}C{-}CH_2CH(NH_2)COOH$ (imidazole ring: N=CH–NH)

L-Histidine **A**

L-Histidine carboxy-lyase
Histidine decarboxylase
4.1.1.22
→ CO_2

$HC{=}C{-}CH_2CH_2NH_2$ (imidazole ring: N=CH–NH)

Histamine **B**

Diamine: oxygen oxidoreductase (deaminating)
Diamine oxidase
Histaminase
1.4.3.6
H_2O, O_2 → NH_3, H_2O_2

$HC{=}C{-}CH_2CHO$ (imidazole ring: N=CH–NH)

Imidazole acetaldehyde **C**

Xanthine: oxygen oxidoreductase
Xanthine oxidase
1.2.32
(Also by aldehyde dehydrogenase (1.2.1.3) with $NADH+H^+$)
H_2O, O_2, FP+Mo → H_2O_2

$HC{=}C{-}CH_2COOH$ (imidazole ring: N=CH–NH)

Imidazole acetate (Excreted as the ribotide by mammals) **D**

Metabolism of Imidazole acetate in *Pseudomonas*

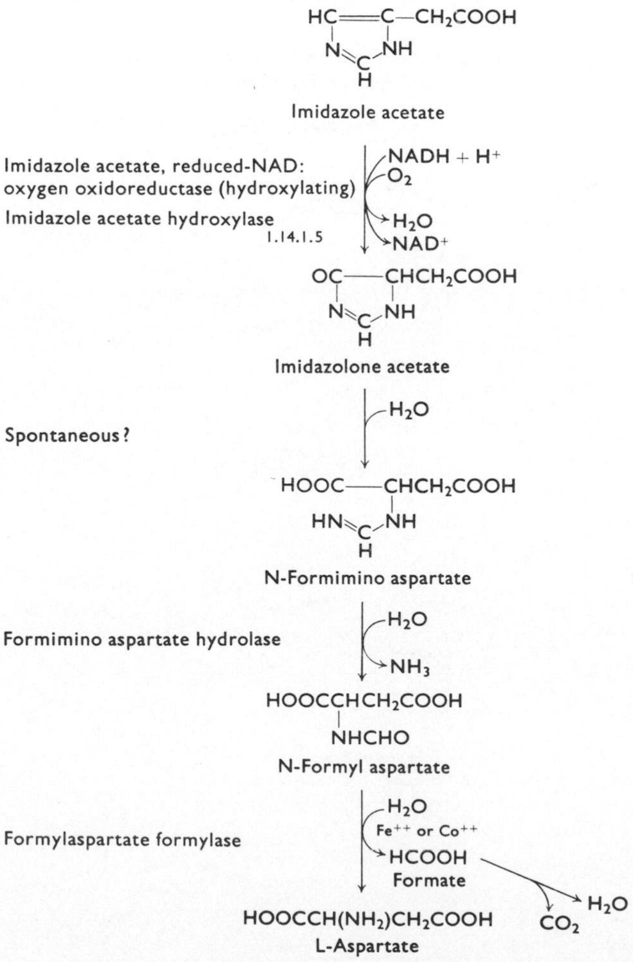

$HC{=}C{-}CH_2COOH$ (imidazole ring: N=CH–NH)

Imidazole acetate **E**

Imidazole acetate, reduced-NAD: oxygen oxidoreductase (hydroxylating)
Imidazole acetate hydroxylase
1.14.1.5
$NADH+H^+$, O_2 → H_2O, NAD^+

$OC{-}CHCH_2COOH$ (ring: N=CH–NH)

Imidazolone acetate **F**

Spontaneous?
H_2O

$HOOC{-}CHCH_2COOH$ (HN=CH–NH)

N-Formimino aspartate **G**

Formimino aspartate hydrolase
H_2O → NH_3

$HOOCCHCH_2COOH$ with $NHCHO$

N-Formyl aspartate **H**

Formylaspartate formylase
H_2O, Fe^{++} or Co^{++} → HCOOH Formate → CO_2 + H_2O

$HOOCCH(NH_2)CH_2COOH$

L-Aspartate **J**

Metabolism of Hydantoin-5-propionate (Bacteria)

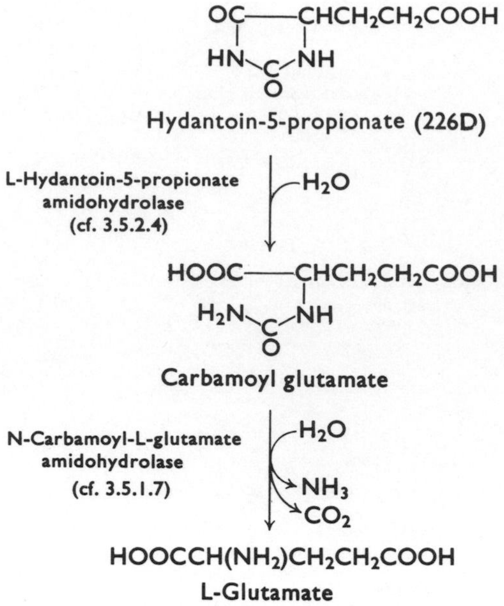

OC——CHCH$_2$CH$_2$COOH
HN–C(=O)–NH

Hydantoin-5-propionate (226D) **A**

L-Hydantoin-5-propionate amidohydrolase (cf. 3.5.2.4)
↓ $-H_2O$

HOOC——CHCH$_2$CH$_2$COOH
H$_2$N–C(=O)–NH

Carbamoyl glutamate **B**

N-Carbamoyl-L-glutamate amidohydrolase (cf. 3.5.1.7)
↓ $-H_2O$, → NH_3, → CO_2

HOOCCH(NH$_2$)CH$_2$CH$_2$COOH
L-Glutamate **C**

Imidazolone propionate may be hydrolysed non-enzymically to form N-formyl isoglutamine

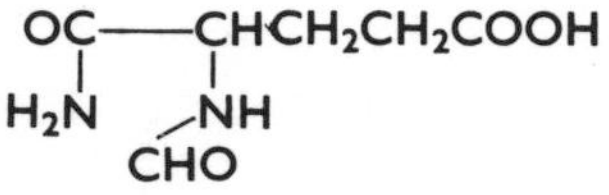

OC——CHCH$_2$CH$_2$COOH
H$_2$N   NH–CHO

D

It may also be oxidized to 4-oxoglutaramate ammonia and formate

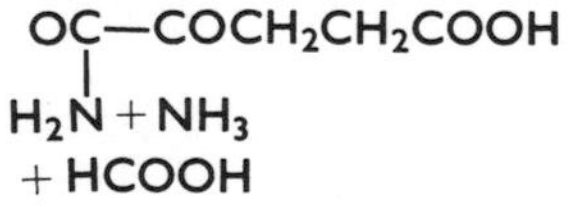

OC—COCH$_2$CH$_2$COOH
H$_2$N + NH$_3$
+ HCOOH

E

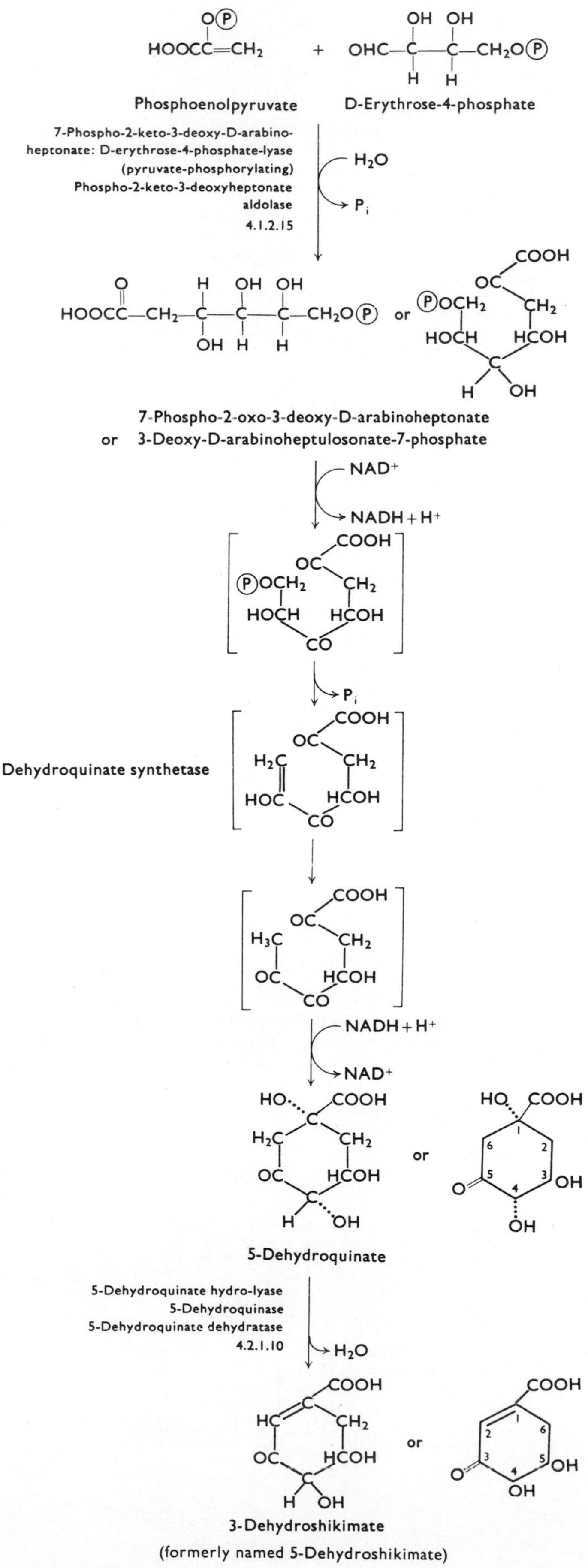

Phosphoenolpyruvate + D-Erythrose-4-phosphate

AB

7-Phospho-2-keto-3-deoxy-D-arabinoheptonate: D-erythrose-4-phosphate-lyase (pyruvate-phosphorylating)
Phospho-2-keto-3-deoxyheptonate aldolase
4.1.2.15

H_2O → P_i

7-Phospho-2-oxo-3-deoxy-D-arabinoheptonate
or 3-Deoxy-D-arabinoheptulosonate-7-phosphate

C

NAD^+ → $NADH + H^+$

P_i

Dehydroquinate synthetase

$NADH + H^+$ → NAD^+

5-Dehydroquinate

D

5-Dehydroquinate hydro-lyase
5-Dehydroquinase
5-Dehydroquinate dehydratase
4.2.1.10

H_2O

3-Dehydroshikimate
(formerly named 5-Dehydroshikimate)

E

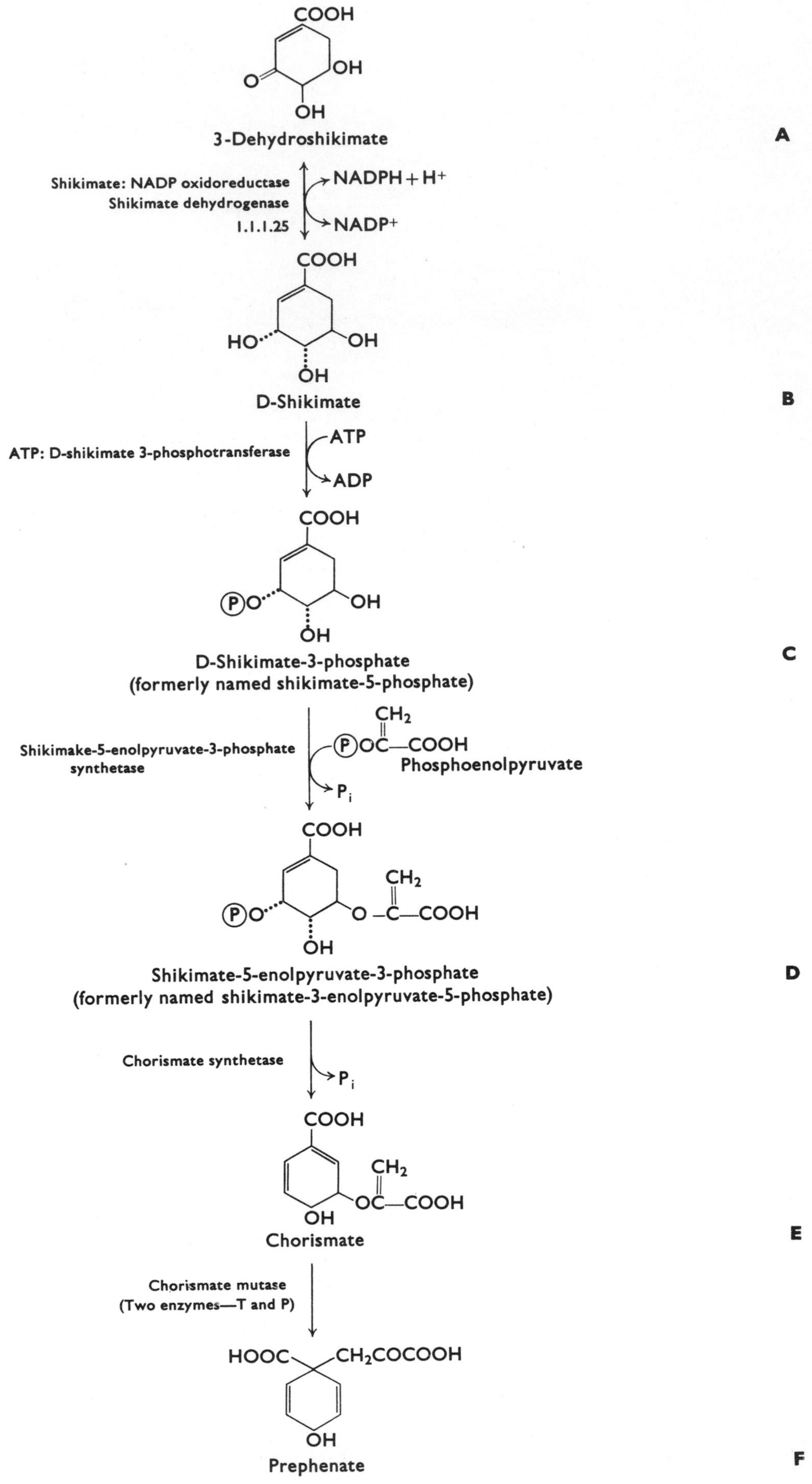
COOH
OH
O
OH
3-Dehydroshikimate
A
Shikimate: NADP oxidoreductase
Shikimate dehydrogenase
1.1.1.25
NADPH + H+
NADP+
COOH
HO
OH
OH
D-Shikimate
B
ATP: D-shikimate 3-phosphotransferase
ATP
ADP
COOH
P O
OH
OH
D-Shikimate-3-phosphate
(formerly named shikimate-5-phosphate)
C
Shikimake-5-enolpyruvate-3-phosphate
synthetase
CH2
P OC—COOH
Phosphoenolpyruvate
Pi
COOH
CH2
P O
O —C—COOH
OH
Shikimate-5-enolpyruvate-3-phosphate
(formerly named shikimate-3-enolpyruvate-5-phosphate)
D
Chorismate synthetase
Pi
COOH
CH2
OC—COOH
OH
Chorismate
E
Chorismate mutase
(Two enzymes—T and P)
HOOC
CH2COCOOH
OH
Prephenate
F

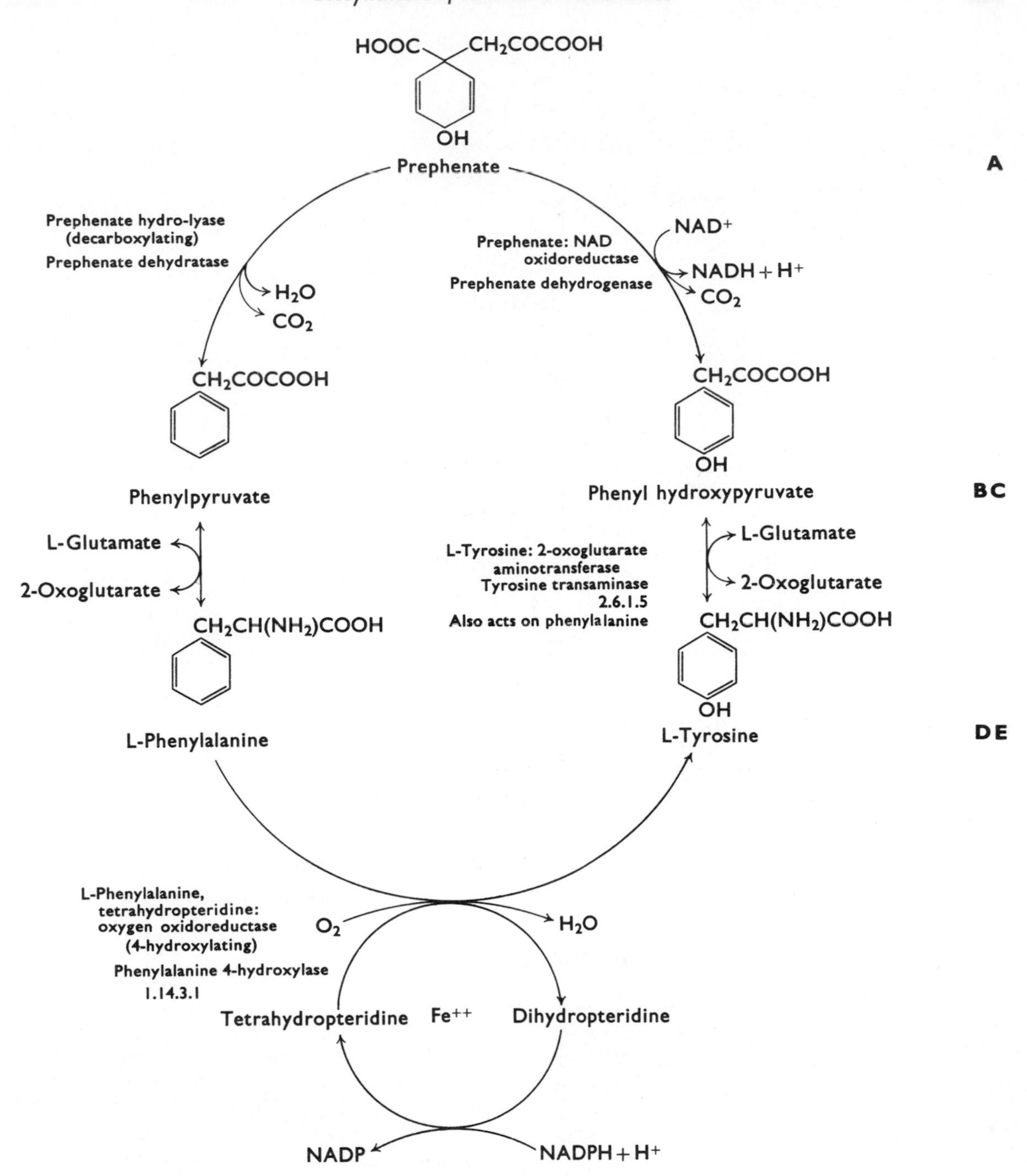
HOOC CH2COCOOH
OH
Prephenate
A
Prephenate hydro-lyase (decarboxylating)
Prephenate dehydratase
H2O
CO2
Prephenate: NAD oxidoreductase
Prephenate dehydrogenase
NAD+
NADH + H+
CO2
CH2COCOOH
Phenylpyruvate
CH2COCOOH
OH
Phenyl hydroxypyruvate
BC
L-Glutamate
2-Oxoglutarate
CH2CH(NH2)COOH
L-Phenylalanine
L-Tyrosine: 2-oxoglutarate aminotransferase
Tyrosine transaminase
2.6.1.5
Also acts on phenylalanine
L-Glutamate
2-Oxoglutarate
CH2CH(NH2)COOH
OH
L-Tyrosine
DE
L-Phenylalanine, tetrahydropteridine: oxygen oxidoreductase (4-hydroxylating)
Phenylalanine 4-hydroxylase
1.14.3.1
O2
H2O
Tetrahydropteridine
Fe++
Dihydropteridine
NADP
NADPH + H+

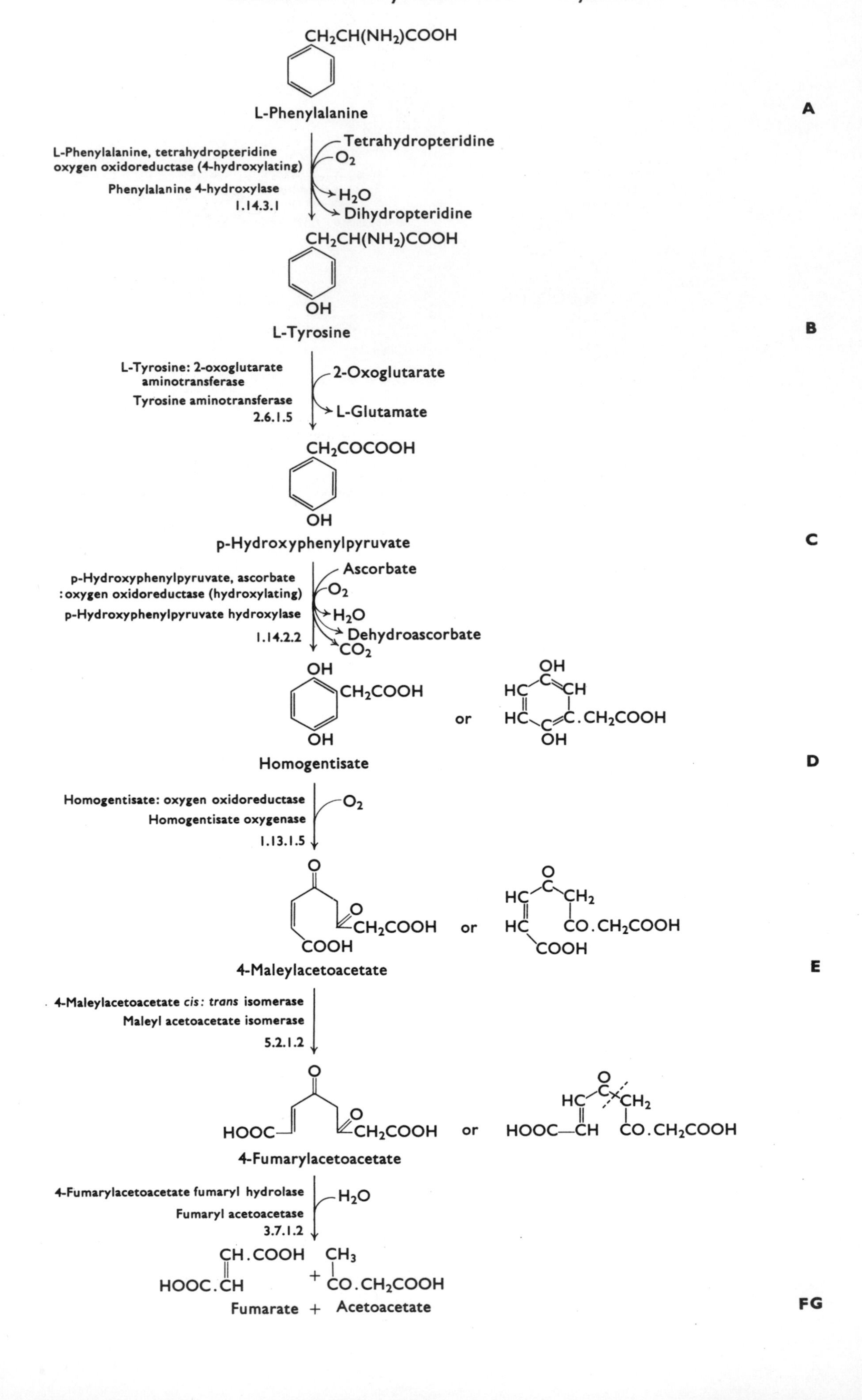

CH₂CH(NH₂)COOH
L-Phenylalanine
A
L-Phenylalanine, tetrahydropteridine oxygen oxidoreductase (4-hydroxylating)
Phenylalanine 4-hydroxylase
1.14.3.1
Tetrahydropteridine
O2
H2O
Dihydropteridine
CH2CH(NH2)COOH
OH
L-Tyrosine
B
L-Tyrosine: 2-oxoglutarate aminotransferase
Tyrosine aminotransferase
2.6.1.5
2-Oxoglutarate
L-Glutamate
CH2COCOOH
OH
p-Hydroxyphenylpyruvate
C
p-Hydroxyphenylpyruvate, ascorbate :oxygen oxidoreductase (hydroxylating)
p-Hydroxyphenylpyruvate hydroxylase
1.14.2.2
Ascorbate
O2
H2O
Dehydroascorbate
CO2
OH
CH2COOH
OH
or
HC.CH2COOH
Homogentisate
D
Homogentisate: oxygen oxidoreductase
Homogentisate oxygenase
1.13.1.5
O2
CH2COOH
COOH
or
CO.CH2COOH
4-Maleylacetoacetate
E
4-Maleylacetoacetate cis: trans isomerase
Maleyl acetoacetate isomerase
5.2.1.2
HOOC
CH2COOH
or
HOOC—CH CO.CH2COOH
4-Fumarylacetoacetate
4-Fumarylacetoacetate fumaryl hydrolase
Fumaryl acetoacetase
3.7.1.2
H2O
CH.COOH
CH3
HOOC.CH
CO.CH2COOH
Fumarate + Acetoacetate
FG

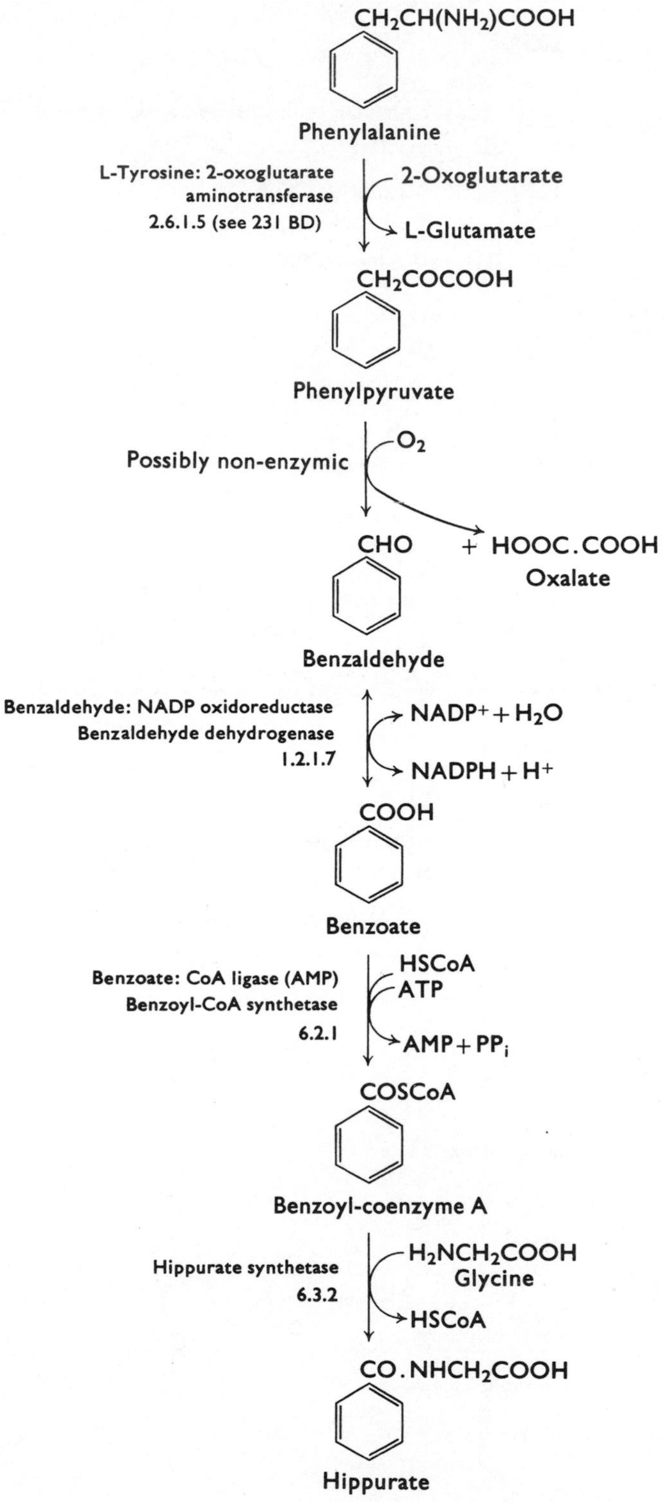

$CH_2CH(NH_2)COOH$

Phenylalanine **A**

L-Tyrosine: 2-oxoglutarate aminotransferase 2.6.1.5 (see 231 BD)

→ 2-Oxoglutarate → L-Glutamate

$CH_2COCOOH$

Phenylpyruvate **B**

Possibly non-enzymic

→ O_2

CHO + HOOC.COOH

Oxalate

Benzaldehyde **C**

Benzaldehyde: NADP oxidoreductase
Benzaldehyde dehydrogenase
1.2.1.7

$NADP^+ + H_2O$ ⇌ $NADPH + H^+$

COOH

Benzoate **D**

Benzoate: CoA ligase (AMP)
Benzoyl-CoA synthetase
6.2.1

→ HSCoA, ATP → $AMP + PP_i$

COSCoA

Benzoyl-coenzyme A **E**

Hippurate synthetase
6.3.2

→ H_2NCH_2COOH Glycine → HSCoA

$CO.NHCH_2COOH$

Hippurate **F**

Formation of Melanin

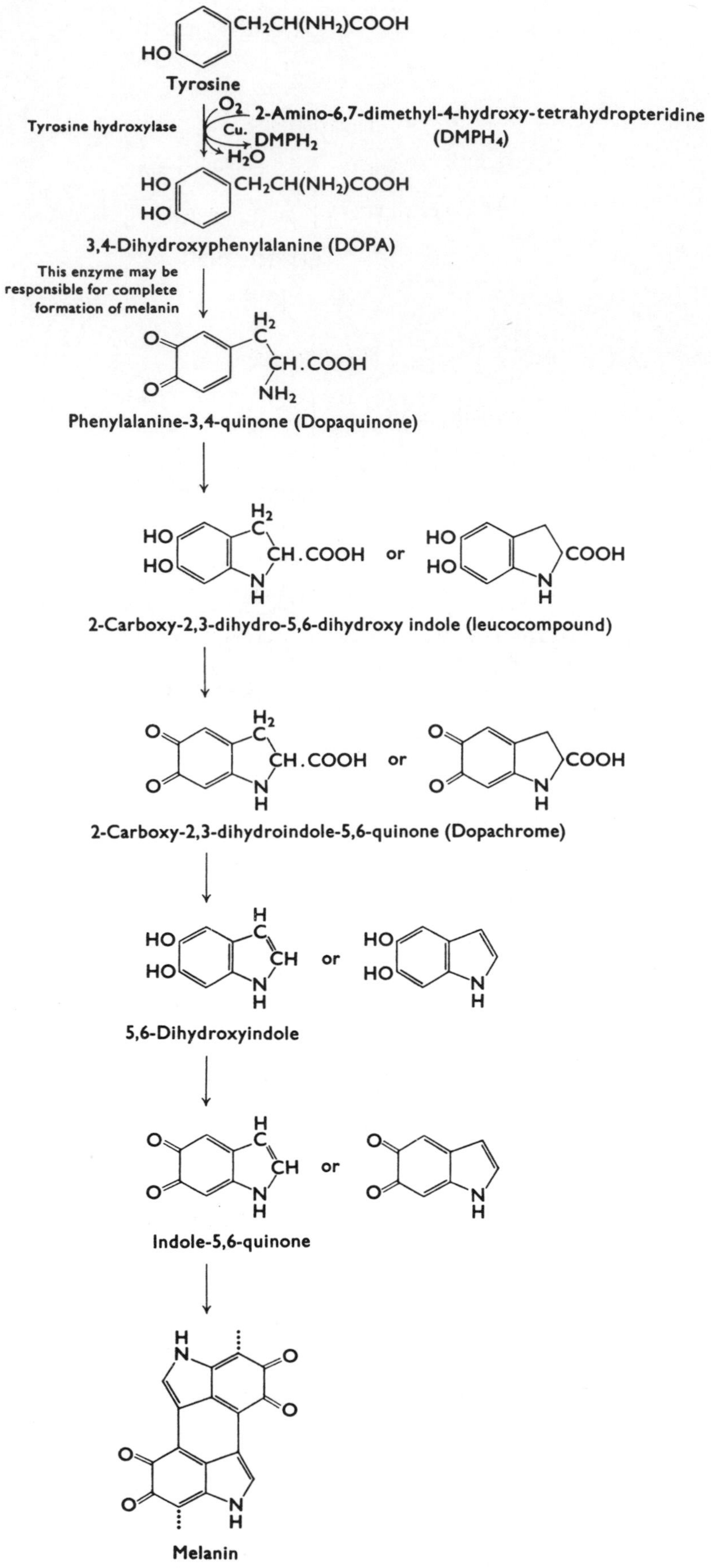

$HO-C_6H_4-CH_2CH(NH_2)COOH$

Tyrosine

A

Tyrosine hydroxylase

O_2, Cu., 2-Amino-6,7-dimethyl-4-hydroxy-tetrahydropteridine ($DMPH_4$) → $DMPH_2$, H_2O

$(HO)_2C_6H_3-CH_2CH(NH_2)COOH$

3,4-Dihydroxyphenylalanine (DOPA)

B

This enzyme may be responsible for complete formation of melanin

H_2 C CH.COOH NH_2

Phenylalanine-3,4-quinone (Dopaquinone)

C

HO HO H_2 C CH.COOH N H or HO HO COOH N H

2-Carboxy-2,3-dihydro-5,6-dihydroxy indole (leucocompound)

D

O O H_2 C CH.COOH N H or O O COOH N H

2-Carboxy-2,3-dihydroindole-5,6-quinone (Dopachrome)

E

HO HO H C CH N H or HO HO N H

5,6-Dihydroxyindole

F

O O H C CH N H or O O N H

Indole-5,6-quinone

G

H N O O O O N H

Melanin

H

The dotted lines indicate polymerization to give a pigment of undetermined size.

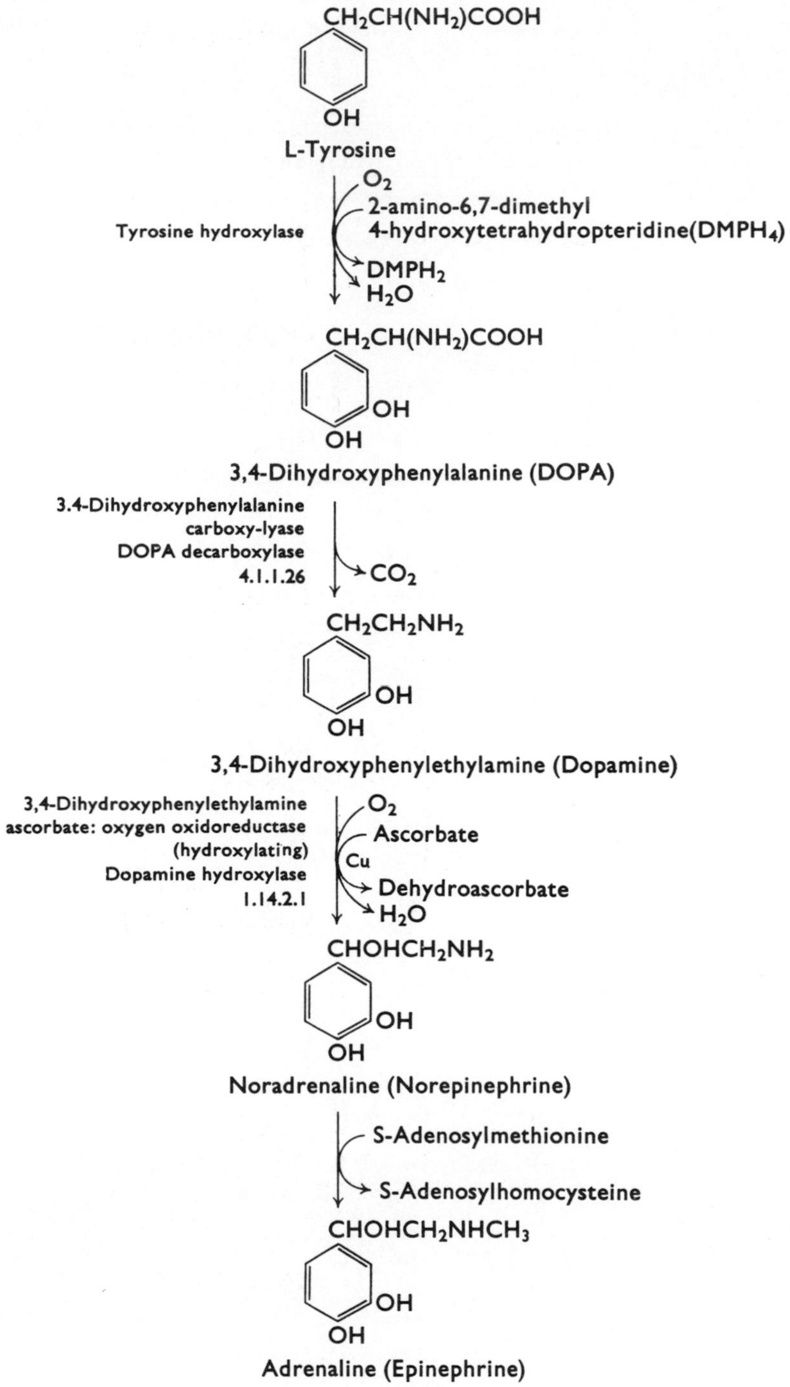

$CH_2CH(NH_2)COOH$

OH

L-Tyrosine **A**

Tyrosine hydroxylase

O_2

2-amino-6,7-dimethyl
4-hydroxytetrahydropteridine($DMPH_4$)

$DMPH_2$
H_2O

$CH_2CH(NH_2)COOH$

OH
OH

3,4-Dihydroxyphenylalanine (DOPA) **B**

3.4-Dihydroxyphenylalanine
carboxy-lyase
DOPA decarboxylase
4.1.1.26

CO_2

$CH_2CH_2NH_2$

OH
OH

3,4-Dihydroxyphenylethylamine (Dopamine) **C**

3,4-Dihydroxyphenylethylamine
ascorbate: oxygen oxidoreductase
(hydroxylating)
Dopamine hydroxylase
1.14.2.1

O_2
Ascorbate
Cu
Dehydroascorbate
H_2O

$CHOHCH_2NH_2$

OH
OH

Noradrenaline (Norepinephrine) **D**

S-Adenosylmethionine

S-Adenosylhomocysteine

$CHOHCH_2NHCH_3$

OH
OH

Adrenaline (Epinephrine) **E**

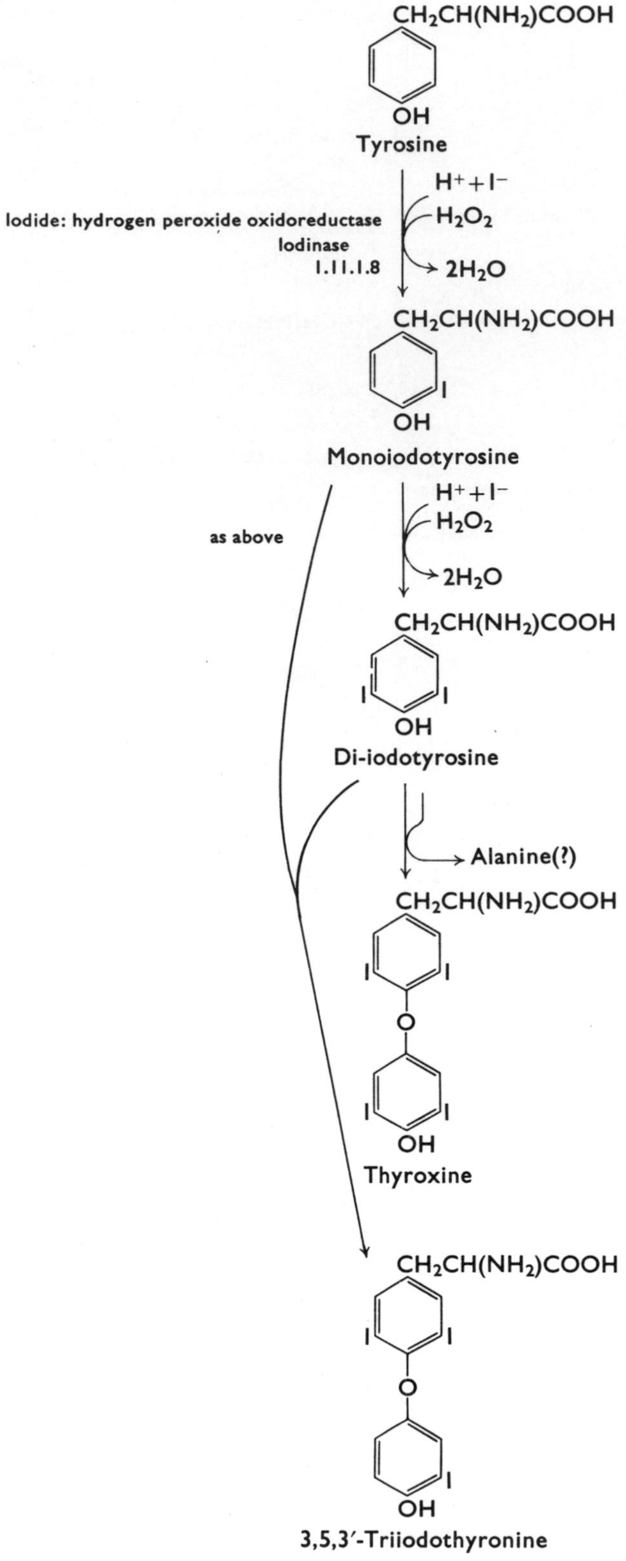

Formation of Thyroxine

$CH_2CH(NH_2)COOH$

OH

Tyrosine

A

$H^+ + I^-$

Iodide: hydrogen peroxide oxidoreductase
Iodinase
1.11.1.8

H_2O_2

$2H_2O$

$CH_2CH(NH_2)COOH$

I

OH

Monoiodotyrosine

B

as above

$H^+ + I^-$

H_2O_2

$2H_2O$

$CH_2CH(NH_2)COOH$

I I

OH

Di-iodotyrosine

C

Alanine(?)

$CH_2CH(NH_2)COOH$

I I

O

I I

OH

Thyroxine

D

$CH_2CH(NH_2)COOH$

I I

O

I

OH

3,5,3′-Triiodothyronine

E

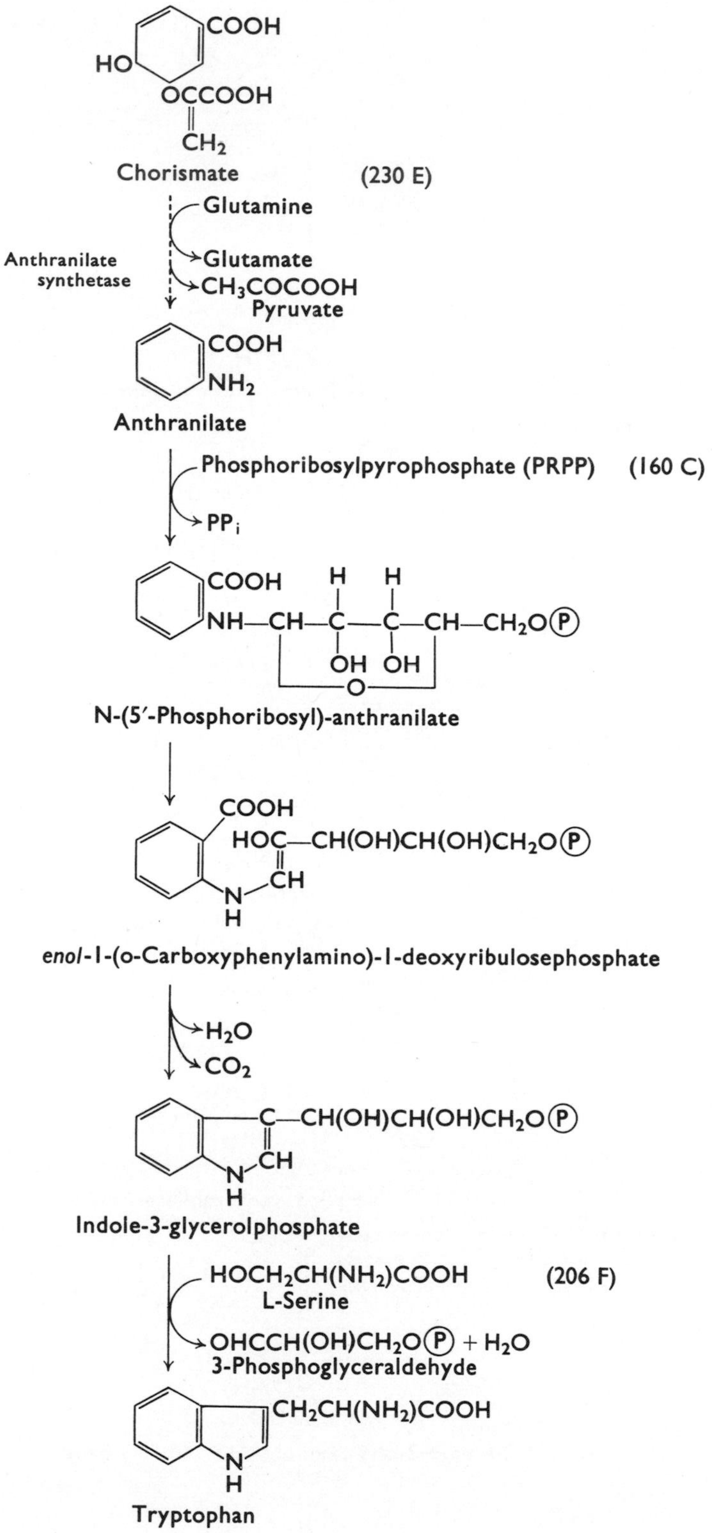

HO — $COOH$ — $OCCOOH$ / $\|$ / CH_2

Chorismate (230 E) **A**

Anthranilate synthetase

Glutamine → Glutamate

$CH_3COCOOH$ Pyruvate

$COOH$ — NH_2

Anthranilate **B**

Phosphoribosylpyrophosphate (PRPP) (160 C)

→ PP_i

$COOH$ H H

$NH—CH—C—C—CH—CH_2O$Ⓟ

OH OH

O

N-(5′-Phosphoribosyl)-anthranilate **C**

$COOH$

$HOC—CH(OH)CH(OH)CH_2O$Ⓟ

CH

N

H

enol-1-(o-Carboxyphenylamino)-1-deoxyribulosephosphate **D**

→ H_2O

→ CO_2

$C—CH(OH)CH(OH)CH_2O$Ⓟ

CH

N

H

Indole-3-glycerolphosphate **E**

$HOCH_2CH(NH_2)COOH$ (206 F)

L-Serine

→ $OHCCH(OH)CH_2O$Ⓟ $+ H_2O$

3-Phosphoglyceraldehyde

$CH_2CH(NH_2)COOH$

N

H

Tryptophan **F**

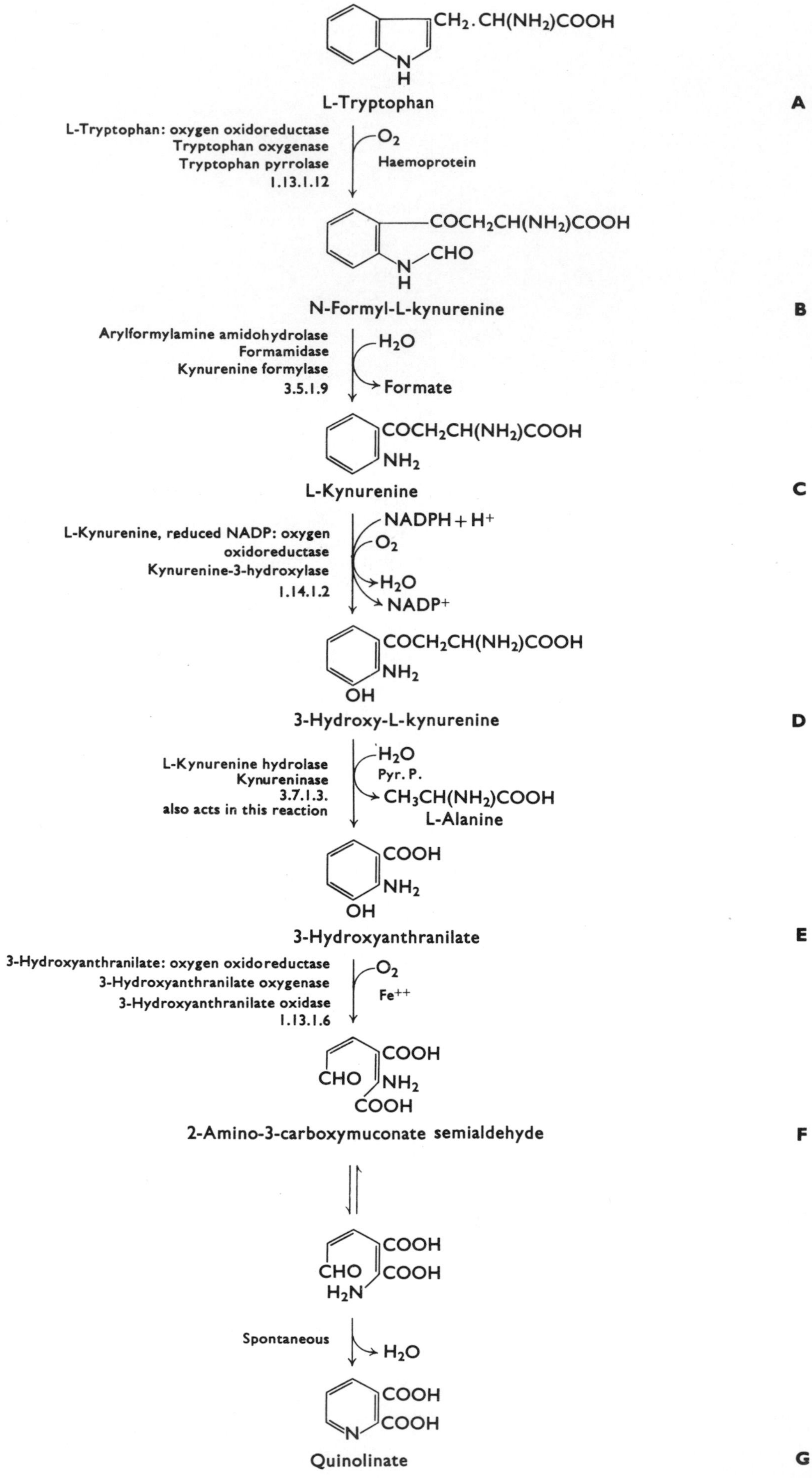

$CH_2.CH(NH_2)COOH$

L-Tryptophan **A**

L-Tryptophan: oxygen oxidoreductase
Tryptophan oxygenase
Tryptophan pyrrolase
1.13.1.12

O_2
Haemoprotein

$COCH_2CH(NH_2)COOH$
CHO

N-Formyl-L-kynurenine **B**

Arylformylamine amidohydrolase
Formamidase
Kynurenine formylase
3.5.1.9

H_2O
Formate

$COCH_2CH(NH_2)COOH$
NH_2

L-Kynurenine **C**

L-Kynurenine, reduced NADP: oxygen oxidoreductase
Kynurenine-3-hydroxylase
1.14.1.2

$NADPH + H^+$
O_2
H_2O
$NADP^+$

$COCH_2CH(NH_2)COOH$
NH_2
OH

3-Hydroxy-L-kynurenine **D**

L-Kynurenine hydrolase
Kynureninase
3.7.1.3.
also acts in this reaction

H_2O
Pyr. P.
$CH_3CH(NH_2)COOH$
L-Alanine

COOH
NH_2
OH

3-Hydroxyanthranilate **E**

3-Hydroxyanthranilate: oxygen oxidoreductase
3-Hydroxyanthranilate oxygenase
3-Hydroxyanthranilate oxidase
1.13.1.6

O_2
Fe^{++}

COOH
CHO NH_2
COOH

2-Amino-3-carboxymuconate semialdehyde **F**

COOH
CHO COOH
H_2N

Spontaneous

H_2O

COOH
COOH

Quinolinate **G**

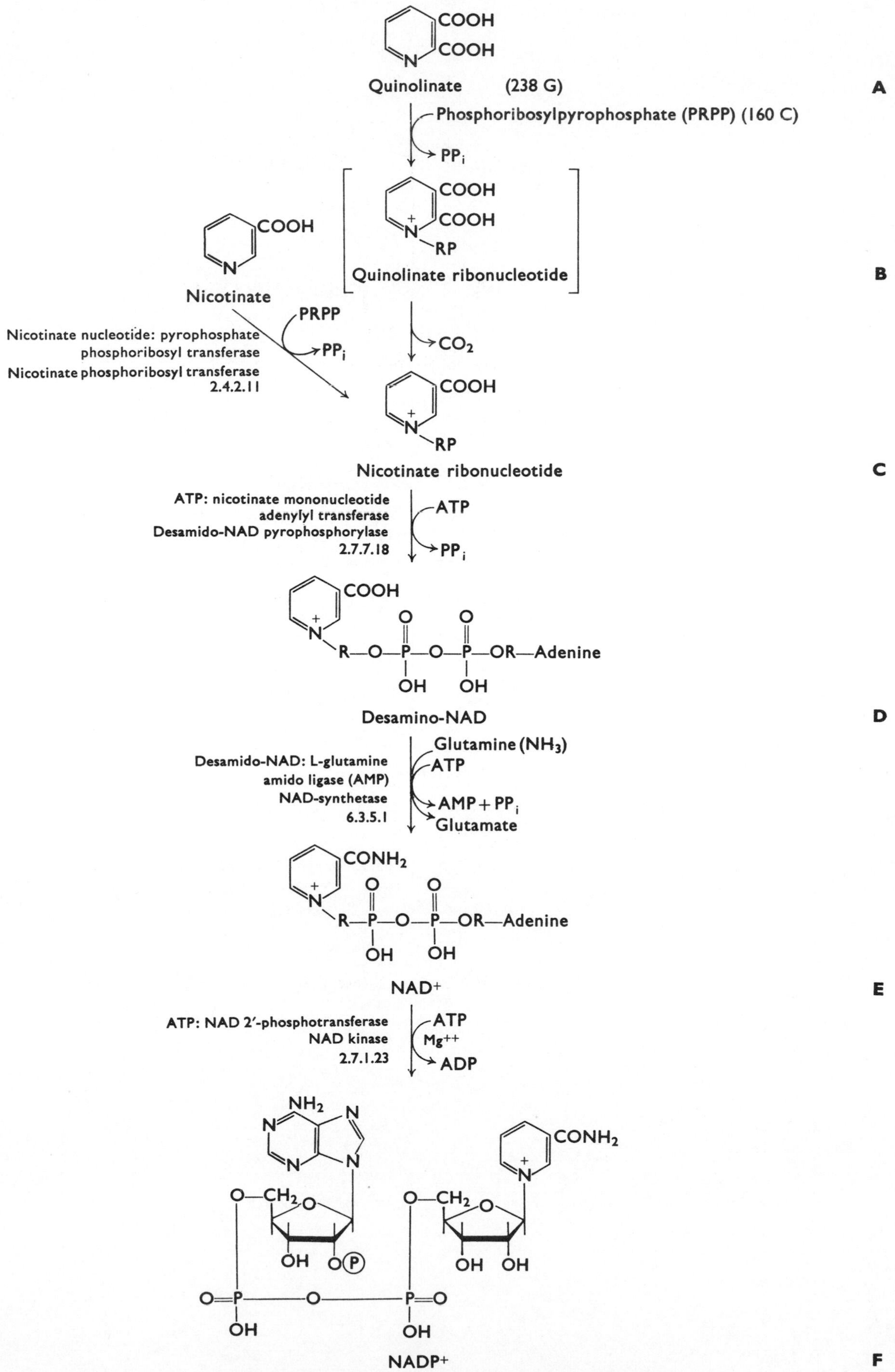

COOH
COOH
N

Quinolinate (238 G) **A**

Phosphoribosylpyrophosphate (PRPP) (160 C)

PP_i

COOH
COOH
$\overset{+}{N}$
RP

[Quinolinate ribonucleotide] **B**

COOH
N

Nicotinate

PRPP

PP_i

Nicotinate nucleotide: pyrophosphate phosphoribosyl transferase
Nicotinate phosphoribosyl transferase
2.4.2.11

CO_2

COOH
$\overset{+}{N}$
RP

Nicotinate ribonucleotide **C**

ATP: nicotinate mononucleotide adenylyl transferase
Desamido-NAD pyrophosphorylase
2.7.7.18

ATP

PP_i

COOH

$\overset{+}{N}$—R—O—P(=O)(OH)—O—P(=O)(OH)—OR—Adenine

Desamino-NAD **D**

Desamido-NAD: L-glutamine amido ligase (AMP)
NAD-synthetase
6.3.5.1

Glutamine (NH_3)
ATP

AMP + PP_i
Glutamate

$CONH_2$

$\overset{+}{N}$—R—P(=O)(OH)—O—P(=O)(OH)—OR—Adenine

NAD^+ **E**

ATP: NAD 2′-phosphotransferase
NAD kinase
2.7.1.23

ATP
Mg^{++}
ADP

NH_2
N N N N
O—CH_2 O
OH O(P)
O=P—O—P=O
OH OH
O—CH_2 O
OH OH
$CONH_2$
$\overset{+}{N}$

$NADP^+$ **F**

R represents the Ribosyl group and RP ribosyl phosphate

Formation of Serotonin and Melatonin

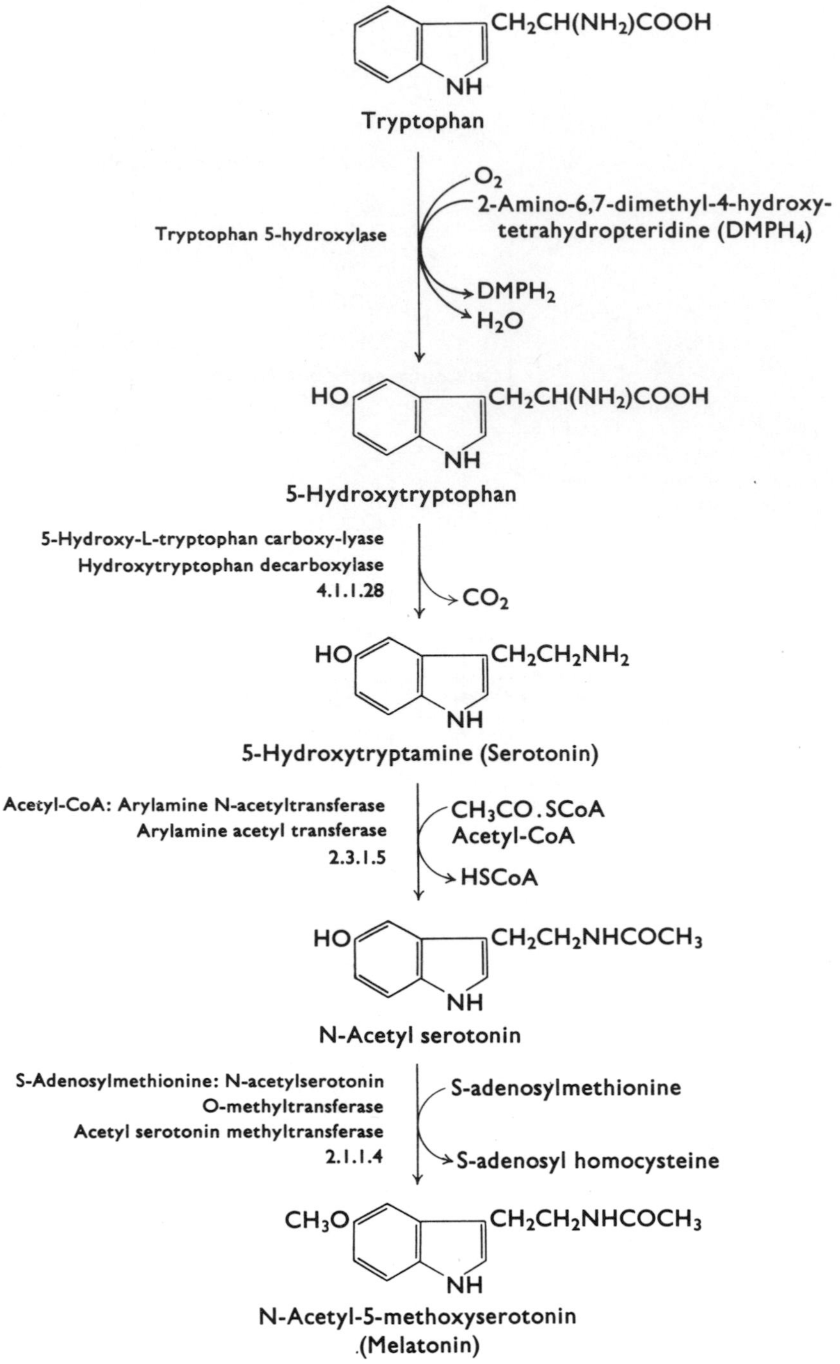

$CH_2CH(NH_2)COOH$

NH

Tryptophan

A

Tryptophan 5-hydroxylase

O_2

2-Amino-6,7-dimethyl-4-hydroxy-tetrahydropteridine ($DMPH_4$)

$\rightarrow DMPH_2$

$\rightarrow H_2O$

HO $CH_2CH(NH_2)COOH$

NH

5-Hydroxytryptophan

B

5-Hydroxy-L-tryptophan carboxy-lyase
Hydroxytryptophan decarboxylase
4.1.1.28

$\rightarrow CO_2$

HO $CH_2CH_2NH_2$

NH

5-Hydroxytryptamine (Serotonin)

C

Acetyl-CoA: Arylamine N-acetyltransferase
Arylamine acetyl transferase
2.3.1.5

$CH_3CO.SCoA$
Acetyl-CoA

$\rightarrow$ HSCoA

HO $CH_2CH_2NHCOCH_3$

NH

N-Acetyl serotonin

D

S-Adenosylmethionine: N-acetylserotonin O-methyltransferase
Acetyl serotonin methyltransferase
2.1.1.4

S-adenosylmethionine

$\rightarrow$ S-adenosyl homocysteine

CH_3O $CH_2CH_2NHCOCH_3$

NH

N-Acetyl-5-methoxyserotonin
(Melatonin)

E

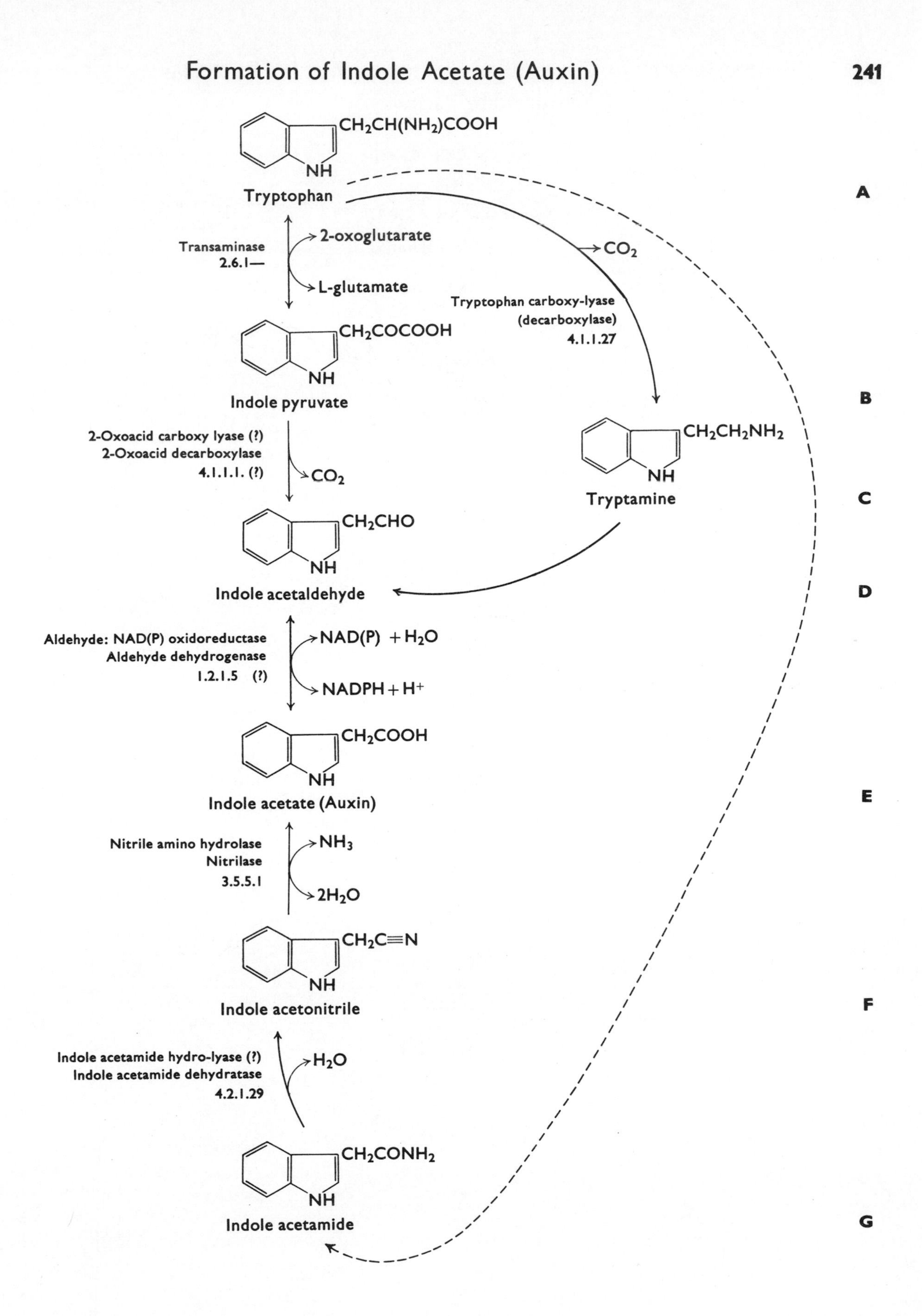

$CH_2CH(NH_2)COOH$
NH
Tryptophan
A
Transaminase
2.6.1—
→2-oxoglutarate
→L-glutamate
→CO_2
Tryptophan carboxy-lyase
(decarboxylase)
4.1.1.27
$CH_2COCOOH$
NH
Indole pyruvate
B
$CH_2CH_2NH_2$
NH
Tryptamine
2-Oxoacid carboxy lyase (?)
2-Oxoacid decarboxylase
4.1.1.1. (?)
CO_2
C
CH_2CHO
NH
Indole acetaldehyde
D
Aldehyde: NAD(P) oxidoreductase
Aldehyde dehydrogenase
1.2.1.5 (?)
NAD(P) + H_2O
NADPH + H^+
CH_2COOH
NH
Indole acetate (Auxin)
E
Nitrile amino hydrolase
Nitrilase
3.5.5.1
NH_3
$2H_2O$
$CH_2C{\equiv}N$
NH
Indole acetonitrile
F
Indole acetamide hydro-lyase (?)
Indole acetamide dehydratase
4.2.1.29
H_2O
CH_2CONH_2
NH
Indole acetamide
G

Catabolism of Tryptophan by Bacteria: Kynurenate Pathway

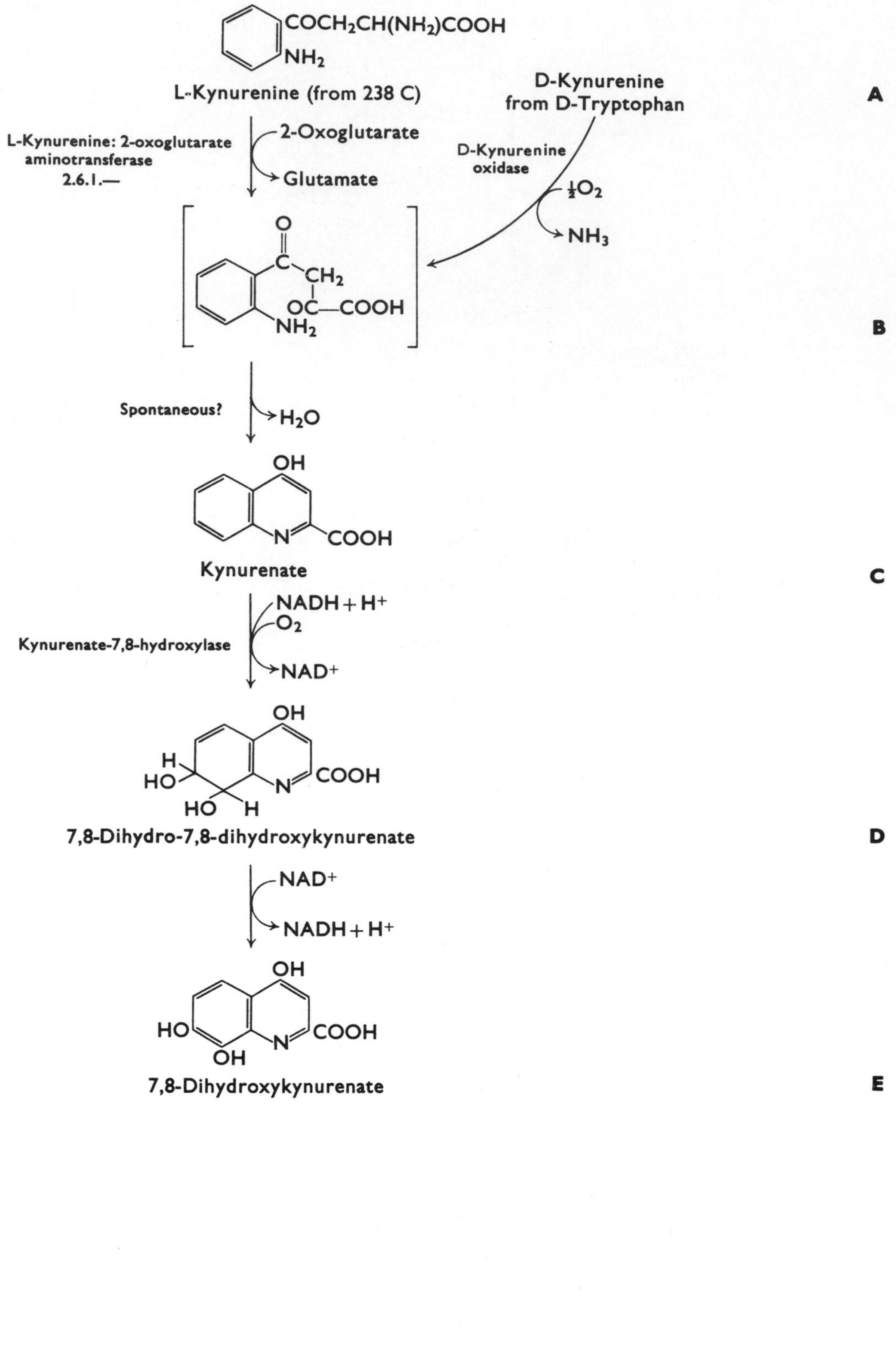

$COCH_2CH(NH_2)COOH$
NH_2
L-Kynurenine (from 238 C)

D-Kynurenine
from D-Tryptophan

A

L-Kynurenine: 2-oxoglutarate aminotransferase 2.6.1.—

2-Oxoglutarate → Glutamate

D-Kynurenine oxidase

$\frac{1}{2}O_2$ → NH_3

O
C
CH_2
OC—COOH
NH_2

B

Spontaneous?

→ H_2O

OH
N COOH
Kynurenate

C

Kynurenate-7,8-hydroxylase

$NADH + H^+$
O_2
→ NAD^+

OH
H
HO
N COOH
HO H
7,8-Dihydro-7,8-dihydroxykynurenate

D

NAD^+ → $NADH + H^+$

OH
HO N COOH
OH
7,8-Dihydroxykynurenate

E

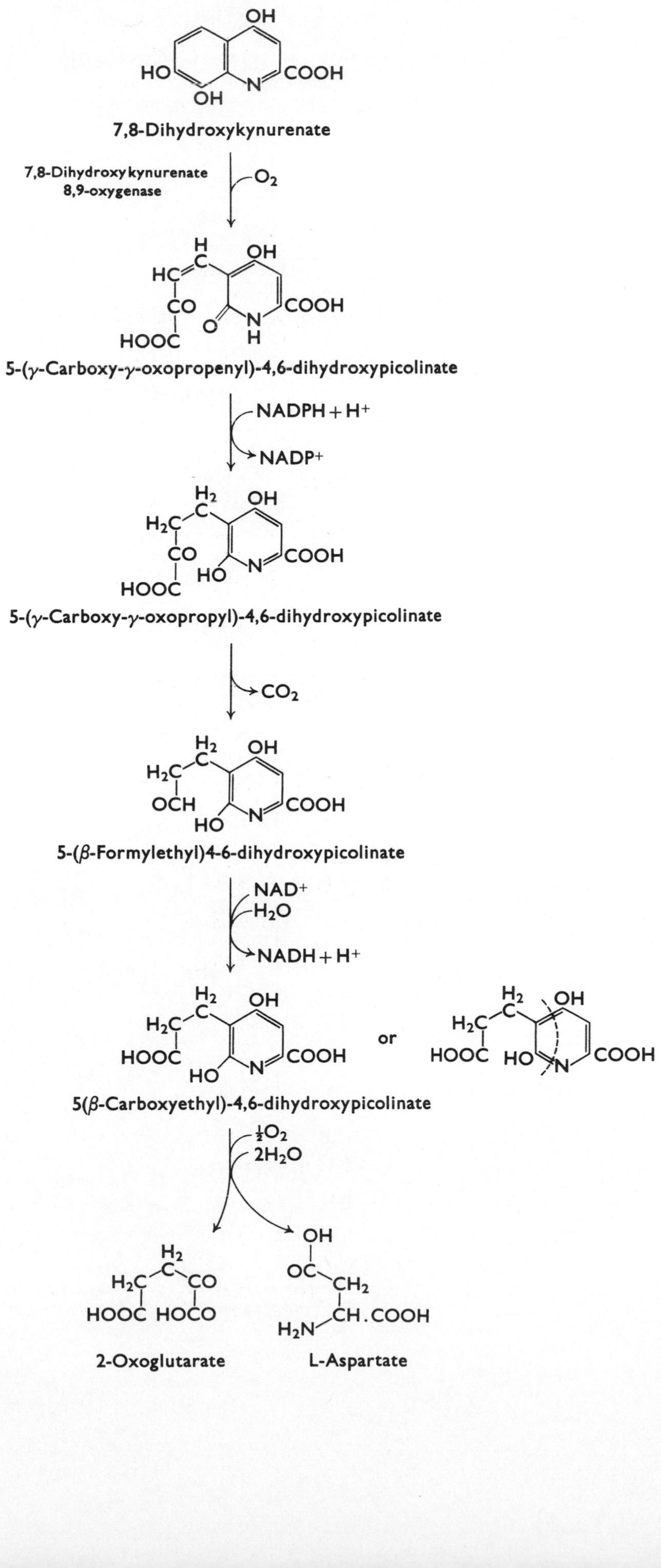

7,8-Dihydroxykynurenate — A

7,8-Dihydroxykynurenate 8,9-oxygenase; O_2

5-(γ-Carboxy-γ-oxopropenyl)-4,6-dihydroxypicolinate — B

NADPH + H^+ → $NADP^+$

5-(γ-Carboxy-γ-oxopropyl)-4,6-dihydroxypicolinate — C

→ CO_2

5-(β-Formylethyl)4-6-dihydroxypicolinate — D

NAD^+, H_2O → NADH + H^+

5(β-Carboxyethyl)-4,6-dihydroxypicolinate — E

$\frac{1}{2}O_2$, $2H_2O$

2-Oxoglutarate   L-Aspartate — F G

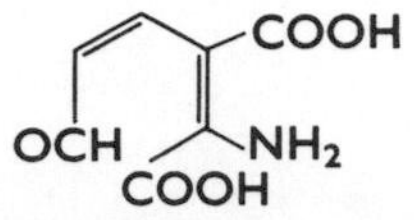

from 238 F

OCH–CH=CH–C(COOH)=C(NH$_2$)–COOH

1-Amino-4-formylbutadiene-1,2-dicarboxylate
2-Acrolyl-3-aminofumarate
2-Amino-3-carboxymuconate semialdehyde **A**

↓ → CO_2

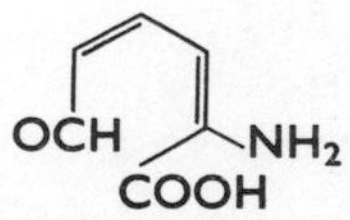

OCH–CH=CH–CH=C(NH$_2$)–COOH

2-Aminomuconate-6 semialdehyde **B**

↓ 2-Hydroxymuconate semialdehyde dehydrogenase; $NAD^+ + H_2O$ → $NADH + H^+$

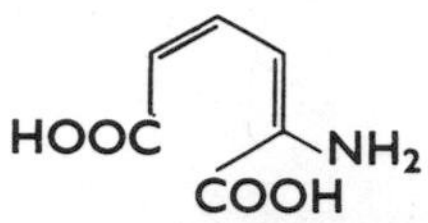

HOOC–CH=CH–CH=C(NH$_2$)–COOH

2-Aminomuconate **C**

↓ H_2O, $NADPH(NADH) + H^+$ → $NADP^+(NAD^+)$, NH_3

$HOOCCH_2CH_2CH_2COCOOH$
2-Oxoadipate **D**

↓ 2-Oxoglutarate dehydrogenase; NAD^+, HSCoA, FAD, TPP, lipoate → CO_2, $NADH + H^+$

$HOOCCH_2CH_2CH_2COSCoA$
Glutaryl-coenzyme A **E**

↓ FAD → $FADH_2$

[$HOOCCH_2CH{=}CHCOSCoA$
Glutaconyl-coenzyme A] **F**

↓ → CO_2

$CH_3CH{=}CH{-}COSCoA$
Crotonyl coenzyme A **G**

⋮

2 $CH_3COSCoA$
Acetyl coenzyme A **H**

Catabolism of Tryptophan and other Aromatic Compounds: By the Catechol Pathway (in Bacteria)

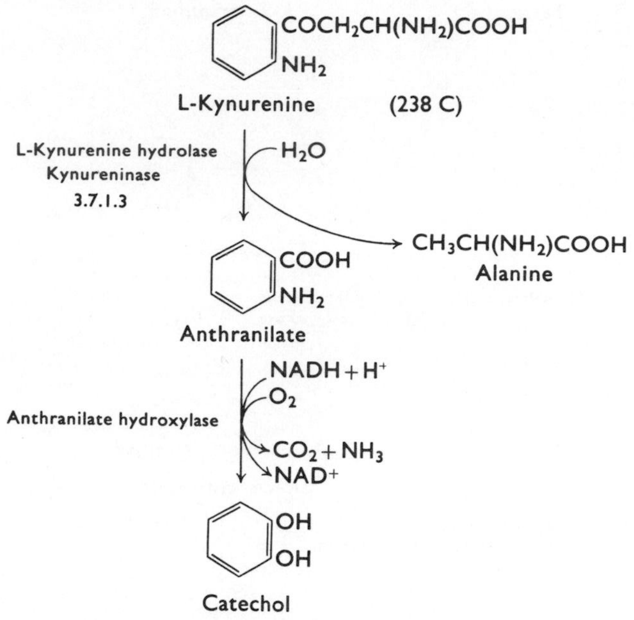

$COCH_2CH(NH_2)COOH$
NH_2
L-Kynurenine (238 C)

B

L-Kynurenine hydrolase
Kynureninase
3.7.1.3

H_2O

$CH_3CH(NH_2)COOH$
Alanine

COOH
NH_2
Anthranilate

C

Anthranilate hydroxylase

$NADH + H^+$
O_2
$CO_2 + NH_3$
NAD^+

OH
OH
Catechol

D

Catabolism of Catechol by Bacteria
By *meta* fission

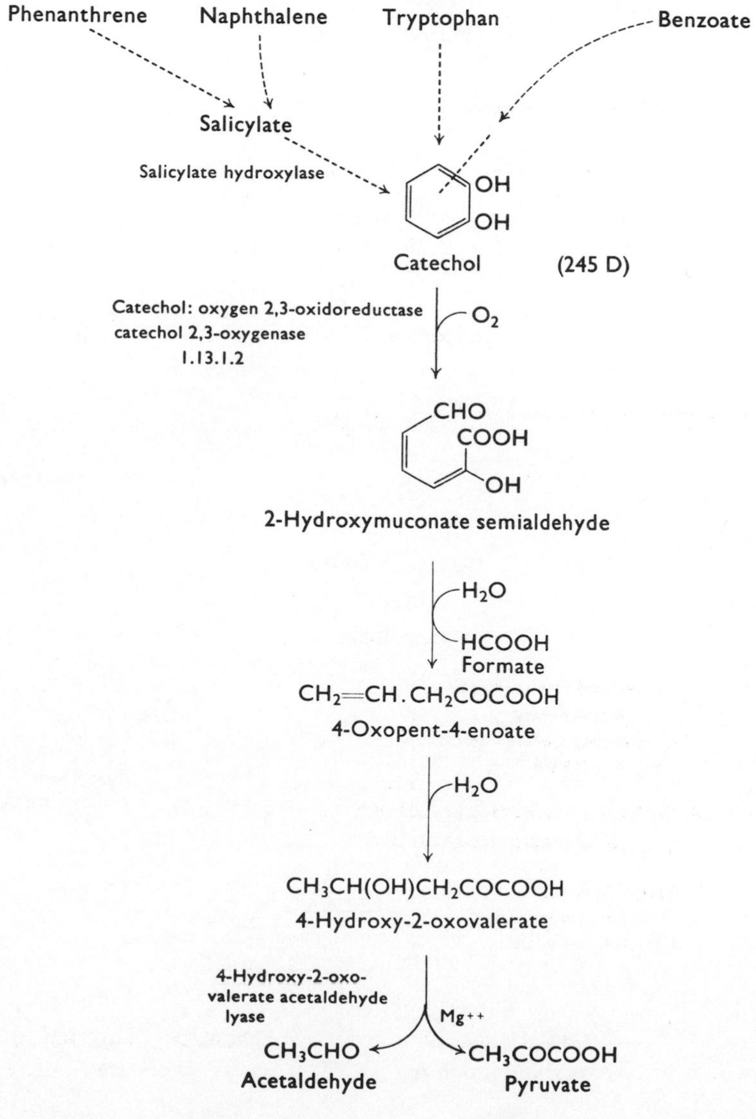

Phenanthrene Naphthalene Tryptophan Benzoate

EFGH

Salicylate

Salicylate hydroxylase

OH
OH
Catechol (245 D)

J

Catechol: oxygen 2,3-oxidoreductase
catechol 2,3-oxygenase
1.13.1.2

O_2

CHO
COOH
OH
2-Hydroxymuconate semialdehyde

K

H_2O
HCOOH
Formate

$CH_2{=}CH.CH_2COCOOH$
4-Oxopent-4-enoate

L

H_2O

$CH_3CH(OH)CH_2COCOOH$
4-Hydroxy-2-oxovalerate

M

4-Hydroxy-2-oxo-valerate acetaldehyde lyase

Mg^{++}

CH_3CHO
Acetaldehyde

$CH_3COCOOH$
Pyruvate

NO

Catabolism of Catechol by Bacteria: By *ortho* fission

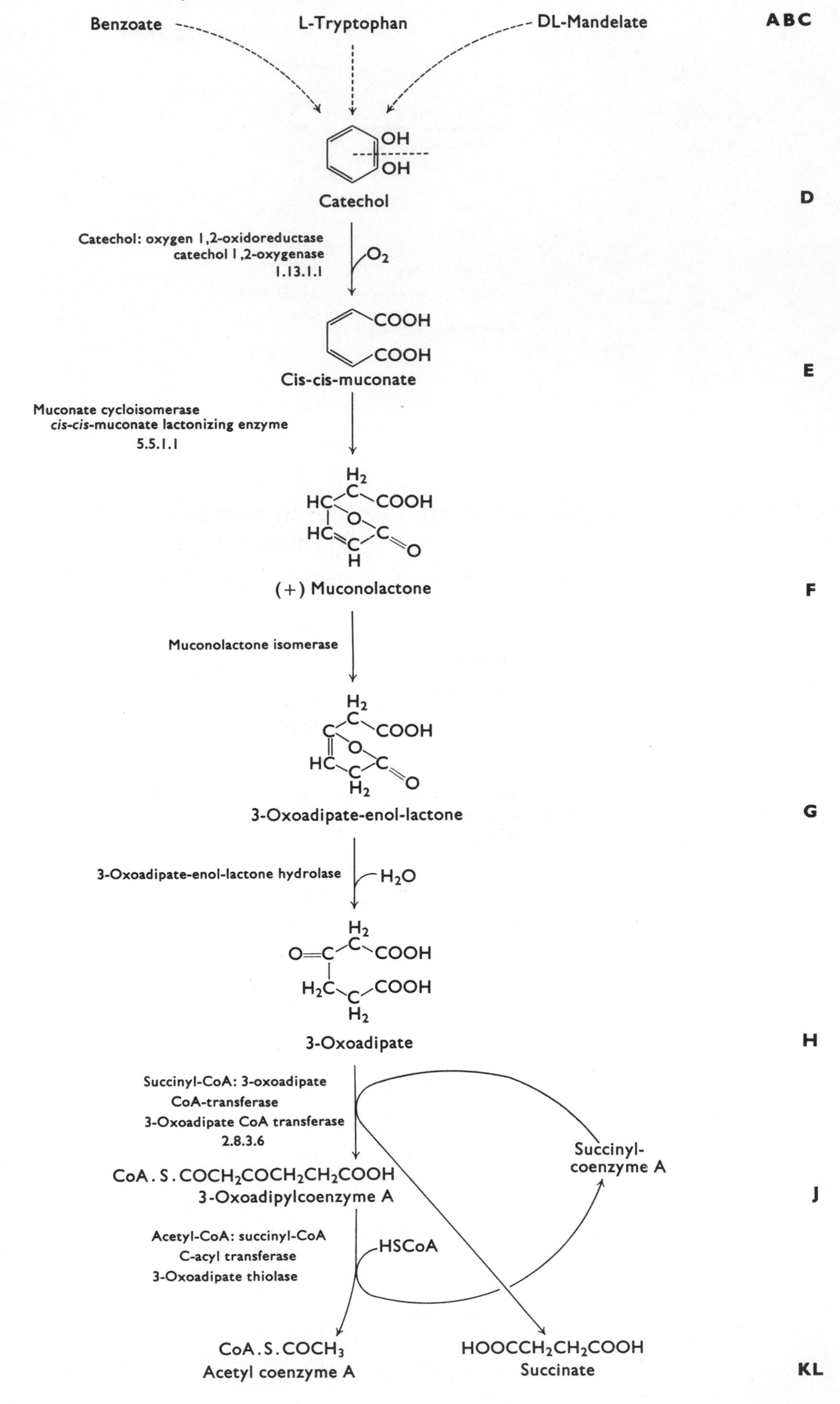

Benzoate
L-Tryptophan
DL-Mandelate
ABC
OH
OH
Catechol
D
Catechol: oxygen 1,2-oxidoreductase
catechol 1,2-oxygenase
1.13.1.1
O2
COOH
COOH
Cis-cis-muconate
E
Muconate cycloisomerase
cis-cis-muconate lactonizing enzyme
5.5.1.1
H2
HC
C
COOH
O
HC
C
C
H
O
(+) Muconolactone
F
Muconolactone isomerase
H2
C
COOH
O
HC
C
H2
O
3-Oxoadipate-enol-lactone
G
3-Oxoadipate-enol-lactone hydrolase
H2O
H2
O=C
C
COOH
H2C
C
COOH
H2
3-Oxoadipate
H
Succinyl-CoA: 3-oxoadipate
CoA-transferase
3-Oxoadipate CoA transferase
2.8.3.6
Succinyl-
coenzyme A
CoA.S.COCH2COCH2CH2COOH
3-Oxoadipylcoenzyme A
J
Acetyl-CoA: succinyl-CoA
C-acyl transferase
3-Oxoadipate thiolase
HSCoA
CoA.S.COCH3
Acetyl coenzyme A
HOOCCH2CH2COOH
Succinate
KL

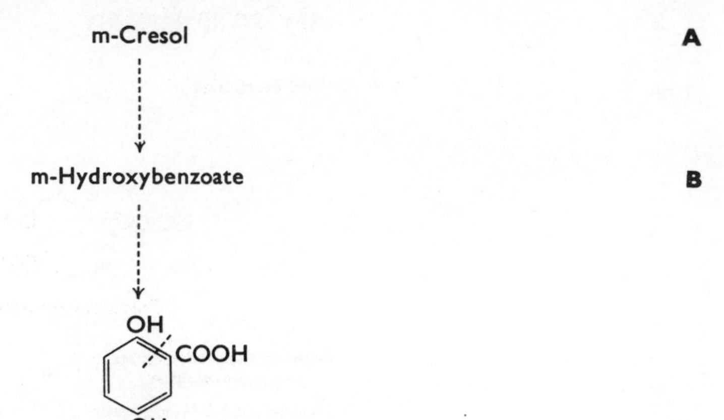

OH
COOH
OH
Gentisate
C

Gentisate oxygenase
O_2

COOH
HC
CO.COOH
HC
C
CH
OH

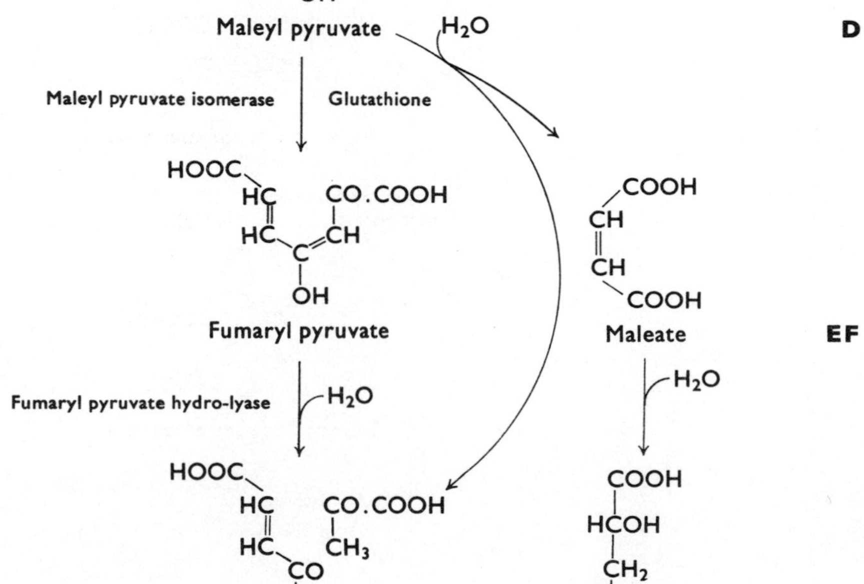

Catabolism of Protocatechuic Acid by Bacteria: By *ortho* fission

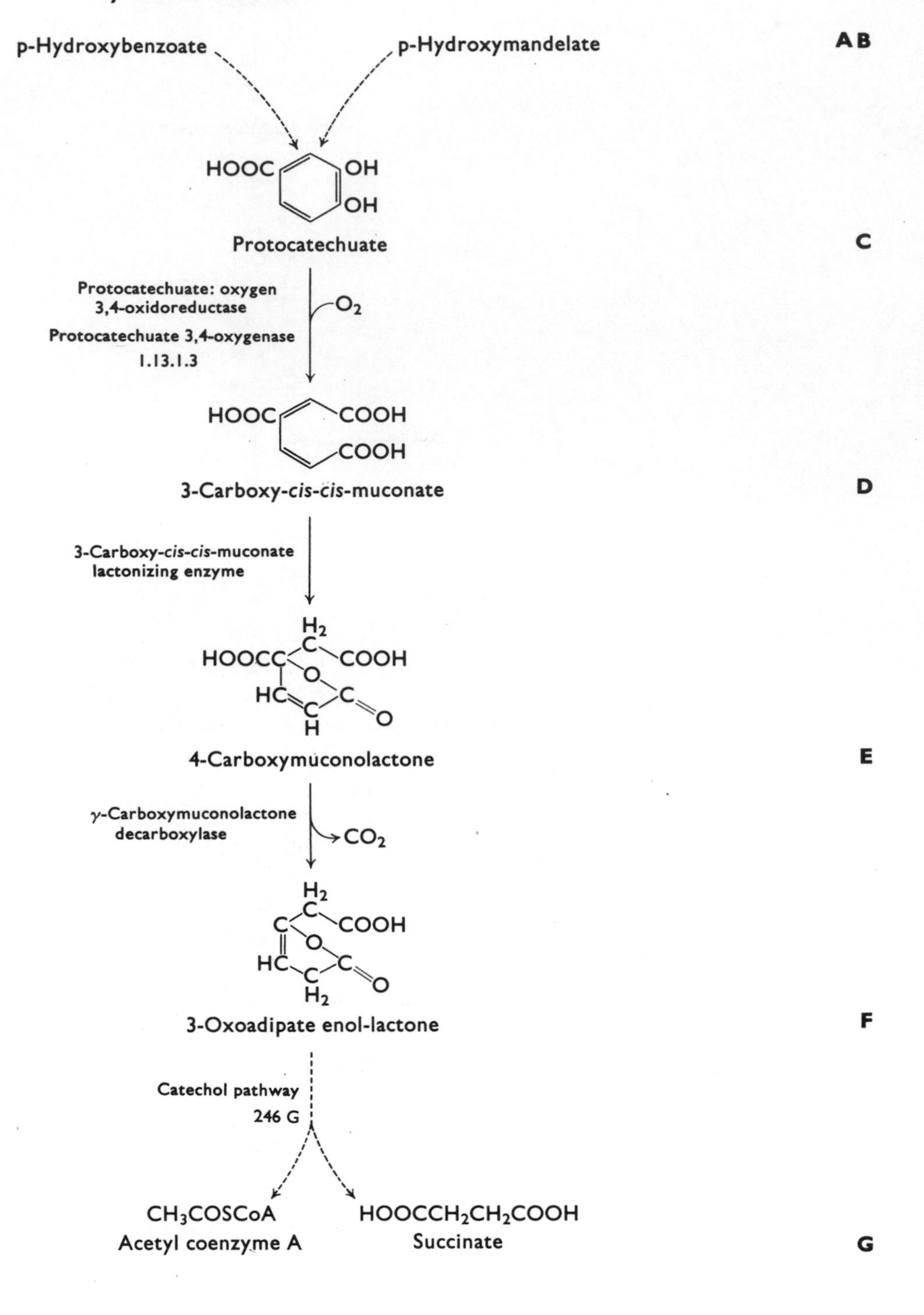

Catabolism of Protocatechuic Acid by Bacteria: By *meta* fission

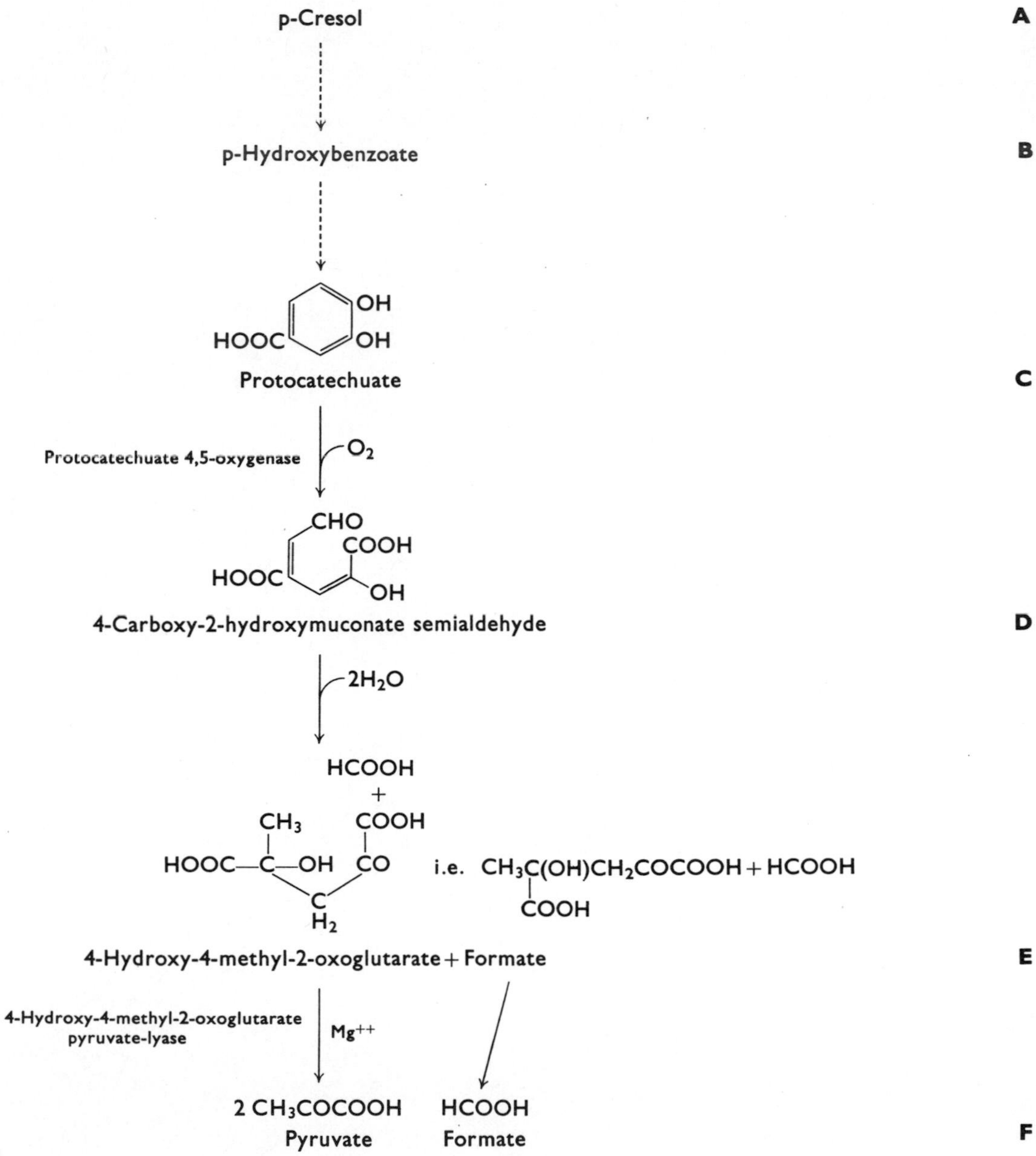

p-Cresol **A**

↓

p-Hydroxybenzoate **B**

↓

HOOC–(benzene ring)–OH, OH
Protocatechuate **C**

↓ Protocatechuate 4,5-oxygenase; O_2

CHO, COOH, HOOC, OH
4-Carboxy-2-hydroxymuconate semialdehyde **D**

↓ $2H_2O$

HCOOH + HOOC–C(CH_3)(OH)–CH_2–CO–COOH

i.e. $CH_3C(OH)(COOH)CH_2COCOOH + HCOOH$

4-Hydroxy-4-methyl-2-oxoglutarate + Formate **E**

↓ 4-Hydroxy-4-methyl-2-oxoglutarate pyruvate-lyase; Mg^{++}

$2\ CH_3COCOOH$ Pyruvate; HCOOH Formate **F**

References

1. The Determination of Metabolic Pathways

4

1 Krebs H.A. (1951) *Radioisotopic Techniques*, vol. 1. H.M. Stationery Office.

2 Lewis G.N. (1926) *The Anatomy of Science*. Yale.

3 Lehninger A.L. (1965) *Bioenergetics*. Benjamin, New York.

4 Krebs H.A. and Kornberg H.L. (1957) *Energy Transformations in Living Matter*. Springer, Berlin.

8

1 Dagley S. and Walker J.R.L. (1956) Accumulation of citrate and pyruvate during growth of a vibrio in the presence of fluoroacetate. *Biochim. biophys. Acta*, **21**, 441.

2 Dixon G.H., Kornberg H.L. and Lund P. (1960) Purification and properties of malate synthetase. *Biochim. biophys. Acta*, **41**, 217.

3 Callely A.G. and Dagley S. (1959) A possible lethal synthesis of monofluoromalate. *Biochim. biophys. Acta*, **35**, 256.

4 Ichihara A., Adachi K., Hosokawa K. and Takeda Y. (1962) The enzymatic hydroxylation of aromatic carboxylic acids; substrate specificities of anthranilate and benzoate oxidases. *J. biol. Chem.* **237**, 2296.

5 Dagley S. and Patel M.D. (1957) Oxidation of *p*-cresol and related compounds by a *pseudomonas*. *Biochem. J.* **66**, 227.

9

1 Dagley S. and Patel M.D. (1955) Excretion of α-ketoglutaric acid during oxidation of acetate by a vibrio. *Biochim. biophys. Acta*, **16**, 418.

2 Woods D.D. (1940) *J. exptl. Pathol.* **21**, 74.

3 Blanchard K.C. (1941) The isolation of *p*-aminobenzoic acid from yeast. *J. biol. Chem.* **140**, 919.

10

1 Chance B. and Williams G.R. (1956) The respiratory chain and oxidative phosphorylation. *Advan. Enzymol.* **17**, 65.

2 Griffiths D.E. (1965) Oxidative phosphorylation. *Essays in Biochemistry*, vol. 1. Academic Press, London.

3 Baldwin E. (1952) *Dynamic Aspects of Biochemistry*, 2nd ed. Cambridge University Press, p. 199.

11

1 Hems R., Ross B.D., Berry M.N. and Krebs H.A. (1966) Gluconeogenesis in the perfused rat liver. *Biochem. J.* **101**, 284.

13

1 Lehninger A.L. (1964) *The Mitochondrion*. Benjamin, New York.

2 Hogeboom G.H., Schneider W.C. and Pallade G.E. (1948) Cytochemical studies of mammalian tissues. Isolation of intact mitochondria from rat liver; some biochemical properties of mitochondria and submicroscopic particulate material. *J. biol. Chem.* **172**, 619.

3 Kennedy E.P. and Lehninger A.L. (1949) Oxidation of fatty acids and tricarboxylic acid cycle intermediates by isolated rat liver mitochondria. *J. biol. Chem.* **179**, 957.

4 Drysdale G.R. and Lardy H.A. (1953) Fatty acid oxidation by a soluble enzyme system from mitochondria. *J. biol. Chem.* **202**, 119.

14

1 Lynen F. and Reichert E. (1952) *Z. Angew. Chem.* **64**, 687.

2 Ochoa S., Stern J.R. and Schneider M.C. (1951) Enzymatic synthesis of citric acid—crystalline condensing enzyme. *J. biol. Chem.* **193**, 691.

3 Barker H.A. (1951) in *Phosphorus Metabolism, vol. I*. Ed. by W.D. McElroy and B. Glass. Johns Hopkins Press, Baltimore.

4 Callely A.G., Dagley S. and Hodgson B. (1958) Oxidation of fatty acids by cell-free extracts of vibrio. *Biochem. J.* **69**, 173.

15

1 Lynen F. (1954) Participation of coenzyme A in the oxidation of fat. *Nature*, **174**, 962.

2 Lynen F. and Ochoa S. (1953) Enzymes of fatty acid metabolism. *Biochim. biophys. Acta*, **12**, 299.

3 Green D.E. (1954) Fatty acid oxidation in soluble systems of animal tissue. *Biological Rev.* **29**, 330.

16

1 Krebs H.A. (1948) The tricarboxylic acid cycle. *The Harvey Lectures, Series 44*. Thomas, Springfield, Illinois.

2 Krebs H.A. (1964) An address presented before *The Robert A. Welch Foundation Conferences on Chemical Research, No. 8*. Held in Houston, Texas.

19

1 Ogston A.G. (1948) Interpretation experiments on metabolic processes, using isotopic tracer elements. *Nature*, **162**, 963.

2 Keech D.B. and Utter M.F. (1963) Pyruvate carboxylase. *J. biol. Chem.* **238**, 2603.

3 Birch A.J. and Donovan F.W. (1953) Some possible routes to derivatives of orcinol and phloroglucinol. *Australian J. Chem.* **6**, 360.

4 Richards J.H. and Hendrickson J.B. (1964) *The Biosynthesis of steroids, terpenes and acetogenins*. Benjamin, New York.

20

1 Ijichi H., Ichiyama A. and Hayaishi O. (1966) Studies on the biosynthesis of NAD—comparative *in vivo* studies on nicotinic acid, nicotinamide and quinolinic acid as precursors of NAD. *J. biol. Chem.* **241**, 3701.

2 Kaplan N.O., Goldin A., Humphreys S.R., Ciotti M.M. and Stolzenbach F.E. (1956) Pyridine nucleotide synthesis in the mouse. *J. biol. Chem.* **219**, 287.

3 Bassham J.A. and Calvin M. (1957) *The Path of Carbon in Photosynthesis*. Prentice-Hall, Englewood Cliffs, N.J.

25

1 Keech D.B. and Utter M.F. (1963) Pyruvate carboxylase—properties. *J. biol. Chem.* **238**, 2609.

2 Shrago E. and Lardy H.A. (1966) Paths of carbon in gluconeogenesis and lipogenesis. *J. biol. Chem.* **241**, 663.

3 Kornberg H.L. (1966) Anaplerotic sequences and their role in metabolism. *Essays in Biochemistry*, vol. 2. Academic Press, London.

26

1 Theodore T.S. and Englesberg E. (1964) Mutant of *Salmonella typhimurium* deficient in the CO_2-fixing enyxme phosphoenolpyruvic carboxylase. *J. Bacteriol.* **88**, 946.

2 Cánovas J.L. and Kornberg H.L. (1965) Fine control of phoshopyruvate carboxylase activity in *E. coli*. *Biochim. biophys. Acta*, **96**, 169.

3 Hsie A.W. and Rickenberg H.V. (1966) A mutant of *E. coli* deficient in phosphoenolpyruvate carboxykinase activity. *Biochem. biophys. Res. Commun.* **25**, 676.

27

1 Watson J.D. (1965) *Molecular Biology of the Gene*. Benjamin, New York, p.74.

2 Wood H.G. and Werkman C.H. (1936) The utilisation of CO_2 in the dissimilation of glycerol by the propionic acid bacteria. *Biochem. J.* **30**, 48.

28

1 Stephenson M. (1930) *Bacterial Metabolism*. Longmans, London.

2 De Ley J. and Stouthamer A.J. (1959) The mechanism and localisation of hexonate metabolism in *acetobacter suboxydans* and *acetobacter melanogenum*. *Biochim. biophys. Acta*, **34**, 171.

29

1 Stanier R.Y. (1947) Simultaneous adaptation: a new technique for the study of metabolic pathways. *J. Bacteriol.* **54**, 339.

30

1 Cohn M., Monod J., Pollock M.R., Spiegleman S. and Stanier R.Y. (1953) Terminology of enzyme formation. *Nature*, **172**, 1096.

31

1 Stone R.W. and Wilson P.W. (1952) Respiratory activity of cell-free extracts from *azotobacter*. *J. Bacteriol.* **63**, 605.

2 Mitchell P. (1962) Metabolism transport and morphogenesis: which drives which? *J. Gen. Microbiol.* **29**, 25.

32

1 Saz H.J. and Krampitz L.O. (1954) The oxidation of acetate by *Micrococcus lysodeikticus*. *J. Bacteriol.* **67**, 409.

2 Swim H.E. and Krampitz L.O. (1954) Acetic acid oxidation by *E. coli*: evidence for the occurrence of a tricarboxylic acid cycle. *J. Bacteriol.* **67**, 419.

33 1 Ornston L.N. and Stanier R.Y. (1966) The conversion of catechol and protocatechuate to β-keto adipate by *Pseudomonas putida*—Biochemistry. *J. biol. Chem.* **241**, 3776.

2 Ornston L.N. (1966) The conversion of catechol and protocatechuate to β-keto adipate by *Pseudomonas putida*—enzymes of the catechol pathway. *J. biol. Chem.* **241**, 3795.

3 Dagley S., Evans W.C. and Ribbons D.W. (1960) New pathways in the oxidative metabolism of aromatic compounds by micro-organisms. *Nature*, **188**, 560.

34 1 Ornston L.N. (1966) The conversion of catechol and protocatechuate to α-keto adipate by *Pseudomonas Putida*—Regulation. *J. biol. Chem.* **241**, 3800.

2 Dagley S., Chapman P.J. and Gibson D.T. (1965) The metabolism of β-Phenylpropionic acid by an *achromobacter*. *Biochem J.* **97**, 643.

35 1 Roberts R.B., Abelson P.H., Cowie D.B., Bolton E.T. and Britten R.J. (1955) Studies of biosynthesis in *Escherichia coli*. *Carnegie Institution of Washington*, Publication no. 607.

37 1 Calvin M. and Bassham J.A. (1963) *The Photosynthesis of Carbon Compounds*. Benjamin, New York.

2 Bassham J.A. (1962) *The Path of Carbon in Photosynthesis*. Scientific American, June.

39 1 Hatch M.D. and Slack C.R. (1966) Photosynthesis by sugar cane leaves—a new carboxylation reaction and the pathway of sugar formation. *Biochem. J.* **101**, 103.

40 1 Wood H.G. and Utter M.F. (1965) The role of carbon dioxide fixation in metabolism. *Essays in biochemistry*, vol. 1. Academic Press, London.

2 Dagley S. and Hinshelwood C.N. (1938) Physiochemical aspects of bacterial growth. Quantitative dependence of the growth rate of *Bact. lactis aerogenes* on the CO_2 content of the gas atmosphere. *J. Chem. Soc.* 1936.

3 Kornberg H.L. (1958) Metabolism of C_2-compounds in micro-organisms. The incorporation of [$2^{14}C$] acetate by *Pseudomonas fluorescens* and by a *corynebacterium* grown on ammonium acetate. *Biochem J.* **68**, 535.

41 1 Campbell J.J.R., Smith R.A. and Eagles B.A. (1953) A deviation from the conventional TCA cycle in *Pseudomonas aeruginosa*. *Biochim. biophys. Acta*, **11**, 594.

2 Wong, D.T.O. and Ajl S.J. (1956) Conversion of acetate and glyoxalate to malate. *J. Amer. chem. Soc.* **78**, 3230.

3 Dagley S., Trudgill P.W. and Callely A.B. (1961) Synthesis of cell constituents from glycine by a *Pseudomonas*. *Biochem. J.* **81**, 623.

42 1 Kornberg H.L. and Gotto A.M. (1961) Metabolism of C_2 compounds in micro-organisms. Synthesis of cell constituents from glycollate by *Pseudomonas* spp. *Biochem. J.* **78**, 69.

2 Quayle J.R. (1960) Carbon assimilation by *Pseudomonas oxalaticus* (Ox/I)-oxalate utilisation during growth on oxalate. *Biochem. J.* **75**, 515.

3 Gotto A.M. and Kornberg H.L. (1961) Metabolism of C_2 compounds in micro-organisms—preparation and properties of crystalline tartronic semialdehyde reductase. *Biochem. J.* **81**, 273.

44 1 Quayle J.R. and Keech D.B. (1959) Carbon assimilation by *pseudomonas oxalaticus* (Ox/I)—formate and CO_2 utilisation by cell-free extracts of the organism grown on formate. *Biochem. J.* **72**, 631.

2 Baeyer A. (1870) Ueber die Wasserentziehung und ihre Bedeutung für das Pflanzenleben und die Gährung. *Ber. Dtsch. chem. Ges.* **3**, 63.

3 Johnson P.A. and Quayle J.R. (1965) Microbial growth on C_1 compounds—synthesis of cell constituents by methane and methanol grown *Pseudomonas methanica*. *Biochem. J.* **95**, 859.

4 Kemp M.B. and Quayle J.R. (1966) Microbial growth on C_1

compounds. Incorporation of C_1 units into allulose phosphate by extracts of *Pseudomonas methanica*. *Biochem. J.* **99**, 41.

47 1 Novelli G.D. and Lipmann F. (1950) The catalytic function of coenzyme A in citric acid synthesis. *J. biol. Chem.* **182**, 213.

49 1 Bonner D.M., Tatum E.L. and Beadle G.W. (1943) The genetic control of biochemical reactions in *neurospora*; a mutant requiring isoleucine and valine. *Arch. Biochem.* **3**, 71.

2 Sjolander J.R.S., Folkers K., Adelberg E.A. and Tatum E.L. (1954) α,β-Dihydroxyisovaleric acid and α,β-dihydroxy-β-methyl valeric acid, precursors of valine and isoleucine. *J. Amer. chem. Soc.* **76**, 1085.

3 Strassman M., Thomas A.J. and Weinhouse S. (1953) Valine biosynthesis in *Torulopsis utilis*. *J. Amer. chem. Soc.* **75**, 5135.

50 1 Halpern Y.S. and Umbarger H.E. (1959) Evidence for two distinct enzyme systems forming acetolactate in aerobacter aerogenes. *J. biol. Chem.* **234**, 3067.

51 1 Armstrong F.B. and Wagner R.P. (1961) Biosynthesis of valine and isoleucine—electrophoretic studies on the reductoisomerase and reductase of *Salmonella*. *J. biol. Chem.* **236**, 3252.

2 Wixom R.L., Wikman J.H. and Howell G.B. (1961) Studies in valine biosynthesis. Biological distribution of a dihydroxy acid dehydrase. *J. biol. Chem.* **236**, 3257.

3 Satyanarayana T. and Radhakrishran A.N. (1962) Biosynthesis of valine and isoleucine in plants. *Biochim. biophys. Acta*, **56**, 197.

52 1 Abramsky T. and Shemin D. (1965) The formation of isoleucine from β-methyl aspartic acid in *Escherichia coli* W. *J. biol. Chem.* **240**, 2971.

2 Watson J.D. (1965) *Molecular Biology of the Gene*. Benjamin, New York, p. 99.

Regulation of Metabolism

55 1 Dean A.C.R. and Hinshelwood C.N. (1963) Integration of cell reactions. Some basic aspects of cell regulation. *Nature*, **199**, 7; *ibid* (1964) **201**, 232.

57 1 Krebs H.A., Gascoyne T. and Notton B.M. (1967) Generation of extra-mitochondrial reducing power in gluconeogenesis. *Biochem. J.* **102**, 275.

2 Chance B. and Williams G.R. (1956) The respiratory chain and oxidative phosphorylation. *Advanc. Enzymol.* **17**, 65.

3 Lehninger A.L. (1964) *The Mitochondrion*. Benjamin, New York.

58 1 Klingenberg M. and Schollmeyer P. (1961) ATP-abhängige Atmungs-kontrolle und DPN. *Biochem.Z.* **333**, 337.

2 Mitchell P. (1963) *Biochemical Society Symposia*, vol. 22, p. 142.

59 1 Boxer G.E. and Devlin T.M. (1961) Pathways of intracellular hydrogen transport. *Science*, **134**, 1495.

60 1 Stephenson M. (1949) *Bacterial Metabolism*, 3rd ed. Longmans, London.

2 Kornberg H.L. (1965) *Symp. Soc. Gen. Microbiol.* **15**, 8.

3 Roberts R.B., Abelson P.H., Cowie D.B., Bolton E.T. and Britten R.J. (1955) Studies of biosynthesis in *Escherichia coli*. Carnegie Institution of Washington Publication, no. 607.

61 1 Dische Z. (1940) *Bull. Soc. Biochim. France*, **23**, 1140.

2 Umbarger H.E. (1956) Evidence for a negative feedback mechanism in the biosynthesis of isoleucine. *Science*, **123**, 848.

3 Yates R.A. and Pardee A.B. (1956) Control of pyrimidine biosynthesis in *Escherichia coli* by a feedback mechanism. *J. biol. Chem.* **221**, 757.

62 1 Koshland D.E., Jr. (1963) The role of flexibility in enzyme action. *Cold Spring Harbor Symp. Quant. Biol.* **28**, 473.

63 1 Monod J., Changeux J-P. and Jacob F. (1963) Allosteric proteins and cellular control systems. *J. molec. Biol.* **6**, 306.

2 Monod J., Wyman J. and Changeux J-P. (1965) On the nature of allosteric transitions: a plausible model. *J. molec. Biol.* **12**, 88.

65 1 Gerhart J.C. and Schachman H.K. (1965) Distinct subunits for the regulation and catalytic activity of aspartate transcarbamylase. *Biochemistry*, **4**, 1054.

2 Weber K. (1968) Aspartate transcarbamylase from *E. coli* characterisation of the peptide chains by molecular weight, amino acid composition and amino terminal residues. *J. biol. Chem.* **243**, 543.

66 1 Martin R.G. (1963) The first enzyme in histidine biosynthesis: the nature of feedback inhibition by histidine. *J. biol. Chem.* **238**, 257.

67 1 Cohen G. (1965) Regulation of enzyme activity in micro-organisms. *Ann. Rev. Microbiol.* **19**, 105.

68 1 Hess B., Haeckel R. and Brand K. (1966) FDP-activation of yeast pyruvate kinase. *Biochem. biophys. Res. Commun.* **24**, 824.

2 Taylor C.B. and Bailey E. (1967) Activation of liver pyruvate kinase by fructose 1-6,diphosphate. *Biochem. J.* **102**, 320.

3 Krebs E.G. and Fischer E.H. (1962) Molecular properties and transformations of glycogen phosphorylase in animal tissues. *Advanc. Enzymol.* **24**, 263.

69 1 Rosell-Perez M., Villar-Palasi C. and Larner J. (1962) Studies on UDPG-glycogen transglucosylase. 1. Preparation and differentiation of two activities of UDPG-glycogen transglucosylase from rat skeletal muscle. *Biochemistry* **1**, 763.

2 Hizukuri S. and Larner J. (1964) Studies on UDPG: α-1,4-glucan α-4-glucosyltransferase. VII. Conversion of the enzyme from glucose-6-phosphate-dependent to independent form in liver. *Biochemistry*, **3**, 1783.

70 1 Lynen F. (1967) The role of biotin-dependent carboxylations in biosynthetic reactions. *Biochem. J.* **102**, 381.

2 Brady R.O. and Gurin S. (1952) Biosynthesis of fatty acids by cell-free or water-soluble enzyme systems. *J. biol. Chem.* **199**, 421.

3 Gregolin C., Ryder E., Kleinschmidt A.K., Warner R.C. and Lane M.D. (1966) Molecular characteristics of liver acetyl CoA carboxylase. Liver acetyl CoA carboxylase: the dissociation-reassociation process and its relation to catalytic activity. *Proc. Nat. Acad. Science. U.S.* **56**, 148; ibid **56**, 1751.

71 1 Krebs H.A. (1961) The physiological role of the ketone bodies. *Biochem J.* **80**, 225.

72 1 Wieland O., Weiss L. and Eger-Neufeldt I. (1964) Enzymatic regulation of liver acetyl CoA metabolism in relation to ketogenesis. General discussion (citrate synthetase). *Advanc. in Enzyme Regulation*, **2**, 85; (1966) ibid **4**, 281.

2 Hathaway J.H. and Atkinson D.E. (1965) Kinetics of regulatory enzymes: Effect of adenosine triphosphate on citrate synthase *Biochem. biophys. Res. Commun.* **20**, 661; (1963) The effect of adenylic acid on yeast nicotinamide adenine dinucleotide isocitrate dehydrogenase, a possible control mechanism. *J. biol. Chem.* **238**, 2875.

73 1 Monod J. and Cohn M. (1952) La biosynthèse induite des enzymes. *Advanc. Enzymol* **13**, 67.

2 Ames B.N. and Garry B. (1959) Coordinate repression of the synthesis of four histidine biosynthetic enzymes by histidine. *Proc. Nat. Acad. Science. U.S.* **45**, 1453.

3 Shepherdson M. and Pardee A.B. (1960). Production and crystallisation of asparate transcarbamylase *J. biol. Chem.* **235**, 3233.

74 1 Ornston L.N. (1966) The conversion of catechol and protocatechuate to β-ketoadipate by *Pseudomonas putida*. *J. biol. Chem.* **241**, 3800.

75 1 Cánovas J.L., Ornston L.N. and Stanier R.Y. (1967) Evolutionary significance of metabolic control systems. *Science*, **156**, 1695.

77 1 Freundlich M., Burns R.O. and Umbarger H.E. (1962) Control of isoleucine, valine and leucine biosynthesis. 1. Multivalent repression. *Proc. Nat. Acad. Science, U.S.* **48**, 1804.

2 Stalon V., Ramos F., Piérard A. and Wiame J.M. (1967) The occurrence of a catabolic and an anabolic ornithine carbamoyl transferase in *Pseudomonas*. *Biochim. biophys. Acta*, **139**, 91.

78 1 Bernlohr R.W. (1966) Ornithine transcarbamylase enzymes: occurrence in *Bacillus licheniformis*. *Science*, **152**, 87.

2 Fox C.F., Robinson, W.S., Haselkorn, R. and Weiss S.B. (1964) Enzymatic synthesis of ribonucleic acid. The ribonucleic acid-primed synthesis of ribonucleic acid with *Micrococcus lysodeikticus* ribonucleic acid polymerase. *J. biol. Chem.* **239**, 186.

3 Raina A. and Cohen S.S. (1966) Polyamines and RNA synthesis in a polyauxotrophic strain of *E. coli*. *Proc. Nat. Acad. Science, U.S.* **55**, 1587.

4 Morris D.R. and Pardee A.B. (1966) Multiple pathways of putrescine biosynthesis in *Escherichia coli*. *J. biol. Chem.* **241**, 3129.

5 Watson J.D. (1965) *Molecular Biology of the Gene*. Benjamin, New York, Chapters 13 and 14.

80 1 Ghosh H.P., Söll D. and Khorana H.G. (1967) Studies on polynucleotides LXVII. Initiation of protein synthesis *in vitro* as studied by using ribonucleotides with repeating nucleotide sequences as messengers. *J. molec. Biol.* **25**, 275.

2 Jacob F. and Monod J. (1961) Genetic regulatory mechanisms in the synthesis of proteins. *J. molec. Biol.* **3**, 318.

82 1 Gilbert W. & Müller-Hill B. (1966) Isolation of the lac repressor. *Proc. Nat. Acad. Science, U.S.* **56**, 1891.

2 Ptashne M. (1967) Specific binding of the λ-phage repressor to λ-DNA. *Nature*, **214**, 232.

83 1 Zabin I. (1963) Proteins of the lactose system. *Cold Spring Harbor Symp. Quant. Biol.* **28**, 431.

2 Jacob F. and Monod J. (1961) On the regulation of gene activity. *Cold Spring Harbor Symp. Quant. Biol.* **26**, 193.

3 Franklin N. and Luria S. (1961) Transduction by bacteriophage P_i and the properties of the lac genetic region in *E. coli* and *S. dysenteriae*. *Virology*, **15**, 299.

4 Ames B.N. and Martin R.G. (1964) Biochemical aspects of genetics: the operon. *Ann. Rev. Biochem.* **33**, 235.

5 Martin R.G., Silbert D.F., Smith D.W.E. and Whitfield H.G. (Jr.) (1966) Polarity in the histidine operon. *J. molec. Biol.* **21**, 357.

84 1 Magasanik B. (1961) Catabolite repression. *Cold Spring Harbor Symp. Quant. Biol.* **26**, 249.

85 1 Moses V. and Prevost C. (1966) Catabolite repression of β-galactosidase synthesis in *Escherichia coli*. *Biochem. J.* **100**, 336.

2 Palmer J. and Moses V. (1967) Involvement of the lac regulatory genes in catabolite repression in *Escherichia coli*. *Biochem. J.* **103**, 358.

86 1 Tyler B., Loomis W.F. and Magasanik B. (1967) Transient repression of the lac operon. *J. Bacteriol.* **94**, 2001.

87 1 Maaløe O. and Kjeldgaard N.O. (1966) Control of macromolecular synthesis. Benjamin, New York.

2 Sykes J. (1966) A mechanism for the regulation of ribosomal ribonucleic acid synthesis in bacteria. *J. theor. Biol.* **12**, 373.

88 1 Piras, M. and Knox W.E. (1967) Tryptophan pyrrolase of liver. The activating reactions in crude preparations from rat liver. *J. biol. Chem.* **242**, 2952.

2 Yudkin J. (1938) *Biol. Rev. Cambridge Phil. Soc.* **13**, 93.

3 Weber G., Singhal R.L. and Srivastava S.K. (1965) Regulation of RNA metabolism and amino acid level in hepatomas of different growth rate. *Advanc. Enzyme Regulation* **3**, 369; (1966) Synchronous behaviour pattern of key glycolytic enzymes: glucokinase, phosphofructokinase, and pyruvate kinase. ibid **4**, 59.

89 1 Korner A. (1967) Ribonucleic acid and hormonal control of protein synthesis. *Progress in Biophysics*, **17**, 61.

93 GEN Kaplan N.O. (1960) The pyridine coenzymes. *The Enzymes*, 2nd ed., vol. 3. Academic Press, New York, p. 105.

GEN Plant G.W.E. (1961) Water soluble vitamins. Part II. *Ann. Rev. Biochem.* **30**, 409.

GEN Cheldelin V.H. and Baich A. (1962) Biosynthesis of the water soluble vitamins. *Biogenesis of Natural Compounds*. Pergamon Press, Oxford, p. 509.

GEN Stokstad E.L.R. (1962) The biochemistry of the water soluble vitamins. *Ann. Rev. Biochem.* **31**, 451.

GEN Strittmatter P. (1966) Dehydrogenases and flavoproteins. *Ann. Rev. Biochem.* **35**, 125.

GEN Chaykin S. (1967) Nicotinamide coenzymes. *Ann. Rev. Biochem.* **36**, 149

94 GEN Beinert H. (1960) Flavin coenzymes. *The Enzymes*, 2nd ed., vol. 2. Academic Press, New York, p. 339.

GEN Wagner A.F. and Folkers K. (1964) Riboflavine, flavine mononucleotide and flavine-adenine dinucleotide. *Vitamins and coenzymes*. Interscience Publishers, New York, p. 46.

GEN Strittmatter P. (1966) see 93.

95 GEN Gunsalus I.C. (1954) Group transfer and acyl generating functions of lipoic acid derivatives. *Symposium on the Mechanism of Enzyme Action*, p. 545. Johns Hopkins Press, Baltimore.

GEN Gunsalus I.C. (1954) Oxidative and transfer reactions of lipoic acid. *Fed. Proc.* **13**, 715.

GEN Reed L.J. (1957) The chemistry and functions of lipoic acid. *Advanc. Enzymol.* **18**, 319.

GEN Reed L.J. (1960) Lipoic acid. *The Enzymes*, 2nd ed., vol. 3. Academic Press, New York, p. 195.

GEN Massey V. (1963) Lipoyl dehydrogenase. *The Enzymes*. 2nd ed., vol. 7. Academic Press, New York, p. 275.

GEN Wagner A.F. and Folkers K. (1964) Lipoic acid. *Vitamins and Coenzymes*. Interscience Publishers, New York, Chapter II.

GEN Schmidt U., Grafen P., Altland K. and Goedde H.W. (1969) Biochemistry and chemistry of lipoic acid. *Advanc. Enzymol.* **32**, 423.

96 GEN Crane F.L., Hatefi Y., Lester R.L. and Widmer C. (1957) Isolation of a quinone from beef heart mitochondria. *Biochim. biophys. Acta*, **25**, 220.

Crane F.L. (1962) Quinones in lipoprotein electron transport systems. *Biochemistry*, **1**, 510.

GEN Hatefi Y. (1963) Coenzyme Q (ubiquinone). *Advanc. Enzymol.* **25**, 275.

97 GEN Du Vigneaud V. (1942) The structure of biotin. *Science*, **96**, 455.

GEN Wright L.D., Cresson E.L., Skeggs H.R., Peck R.L., Wolf D.E., Wood T.R., Valiant J. and Folkers K. (1951) The elucidation of biocytin. *Science* **114**, 635.

GEN Lardy H.A. and Peanansky R. (1953) Metabolic functions of biotin. *Physiol. Rev.* **33**, 560.

GEN Lynen F., Knappe F.J., Lorch E., Jutting G. and Ringelmann E. (1958) Die biochemische Funktion des Biotins. *Angew. Chem.* **71**, 481.

GEN Kosow D.P. and Lane M.D. (1962) Propionyl holocarboxylase formation. Covalent bonding of biotin to apocarboxylase lyase ε-amino groups. *Biochem. biophys. Res. Commun.* **7**, 439.

GEN Vagelos P.R. (1964) Lipid metabolism. *Ann. Rev. Biochem.* **33**, 139.

GEN Lane M.D., Young D.L. and Lynen F. (1964) The enzymatic synthesis of holotranscarboxylase from apotranscarboxylase and (+)-biotin. Purification of the apoenzyme and synthesis: characteristics of the reaction. *J. biol. Chem.* **239**, 2858.

GEN Lane M.D., Rominger K.L., Young D.L. and Lynen F. (1964) The enzymatic synthesis of holotranscarboxylase from apotranscarboxylase and (+)-biotin. *J. biol. Chem.* **239**, 2865.

GEN Lynen F. (1967) The role of biotin-dependent carboxylations in biosynthetic reactions. *Biochem. J.* **102**, 381.

GEN Sebrell W.H., Jr. and Harris R.S. (Eds.) (1969) *Biotin in the Vitamins*, 2nd ed., vol. II. Academic Press, New York, p. 262.

98–99 GEN Ingraham L.L. and Green D.E. (1958) Role of magnesium in enzyme catalysed syntheses involving adenosine triphosphate. *Science*, **128**, 310.

GEN Bock R.M. (1960) Adenine nucleotides and properties of pyrophosphate compounds. *The Enzymes*, 2nd ed., vol. 2. Academic Press, New York, p. 3.

GEN George P. and Rutman R.J. (1960) The high energy phosphate bond concept. *Concept. progs. Biophys.* **10**, 2.

GEN Atkinson M.R. and Morton R.K. (1960) Free energy and the biosynthesis of phosphates. *Comparative Biochemistry*, vol. 2. Academic Press, New York, p. 1.

GEN Cohn M. (1963) Magnetic resonance studies of metal activation of enzymic reactions of nucleotides and other phosphate substrates. *Biochemistry*, **2**, 623.

99 H Robison G.A., Butcher R.W. and Sutherland E.W. (1968) Cyclic AMP. *Ann. Rev. Biochem.* **37**, 149.

H Krishna G., Hynies S. and Brodie B.B. (1968) Effects of thyroid hormones on adenyl cyclase in adipose tissue and free fatty acid mobilisation. *Proc. Nat. Acad. Sci. U.S.* **59**, 884.

100 GEN Braunstein A.E. (1960) Pyridoxal phosphate. *The Enzymes*, 2nd ed., vol. 2. Academic Press, New York, p. 113.

GEN Snell E.E., Fasella P.M., Braunstein A. and Rossi-Fannelli A. (Eds.) (1963) *Chemical and Biological Aspects of Pyridoxal Catalysis*. The Macmillan Co. New York.

GEN Wagner A.F. and Folkers K. (1964) *Vitamins and Coenzymes*. John Wiley & Sons, New York, p. 160.

GEN Harris R.S., Wool I.G. and Lorraine J.A. (Eds.) (1964) *Vitamins and Hormones*, vol. 22. International Symposium on Vitamin B.

GEN Fasella P. (1967) Pyridoxal phosphate. *Ann. Rev. Biochem.* **36**, 185.

GEN Sebrell W.H., Jr. and Harris R.S. (Eds.) (1969) Vitamin B_6 Group, in *The Vitamins*, 2nd ed., vol. II, 2. Academic Press, New York.

101 GEN Cammarata P.S. and Cohen P.P. (1950) The scope of the transamination reaction in animal tissues. *J. biol. Chem.* **187**, 439.

GEN Olivard J. and Snell E.E. (1955) Growth and enzymatic activities of Vitamin B_6 analogues. 1. D-Alanine synthesis. *J. biol. Chem.* **213**, 203.

102–103 GEN Breslow R. (1958) On the mechanism of thiamine action: evidence from studies on model systems. *J. Amer. chem. Soc.* **80**, 3719.

GEN Metzler D.E. (1960) Thiamine coenzymes. *The Enzymes*. 2nd ed., vol. 2. Academic Press, New York, p. 295.

GEN Krampitz L.O., Suzuki I. and Greull G. (1961) Mechanism of action of thiamin diphosphate. *Fifth Intern. Cong. Biochem.* 321.

GEN Krampitz L.O., Suzuki I. and Greull G. (1961) Role of thiamin diphosphate in catalysis. *Fed. Proc.* **20**, 971.

GEN Carlson G.L. and Brown G.M. (1961) The natural occurrence, enzymatic formation, and biochemical significance of a hydroxyethyl derivative of thiamine pyrophosphate. *J. biol. Chem.* **236**, 2099.

GEN Goldsmith G. (1964) *Thiamine Nutrition*, vol. 2. Academic Press, New York, p. 109.

GEN Krampitz L.O. (1969) Catalytic functions of thiamine diphosphate. *Ann. Rev. Biochem.* **38**, 213.

A–G Detar D.F. and Westheimer F.H. (1959) The role of thiamin in carboxylase. *J. Amer. chem. Soc.* **81**, 175.

A–H Lipmann F., Jones M.E., Black S. and Flynn R.M. (1953) The mechanism of the ATP-CoA-Acetate reaction. *J. cell. comp. Physiol.* **41**, suppl. 1, 109.

A–H Jones M.E., Lipmann F.L., Hilz H. and Lynen F. (1953) On the enzymatic mechanism of coenzyme A acetylation with adenosine triphosphate and acetate. *J. Amer. chem. Soc.* **75**, 3285.

A–H Hele P. (1954) The acetate activating enzyme of beef heart. *J. biol. Chem.* **206**, 671.

A–H Eisenberg M.A. (1955) The acetate-activating enzyme of *Rhodospirillum rubrum. Biochim. biophys. Acta*, **16**, 58.

A–J Downes J. and Sykes P. (1957) Thiazolium salts as catalysts in the acyloin condensation. *Chem. and Ind.* vol. 2, 1095.

B–K Racker E., de la Haba G. and Leder I.G. (1953) Thiamine pyrophosphate, a coenzyme of transketolase. *J. Amer. chem. Soc.* **75**, 1010.

B–K Datta A.G. and Racker E. (1961) Mechanism of action of transketolase. II. The substrate-enzyme intermediate. *J. biol. Chem.* **236**, 624.

B–K Holzer H., Katteman R. and Busch D. (1962) A thiamine pyrophosphate-glycolaldehyde compound ("active glycolaldehyde") as intermediate in the transketolase reaction. *Biochem. biophys. Res. Commun.* **7**, 167.

104 GEN Lipmann F., Kaplan N.O., Novelli G.D., Tuttle L.C. and Guirard B.M. (1947) Coenzyme for acetylation; a pantothenic acid derivative. *J. biol. Chem.* **167**, 869.

GEN Lipmann F. (1953) On the chemistry and function of coenzyme A. *Bact. Rev.* **17**, 1.

GEN Jaenicke L. and Lynen F. (1960) Coenzyme A, in *The Enzymes*, 2nd ed., vol. 3. Academic Press, New York, p. 3.

GEN Goldman P. and Vagelos P.R. (1964) Acyl-transfer reactions (CoA structure, function) 'Group transfer reactions'. *Comprehensive Biochemistry*, vol. 15. Elsevier, New York, p. 71.

106 GEN Huennekens F.M. and Osborn M.J. (1959) Folic acid coenzymes and one carbon metabolism. *Advanc. Enzymol.* **21**, 369.

GEN Rabinowitz J.C. (1960) Folic acid. *The Enzymes*, 2nd ed., vol. 2. Academic Press, New York, p. 185.

GEN Silverman M. (1962) Folic acid. *Methods in Enzymology*, vol. 5. Academic Press, New York, p. 790.

GEN Jaenicke L. (1964) Vitamin and coenzyme function: vitamin B_{12} and folic acid. *Ann. Rev. Biochem.* **33**, 287.

GEN Zakrzewski S.F. and Sansone A. (1967) Mechanism of reduction of dihydrofolate to tetrahydrofolate. Studies with 7-methyldihydrofolate as a model compound. *J. biol. Chem.* **242**, 5661.

B Osborn M.J., Talbert P.T. and Huennekens F.M. (1960) The structure of 'active formaldehyde'. (N^5, N^{10}-methylene tetrahydrofolic acid. *J. Amer. chem. Soc.* **82**, 4921.

B Schirch L. and Mason N. (1962) Serine transhydroxymethylase: spectral properties of the enzyme bound pyridoxal-5-phosphate. *J. biol. Chem.* **237**, 2578.

D Larrabee A.R., Rosenthal S., Cathou R.E. and Buchanan J.M. (1963) Enzymatic synthesis of the methyl group of methionine. IV. Isolation, characterisation, and role of 5-methyl tetrahydrofolate. *J. biol. Chem.* **238**, 1025.

107 GEN Friedkin M. (1963) Enzymatic aspects of folic acid. *Ann. Rev. Biochem.* **32**, 185.

A–B Bertino J.R., Simmons B. and Donahue D.M. (1962) Purification and properties of the formate-activating enzyme from erythrocytes. *J. biol. Chem.* **237**, 1314.

A–B Rabinowitz J.C. and Pricer W.E. (1962) Formyltetrahydrofolate synthetase. I. Isolation and crystallisation of the enzyme. *J. biol. Chem.* **237**, 2898.

A–B Himes R.H. and Rabinowitz J.C. (1962) Formyltetrahydrofolate synthetase. II. Characteristics of the enzyme and the enzymic reaction. *J. biol. Chem.* **237**, 2903, 2915.

A–B Joyce B.K. and Himes R.H. (1966) Formyltetrahydrofolate synthetase. A study of equilibrium reaction rates. *J. biol. Chem.* **241**, 5716.

A–C Silverman M. (1962) N^5-Formyltetrahydrofolic acid-glutamic acid transformylase from hog liver. *Methods in Enzymology*, vol. 5. Academic Press, New York, p. 790.

B–C Kay L.D., Osborn N.J., Hatefi Y. and Huennekens F.M. (1960) The enzymatic conversion of N^5-formyltetrahydrofolic acid (folinic acid) to N^{10}-formyltetrahydrofolic acid. *J. biol. Chem.* **235**, 195.

F–G Cathou R.E. and Buchanan J.M. (1963) Enzymatic synthesis of the methyl group of methionine. *J. biol. Chem.* **238**, 1746.

F–G Kisliuk R.L. (1963) The source of hydrogen for methionine methyl formation. *J. biol. Chem.* **238**, 397.

108 GEN Tabor H. and Wyngarden L. (1959) The enzymatic formation of formiminotetrahydrofolic acid, 5,10-methenyltetrahydrofolic acid and 10-formyltetrahydrofolic acid in the metabolism of formiminoglutamic acid. *J. biol. Chem:* **234**, 1830.

109 GEN Chance B. and Williams G.R. (1956) The respiratory chain and oxidative phosphorylation. *Advanc. Enzymol.* **17**, 65.

GEN King T.E. (1962) Reconstitution of respiratory-chain enzymes. V. Reversible dissociation and reconstitution of mitochondrial succinate oxidase. *Biochim. biophys. Acta*, **58**, 375.

111 A–B Green A.A. and Cori G.T. (1943) Crystalline muscle phosphorylase. *J. biol. Chem.* **151**, 21.

A–B Cori G.T. and Larner J. (1951) Action of amylo-1,6-glucosidase and phosphorylase on glycogen and amylopectin. *J. biol. Chem.* **188**, 17.

A–B Krebs E.G. and Fischer E.H. (1962) Molecular properties and transformations of glycogen phosphorylase in animal tissues. *Advanc. Enzymol.* **24**, 263.

A–B Lowry O.H., Schulz D.N. and Passoneau J.V. (1964) Effects of adenylic acid on the kinetics of muscle phosphorylase A. *J. biol. Chem.* **239**, 1947.

A–B Wang J.H. and Graves D.J. (1964) The relationship of the dissociation to the catalytic activity of glycogen phosphorylase a. *Biochemistry* **3**, 1437.

B–C Josh J.G. and Handler P. (1964) Phosphoglucomutase. Purification and properties of phosphoglucomutase from *E. coli*. *J. biol. Chem.* **239**, 2741.

B–C Yankeelon J.A., Jr. and Koshland D.E., Jr. (1965) Evidence for conformation changes induced by substrates of phosphoglucomutase. *J. biol. Chem.* **240**, 1593.

B–C Ray W.J., Jr. and Roscelli G.A. (1966) The addition and release of magnesium in the phosphoglucomutase reaction. Kinetic control of alternative pathways. *J. biol. Chem.* **241**, 3499.

B–C Britton H.G. and Clarke J.B. (1968) The mechanism of the phosphoglucomutase reaction. Studies on rabbit muscle phosphoglucomutase with flux techniques. *Biochem J.* **110**, 161.

B–C Passonneau J.V., Lowry O.H., Schulz D.W. and Brown J.G. (1969) Glucose 1,6-diphosphate formation by phosphoglucomutase in mammalian tissues. *J. biol. Chem.* **244**, 902.

112 GEN Caputto R., Barra H.S. and Cumar F.A. (1967) Glycolysis. *Ann. Rev. Biochem.* **36**, 223.

B Harden A. and Young W. (1909) The hexosephosphate formed by yeast-juice from hexose and phosphate. *Proc. Roy. Soc.* **B81**, 528.

A–B Fromm H.J., Silverstein E. and Boyer P.D. (1964) Equilibrium and net reaction rates in relation to the mechanism of yeast hexokinase. *J. biol. Chem.* **239**, 3645.

B–C Topper Y.J. (1957) On the mechanism of action of phosphoglucose isomerase and phosphomannose isomerase. *J. biol. Chem.* **225**, 419.

B–C Tsuboi K.K., Estrada J. and Hudson P.B. (1958) Phosphoglucose isomerase, purification and properties. *J. biol. Chem.* **231**, 19.

B–C Kahana S.E., Lowry O.H., Schulz D.W., Passoneau J.V. and Crawford E. (1960) The kinetics of phosphoglucose isomerase. *J. biol. Chem.* **235**, 2178.

B–C Topper Y.J. (1961) Aldose-ketose transformations. *The Enzymes*, 2nd ed., vol. 5. Academic Press, New York, p. 437.

B–C Noltmann E.A. (1964) Isolation of crystalline phosphoglucose isomerase from rabbit muscle. *J. biol. Chem.* **239**, 1545.

C–D Ling K. and Lardy H.A. (1954) Uridine- and inosine-triphosphates as phosphate donors for phosphohexokinase. *J. Amer. chem. Soc.* **76**, 2842.

C–D Lardy H.A. (1962) Phosphohexokinases. *The Enzymes*, 2nd ed., vol. 6. Academic Press, New York, p. 67.

C–D Gonzalez C.T., Ureta R., Sanchez H. and Niemeyer H. (1964) Multiple molecular forms of ATP: hexose 6-phosphotransferase from rat liver. *Biochem. biophys. Res. Commun.* **16**, 347.

C–D Atkinson D.E. and Walton G.M. (1965) Kinetics of regulatory enzymes. *E. coli*. Phosphofructokinase. *J. biol. Chem.* **240**, 757.

C–D Paetkau V. and Lardy H.A. (1967) Phosphofructokinase. Correlation of physical and enzymatic properties. *J. biol. Chem.* **242**, 2035.

C–D Mansour T.E. and Ahlfors C.E. (1968) Studies on heart phosphofructokinase. Some kinetic and physical properties of the enzyme. *J. biol. Chem.* **243**, 2523.

C–D Froede H.C., Geraci G. and Mansour T.E. (1968) Studies on heart phosphofructokinase. Thiol groups and their relationship to activity. *J. biol. Chem.* **243**, 6021.

C–D Younathan E.S., Paetkau V. and Lardy H.A. (1968) Rabbit muscle phosphofructokinase. Reactivity and function of thiol groups. *J. biol. Chem.* **243**, 1603.

C–D Lindell T.J. and Stellwagen E. (1968) Purification and properties of phosphofructokinase from yeast. *J. biol. Chem.* **243**, 907.

F Grazi E., Cheng T. and Horecker B.L. (1962) The formation of a stable aldolase-dihydroxyacetone phosphate complex. *Biochem. biophys. Res. Commun.* **7**, 250.

D–EF Rose I.A. and Reider S.V. (1955) The mechanism of action of muscle aldolase. *J. Amer. chem. Soc.* **77**, 5764.

D–EF Bloom B. and Topper Y.J. (1956) Mechanism of action of aldolase and phosphotriose isomerase. *Science*, **124**, 982.

D–EF Penansky R.J. and Lardy H.A. (1958) Bovine liver aldolase-isolation, crystallisation and some general properties. *J. biol. Chem.* **233**, 365.

D–EF Rose I.A. (1962) Mechanism of C-H bond cleavage in aldolase and isomerase reactions. *Brookhaven. Symp. Biol.* **15**, 293.

D–EF Grazi E., Meloche H., Martinez G., Wood W.A. and Horecker B.L. (1963) Evidence for Schiff base formation in enzymatic aldol condensations. *Biochem. biophys. Res. Commun.* **10**, 4.

D–EF Horecker B.L., Rowley P.T., Grazi E., Cheng T. and Tchola O. (1963) The mechanism of action of aldolases. Lysine as the substrate binding site. *Biochem. Z.* **338**, 36.

D–EF Rutter W.J. (1964) Evolution of aldolase. *Fed. Proc.* **23**, 1248.

D–EF Dahlquist A. and Crane R.K. (1964) The influence of the method of assay on the apparent specificity of rabbit liver aldolase. *Biochim. biophys. Acta*, **85**, 132.

D–EF Rose I.A., O'Connell E.L. and Mehler A.H. (1965) Mechanism of the aldolase reaction. *J. biol. Chem.* **240**, 1758.

D–EF Morse D., Lai C.F., Horecker B.L., Rajkumar T. and Rutter W.J. (1965) The mechanism of action of aldolases—the combining sites of rabbit liver aldolase. *Biochem. biophys. Res. Commun.* **18**, 679.

D–EF Lai C.Y., Tchola O., Cheng T. and Horecker B.L. (1965) The mechanism of action of aldolases. The number of combining sites in fructose diphosphate aldolase. *J. Biol. Chem.* **240**, 1347.

D–EF Morse D.E. and Horecker B.L. (1968) The mechanism of action of aldolases. *Advanc. Enzymol.* **31**, 125.

113 A–C Velick S.F. and Furfine C.S. (1963) Glyceraldehyde 3-phosphate dehydrogenase. *The Enzymes*, 2nd ed., vol. 7. Academic Press, New York, p. 243.

A–C Hilvers A.G., Vandam K. and Slater E.C. (1964) The reaction mechanism of D-glyceraldehyde-3-phosphate: NAD^+ oxido-reductase (phosphorylating) of rabbit muscle. *Biochim. biophys. Acta*, **85**, 206.

A–C Murdock A.L. and Koeppe O.J. (1964) The content and action of diphosphopyridine nucleotide in triosephosphate dehydrogenase. *J. biol. Chem.* **239**, 1983.

A–C Allison N.S. and Kaplan N.O. (1964) The comparative enzymology of triosephosphate dehydrogenases. *J. biol. Chem.* **239**, 2140.

A–C Mathew E., Meriwether B.P. and Park J.H. (1967) The enzymatic significance of *S*-acetylation and *N*-acetylation of 3-phosphoglyceraldehyde dehydrogenase. *J. biol. Chem.* **242**, 5024.

A–C De Vijlder J.J. and Slater E.C. (1968) The reaction between NAD^+ and rabbit muscle glyceraldehyde phosphate dehydrogenase. *Biochim. biophys. Acta*, **167**, 23.

A–C Singleton R., Jr., Kimmel J.R., and Amelunxen R.E. (1969) The amino acid composition and other properties of thermostable glyceraldehyde 3-phosphate dehydrogenase from *Bacillus stearothermophilus*. *J. biol. Chem.* **244**, 1623.

C–D Harrison W.H., Boyer P.D. and Falcone A.B. (1955) The mechanism of enzymic phosphate transfer reactions. *J. biol. Chem.* **215**, 303.

D–E Sutherland E.W., Posternak T. and Cori C.F. (1949) Mechanism of the phosphoglyceric mutase reaction. *J. biol. Chem.* **181**, 153.

D–E Lowry O.H. and Passonneau J.V. (1964) The relationships between substrates and enzymes of glycolysis in brain. *J. biol. Chem.* **239**, 31.

D–E Grisolia S. and Detter J.C. (1965) Immunological studies on glycerate 2,3-diphosphatase and phosphoglycerate mutase. *Biochem. Z.* **342**, 239.

D–E Torralba A. and Grisolia S. (1966) The purification and properties of phosphoglycerate mutase from chicken breast muscle. *J. biol. Chem.* **241**, 1713.

D–E Jacobs R.J. and Grisolia S. (1966) Phosphoryl intermediates formed with phosphoglycerate mutase. Role and labilisation of 2,3-diphosphoglycerate. *J. biol. Chem.* **241**, 5926.

E–F Malmström B.G. (1961) Enolase. *The Enzymes*, 2nd ed., vol. 5. Academic Press, New York, p. 471.

E–F Winstead J.A. and Wold F. (1964) Studies on rabbit muscle enolase. Chemical evidence for two polypeptide chains in the active enzyme. *Biochemistry*, **3**, 791.

F–G Bücher T. and Pfleiderer G. (1955) Pyruvate kinase from muscle. *Methods in Enzymology*, vol. 6. p. 435.

F–G Hess B., Haeckel R. and Brand K. (1966) FDP-activation of yeast pyruvate kinase. *Biochem. biophys. Res. Commun.* **24**, 824.

F–G Taylor C.B. and Bailey E. (1967) Activation of liver pyruvate kinase by fructose 1,6-diphosphate. *Biochem. J.* **102**, 32c.

F–G Pogson C.I. (1968) Adipose tissue pyruvate kinase. Properties and interconversion of two active forms. *Biochem. J.* **110**, 67.

F–G Carminatti H., De Asuá J.L., Recondo E., Passeron S. and Rozengurt E. (1968) Some kinetic properties of liver pyruvate kinase (Type L). *J. biol. Chem.* **243**, 3051.

114 GEN Cooper R.A. and Kornberg H.L. (1965) Net formation of phosphoenol pyruvate from pyruvate by *E. coli*. *Biochim. biophys. Acta*, **104**, 618.

GEN Hems R., Ross B.D., Berry M.N. and Krebs H.A. (1966) Gluconeogenesis in the perfused fat liver. *Biochem. J.* **101**, 284.

A–B Appleby C.A. and Morton R.K. (1959) Lactic dehydrogenase and cytochrome b_2 of bakers yeast. *Biochem. J.* **71**, 492.

A–B Apella E. and Markert C.L. (1961) Dissociation of lactate dehydrogenase into subunits with guanidine hydrochloride. *Biochem. biophys. Res. Commun.* **6**, 171.

A–B Di Sabato G. and Kaplan N.O. (1963) The role of sulphydryl groups of lactic dehydrogenase. *Biochemistry*, **2**, 776.

A–B Dawson D.M., Goodfriend T.L. and Kaplan N.O. (1964) Lactic dehydrogenases: functions of the two types. *Science*, **143**, 929.

A–B Fondy T.P. (1964) The comparative enzymology of lactic dehydrogenases. Properties of the crystalline HM_3 hybrid from chicken muscle and of the H_2M_2 hybrid and H_4 enzyme from chicken liver. *Biochemistry*, **3**, 522.

A–B Heck H.D'A., McMurray C.H. and Gutfreund H. (1968) The resolution of some steps of the reactions of lactate dehydrogenase with its substrates. *Biochem. J.* **108**, 793.

A–B Tarmy E.M. and Kaplan N.O. (1968) Chemical characterisation of D-lactate dehydrogenase from *E. coli* B. *J. biol. Chem.* **243**, 2579.

C–D Green D.E., Herbert D. and Subrahmanyan V. (1941) Carboxylase. *J. biol. Chem.* **138**, 327.

C–D Singer T.P. and Pensky J. (1952) Isolation and properties of the α-carboxylase of wheat-germ. *J. biol. Chem.* **196**, 375.

C–D Holzer H. and Beauchamp K. (1961) Nachweiss und Charakterisierung von α-Lactyl-thiaminpyrophosphat ('aktives Pyruvat') und α-Hydroxyäthyl-thiaminpyrophosphat ('aktiver Acetaldehyd') als Zwischenprodukte der Decarboxylierung von Pyruvat mit Pyruvatdecarboxylase aus Bierhefe. *Biochem. biophys. Acta*, **46**, 225.

D–E Kägi J.H.R. and Vallee B.L. (1960) The role of zinc in alcohol dehydrogenase. V. The effect of metal-binding agents on the structure of the yeast alcohol dehydrogenase molecule. *J. biol. Chem.* **235**, 3188.

D–E Sund H. and Theorell, H. (1963) Alcohol dehydrogenases. *The Enzymes*, 2nd ed., vol. 7. Academic Press, New York, p. 25.

116 A–B Utter M.F. and Keech D.B. (1960) Formation of oxaloacetate from pyruvate and CO_2. *J. biol. Chem.* **235**, PC 17.

A–B Utter M.F. and Keech D.B. (1963) Pyruvate carboxylase. I. Nature of the reaction. *J. biol. Chem.* **238**, 2603.

A–B Keech D.B. and Utter M.F. (1963) Pyruvate carboxylase. II. Properties. *J. biol. Chem.* **238**, 2609.

A–B Scrutton M.C., Keech D.B. and Utter M.F. (1965) Pyruvate carboxylase. Partial reactions and the locus of activation by acetyl coenzyme A. *J. biol. Chem.* **240**, 574.

A–B Scrutton M.C. and Utter M.F. (1965) Pyruvate carboxylase. V. Interaction of the enzyme with adenosine triphosphate. *J. biol. Chem.* **240**, 3714.

A–B Scrutton M.C., Utter M.F. and Mildvan A.S. (1966) Pyruvate carboxylase. The presence of tightly bound manganese. *J. biol. Chem.* **241**, 3480.

A–C Cooper R.A. and Kornberg H.L. (1965) Net formation of phosphoenolpyruvate from pyruvate by *Escherichia coli*. *Biochim. biophys. Acta*, **104**, 618; (1967) The mechanism of the phosphoenolpyruvate synthetase reaction, ibid, **141**, 211.

B–C Tchen T.T. and Vennesland B. (1955) Enzymatic carbon dioxide fixation into oxalacetate in wheat germ. *J. biol. Chem.* **213**, 533.

B–C Bandurski R.S. (1955) Further studies on the enzymic synthesis of oxalacetate from phosphorylenolpyruvate and carbon dioxide. *J. biol. Chem.* **217**, 137.

B–C Bandurski R.S. and Lipmann F. (1956) Studies on an oxalacetic carboxylase from liver mitochondria. *J. biol. Chem.* **291**, 741.

B–C Graves J.L., Vennesland B., Utter M.F. and Pennington R.J. (1956) The mechanism of the reversible carboxylation of phosphoenolpyruvate. *J. biol. Chem.* **223**, 551.

B–C Siu P.M.L., Wood H.G. and Stjernholm R.L. (1961) Fixation of CO_2 by phosphoenolpyruvic carboxytransphosphorylase. *J. biol. Chem.* **236**, PC 21.

B–C Cánovas J.L. and Kornberg H.L. (1965) Fine control of phosphopyruvate carboxylase activity in *Escherichia coli*. *Biochim. biophys. Acta*, **96**, 169.

B–C Shrago E. and Lardy H.A. (1966) Paths of carbon in gluconeogenesis and lipogenesis. II. Conversion of precursors to phosphoenolpyruvate in liver cytosol. *J. biol. Chem.* **241**, 663.

D–E Mokrasch L.C. and Mcgilvery R.N. (1956) Purification and properties of fructose-1,6-diphosphatase. *J. biol. Chem.* **221**, 909.

D–E Racker E. and Schroeder E.A.R. (1958) The reductive pentose phosphate cycle. II. Specific C-1 phosphatases for fructose-1,6-diphosphate and sedoheptulose 1,7-diphosphate. *Arch. biochem. Biophys.* **74**, 326.

D–E Newsholme E.A. (1963) Properties of fructose 1,6-diphosphatase of rat liver. *Biochem. J.* **89**, 38P.

D–E Luppin B., Tramello S., Wood W.A. and Pontremoli S. (1964) Evidence for two forms of fructose-diphosphatase. *Biochem. biophys. Res. Commun.* **15**, 458.

D–E Pontremoli S., Traniello S., Luppin B. and Wood W.A. (1965) Fructose-diphosphatase from rabbit liver. I. Purification and properties. *J. biol. Chem.* **240**, 3459.

D–E Krebs H.A. and Woodford M. (1965) Fructose 1,6-diphosphatase in striated muscle. *Biochem. J.* **94**, 436.

117 GEN Ginsberg V. (1964) Sugar nucleotides and the synthesis of carbohydrates. *Advanc. Enzymol.* **26**, 35.

A–B Kalckar H.M. (1953) The role of phosphoglycosyl compounds in the biosynthesis of nucleosides and nucleotides. *Biochim. biophys. Acta*, **12**, 250.

A–B Smith E.E.B. and Mills G.T. (1955) The uridyl transferase of mammary gland. *Biochim. biophys. Acta*, **18**, 152.

A–B Ginsburg V. (1958) Purification of uridine diphosphate glucose pyrophosphorylase from mung bean seedlings. *J. biol. Chem.* **232**, 55.

A–B Newell P.C. and Sussman M. (1969) Uridine diphosphate glucose pyrophosphorylase in *dictyostelium discoideum*. *J. biol. Chem.* **244**, 2990.

B–C Leloir L.F. and Cardini C.E. (1957) Biosynthesis of glycogen from uridine diphosphate glucose. *J. Amer. chem. Soc.* **79**, 6340

B–C Leloir L.F. and Goldenberg S.H. (1960) Synthesis of glycogen from uridine diphosphate glucose in liver. *J. biol. Chem.* **235**, 919.

B–C Friedman D.L. and Larner J. (1962) Interconversion of two forms of muscle UDPG-α-glucan transglucosylase by a phosphorylation-dephosphorylation reaction sequence. *Biochim. biophys. Acta*, **64**, 185.

B–C Rosell-Perez M., Villar-Palasi C. and Larner J. (1962) Studies on UDPG—glycogen transglucosylase. Preparation and differentiation of two activities of UDPG-glycogen transglucosylase from rat skeletal muscle. *Biochemistry* **1**, 763.

B–C Hizukuri, S. and Larner J. (1964) Studies on UDPG: α-1,4 glucan, α-4-glucosyltransferase. Conversion of the enzyme from glucose-6-phosphate dependent to independent form in liver. *Biochemistry*, **3**, 1783.

B–C Goldberg N.D. and O'Toole A.G. (1969) The properties of glycogen synthetase and regulation of glycogen biosynthesis in rat brain. *J. biol. Chem.* **224**, 3053.

D–E Crane R.K. (1955) The substrate specificity of liver glucose-6-phosphatase. *Biochim. biophys. Acta*, **17**, 443.

D–E Langdon R.G. and Weakley D.R. (1957) Preparation and some properties of soluble glucose-6-phosphatase. *Fed. Proc.* **16**, 208.

D–E Arion W.J. and Nordlie R.C. (1964) Evidence for the common identity of glucose-6-phosphatase, inorganic pyrophosphatase and pyrophosphate-glucose phosphotransferase. *J. biol. Chem.* **239**, 1680.

D–E Duttera S.M., Byrne W.L. and Ganoza M.C. (1968) Studies on the phospholipid requirement of glucose-6-phosphatase. *J. biol. Chem.* **243**, 2216.

118 B–C Racker E. (1949) Aldehyde dehydrogenase, a diphosphopyridine nucleotide-linked enzyme. *J. biol. Chem.* **177**, 803.

B–C Seegmiller J.E. (1953) Triphosphopyridine nucleotide-linked aldehyde dehydrogenase from yeast. *J. biol. Chem.* **201**, 629.

D–E Burton R.H. and Stadtman E.R. (1953) The oxidation of acetaldehyde to acetyl coenzyme A. *J. biol. Chem.* **202**, 873.

E–F Stadtman E.R. (1952) The purification and properties of phosphotransacetylase. *J. biol. Chem.* **196**, 527.

F–G Rose I.A., Grunberg-Manago M., Korey S. and Ochoa S. (1955) Enzymatic phosphorylation of acetate. *J. biol. Chem.* **211**, 737.

119 B Krampitz L.O., Gruell G. and Miller R. (1958) An active acetaldehyde-thiamine intermediate. *J. Amer. chem. Soc.* **80**, 5893.

A–B Singer T.P. and Pensky J. (1952) Mechanism of acetoin synthesis by α-carboxylase. *Biochim. biophys. Acta*, **9**, 316.

C Watt D. and Krampitz L.O. (1947) α-Acetolactic acid, an intermediate in acetyl methyl carbinol formation. *Fed. Proc.* **6**, 301.

A–C Halpern Y.S. and Umbarger H.E. (1959) Evidence for two distinct enzyme systems forming acetolactate in aerobacter aerogenes. *J. biol. Chem.* **234**, 3067.

A–D Juni E. (1952) The mechanism of formation of acetion by bacteria. *J. biol. Chem.* **195**, 715.

D–E F–G Strecker H.J. and Harary I. (1954) Bacterial butylene glycol dehydrogenase and diacetyl reductase. *J. biol. Chem.* **211**, 263.

120 A–D Sanadi D.R., Langley M. and White F. (1958) On the mechanism of oxidative decarboxylation of α-keto acids. *Biochim. biophys. Acta*, **29**, 218

A–D Koike M. and Reed L.J. (1959) On the mechanism of oxidative decarboxylation of pyruvate. *J. Amer. chem. Soc.* **81**, 505.

A–D Koike M., Reed L.J. and Carroll W.R. (1963) α-keto acid dehydrogenation complexes. Resolution and reconstitution of the *E. coli* pyruvate dehydrogenation complex. *J. biol. Chem.* **238**, 30.

A–D Hayakawa T., Hirashima M., Ide S., Hamada M., Okabe K. and Koike M. (1966) Mammalian α-keto acid dehydrogenase complexes. Isolation, purification and properties of pyruvate dehydrogenase complex of pig heart muscle. *J. biol. Chem.* **241**, 4694.

A–D Kanzaki T., Hayakawa T., Hamada M., Fukuyoshi Y. and Koike M. (1969) Mammalian α-keto acid dehydrogenase complexes. Substrate specificities and kinetic properties of the pig heart pyruvate and 2-oxoglutarate dehydrogenase complexes. *J. biol. Chem.* **244**, 1138.

A–D Hayakawa T., Kanzaki T., Kitamura T., Fukuyoshi Y., Sakurai Y., Koike K., Suematsu T. and Koike M. (1969) Mammalian α-keto dehydrogenase complexes. Resolution and reconstitution studies on the pig heart pyruvate dehydrogenase complex. *J. biol. Chem.* **244**, 3660.

A–D See also Reed and Cox 122 gen.

C–D Korkes S. (1951) Acetyl transfer in the enzymatic oxidation of pyruvic acid. *Phosphorus Metabolism*, vo. 1, p. 259.

C–D Willms C.R., Oliver R.M., Henney H.R., Jr., Mukherjee B.B. and Reed L.J. (1967) α-Keto acid dehydrogenase complexes. Dissociation and reconstitution of the dihydrolipoyl transacetylase of *Escherichia coli*. *J. biol. Chem.* **242**, 889.

121 GEN Krebs H.A. and Johnson W.A. (1937) The role of citric acid in intermediate metabolism in animal tissues. *Enzymologia*, **4**, 148.

GEN Krebs H.A. (1948–49) The tricarboxylic acid cycle. *Harvey Lectures*, **44**, 165.

GEN Martius C. and Lynen F. (1950) Probleme des citronesäure cyklus. *Advanc. Enzymol.* **10**, 167.

GEN Ochoa S. (1954) Enzymic mechanisms in the citric acid cycle. *Advanc. Enzymol.* **15**, 183.

GEN Lowenstein J.M. (1967) The tricarboxylic acid cycle. *Metabolic Pathways*. D. Greenberg, p. 146.

GEN Müller M., Hogg J.F. and De Duve C. (1968) Distribution of tricarboxylic acid cycle enzymes and glyoxalate cycle enzymes between mitochondria and peroxisomes in *tetrahymena pyriformis*. *J. biol. Chem.* **243**, 5385.

122 GEN Reed L.J. and Cox D.J. (1966) Macromolecular organisation of enzyme systems. *Ann. Rev. Biochem.* **35**, 57.

AB–C Novelli G.D. and Lipmann F. (1950) The catalytic function of coenzyme A in citric acid synthesis. *J. biol. Chem.* **182**, 213.

AB–C Ochoa S., Stern J.R. and Schneider M.C. (1951) Enzymatic synthesis of citric acid. II. Crystalline condensing enzyme. *J. biol. Chem.* **193**, 691.

AB–C Stern J.R., Shapiro B., Stadtman E.R. and Ochoa S. (1951) Enzymatic synthesis of citric acid. III. Reversibility and mechanism. *J. biol. Chem.* **193**, 703.

AB–C Stern J.R. (1961) Oxalacetate transacetase (condensing enzyme, citrogenase). *The Enzymes*, 2nd ed., vol. 5, part B. Academic Press, New York, p. 367.

AB–C Srere P.A. and Kosicki G.W. (1961) The purification of the citrate condensing enzyme. *J. biol. Chem.* **236**, 2557.

AB–C Hathaway J.H. and Atkinson D.E. (1965) Kinetics of regulatory enzymes. Effect of adenosine triphosphate on yeast citrate synthetase. *Biochem. biophys. Res. Commun.* **20**, 661.

C–E Morrison J.F. (1954) The purification of aconitase. *Biochem. J.* **56**, 99.

C–E Speyer J.F. and Dickman S.R. (1956) On the mechanism of action of aconitase. *J. biol. Chem.* **220**, 193.

C–E Dickman S.R. (1961) Aconitase. *The Enzymes*, 2nd ed., vol. 5. Academic Press, New York, p. 495.

C–E Gawron O. and Mahajan K.P. (1966) α-Methyl-cis-aconitic acid. Aconitase substrate. II. Substrate properties and aconitase mechanism. *Biochemistry*, **5**, 2343.

C–E Rose I.A. and O'Connell E.L. (1967) Mechanism of aconitase action. The hydrogen transfer reaction. *J. biol. Chem.* **242**, 1870.

C–E Henson C.P. and Cleland W.W. (1967) Purification and kinetic studies of beef liver cytoplasmic aconitase. *J. biol. Chem.* **242**, 3833.

E–F Kornberg A. and Pricer W.E. (1951) Di- and tri-phosphopyridine nucleotide isocitrate dehydrogenases in yeast. *J. biol. Chem.* **189**, 123.

E–F Ramakrishnan C.V. and Martin S.M. (1955) Isocitrate dehydrogenase in *Aspergillus niger*. *Arch. biochem. Biophys.* **55**, 403.

E–F Plaut G.W.E. (1962) Isocitrate dehydrogenase (TPN-linked) from pig heart. *Methods in Enzymology*, vol. 5. Academic Press, New York, p. 645.

E–F Chen R.F. and Plaut G.W.E. (1963) Activation and inhibition of DPN-linked isocitrate dehydrogenase of heart by certain nucleotides. *Biochemistry*, **2**, 1023.

E–F Hathaway J.H. and Atkinson D.E. (1963) The effect of adenylic acid on yeast nicotinamide adenine dinucleotide isocitrate dehydrogenase. A possible metabolic control mechanism. *J. biol. Chem.* **238**, 2875.

E–F Dalziel K. and Londesborough J.C. (1968) The mechanisms of reductive carboxylation reactions. Carbon dioxide or bicarbonate as substrate of nicotinamide-adenine dinucleotide phosphate-linked isocitrate dehydrogenase and malic enzyme. *Biochem. J.* **110**, 223.

E–F Plaut, G.W.E. and Aogaichi T. (1968) Purification and properties of diphosphopyridine nucleotide-linked isocitrate dehydrogenase of mammalian liver. *J. biol. Chem.* **243**, 5572.

E–F Cox G.F. and Davies D.D. (1969) The effects of pH and citrate on the activity of nicotinamide adenine dinucleotide-specific isocitrate dehydrogenase from pea mitochondria. *Biochem. J.* **113**, 813.

F–G Goldberg M. and Sanadi D.R. (1952) Incorporation of labelled carbon dioxide into pyruvate and α-keto glutarate. *J. Amer. chem. Soc.* **74**, 4972.

F–G Breslow R. (1957) Rapid deuterium exchange in thiazolium salts. *J. Amer. chem. Soc.* **79**, 1762.

F–L Sanadi D.R., Littlefield J.W. and Bock R.M. (1952) Studies on α-ketoglutaric oxidase. II. Purification and properties. *J. biol. Chem.* **197**, 851.

F–L Hager L.P., Fortney J.D. and Gunsalus I.C. (1953) Mechanism of pyruvate and α-ketoglutarate dehydrogenase systems. *Fed. Proc.* **12**, 213.

F–L Koike M. and Reed L.J. (1960) α-Keto acid dehydrogenation complexes. *J. biol. Chem.* **235**, 1931.

F–L Koike M., Reed L.J. and Carroll W.R. (1960) α-Keto acid dehydrogenation complexes. *J. biol. Chem.* **235**, 1924.

F–L Mukherjee B.B., Matthews J., Horney D.L. and Reed L.J. (1965) Resolution and reconstitution of the *E. coli* α-ketoglutarate dehydrogenase complex. *J. biol. Chem.* **240**, PC 2268.

F–L *See* Kanzaki *et al.* 120 A–D.

H–K Hager L.P. and Gunsalus I.C. (1953) Lipoic acid dehydrogenase: the function of *E. coli.* fraction B. *J. Amer. chem. Soc.* **75**, 5767.

H–K Gunsalus I.C., Barton L.S. and Gruber W. (1956) Biosynthesis and structure of lipoic acid derivatives. *J. Amer. chem. Soc.* **78**, 1763.

H–K Reed L.J., Leach R.F. and Koike M. (1958) Studies on a lipoic acid-activating system. *J. biol. Chem.* **232**, 123.

H–K Massey V. (1959) The role of diaphorase in ketoglutarate oxidation. *Biochim. biophys. Acta*, **32**, 286.

H–K Reed L.J., Koike M., Levitch M.E. and Leach R.F. (1958) Studies on the nature and reactions of protein-bound lipoic acid. *J. biol. Chem.* **232**, 143.

H–K *See* Reed L.J., 95 GEN.

123 A–B Gergely J., Hele P. and Ramakrishnan C.V. (1952) Succinyl- and acetyl-coenzyme A deacylases. *J. biol. Chem.* **198**, 323.

A–B Sanadi D.R., Gibson D.M., Ayengar P. and Jacob M. (1956) α-Ketoglutaric dehydrogenase. V. Guanosine diphosphate in coupled phosphorylation. *J. biol. Chem.* **218**, 505.

A–B Smith R.A., Frank I.R. and Gunsalus I.C. (1957) Phosphoryl-S-coenzyme A: an intermediate in succinate activation. *Fed. Proc.* **16**, 251.

A–B Menon G.K.K. and Stern J.R. (1960) Enzyme synthesis and metabolism of malonyl coenzyme A and glutaryl coenzyme A. *J. biol. Chem.* **235**, 3393.

A–B Hager L.P. (1962) Succinyl CoA synthetase. *The Enzymes*, 2nd ed., vol. 6. Academic Press, New York, p. 387.

A–B Cha S., Cha C.J.M. and Parks R.E., Jr. (1965) Succinic thiokinase: the occurrence of a non-phosphorylated high energy intermediate of the enzyme. *J. biol. Chem.* **240**, PC 3700.

A–B Hersch L. and Jencks W.P. (1966) Mechanism of action of succinyl coenzyme A-acetoacetate transferase. *Fed. Proc.* **25**, 585.

A–B Ramaley R.F., Bridger W.A., Moyer R.W., Boyer P.D. (1967) The preparation, properties and reactions of succinyl coenzyme A synthetase and its phosphorylated form. *J. biol. Chem.* **242**, 4287.

B–C Neufeld H.A., Scott C.R. and Stotz E. (1956) Purification of heart muscle succinic dehydrogenase. *J. biol. Chem.* **221**, 869.

B–C Singer T.P., Kearney E.B. and Bernath P. (1956) Studies on succinic dehydrogenase. II. Isolation and properties of the dehydrogenase from beef heart. *J. biol. Chem.* **223**, 599.

B–C Massey V. and Singer T.P. (1957) Studies on succinic dehydrogenase. III. The fumaric reductase activity of succinic dehydrogenase. *J. biol. Chem.* **228**, 263.

B–C Tchen T.T. and Milligan H.V. (1960) The stereochemistry of the succinic dehydrogenase reaction. *J. Amer. chem. Soc.* **82**, 4115.

B–C Gawron O., Glaid A.J., Fransisco J. and Fondy T.P. (1963) Mechanism of the succinic dehydrogenase catalysed reaction. *Nature*, **197**, 1270.

B–C Kimura T., Hauber J. and Singer T.P. (1967) Studies on succinate dehydrogenase. Reversible activation of the mammalian enzyme. *J. biol. Chem.* **242**, 4987.

C–D Krebs H.A. (1953) The equilibrium constants of the fumarase and aconitase systems. *Biochem. J.* **54**, 78.

C–D Kanarek L. and Hill R.L. (1964) The preparation and characterisation of fumarase from swine heart muscle. *J. biol. Chem.* **239**, 4202.

C–D Teipel J.W. and Hill R.L. (1968) The number of substrate and inhibitor binding sites of fumarase. *J. biol. Chem.* **243**, 5679.

C–E Fahien L.A. and Strmecki M. (1969) Studies of gluconeogenic mitochondrial enzymes. The conversion of oxaloacetate to fumarate by bovine liver mitochondrial malate dehydrogenase and fumarase. *Arch. biochim. Biophys.* **130**, 478.

D–E Loewus F.A., Tchen T.T. and Vennesland B. (1955) The enzymatic transfer of hydrogen. III. The reaction catalysed by malic dehydrogenase. *J. biol. Chem.* **212**, 787.

D–E Rutter W.J. and Lardy H.A. (1958) Purification and properties of pigeon liver malic enzyme. *J. biol. Chem.* **233**, 374.

D–E Thorne C.J.R. (1960) Characterisation of two malic dehydrogenases from rat liver. *Biochim. biophys. Acta*, **42**, 175.

D–E Thorne C.J.R., Grossman L.I. and Kaplan N.O. (1963) Starch gel electrophoresis of malate dehydrogenase. *Biochim. biophys. Acta*, **73**, 193.

D–E Murphey W.H., Barnaby C., Lin, F.J. and Kaplan N.O. (1967) Malate dehydrogenases. Purification and properties of *Bacillus subtilis*, *Bacillus stearothermophilus* and *Escherichia coli* malate dehydrogenases. *J. biol. Chem.* **242**, 1548.

D–E See Dalziel K. *et al.* 122 EF.

D–E Guha A., Englard S. and Listowsky I. (1968) Beef heart malic dehydrogenases. Reactivity of sulphydryl groups and conformation of the supernatant. *J. biol. Chem.* **243**, 609.

D–E Heyde E. and Ainsworth S. (1968) Kinetic studies on the mechanism of the malate dehydrogenase reaction. *J. biol. Chem.* **243**, 2413.

124 A–BC Srere P.A. (1959) The citrate cleavage enzyme. I. Distribution and purification. *J. biol. Chem.* **234**, 2544.

A–BC Walsh C.T., Jr., Spector L.B. (1968) Citryl-1-phosphate and the citrate cleavage reaction. *J. biol. Chem.* **243**, 446.

D–EF Dagley S. and Dawes E.A. (1955) Citridesmolase: its properties and mode of action. *Biochem. biophys. Acta*, **17**, 177.

D–EF Wheat R.W. and Ajl S.J. (1955) Citratase, the citrate splitting enzyme from *E. coli*. I. Purification and properties. *J. biol. Chem.* **217**, 897.

GEN Wood H.G. (1955) Significance of alternate pathways in the metabolism of glucose. *Physiol. Rev.* **35**, 841.

125 GEN Touster O. (1959) Pentose metabolism and pentosuria. *Amer. J. Med.* **26**, 724.

GEN Wood H.G., Katz J. and Landau B.R. (1963) Estimation of pathways of carbohydrate metabolism. *Biochem. Z.* **338**, 809.

GEN Axelrod B. (1967) Other pathways of carbohydrate metabolism. *Metabolic Pathways*, vol. 1. Greenberg, p. 271.

GEN Novello F. and Mclean P. (1968) The pentose phosphate pathway of glucose metabolism. Measurement of the non-oxidative reactions of the cycle. *Biochem. J.* **107**, 775.

126 GEN Muntz J.A. and Murphy J.R. (1957) The metabolism of variously labelled glucose in rat liver, *in vivo*. *J. biol. Chem.* **224**, 971.

C–D Cori O. and Lipmann F. (1952) The primary oxidation product of enzymatic glucose-6-phosphate oxidation. *J. biol. Chem.* **194**, 417.

C–D Scott D.B.M. and Cohen S.S. (1953) The oxidative pathways of carbohydrate metabolism in *Escherichia coli*. I. The isolation and properties of glucose-6-phosphate dehydrogenase and 6-phosphogluconate dehydrogenase. *Biochem. J.* **55**, 23.

C–D Horecker B.L. and Smyrniotis P.Z. (1953) Reversibility of glucose-6-phosphate oxidation. *Biochim. biophys. Acta*, **12**, 98.

C–D Glaser L. and Brown D.H. (1955) Purification and properties of D-glucose-6-phosphate dehydrogenase. *J. biol. Chem.* **216**, 67.

C–D Hochster R.M. (1957) Pyridine nucleotide specificities and rates of formation of glucose-6-phosphate and of 6-phosphogluconate dehydrogenases in *Aspergillus flavus-oryzae*. *Arch. biochem. Biophys*, **66**, 499.

C–D Noltmann E.A., Gubler C.J. and Kuby S.A. (1961) Glucose-6-phosphate dehydrogenase (zwischenferment). I. Isolation of the crystalline enzyme from yeast. *J. biol. Chem.* **236**, 1225.

C–D Julian G.R., Wolfe R.G. and Reithel F.J. (1961) The enzymes of mammary gland. II. The preparation of glucose-6-phosphate dehydrogenase. *J. biol. Chem.* **236**, 754.

C–D Beutler E. and Morrison M. (1967) Localisation and characterisation of hexose 6-phosphate dehydrogenase (glucose dehydrogenase). *J. biol. Chem.* **242**, 5289.

D–E Brodie A.F. and Lipmann F. (1955) Identification of gluconolactonase. *J. biol. Chem.* **212**, 677.

D–E Kawada M., Yagawa Y., Takiguchi H. and Shimazono N. (1962) Purification of 6-phosphogluconolactonase from rat liver and yeast; its separation from gluconolactonase. *Biochim. biophys. Acta*, **57**, 404.

F Scott D.B.M. and Cohen S.S. (1957) The oxidative pathways of carbohydrate metabolism in *Escherichia coli*. 5. Isolation and identification of ribulose phosphate produced from 6-phosphogluconate by the dehydrogenase of *E. coli*. *Biochem. J.* **65**, 686.

E–F Pontremoli S., De Flora A., Grazi E., Mangiarotti G., Bonsignore A. and Horecker B.L. (1961) Crystalline D-gluconate 6-phosphate dehydrogenase. *J. biol. Chem.* **236**, 2975.

E–F Leinhard G.E. and Rose I.A. (1964) The mechanism of action of 6-phosphogluconate dehydrogenase. *Biochemistry* **3**, 190.

E–F Pontremoli S. and Grazi E. (1966) 6-phosphogluconate dehydrogenase-crystalline. *Meths. Enzymol.* **9**, 137.

E–H Horecker B.L. and Smyrniotis P.Z. (1950) The enzymatic production of ribose 5-phosphate from 6-phosphogluconate. *Arch. Biochem.* **29**, 232.

G Srere P.A., Cooper J.R., Klybas V. and Racker E. (1955) Xylulose-5-phosphate, a new intermediate in the pentose phosphate cycle. *Arch. biochem. Biophys.* **59**, 535.

F–G Hurwitz J. and Horecker B.L. (1956) The purification of phosphoketopentoepimerase from *Lactobacillus pentosus* and the preparation of xylulose 5-phosphate. *J. biol. Chem.* **223**, 993.

F–H Axelrod B. and Jang R. (1954) Purification and properties of phosphoribo-isomerase from alfalfa. *J. biol. Chem.* **209**, 847.

F–GH Dickens F. and Williamson D.H. (1956) Pentose phosphate isomerase and epimerase from animal tissues. *Biochem. J.* **64**, 567.

127 AB–CD Horecker B.L., Smyrniotis P.Z. and Klenow, H. (1953) The formation of sedoheptulose phosphate from pentose phosphate. *J. biol. Chem.* **205**, 661.

AB–CD Racker E., de la Haba G. and Leder G. (1954) Transketolase catalysed utilisation of fructose-6-phosphate and its significance in a glucose-6-phosphate oxidation cycle. *Arch. biochem. Biophys.* **48**, 238.

AB–CD Horecker B.L., Hurwitz J. and Smyrniotis P.Z. (1956) Xylulose 5-phosphate and the formation of sedoheptulose 7-phosphate with liver transketolase. *J. Amer. chem. Soc.* **78**, 692.

AB–CD Heath E.C., Hurwitz J., Horecker B.L. and Ginsburg A. (1958) Pentose fermentation by *Lactobacillus plantarum*. I. The cleavage of xylulose 5-phosphate by phosphoketolase. *J. biol. Chem.* **231**, 1009.

AB–CD Horecker, B.L., Florkin M. and Stotz E.H. (1964) Transketolase and transaldolase 'group transfer reactions' *Comprehensive Biochemistry*, vol. 15, p. 58.

AB–CD See Krampitz L.O. (1969) p. 103.

CD–EF Horecker B.L. and Smyrniotis P.Z. (1955) Purification and properties of yeast transaldolase. *J. biol. Chem.* **212**, 811.

CD–EF Venkataraman R. and Racker E. (1961) Mechanism of action of transaldolase. I. Crystallisation and properties of yeast enzyme. *J. biol. Chem.* **236**, 1876.

128 GEN Entner N. and Doudoroff M. (1952) Glucose and gluconic acid oxidation of *Pseudomonas saccharophila*. *J. biol. Chem.* **196**, 853.

DE Kovachevich R. and Wood W.A. (1955) Carbohydrate metabolism of *Pseudomonas fluorescens*. Purification and properties of a 6-phosphogluconate dehydrase. *J. biol. Chem.* **213**, 745.

D–E Meloche H.P. and Wood W.A. (1964) The mechanism of 6-phosphogluconic dehydrase. *J. biol. Chem.* **239**, 3505.

E–F Kovachevich R. and Wood W.A. (1955) Carbohydrate metabolism by *Pseudomonas fluorescens*. IV. Purification and properties of 2-keto-3-deoxy-6-phosphogluconate aldolase. *J. biol. Chem.* **213**, 757.

E–F Meloche H.P. and Wood W.A. (1964) The mechanism of 2-keto-3-deoxy-6-phosphogluconate aldolase. *J. biol. Chem.* **239**, 3511.

E–F Ingram J.M. and Wood W.A. (1966) Mechanism of 2-keto-3-deoxy-6-phosphogluconic aldolase. *Fed. Proc.* **25**, 586.

129 GEN Touster O. (1960) Essential pentosuria and the glucuronate-xylulose pathway. *Fed. Proc.* **19**, 977.

130 C Dutton G.J. and Storey I.D.E. (1953) The isolation of a compound of uridine diphosphate and glucuronic acid from liver. *Biochem. J.* **53**, 37.

C Storey I.D.E. and Dutton G.J. (1955) Uridine compounds in glucuronic acid metabolism. 2. The isolation and structure of uridine-diphosphate-glucuronic acid. *Biochem. J.* **59**, 279.

C Feingold D.S., Neufeld E.F. and Hassid W.Z. (1958) Enzymic synthesis of uridine diphosphate glucuronic acid and uridine diphosphate galacturonic acid with extracts from *Phaseolus aureus* seedlings. *Arch. biochem. Biophys.* **78**, 401.

B–C Strominger J.L., Kalckar H.M., Axelrod J. and Maxwell E.S. (1954) Enzymatic oxidation of uridine diphosphate glucose to uridine diphosphate glucuronic acid. *J. Amer. chem. Soc.* **76**, 6411.

B–C Maxwell E.S., Kalckar H.M. and Strominger J.L. (1956) Some properties of uridine diphosphoglucose dehydrogenase. *Arch. biochem. Biophys.* **65**, 2.

B–C Strominger J.L., Maxwell E.S., Axelrod J. and Kalckar H.M. (1957) Enzymatic formation of uridine diphosphoglucuronic acid. *J. biol. Chem.* **224**, 79.

B–C Strominger J.L. and Mapson L.W. (1957) Uridine diphosphoglucose dehydrogenase of pea seedlings. *Biochem. J.* **66**, 567.

C–E Ginsburg, V., Weissbach, A. and Maxwell E.S. (1958) Formation of glucuronic acid from uridine diphosphate glucuronic acid. *Biochem. biophys. Acta*, **28**, 649.

131 A–B Charalampous F.C. and Lyras C. (1957) Biochemical studies on inositol. IV. Conversion of inositol to glucuronic acid by rat kidney extracts. *J. biol. Chem.* **228**, 1.

A–B Charalampous F.C., Bumiller S. and Graham S. (1958) The site of cleavage of myo-inositol by purified enzyme of rat kidney. *J. Amer. chem. Soc.* **80**, 2022.

A–B Charalampous F.C. (1959) Biochemical studies on inositol. V. Purification and properties of the enzyme that cleaves inositol to D-glucuronic acid. *J. biol. Chem.* **234**, 220.

132 GEN Dayton P.G., Eisenberg F. and Burns J.S. (1958) Conversion of D-glucuronolactone, L-gulonolactone and L-ascorbic acid to glucose. *Fed. Proc.* **17**, 209.

GEN Eisenberg F. (Jr.), Dayton P.G. and Burns J.S. (1959) Studies on the glucuronic acid pathway of glucose metabolism. *J. biol. Chem.* **234**, 250.

GEN Hollmann S. and Touster O. (1964) Metabolism of glucuronic acid and ascorbic acid. *Non Glycolytic Pathways of Metabolism of Glucose*, chap. 6. Academic Press, New York, p. 83.

A–B Mano Y., Yamada K., Suzuki K. and Shimazono N. (1959) Formation of L-gulonolactone from D-glucurono-lactone with TPN L-gulonic dehydrogenase. *Biochim. biophys. Acta*, **34**, 563.

A–B Shimazono N. and Mano Y. (1961) Enzymatic studies on the metabolism of uronic and aldonic acids related to L-ascorbic acid in animal tissues. *Ann. N.Y. Acad. Sci.* **92**, 91.

D Burns J.J. and Kanfer J. (1957) Formation of L-xylulose from L-gulonolactone in rat kidney. *J. Amer. chem. Soc.* **79**, 3604.

A–D Touster U., Mayberry R.H. and McCormick D.B. (1957) Conversion of $1^{13}C$-D glucuronolactone to $5^{13}C$ L-xylulose in a pentosuric human. *Biochim. biophys. Acta*, **25**, 196.

C Smiley J.D. and Ashwell G. (1961) Purification and properties of β-L-hydroxy acid dehydrogenase. II. Isolation of β-keto-L-gulonic acid, an intermediate in L-xylulose biosynthesis. *J. biol. Chem.* **236**, 357.

B–C York J.L., Grollman A.P. and Bublitz C. (1961) NADP-L-gulonate dehydrogenase. *Biochim. biophys. Acta*, **47**, 298.

D–E Touster O., Reynolds V.H. and Hutcheson R.M. (1956) The reduction of L-xylulose to xylitol by guinea pig liver mitochondria. *J. biol. Chem.* **221**, 697.

D–E Hollman S. and Touster O. (1957) The L-xylulose-xylitol enzyme and other polyol dehydrogenases of guinea pig liver mitochondria. *J. biol. Chem.* **225**, 87.

D–E Arsenis C. and Touster O. (1969) NADP-linked xylitol dehydrogenase in guinea pig liver cytosol. *J. biol. Chem.* **244**, 3895.

D–G Hollman S. and Touster O. (1956) The enzymatic pathway from L-xylulose to D-xylulose. *J. Amer. chem. Soc.* **78**, 3544.

D–G Hickman J. and Ashwell G. (1956) Enzymatic phosphorylation of D-xylulose in liver. *J. Amer. chem. Soc.* **78**, 6209.

D–G Hickman J. and Ashwell G. (1959) A sensitive and stereospecific enzymatic assay for xylulose. *J. biol. Chem.* **234**, 758.

E–G McCormick D.B. and Touster O. (1957) Conversion *in vivo* of xylitol to glycogen via the pentose phosphate pathway. *J. biol. Chem.* **229**, 451.

F–G Hickman J. and Ashwell G. (1958) Purification and properties of D-xylulokinase in liver. *J. biol. Chem.* **232**, 737.

133 GEN Kilgore W.W. and Starr M.P. (1959) Catabolism of galacturonic acid and glucuronic acid by *Erwinia carotovora*. *J. biol. Chem.* **234**, 2227.

A–B Ashwell G., Wahba A.J. and Hickman J. (1960) Uronic acid metabolism in bacteria. I. Purification and properties of uronic acid isomerase in *Escherichia coli*. *J. biol. Chem.* **235**, 1559.

A–C Kilgore W.W. and Starr M.P. (1958) Metabolism of hexuronates and 5-ketohexonates by *Erwinia carotovora*. *Biochim. biophys. Acta*, **30**, 652.

B–C Hickman J. and Ashwell G. (1960) Uronic acid metabolism in bacteria. II. Purification and properties of D-altronic acid and D-mannonic acid dehydrogenases in *Escherichia coli*. *J. biol. Chem.* **235**, 1566.

C–D Smiley J.D. and Ashwell G. (1960) Uronic acid metabolism in bacteria. III. Purification and properties of D-altronic acid and D-mannonic acid dehydrases in *Escherichia coli*. *J. biol. Chem.* **235**, 1571.

D–E Cynkin M.A. and Ashwell G. (1960) Uronic acid metabolism in bacteria. IV. Purification and properties of 2-keto-3-deoxy-D-gluconokinase. *J. biol. Chem.* **235**, 1576.

E–FG Kovachevich R. and Wood W.A. (1955) Carbohydrate metabolism by *Pseudomonas fluorescens*. IV. Purification and properties of 2-keto-3-deoxy-6-phosphogluconate aldolase. *J. biol. Chem.* **213**, 757.

134 GEN Burns J.J. and Evans C. (1956) The synthesis of L-ascorbic acid in the rat from D-glucuronolactone and L-gulonolactone. *J. biol. Chem.* **223**, 897.

GEN Grollman A.P. and Lehninger A.L. (1957) Enzymic synthesis of L-ascorbic acid in different animal species. *Arch. biochem. Biophys.* **69**, 458.

GEN Kanfer J., Burns J.J. and Ashwell G. (1959) L-Ascorbic acid synthesis in a soluble enzyme system from rat liver microsomes. *Biochim. biophys. Acta*, **31**, 556.

GEN See Touster O., 125 GEN.

GEN Chatterjee I.B., Chatterjee G.C., Ghosh W.C., Ghosh J.J. and Guha B.C. (1960) Biological synthesis of L-ascorbic acid in animal tissues. Conversion of D-glucuronolactone and L-gulonolactone into L-ascorbic acid. *Biochem. J.* **76**, 279.

GEN Ashwell G., Kanfer J., Smiley J.D. and Burns J.J. (1961) Metabolism of ascorbic acid and related uronic acids, aldonic acids and pentoses. *Ann. N.Y. Acad. Sci.* **92**, 105.

GEN See Hollmann S. and Touster O. 132 GEN.

GEN Wagner A.F. and Folkers K. (1964) L-Ascorbic acid. *Vitamins and Coenzymes*. Interscience Publishers, New York, p. 308.

GEN Burns J.J. (1967) Ascorbic acid. *Metabolic Pathways*, vol. VI. Greenberg, p. 394.

A–B Winkelman J. and Lehninger A.L. (1958) Aldono- and uronolactonases of animal tissues. *J. biol. Chem.* **233**, 794.

A–B Yamada K., Ishikawa S. and Shimazono N. (1959) On the microsomal and soluble lactonases. *Biochim. biophys. Acta*, **32**, 253.

B–C Eliceiri G.L., Lai E.K. and McCay P.B. (1969) Gulonolactone oxidase. Solubilisation, properties and partial purification. *J. biol. Chem.* **244**, 2641.

B–D Burns J.J., Peyser P. and Moltz A. (1956) Missing step in guinea pigs required for the biosynthesis of L-ascorbic acid. *Science*, **124**, 1148.

135 GEN Chan P.C., Babineau L.M., Becker R.R. and King C.G. (1957) Ascorbic acid metabolism. *Fed. Proc.* **16**, 163.

GEN Kagawa Y., Mano Y. and Shimazono N. (1960) Biodegradation of dehydro-L-ascorbic acid: 2,3-diketo-aldonic acid decarboxylase from rat liver. *Biochim. biophys. Acta*, **43**, 348.

A–B Staric G.R. and Dawson C.R. (1963) Ascorbic acid oxidase. *The Enzymes*, 2nd ed., vol. 8. Academic Press, New York, p. 297.

A–D Chan P.C., Babineau L.M., Becker R.R. and King C.G. (1957) Ascorbic acid metabolism. *Fed. Proc.* **16**, 163.

A–E
A–F Burns J.J., Kanfer J. and Dayton P.G. (1958) Metabolism of L-ascorbic acid in rat kidney. *J. biol. Chem.* **232**, 107.

A–EF Kanfer J., Ashwell G. and Burns J.J. (1960) Formation of L-lyxonic and L-xylonic acids from L-ascorbic acid in rat kidney. *J. biol. Chem.* **235**, 2518.

B–D Chan P.C., Becker R.R. and King C.G. (1958) Metabolic products of L-ascorbic acid. *J. biol. Chem.* **231**, 231.

C–F See Shimazono *et al.* 132 ABC.

E Kanfer J., Ashwell G. and Burns J.J. (1959) Formation of L-lyxonic acid from L-ascorbic acid in rat kidney. *Fed. Proc.* **18**, 256.

136 A–B Cori G.T., Ochoa S., Slein M.W. and Cori C.F. (1951) The metabolism of fructose in liver. Isolation of fructose-1-phosphate and inorganic pyrophosphate. *Biochim. biophys. Acta*, **7**, 304.

A–B Parks R.E., Ben-Gershom E. and Lardy H.A. (1957) Liver fructokinase. *J. biol. Chem.* **227**, 231.

A–B Hers H.G. (1957) Le metabolisme du fructose (Brussels, Arscia).

A–C Bueding E. and Mackinnon J.A. (1955) Hexokinases of *Schistosoma mansoni*. *J. biol. Chem.* **215**, 495.

A–C Medina A. and Sols A. (1956) A specific fructokinase in peas. *Biochim. biophys. Acta*, **19**, 378.

F–DG *See* 112 D–EF.

B–D Jagannathan V., Singh K. and Damodaran M. (1956) Purification and properties of aldolase from *Aspergillus nidulans*. *Biochem. J.* **63**, 94.

B–DE Rose I.A., O'Connell E.L. and Rieder S.V. (1958) Studies on the mechanism of the aldolase reaction. Isotope exchange reactions of muscle and yeast aldolase. *J. biol. Chem.* **231**, 315.

B–DE Peansky R.J. and Lardy H.A. (1958) Isolation, crystallisation and some general properties of bovine liver aldolase. *J. biol. Chem.* **233**, 365

B–DE *See* Dahlquist A. and Crane R.K. (1964) p. 110 DE.

C–F Axelrod B., Saltman P., Bandurski R.S. and Baker R.S. (1952) Phosphohexokinase in higher plants. *J. biol. Chem.* **197**, 89.

C–F Ling K.H. and Lardy H.A. (1954) Uridine and inosine triphosphates as phosphate donors for phosphohexokinase. *J. Amer. chem. Soc.* **76**, 2842.

B–DE Hers H.G. and Kusaka T. (1953) Le metabolisme du fructose-1-phosphate dans le foie. *Biochim. biophys. Acta*, **11**, 427.

137 A–B *See* Bueding, E. 136 A–C.

B–C Slein M.W. (1950) Phosphomannose isomerase. *J. biol. Chem.* **186**, **753**.

B–C Bruns F.H., Noltmann E. and Willemson A. (1958) Phosphomannose isomerase. *Biochem. Z.* **330**, 411.

B–C Noltmann E and Bruns F.H. (1958) Phosphomannose isomerase. *Biochem. Z.* **330**, 514.

B–C Hollmann S. and Touster O. (1964) Transformations of the primary reaction products of glucose. *Non-Glycolytic Pathways of Metabolism of Glucose*, chap. 3. Academic Press, New York, p. 41.

138 GEN Kalckar H.M. and Maxwell E.S. (1956) Some considerations concerning the nature of the enzymic galactose-glucose conversion. *Biochim. biophys. Acta*, **22**, 588.

GEN Hollmann S. and Touster O. (1964) Relationships between galactose and glucose metabolism. *Non-Glycolytic Pathways of Metabolism of Glucose*, chap. 8. Academic Press, New York, p. 132.

A–B Trucco R.E., Caputto R., Leloir L.F. and Mittelman N. (1948) Galactokinase. *Arch. Biochem.* **18**, 137.

A–B Cardini C.E. and Leloir L.F. (1953) Enzymic phosphorylation of galactosamine and galactose. *Arch. biochem. Biophys.* **45**, 55.

A–B Neufeld E.F., Feingold D.S. and Hassid W.Z. (1960) Phosphorylation of D-galactose and L-arabinose by extracts from *Phaseolus aureus* seedlings. *J. biol. Chem.* **235**, 906.

B–D Isselbacher K.J. (1958) A mammalian uridinediphosphate galactose pyrophosphorylase. *J. biol. Chem.* **232**, 429.

B–D Abraham H.D. and Howell R. (1969) Human hepatic uridinediphosphate galactose pyrophosphorylase. Its characterisation and activity during development. *J. biol. Chem.* **244**, 545.

B–E Kalckar H.M. (1953) The role of phosphoglycosyl compounds in the biosynthesis of nucleosides and nucleotides. *Biochim. biophys. Acta*, **12**, 250.

B–E Kalckar H.M., Braganca B. and Munch-Petersen A. (1953) Uridyl transferase and the formation of uridine diphosphogalactose. *Nature*, **172**, 1038.

B–E Smith E.E.B. and Mills G.T. (1955) The uridyl transferase of mammary gland. *Biochim. biophys. Acta*, **18**, 152.

B–E Kurahashi K. and Sugimura A. (1960) Purification and properties of galactose-1-phosphate uridyl transferase from *Escherichia coli*. *J. biol. Chem.* **235**, 940.

D Leloir L.F. (1958) Uridine diphospho-galactose: metabolism, enzymology and biology. *Advanc. Enzymol.* **20**, 3.

C–D Leloir L.F. (1951) The enzymatic transformation of uridine diphosphate glucose into a galactose derivative. *Arch. biochem. Biophys.* **33**, 186.

C–D Caputto R. and Trucco R.E. (1952) A new galactose-containing compound from mammary glands. *Nature, Lond.* **169**, 1061.

C–D Maxwell E.S. (1957) Enzymic interconversion of uridine diphospho-galactose and uridine diphosphoglucose. *J. biol. Chem.* **229**, 139.

C–D Maxwell E.S. and De Robichon-Szulmajster H.J. (1960) Purification of uridine diphosphate galactose-4-epimerase from yeast, and the identification of protein bound diphosphopyridine nucleotide. *J. biol. Chem.* **235**, 308.

C–D Wilson D.B. and Hogness D.S. (1964) The enzymes of the galactose operon in *E. coli*. *J. biol. Chem.* **239**, 2469.

C–E Ginsberg V. (1958) Purification of uridine diphosphate glucose pyrophosphorylase from mung bean seedlings. *J. biol. Chem.* **232**, 55.

C–E Kalckar H.M. and Maxwell E.S. (1958) Biosynthesis and metabolic function of uridine diphosphoglucose in mammalian organisms and its relevance to certain inborn errors. *Physiol. Rev.* **38**, 77.

139 BC–D Watkins W.M. and Hassid W.Z. (1962) The synthesis of lactose by particulate enzyme preparations from guinea pig and bovine mammary glands. *J. biol. Chem.* **237**, 1432.

BC–D Brodbeck U. and Ebner K.E. (1966) Resolution of a soluble lactose synthetase into two protein components and solubilisation of microsomal lactose synthetase. *J. biol. Chem.* **241**, 762.

BC–D Bartley J.C., Abraham S. and Chaikoff I.L. (1966) Biosynthesis of lactose by mammary gland slices from the lactating rat. *J. biol. Chem.* **241**, 1132.

BC–D Baldwin R.L. and Milligan L.P. (1966) Enzymatic changes associated with the initiation and maintenance of lactation in the rat. *J. biol. Chem.* **241**, 2058.

BC–D Babad H. and Hassid W.Z. (1966) Soluble UDP-D-galactose: D-glucose β-4-D-galactosyltransferase from bovine milk. *J. biol. Chem.* **241**, 2672.

D–EF Landman O.C. (1957) Properties and induction of β-galactosidase in *Bacillus megaterium*. *Biochim. biophys. Acta*, **23**, 558.

D–EF Wallenfels K., Zarnitz M.L., Laule G., Bender H. and Keser M. (1959) Untersuchungen über milchzuckerspaltende Enzyme. III. Reinigung, Kristallisation und Eigenschaften der β-Galaktosidase von *E. coli*. ML 309. *Biochem. Z.* **331**, 459.

D–EF Conchie J., Findlay J. and Levvy G.A. (1959) Mammalian glycosidases. Distribution in the body. *Biochem. J.* **71**, 318.

140 GEN Ishihara H. and Heath E.C. (1968) The metabolism of L-fucose. The biosynthesis of guanosine diphosphate L-fucose in porcine liver. *J. biol. Chem.* **243**, 1110.

141 GEN Horecker B.L. (1962) Pentose metabolism in bacteria. *CIBA*. John Wiley.

GEN Hollmann S. and Touster O. (1964) Metabolism of pentoses. *Non-Glycolytic Pathways of Metabolism of Glucose*, chap. 7. Academic Press, New York, p. 115.

GEN Mortlock R.P., Fossitt D.D. and Wood W.A. (1965) A basis for utilisation of unnatural pentoses and pentitols by *aerobacter aerogenes*. *Proc. Nat. Acad. Sci. U.S.* **54**, 572.

B Chiang C. and Knight S.G. (1959) D-Xylulose metabolism by cell free extracts of *Penicillium chrysogenum*. *Biochim. biophys. Acta*, **35**, 454.

B–J Hochster R.M. and Watson R.W. (1954) Enzymic isomerisation of D-xylulose to D-xylose. *Arch. biochem. Biophys.* **48**, 120.

B–J Slein M.W. (1955) Xylose isomerase from *Pasteurella pestis*, strain A–1122. *J. Amer. chem. Soc.* **77**, 1663.

D–J Paleroni N.J. and Doudoroff M. (1956) Mannose isomerase of *Pseudomonas saccharophila*. *J. biol. Chem.* **218**, 535.

D–J Anderson R.L. and Allison D.P. (1965) Purification and characterisation of D-lyxose isomerase. *J. biol. Chem.* **240**, 2367.

E Touster O. and Harwell S.O. (1958) Isolation of L-arabitol from pentosuria urine. *J. biol. Chem.* **230**, 1031.

E & G Fossitt D., Mortlock R.P., Anderson R.L. and Wood W.A. (1964) Pathways of L-arabitol and xylitol metabolism in *Aerobacter aerogenes*. *J. biol. Chem.* **239**, 2110.

G See McCormick D.B. 132 E–G.

G Horwitz S.B. and Kaplan N.O. (1965) Hexitol dehydrogenases of *Bacillus subtilis*. *J. biol. Chem.* **239**, 830.

H–K Anderson R.L. and Wood W.A. (1962) Purification and properties of L-xylulokinase. *J. biol. Chem.* **237**, 1029.

J Hickman J. and Ashwell G. (1959) A sensitive and stereospecific enzymatic assay for xylulose. *J. biol. Chem.* **234**, 758.

J–G van Heyningen R. (1959) Metabolism of xylose by the lens. II. Rat lens *in vivo* and *vitro*. *Biochem. J.* **73**, 197.

L Stumpf P.K. and Horecker B.L. (1956) The role of xylulose-5-phosphate in xylose metabolism of *Lactobacillus pentosus*. *J. biol. Chem.* **218**, 753.

J–L See Hickman J. *et al.* 132 FG.

L–P Dickens F. and Williamson D.H. (1956) Pentose phosphate isomerase and epimerase from animal tissues. *Biochem. J.* **64**, 567.

L–P Ashwell G. and Hickman J. (1957) Enzymatic formation of xylulose 5-phosphate from ribose 5-phosphate in spleen. *J. biol. Chem.* **226**, 65.

L–P Wolin M.J., Simpson F.J. and Wood W.A. (1958) Degradation of L-arabinose by *Aerobacter aerogenes*. Identification and properties of L-ribulose-5-phosphate 4-epimerase. *J. biol. Chem.* **232**, 559.

L–MN Heath E.C., Hurwitz J., Horecker B.L. and Ginsburg A. (1958) Pentose fermentation by *Lactobacillus plantarum*. I. The cleavage of xylulose 5-phosphate by phosphoketolase. *J. biol. Chem.* **231**, 1009.

O–S Simpson F.J. and Wood W.A. (1958) Degradation of L-arabinose by *Aerobacter aerogenes*. II. Purification and properties of L-ribulokinase. *J. biol. Chem.* **230**, 473.

P–T Burma D.P. and Horecker B.L. (1958) Pentose fermentation by *Lactobacillus Plantarum* Ribulokinase. *J. biol. Chem.* **231**, 1039.

P–T Fromm H.J. (1959) D-Ribulokinase from *Aerobacter aerogenes*. *J. biol. Chem.* **234**, 3097.

L–P–Q Tabachnick M., Srere P.A., Cooper J. and Racker E. (1958) The oxidative pentose phosphate cycle. II. The interconversion of ribose 5-phosphate, ribulose 5-phosphate and xylulose 5-phosphate. *Arch. biochem. Biophys.* **74**, 315.

Q–R Long C. (1955) Studies involving enzymic phosphorylation. IV. The conversion of D-ribose into D-ribose 5-phosphate by extracts of *Escherichia coli*. *Biochem. J.* **59**, 322.

S–V Heath E.C., Horecker B.L., Smyrniotis P.Z. and Takagi Y. (1958) Pentose fermentation by *Lactobacillus plantarum*. II. L-Arabinose isomerase. *J. biol. Chem.* **231**, 1031.

T–W Cohen S.S. (1953) Studies on D-ribulose and its enzymatic conversion to D-arabinose. *J. biol. Chem.* **201**, 71.

T–V Fromm H.J. (1958) Ribitol dehydrogenase. I. Purification and properties of the enzyme. *J. biol. Chem.* **233**, 1049.

142 A–E Weimberg R. and Doudoroff M. (1955) The oxidation of L-arabinose by *Pseudomonas saccharophila*. *J. biol. Chem.* **217**, 607.

A–H Weimberg R. (1959) L-2-Keto-4,5-dihydroxyvaleric acid: an intermediate in the oxidation of L-arabinose by *Pseudomonas saccharophila*. *J. biol. Chem.* **234**, 727.

B Trudgill P.W. and Widdus R. (1966) D-Glucarate catabolism by *Pseudomonadaceae* and *Enterobacteriaceae*. Nature, **211**, 1097.

B–H Dagley S. and Trudgill P.W. (1965) The metabolism of galactarate, D-glucarate and various pentoses by species of *Pseudomonas*. *Biochem. J.* **95**, 48.

F Palleroni N.J. and Doudoroff M. (1956) Characterisation and properties of 2-keto-3-deoxy-D-arabonic acid. *J. biol. Chem.* **223**, 499.

F–G Portsmouth D., Stoolmiller A.C. and Abeles R.H. (1967) Studies on the mechanism of action of 2-keto-3-deoxy-L-arabonate dehydratase. *J. biol. Chem.* **242**, 2751.

G–H Adams E. and Rosso G. (1967) α-Ketoglutaric semialdehyde dehydrogenase of *Pseudomonas*. Properties of the purified enzyme induced by hydroxyproline and of the glucarate-induced and constitutive enzyme. *J. biol. Chem.* **242**, 1802.

143 GEN Blumenthal H.J. and Fish D.C. (1963) Bacterial conversion of D-glucarate to glycerate and pyruvate. *Biochem. biophys. Res. Commun.* **11**, 239.

GEN *See* Trudgill P.W. and Widdus R. 142 B.

C–D Gotto A.M. and Kornberg H.L. (1961) Metabolism of C_2 compounds in microorgansims. Preparation and properties of crystalline tartronic semialdehyde reductase. *Biochem. J.* **81**, 273.

144 GEN Chen I.W. and Charalampous F.C. (1966) Biochemical studies on inositol. IX. D-Inositol-1-phosphate as intermediate in the biosynthesis of inositol from glucose-6-phosphate, and characteristics of two reactions in this biosynthesis. *J. biol. Chem.* **241**, 2194.

B Eisenberg F. and Bolden A.H. (1965) D-Myo-inositol-1-phosphate, an intermediate in the biosynthesis of inositol in the mammal. *Biochem. biophys. Res. Commun.* **21**, 100.

145 GEN Hill R. (1951) Reduction by chloroplasts. *Sym. Soc. Exptl. Biol.* **5**, 222.

GEN Hill R. and Whittingham C.P. (1955) *Photosynthesis*. Methuen, London.

GEN Vishniac W., Horecker B.L. and Ochoa S. (1957) Enzymic aspects of photosynthesis. *Advanc. Enzymol.* **19**, 1.

GEN Arnon D. (1959) Conversion of light into chemical energy in photosynthesis. *Nature*, **184**, 10.

GEN Hill R. and Bendall F. (1960) Function of the two cytochrome components in chloroplasts: a working hypothesis. *Nature*, **186**, 136.

GEN Kamen M.D. (1963) *Primary Processes in Photosynthesis*. Academic Press, New York.

GEN Bassham J.A. (1963) Photosynthesis: energetics and related topics. *Advanc. Enzymol.* **25**, 39.

GEN Wassink E.C. (1963) Photosynthesis. *Comparative Biochemistry*, vol. 5. Academic Press, New York, p. 347.

GEN Rose A.H. and Wilkinson J.F. (1968) *Advances in Microbial Physiology*, vol. 2.

147 GEN Calvin M. (1956) The photosynthetic carbon cycle. *J. chem. Soc.* 1895.

GEN Bassham J.A. and Calvin M. (1957) *The pathway of Carbon in Photosynthesis*. Prentice Hall, New Jersey, pp. 39–66.

GEN Bassham J.A. (1962) *The Path of Carbon in Photosynthesis*. Scientific American. June.

GEN Calvin M. and Bassham J.A. (1962) *The Photosynthesis of Carbon Compounds*. Benjamin, New York.

GEN Bassham J.A. (1964) Kinetic studies on the photosynthetic carbon[1,2,3 ... 90] reduction cycle. *Ann. Rev. Plant Physiol.* **15**, 101.

GEN Vernon C.P. and Auron M. (1965) Photosynthesis. *Ann. Rev. Biochem.* **34**, 269.

GEN San Pietro A. and Black C.C. (1965) Enzymology of energy conversion in photosynthesis. *Ann. Rev. Plant Physiol.* **16**, 155.

GEN Hatch M.D and Slack C.R. (1966) Photosynthesis by sugar cane leaves. A new carboxylation reaction and the pathway of sugar formation. *Biochem. J.* **101**, 103.

GEN Pfennig N. (1967) Photosynthetic bacteria. *Ann. Rev. Microbiol.* **21**, 285.

GEN Gibbs M. (1967) Photosynthesis. *Ann. Rev. Biochem.* **36**, 757.

148 A–B Quayle J.R., Fuller R.C., Benson A.A. and Calvin M.J. (1954) Enzymatic carboxylation of ribulose diphosphate. *J. Amer. chem. Soc.* **76**, 3610.

A–B Jakoby W.B., Brummond D.O. and Ochoa S. (1956) Formation of 3-phosphoglyceric acid by carbon dioxide fixation in spinach leaf extracts. *J. biol. Chem.* **218**, 811.

A–B Müllhofer G. and Rose I.A. (1965) The position of carbon–carbon bond cleavage in the ribulose diphosphate carboxydismutase reaction. *J. biol. Chem.* **240**, 1341.

149 H–J Hurwitz J., Weissbach A., Horecker B.L. and Smyrniotis P.Z. (1956) Spinach phosphoribulokinase. *J. biol. Chem.* **218**, 769.

H–J Racker E. (1957) The reductive pentose phosphate cycle. I. Phosphoribulokinase and ribulose diphosphate carboxylase. *Arch. biochem. Biophys.* **69**, 300.

150 GEN Bachofen R., Buchanan B.B. and Arnon D.I. (1964) Ferredoxin as a reductant in pyruvate synthesis by a bacterial extract. *Proc. Nat. Acad. Sci. U.S.* **51**, 690.

GEN Buchanan B.B., Bachofen R. and Arnon D.I. (1964) Role of ferredoxin in the reductive assimilation of CO_2 and acetate by extracts of the photosynthetic *Bacterium chromatium*. *Proc. Nat. Acad. Sci. U.S.* **52**, 839.

GEN Evans M.C.W. and Buchanan B.B. (1965) Photoreduction of ferredoxin and its use in carbon dioxide fixation by a subcellular system from a photosynthetic bacterium. *Proc. Nat. Acad. Sci. U.S.* **53**, 1420.

GEN Buchanan B.B. and Evans M.C.W. (1965) The synthesis of α-ketoglutarate from succinate and carbon dioxide by a subcellular preparation of a photosynthetic bacterium. *Proc. Nat. Acad. Sci. U.S.* **54**, 1212.

GEN Evans M.C.W., Buchanan B.B. and Arnon D.I. (1966) A new ferredoxin-dependent carbon reduction cycle in a photosynthetic bacterium. *Proc. Nat. Acad. Sci. U.S.* **55**, 928.

151 GEN Dagley S. and Patel M.D. (1955) Excretion of α-ketoglutaric acid during oxidation of acetate by a vibrio. *Biochim. biophys. Acta*, **16**, 418.

GEN Kornberg H.L. and Masden N.B. (1955) Synthesis of C_2-dicarboxylic acids from acetate by a 'glyoxalate bypass' of the tricarboxylic acid cycle. *Biochim. biophys. Acta*, **24**, 651.

GEN Kornberg H.L. and Beevers H. (1957) The glyoxalate cycle as a stage in the conversion of fat to carbohydrate in castor beans. *Biochim. biophys. Acta*, **26**, 531.

GEN Kornberg H.L. and Krebs H.A. (1957) Synthesis of cell constituents from C_2 units by a modified tricarboxylic cycle. *Nature*, **179**, 988.

GEN Kornberg H.L. (1958) The metabolism of C_2 compounds in microorganisms. I. The incorporation of [$2^{14}C$]acetate by *Pseudomonas fluorescens*, and by a corynebacterium, grown on ammonium acetate. *Biochem. J.* **68**, 535.

GEN Kornberg H.L. and Elsden S.R. (1961) The metabolism of 2-carbon compounds by microorganisms. *Advanc. Enzymol.* **23**, 401.

GEN Kornberg H.L. (1961) Selective utilisation of metabolic routes by *E. coli*. *Cold Spring Harbor Sym. Quant. Biol.* **26**, 257.

GEN Kornberg H.L. (1966) Anaplerotic sequences and their role in metabolism. *Essays in Biochemistry*, vol. 2. Academic Press, London.

GEN *See* Müller M. *et al*. 121 GEN.

AH–J Wong D.T.O. and Ajl S.J. (1956) Conversion of acetate and glyoxylate to malate. *J. Amer. chem. Soc.* **78**, 3230.

AH–J Wong D.T.O. and Ajl S.J. (1957) Significance of the malate synthetase reaction in bacteria. *Science*, **126**, 1013.

AH–J Kornberg H.L. and Masden N.B. (1958) The metabolism of C_2 compounds in microorganisms. III. Synthesis of malate from acetate via the glyoxalate cycle. *Biochem. J.* **68**, 549.

AH–J Dixon G.H., Kornberg H.L. and Lund P. (1960) Purification and properties of malate synthetase. *Biochim. biophys. Acta*, **41**, 217.

AH–J Yamamoto Y. and Beevers H. (1961) Purification and properties of malate synthetase from castor beans. *Biochim. biophys. Acta*, **48**, 20.

AH–J Falmagne P., Vanderwinkel E. and Wiame J.M. (1965) Mise en evidence de deux malate synthetases chez *Escherichia coli*. *Biochim. biophys. Acta*, **99**, 246.

F–GH Campbell J.J.R., Smith R.A. and Eagles B.A. (1953) A deviation from the conventional tricarboxylic acid cycle in *Pseudomonas aeruginosa*. *Biochim. biophys. Acta*, **11**, 594.

F–GH Saz H.J. and Hillary E.P. (1956) Formation of glyoxylate and succinate from tricarboxylic acids by *Pseudomonas aeruginosa*. *Biochem. J.* **62**, 563.

F–GH Smith R.A. and Gunsalus I.C. (1957) Isocitratase: enzyme properties and reaction equilibrium. *J. biol. Chem.* **229**, 305.

F–GH Olson J.A. (1959) The purification and properties of yeast isocitric lyase. *J. biol. Chem.* **234**, 5.

F–GH Carpenter W.D. and Beevers H. (1959) Distribution and properties of isocitratase in plants. *Plant. Physiol.* **34**, 403.

F–GH Ashworth J.M. and Kornberg H.L. (1964) The role of isocitrate lyase in *E. coli*. *Biochim. biophys. Acta*, **89**, 383.

152 GEN Dagley S., Trudgill P.W. and Callely A.G. (1961) Synthesis of cell constituents from glycine by a *Pseudomonas*. *Biochem. J.* **81**, 623.

A–B Krakow G., Barkulis S.S. and Hayashi J.A. (1961) Glyoxylic acid carboligase: an enzyme present in glycollate grown *Escherichia coli*. *J. Bacteriol.* **81**, 509.

A–B Gupta N.K. and Vennesland B. (1964) Glyoxylate carboligase of *E. coli*: a flavoprotein. *J. biol. Chem.* **239**, 3787.

A–B *See* Krampitz L.O. (1969) p. 103.

A–B Hall R.L., Vennesland B. and Kézdy F.J. (1969) Glyoxylate carboligase of *E. coli* Identification of carbon dioxide as the primary reaction product. *J. biol. Chem.* **244**, 3991.

B–C Gotto A.M. and Kornberg H.L. (1961) The metabolism of C_2 compounds in micro-organisms. 7. Preparation and properties of crystalline tartronic semialdehyde reductase. *Biochem J.* **81**, **273**.

B–C Kohn L.D. (1968) Tartaric acid metabolism. Crystalline tartronic semialdehyde reductase. *J. biol. Chem.* **243**, 4426.

C–D Black S. and Wright N.G. (1956) Enzymatic formation of glyceryl and phosphoglyceryl methylthiol esters. *J. biol. Chem.* **221**, 171.

C–D Ichihara A. and Greenberg D.M. (1957) Studies on the purification and properties of D-glyceric acid kinase of liver. *J. biol. Chem.* **225**, 949.

153 GEN Quayle J.R. and Keech D.B. (1960) Carbon assimilation by *Pseudomonas oxalaticus* (Ox I) 3. Oxalate utilisation during growth on oxalate. *Biochem. J.* **75**, 515.

B Quayle J.R. (1963) Carbon assimilation by *Pseudomonas oxalaticus* (Ox I) 6. Reaction of oxalyl-coenzyme A. *Biochem. J.* **87**, 368.

A–B Quayle J.R., Keech D.B. and Taylor G.A. (1961) Carbon assimilation by *Pseudomonas oxalaticus*. Metabolism of oxalate in cell-free extracts of the organism grown on oxalate. *Biochem. J.* **78**, 225.

B–C Quayle J.R. (1963) Carbon assimilation by *Pseudomonas oxalaticus* (Ox I). 7. Decarboxylation of oxalyl-coenzyme A to formyl-coenzyme A. *Biochem. J.* **89**, 492.

D Quayle J.R. and Keech D.B. (1959) Carbon assimilation by *Pseudomonas oxalaticus* (Ox I). 2. Formate and carbon dioxide utilisation by cell-free extracts of the organism grown on formate. *Biochem. J.* **72**, 631.

154 GEN Dagley S. and Trudgill P.W. (1963) The metabolism of tartaric acid by a *Pseudomonas:* a new pathway. *Biochem. J.* **89**, 22.

GEN Kohn L.D. and Jakoby W.B. (1968) Tartaric acid metabolism. III. The formation of glyceric acid. *J. biol. Chem.* **243**, 2465.

AB–C Shilo M. (1957) The enzymatic conversion of tartaric acids to oxaloacetate. *J. Gen. Microbiol.* **16**, 472.

C–D Corwin L.M. (1959) Oxaloacetic decarboxylase from rat liver mitochondria. *J. biol. Chem.* **234**, 1338.

EF–G Kohn L.D. and Jakoby W.B. (1966) Tartaric acid metabolism. II. Crystalline protein converting meso-tartrate and dihydroxy-fumarate to glycerate. *Biochem. biophys. Res. Commun.* **22**, 33.

155 GEN Wood H.G. and Stjernholm R. (1962) Formation of propionate by propionibacteria. *The Bacteria*, vol. 3. Academic Press, New York, p. 81.

GEN Kaziro Y. and Ochoa S. (1964) The metabolism of propionic acid. *Advanc. Enzymol.* **26**, 283.

GEN Galivan J.H. and Allen S.H.G. (1968) Methylmalonyl coenzyme A decarboxylase. Its role in succinate decarboxylation by *Micrococcus latilyticus. J. biol. Chem.* **243**, 1253.

AB–CD Swick R.W. and Wood H.G. (1960) The role of transcarboxylation in propionic acid fermentation. *Proc. Nat. Acad. Sci. U.S.* **46**, 28.

AB–CD Wood H.G. and Allen S.H.G. (1963) Transcarboxylase. III. Purification and properties of methyl malonyl oxalacetic transcarboxylase. *J. biol. Chem.* **238**, 547.

156 GEN Leloir L.F. and Cardini C.E. (1956) Enzymes acting on glucosamine phospates. *Biochim. biophys. Acta*, **20**, 33.

GEN Jeanloz R.W. (1963) Recent developments in the biochemistry of amino sugars. *Advanc. Enzymol.* **25**, 433.

GEN Dorfman A. (1963) Polysaccharides of connective tissue. *J. Histochem. Cytochem.* **11**, 2.

GEN Hollmann S. and Touster O. (1964) *Non-Glycolytic Pathways of Metabolism of Glucose*, chap. 9. Academic Press, New York.

GEN Salton M.R.J. (1965) Chemistry and function of amino sugars and derivatives. *Ann. Rev. Biochem.* **34**, 143.

C Distler J.J., Merrick J.M. and Roseman S. (1958) Glucosamine metabolism. Preparation and *N*-acetylation of crystalline D-glucosamine and D-galactosamine 6-phosphoric acids. *J. biol. Chem.* **230**, 497.

B–C Pogell B.M. and Gryder R.M. (1957) Enzymic synthesis of glucosamine 6-phosphate in rat liver. *J. biol. Chem.* **228**, 701.

B–C Wolfe J.B. and Nakada H.I. (1956) Glucosamine degradation by *E. coli*. The isomeric conversion of glucosamine-6-P to fructose-6-P and ammonia. *Arch. biochem. Biophys.* **64**, 489.

B–C Comb D.G. and Roseman S. (1958) Glucosamine metabolism. Glucosamine 6-phosphate deaminase. *J. biol. Chem.* **232**, 807.

B–C Gryder R.M. and Pogell B.M. (1960) Further studies on glucosamine 6-phosphate synthesis by rat liver. *J. biol. Chem.* **235**, 558.

B–C Ghosh S., Blumenthal H.J., Davidson E. and Roseman S. (1960) Glucosamine metabolism. Enzymic synthesis of glucosamine-6-phosphate. *J. biol. Chem.* **235**, 1265.

B–E Leloir L.F. and Cardini C.E. (1953) Biosynthesis of glucosamine. *Biochim. biophys. Acta*, **12**, 15.

C–D Brown D.H. (1955) The D-glucosamine-6-phosphate *N*-acetylase of yeast. *Biochim. biophys. Acta*, **16**, 429.

C–D Davidson E.A., Blumenthal H.J. and Roseman S. (1957) Glucosamine metabolism. Studies on glucosamine 6-phosphate *N*-acetylase. *J. biol. Chem.* **226**, 125.

D–E Reissig J.L. (1956) Phosphoacetylglucosamine mutase of *Neurospora*. *J. biol. Chem.* **219**, 753.

157 A–B Smith E.E.B. and Mills G.T. (1954) Uridine nucleotide compounds of liver. *Biochim. biophys. Acta*, **13**, 386.

A–B Mills G.T., Ondarz A. and Smith E.E.B. (1954) The uridyl transferase of liver. *Biochim. biophys. Acta*, **14**, 159.

A–B Maley F., Maley G.F. and Lardy H.A. (1956) The synthesis of α-D-glucosamine-1-phosphate and *N*-acetyl-α-D-glucosamine-1-phosphate. *J. Amer. chem. Soc.* **78**, 5303.

A–B Maley F. and Lardy H.A. (1956) Formation of UDP-glucose and related compounds by the soluble fraction of the liver. *Science* **124**, 1207.

A–B Strominger J.L. and Smith M.S. (1959) Uridine acetyl glucosamine diphosphorylase. *J. biol. Chem.* **234**, 1822.

B–C Maley F. and Maley G.F. (1959) UDPN-acetylglucosamine-4-epimerase from *B. subtilis*. *Biochim. biophys. Acta*, **31**, 575.

B–D Comb D.G. and Roseman S. (1958) Enzymic synthesis of *N*-acetyl-D-mannosamine. *Biochim. biophys. Acta*, **29**, 653.

B–D Glaser L. (1959) The biosynthesis of *N*-acetyl glucosamine. *J. biol. Chem.* **234**, 2801.

B–D Ghosh S. and Roseman S. (1965) The sialic acids. IV. *N*-acyl-D-glucosamine-6-phosphate 2-epimerase. V. *N*-Acyl-D-glucosamine-2-epimerase. *J. biol. chem.* **240**, 1525, 1531.

158 GEN Strange R.E. and Dark F.A. (1956) An unidentified amino sugar present in cell walls and spores of various bacteria. *Nature*, **177**, 186.

GEN Strange R.E. and Kent L.H. (1959) The isolation characterisation and chemical synthesis of muramic acid. *Biochem. J.* **71**, 333.

GEN Richmond M.H. and Perkins H.R. (1960) Possible precursors for the synthesis of muramic acid by *Staphylococcus aureus* 524. *Biochem. J.* **76**, 1P.

GEN Perkins H.R. (1963) Chemical structure and biosynthesis of bacterial cell walls. *Bact. Rev.* **27**, 18.

GEN Martin H.H. (1966) Biochemistry of bacterial cell walls. *Ann. rev. Biochem.* **35**, 457.

GEN Gunetileke K.G. and Anwar R.A. (1966) Biosynthesis of uridine diphosphate-*N*-acetyl muramic acid. *J. biol. Chem.* **241**, 5740.

GEN Mirelman D. and Sharon N. (1968) Isolation and characterisation of the disaccharide *N*-acetyl glucosaminyl-B (1→4) *N*-acetyl muramic acid and two tripeptide derivatives of this disaccharide from lysozyme digests of *Bacillus licheniformis*. *J. biol. Chem.* **243**, 2279.

GEN Gunetileke K.G. and Anwar R.A. (1968) Biosynthesis of uridine diphospho-*N*-acetylmuramic acid. Purification and properties of pyruvate-uridine diphospho-*N*-acetyl glucosamine transferase and characterisation of uridine diphospho-*N*-acetylenolpyruvalglucosamine. *J. biol. Chem.* **243**, 5770.

159 AB–C Roseman S., Jourdian G.G., Watson W. and Rood R. (1961) Enzymic synthesis of sialic acid 9-phosphates. *Proc. Nat. Acad. Sci. U.S.* **47**, 958.

AB–C Warren L. and Felsenfeld H. (1961) *N*-acetylmannosamine-6-phosphate and *N*-acetyl-neuraminic acid 9-phosphate as intermediates in sialic acid synthesis. *Biochem. biophys. Res. Commun.* **5**, 185.

D Comb D.G. and Roseman S. (1960) The sialic acids: the structure and enzymic synthesis of *N*-acetylneuraminic acid. *J. biol. Chem.* **235**, 2529.

A–D Warren L. and Felsenfeld H. (1962) The biosynthesis of sialic acids. *J. biol. Chem.* **237**, 1421.

C–D Blix F.G., Gottschalk A. and Klenk E. (1957) Proposed nomenclature of neuraminic and sialic acids. *Nature*, **179**, 1088.

D–E Comb D.G., Shimizu F. and Roseman S. (1959) Isolation of cytidine-5′-monophospho-*N*-acetylneuraminic acid. *J. Amer. chem. Soc.* **81**, 5513.

D–E Roseman S. (1962) Enzymatic synthesis of cytidine-5′-monophospho-sialic acids. *Proc. Nat. Acad. Sci. U.S.* **48**, 437.

160 GEN Brown G.B., Roll P.M. and Neinfeld H. (1952) Origins of the individual atoms of the purines. *Phosphorous Metabolism*, **2**, 389.

GEN Hartman S.C. and Buchanan J.M. (1959) Nucleic acids, purines, pyrimidines (Nucleotide synthesis). *Ann. rev. Biochem.* **28**, 365.

GEN Buchanan J.M. and Hartman S.C. (1959) Enzymatic reactions in the synthesis of the purines. *Advanc. Enzymol.* **21**, 199.

GEN Buchanan J.M. (1960) The enzymatic synthesis of the purine nucleotides. *Harvey Lectures, series 54*, 104.

A–B Heald K. and Long C. (1955) Studies involving enzymatic phosphorylation. The phosphorylation of D-ribose by extracts of *Escherichia coli*. *Biochem. J.* **59**, 316.

A–B Long C. (1955) Studies involving enzymic phosphorylation. 4. The conversion of D-ribose into D-ribose 5-phosphate by extracts of *E. coli*. *Biochem. J.* **59**, 322.

A–B Agranoff B.W. and Brady R.O. (1956) Purification and properties of calf liver ribokinase. *J. biol. Chem.* **219**, 221.

A–E Westby C.A. and Gots J.S. (1969) Genetic blocks and unique features in the biosynthesis of 5′-phosphoribosyl-*N*-formyl glycinamide in *Salmonella typhimurium*. *J. biol. Chem.* **244**, 2095.

B–C Remy C.N., Remy W.T. and Buchanan J.M. (1955) Biosynthesis of the purines. VIII. Enzymatic synthesis and utilisation of α-5-phosphoribosylpyrophosphate. *J. biol. Chem.* **217**, 885.

B–C Khorana H.G., Fernandes J.F. and Kornberg A. (1958) Phosphorylation of ribose-5-phosphate in the enzymatic synthesis of 5-phosphorylribose 1-pyrophosphate. *J. biol. Chem.* **230**, 941.

B–C Switzer R.L. (1969) Regulation and mechanism of phosphoribosylpyrophosphate synthetase. *J. biol. Chem.* **244**, 2854.

C–D Hartman S.C. and Buchanan J.M. (1958) Biosynthesis of the purines. XXI. 5-Phosphoribosylpyrophosphate amidotransferase. *J. biol. Chem.* **233**, 451.

C–D Wyngaarden J.B. and Ashton D.M. (1959) The regulation of activity of phosphoribosylpyrophosphate amidotransferase by purine ribonucleotides: a potential feedback control of purine biosynthesis. *J. biol. Chem.* **234**, 1492.

D–E Goldthwait D.A., Peabody R.A. and Greenberg G.R. (1956) On the mechanism of synthesis of glycinamide ribotide and its formyl derivative. *J. biol. Chem.* **221**, 569.

D–E Hartman S.C. and Buchanan J.M. (1958) Biosynthesis of the purines. 2-amino-*N*-ribosylacetamide-5′-phosphate kinosynthetase. *J. biol. Chem.* **233**, 456.

161 A–B Levenberg B. and Buchanan J.M. (1957) Structure, enzymatic synthesis and metabolism of (α-*N*-formyl)-glycinamide ribotide. *J. biol. Chem.* **224**, 1019.

A–B Warren L. and Buchanan J.M. (1957) Biosynthesis of the purines. XIX. 2-Amino-*N*-ribosylacetamide 5′-phosphate (glycinamide ribotide) transformylase. *J. biol. Chem.* **229**, 613.

A–B Hartman S.C. and Buchanan J.M. (1959) Biosynthesis of the purines. XXVI. The identification of the formyl donors of the transformylation reactions. *J. biol. Chem.* **234**, 1812.

A–B Mizobuchi K. and Buchanan J.M. (1968) Biosynthesis of the purines. Isolation and characterisation of formylglycinamide ribonucleotide amidotransferase-glutaryl complex. *J. biol. Chem.* **243**, 4853.

B–C Melnick I. and Buchanan J.M. (1957) Biosynthesis of the purines. XIV. Conversion of (α-*N*-formyl) glycinamide ribotide to (α-*N*-formyl) glycinamidine ribotide; purification and requirements of the enzyme system. *J. biol. Chem.* **225**, 157.

C–D Levenberg B. and Buchanan J.M. (1957) Biosynthesis of the purines. XII. Structure, enzymic synthesis, and metabolism of 5-amino-imidazole ribotide. *J. biol. Chem.* **224**, 1005.

D–E Lukens L.N. and Buchanan J.M. (1959) Biosynthesis of the purines. XXIV. The enzymatic synthesis of 5-amino-1-ribosyl-4-imidazole-carboxylic acid 5′-phosphate from 5-amino-1-ribosylimidazole 5′-phosphate and carbon dioxide. *J. biol. Chem.* **234**, 1799.

162 A–B Lukens L.N. *et al.* as above.

B–C Miller R.W., Lukens L.N. and Buchanan J.M. (1959) Biosynthesis of the purines. XXV. The enzymatic cleavage of *N*-(5-amino-ribosyl-4-imidazolylcarboxy)-L-aspartic acid 5′–phosphate. *J. biol. Chem.* **234**, 1806.

C–D Hartman S.C. and Buchanan J.M. (1959) Biosynthesis of purines. XXVI. The identification of the formyl donors of the trans-formylation reactions. *J. biol. Chem.* **234**, 1812.

C–E Flaks J.G., Erwin M.J. and Buchanan J.M. (1957) Biosynthesis of the purines. XVIII. 5-Amino-1-ribosyl-4-imidazolecarboxamide 5′-phosphate transformylase and inosinicase. *J. biol. Chem.* **229**, 603.

163 A–B Carter C.E. (1956) Synthesis of 6-succinyl aminopurine and structure of adenylosuccinic acid. *Fed. proc.* **15**, 230.

A–B Davey C.L. (1959) Synthesis of adenylosuccinic acid in preparations of mammalian skeletal muscle. *Nature*, **183**, 995.

A–C Lieberman I. (1956) Enzymatic synthesis of adenosine-5′-phosphate from inosine-5′-phosphate. *J. biol. Chem.* **223**, 327.

B–C Carter C.E. and Cohen L.H. (1956) The preparation and properties of adenylosuccinase and adenylosuccinic acid. *J. biol. Chem.* **222**, 17.

D–E Magasanik B., Moyed H.S. and Gehring L.B. (1957) Enzymes essential for the biosynthesis of nucleic acid guanine; inosine 5′-phosphate dehydrogenase of *Aerobacter aerogenes*. *J. biol. Chem.* **226**, 339.

D–E Turner J.F. and King J.E. (1961) Inosine 5′-phosphate dehydrogenase of pea seeds. *Biochem. J.* **79**, 147.

D–E Yefimochkina E.F. and Braunstein A.E. (1959) The amination of inosinic acid to adenylic acid in muscle extracts. *Arch. biochem. Biophys.* **83**, 350.

E–F Moyed H.S. and Magasanik B. (1957) Enzymes essential for the biosynthesis of nucleic acid guanine; xanthosine 5′-phosphate aminase of *Aerobacter aerogenes*. *J. biol. Chem.* **226**, 351.

164 GEN Kalckar H.M. (1947) Differential spectrophotometry of purine compounds by means of specific enzymes. III. Studies of the enzymes of purine metabolism. *J. biol. Chem.* **167**, 461.

A–B Weil-Malherbe H. and Green R.H. (1955) Ammonia formation in brain. 2. Brain adenylic deaminase. *Biochem. J.* **61**, 218.

A–B Nikiforuk G. and Colowick S.P. (1956) The purification and properties of 5-adenylic deaminase from muscle. *J. biol. Chem.* **219**, 119.

A–B Lee Y.P. (1957) 5′-Adenylic acid deaminase. I. Isolation of the crystalline enzyme from rabbit skeletal muscle. *J. biol. Chem.* **227**, 987.

A–B Mendicino J. and Muntz J.A. (1958) Activating effect of adenosine triphosphate on brain adenylic deaminase. *J. biol. Chem.* **233**, 178.

A–B Turner D.H. and Turner J.F. (1961) Adenylic deaminase of pea seeds. *Biochem. J.* **79**, 143.

B–C Kornberg A., Lieberman I. and Simms E.S. (1955) Enzymatic synthesis of purine nucleotides. *J. biol. Chem.* **215**, 417.

B–C Remy C.N., Remy W.T. and Buchanan J.M. (1955) Biosynthesis of the purines. VIII. Enzymatic synthesis and utilisation of α-5-phosphoribosylpyrophosphate. *J. biol. Chem.* **217**, 885.

B–C Lukens L.N. and Herrington K.A. (1957) Enzymic formation of 6-mercaptopurine ribotide. *Biochim. biophys. Acta*, **24**, 432.

C–DE Mackler B., Mahler H.R. and Green D.E. (1954) Studies on metalloflavoproteins. I. Xanthine oxidase, a molybdoflavoprotein. *J. biol. Chem.* **210**, 149.

C–DE Villela G.G., Affonso O.R. and Mitidieri E. (1955) Xanthine oxidase in *Lactobacillus casei*. *Arch. biochem. Biophys.* **59**, 532.

C–DE Remy C.N., Richert D.A., Doisy R.J., Wells I.C. and Westerfeld, W.W. (1955) Purification and characterisation of chicken liver xanthine dehydrogenase. *J. biol. Chem.* **217**, 293.

C–DE Avis P.G., Bergel F. and Bray R.C. (1955) Cellular constituents: the chemistry of xanthine oxidase. *J. chem. Soc.* 1100.

C–DE De Renzo E.C. (1956) Chemistry and biochemistry of xanthine oxidase. *Advanc. Enzymol.* **17**, 293.

C–DE Dikstein S., Bergmann F. and Henis Y. (1957) Studies on uric acid and related compounds. IV. The specificity of bacterial xanthine oxidases. *J. biol. Chem.* **224**, 67.

C–DE Rajagopalan K.V. and Handler P. (1967) Purification and properties of chicken liver xanthine dehydrogenase. *J. biol. Chem.* **242**, 4097.

F–G Leone E. (1953) Preparation of uricase. *Biochem. J.* **54**, 393.

F–G Mahler H.R., Hübscher G. and Baum H. (1955) Studies on uricase. I. Preparation, purification and properties of a cuproprotein. *J. biol. Chem.* **216**, 625.

F–G Robbins K.C., Barnett E.L. and Grant N.H. (1955) Partial purification of porcine liver uricase. *J. biol. Chem.* **216**, 27.

F–G London M. and Hudson P.B. (1956) Purification and properties of solubilised uricase. *Biochim. biophys. Acta*, **21**, 290.

F–G Roush A.H. and Domnas A.J. (1956) Induced biosynthesis of uricase in yeast. *Science*, **124**, 125.

F–G Hübscher G., Baum H. and Mahler H.R. (1957) Studies on uricase. IV. The nature and composition of some stable reaction products. *Biochim. biophys. Acta*, **23**, 43.

H–JK Florkin M. and Duchateau-Bosson G. (1940) Microdosage photométrique de l'allantoïne en solution pures et dans l'urine. *Enzymologia*, **9**, 5.

166 GEN Kaplan N.O., Goldin A., Humphreys S.R., Ciotti M.M. and Stolzenbach F.E. (1956) Pyrimidine nucleotide synthesis in the mouse. *J. biol. Chem.* **219**, 287.

A–C Hager S.E. and Jones M.E. (1967) A glutamate-dependent enzyme for the synthesis of carbamyl phosphate for pyrimidine biosynthesis in fetal rat liver. *J. biol. Chem.* **242**, 5674.

A–C Glasziou K.T. (1956) The metabolism of arginine in *Serratia marcescens*. II. Carbamyl-adenosine diphosphate phosphoferase. *Aust. J. biol. Sci.* **9**, 253.

B–C Jones M.E. and Spector L. (1960) The pathway of carbonate in the biosynthesis of carbamyl phosphate. *J. biol. Chem.* **235**, 2897.

B–C Marshall M., Metzenberg R.L. and Cohen P.P. (1961) Physical and kinetic properties of carbamyl phosphate synthetase from frog liver. *J. biol. Chem.* **236**, 2229.

C Jones M.E., Spector L. and Lipmann F. (1955) Carbamyl phosphate: the carbamyl donor in enzymatic citrulline synthesis. *J. Amer. chem. Soc.* **77**, 819.

E Lowenstein J.M. and Cohen P.P. (1956) Studies on the biosynthesis of carbamylaspartic acid. *J. biol. Chem.* **220**, 57.

CD–E Reichard P. and Hanshoff G. (1956) Aspartate carbamyl transferase from *Escherichia coli*. *Acta chem. Scand.* **10**, 548.

CD–E Shepherdson M. and Pardee A.B. (1960) Production and crystallisation of aspartate transcarbamylase. *J. biol. Chem.* **235**, 3233.

CD–E Gerhart J.C. and Schachman H.K. (1965) Distinct subunits for the regulation and catalytic activity of aspartate transcarbamylase. *Biochemistry*, **4**, 1054.

CD–E Schmidt P.G., Stark G.R. and Baldeschweiler J.D. (1969) Aspartate transcarbamylase. *J. biol. Chem.* **244**, 1860.

CD–E Porter R.W., Modebe M.O. and Stark G.R. (1969) Aspartate transcarbamylase. Kinetic studies of the catalytic subunit. *J. biol. Chem.* **244**, 1846.

E–F Cooper R.C. and Wilson D.W. (1954) Biosynthesis of pyrimidines. *Fed. Proc.* **13**, 194.

E–F Lieberman I. and Kornberg A. (1954) Enzymatic synthesis and breakdown of a pyrimidine, orotic acid. II. Dihydroorotic acid, ureidosuccinic acid, and 5-carboxymethylhydantion. *J. biol. Chem.* **207**, 911.

F–G Friedmann H.C. and Vennesland B. (1960) Crystalline dihydro-orotic dehydrogenase. *J. biol. Chem.* **235**, 1526.

G–J Kornberg A., Lieberman I. and Simms E.S. (1954) Enzymatic synthesis of pyrimidine and purine nucleotides. II. Orotidine-5′-phosphate pyrophosphorylase and decarboxylase. *J. Amer. chem. Soc.* **76**, 2844.

G–J Lieberman I., Kornberg A. and Simms E.S. (1955) Enzymatic synthesis of pyrimidine nucleotides. Orotidine-5′-phosphate and uridine-5′-phosphate. *J. biol. Chem.* **215**, 403.

G–J Hurlbert R.B. and Reichard P. (1955) The conversion of orotic acid to uridine nucleotides *in vitro*. *Acta chem. Scand.* **9**, 251.

G–J Crawford I., Kornberg A. and Simms E.S. (1957) Conversion of uracil and orotate to uridine 5′-phosphate by enzymes in *Lactobacilli*. *J. biol. Chem.* **226**, 1093.

H–J Creasy W.A. and Handschumacher R.E. (1961) Purification and properties of orotidylate decarboxylases from yeast and rat liver. *J. biol. Chem.* **236**, 2058.

167 A–C Berg P. and Joklik W.K. (1954) Enzymatic phosphorylation of nucleoside diphosphates. *J. biol. Chem.* **210**, 657.

A–C Lieberman I., Kornberg A. and Simms E.S. (1955) Enzymatic synthesis of nucleoside diphosphates and triphosphates. *J. biol. Chem.* **215**, 429.

A–C Gibson D.M., Ayengar, P. and Sanadi D.R. (1956) Transphosphorylations between nucleoside phosphates. *Biochim. biophys. Acta*, **21**, 86.

A–C Heppel L.A., Strominger J.L. and Maxwell E.S. (1959) Nucleoside monophosphate kinases. II. Transphosphorylation between adenosine monophosphate and nucleoside triphosphates. *Biochim. biophys. Acta*, **32**, 422.

C–D Lieberman I. (1956) Enzymatic amination of uridine triphosphate to cytidine triphosphate. *J. biol. Chem.* **222**, 765.

168 A–B & C–D Kirkland R.J.A. and Turner J.F. (1959) Nucleoside diphosphokinase of pea seeds. *Biochem. J.* **72**, 716.

A–D *See* 167 A–C.

B–C Moore E.C., Reichard P. and Thelander L. (1964) Purification and properties of thioredoxin from *E. coli*. *J. biol. Chem.* **239**, 3445.

B–C Laurent T.C., Moore E.C. and Reichard P. (1964) Enzymatic synthesis of deoxyribonucleotides. IV. Isolation and characterisation of thioredoxin, the hydrogen donor from *Escherichia coli* B. *J. biol. Chem.* **239**, 3436.

C–D Krebs H.A. and Hems R. (1953) Some reactions of adenosine and inosine phosphates in animal tissues. *Biochim. biophys. Acta*, **12**, 172.

C–D Gibson D.M., Ayengar P. and Sanadi D.R. (1956) Transphosphorylations between nucleoside phosphates. *Biochim. biophys. Acta*, **21**, 86.

C–D Ginsburg A. (1959) A deoxyribokinase from *Lactobacillus plantarum*. *J. biol. Chem.* **234**, 481.

169 A–B Hurwitz J. (1959) Enzymatic incorporation of ribonucleotides and poly deoxy nucleotide material. *J. biol. Chem.* **234**, 2351.

B–C Wang T.P., Sable H.Z. and Lampen J.O. (1950) Enzymatic deamination of cytosine nucleoside. *J. biol. Chem.* **184**, 17.

C–D Flaks J.G. and Cohen S.S. (1959) Virus induced acquisition of metabolic function. I. Enzymatic formation of 5-hydroxy-methyl deoxycytidylate. *J. biol. Chem.* **234**, 1501.

C–D Wahba A.J. and Friedkin M. (1962) The enzymatic synthesis of thymidylate. *J. biol. Chem.* **237**, 3794.

C–D Pastore E.J. and Friedkin M. (1962) The enzymatic synthesis of thymidylate. Transfer of tritium from tetrahydrofolate to the methyl group of thymidylate. *J. biol. Chem.* **237**, 3802.

C–D Friedkin M., Crawford E.J., Donovan E. and Pastore E.J. (1962) The enzymatic synthesis of thymidylate. The further purification of thymidylate synthetase and its separation from natural fluorescent inhibitors. *J. biol. Chem.* **237**, 3811.

C–D Friedkin M. (1963) Enzymatic aspects of folic acid. Thymidylate synthetase. *Ann. rev. Biochem.* **32**, 201.

C–D Blakley R.L. (1963) The biosynthesis of thymidylic acid. *J. biol. Chem.* **238**, 2113.

D–E *See* Hurwitz J. 169 A–B.

E–F *See* 167 A–C.

170 GEN Canellakis E.S. (1956) Pyrimidine metabolism. I. Enzymatic pathways of uracil and thymine degradation. *J. biol. Chem.* **221**, 315.

GEN Fink K., Cline R.E., Henderson R.B. and Fink R.M. (1956) Metabolism of thymine (methyl-C^{14} or-2-C^{14}) by rat liver *in vitro*. *J. biol. Chem.* **221**, 425.

A–C *See* Crawford I., Kornberg A. and Simms E.S. 166 G–J.

C–E Campbell L.L. (1957) Reductive degradation of pyrimidines. III. Purification and properties of dihydrouracil dehydrogenase. *J. biol. Chem.* **227**, 693.

C–E Grisolia S. and Cardoso S. (1957) The purification and properties of hydropyrimidine dehydrogenase. *Biochim. biophys. Acta*, **25**, 430.

C–E Fritzson P. (1960) Properties and assay of dihydrouracil dehydrogenase of rat liver. *J. biol. Chem.* **235**, 719.

E–G & F–H Wallach D.P. and Grisolia S. (1957) The purification and properties of hydropyrimidine hydrase. *J. biol. Chem.* **226**, 277.

E–J Fritzson P. and Pihl A. (1957) The catabolism of C^{14}-labelled uracil, dihydrouracil and β-ureidopropionic acid in the intact rat. *J. Biol. Chem.* **226**, 229.

G–J Caravaca J. and Grisolia S. (1958) Enzymatic decarbamylation of carbamyl β-alanine and carbamyl β-aminoisobutyric acid. *J. biol. Chem.* **231**, 357.

G–J Campbell L.L. (1960) Reductive degradation of pyrimidines. V. Enzymatic conversion of *N*-carbamyl-β-alanine, carbon dioxide and ammonia. *J. biol. Chem.* **235**, 2375.

J–L Hayaishi O., Nishizuka Y., Tatibana M., Takeshita M. and Kuno S. (1961) Enzymatic studies on the metabolism of β-alanine. *J. biol. Chem.* **236**, 781.

172 GEN Novelli G.D. (1953) Enzymatic synthesis and structure of CoA. *Fed. Proc.* **12**, 675.

GEN Brown G.M. and Reynolds J.S. (1963) Biogenesis of water soluble vitamins. Review of CoA biosynthesis. *Ann. Rev. Biochem.* **32**, 419.

GEN Goldman P. and Vagelos P.R. (1964) Acyl-transfer reactions (CoA-structure, functions). *Comprehensive Biochemistry*, vol. 15, p. 71.

GEN Cheldelin V.A. and Baich A. (1964) The biosynthesis of water soluble vitamins. *Biogenesis of Natural Compounds*. Pergamon Press, p. 509.

A Maas W.K. and Vogel H.J. (1953) α-Ketoisovaleric acid, a precursor of pantothenic acid in *E. coli*. *J. Bact.* **65**, 388.

A–B Mcintosh E.N., Purko M. and Wood W.A. (1957) Ketopantoate formation by a hydroxymethylation enzyme from *E. coli*. *J. biol. Chem.* **228**, 499.

D Lipmann F., Kaplan N.O., Novelli G.D., Tuttle L.C. and Guirard B.M. (1947) Coenzyme for acetylation. A pantothenic acid derivative. *J. biol. Chem.* **167**, 869.

C–D Maas W.K. (1952) Description of the extracted pantothenate synthesising enzyme of *E. coli*. *J. biol. Chem.* **198**, 23.

C–D Maas W.K. and Novelli G.D. (1953) Synthesis of pantothenic acid by depyrophosphorylation of adenosine triphosphate. *Arch. biochem. Biophys.* **43**, 236.

C–D Ginoza H.S. and Altenbern R.A. (1955) The pantothenate-synthesising enzyme in cell free extracts of *Brucella abortis* strain 19. *Arch. biochem. Biophys.* **56**, 537.

C–D Maas W.K. (1956) Mechanism of the enzymatic synthesis of pantothenate from beta-alanine and pantoate. *Fed. Proc.* **15**, 305.

D–E Pierpont W.S., Hughes D.E., Baddiley J. and Mathias A.P. (1955) The phosphorylation of pantothenic acid by *Lactobacillus arabinosus*. *Biochem. J.* **61**, 368.

D–G Brown G.M. (1959) The metabolism of pantothenic acid. *J. biol. Chem.* **234**, 370.

D–G Hoagland M.B. and Novelli G.D. (1954) Biosynthesis of coenzyme A from phosphopantetheine and of pantetheine from pantothenate. *J. biol. Chem.* **207**, 767.

E–F Brown G.M. (1957) Pantothenylcysteine, a precursor of pantetheine in *Lactobacillus helveticus*. *J. biol. Chem.* **226**, 651.

173 A–B Novelli G.D. (1953) Enzymatic synthesis and structure of CoA. *Fed. Proc.* **12**, 675.

A–C See Hoagland M.B. and Novelli G.D. 172 D–G.

B–C Wang T.P. and Kaplan N.O. (1954) Kinases for the synthesis of coenzyme A and triphosphopyridine nucleotide. *J. biol. Chem.* **206**, 311.

174 GEN Stansly P.G. and Beinert, H. (1953) Synthesis of butyryl CoA by reversal of the oxidative pathway. *Biochim. biophys. Acta*, **11**, 600.

GEN Wakil S.J. (1961) Mechanism of fatty acid synthesis. *J. lipid Res.* **2**, 1.

GEN Lynen F. (1961) Biosynthesis of saturated fatty acids. *Fed. Proc.* **20**, 941.

GEN Dils R. and Popjack G. (1962) Synthesis of fatty acids from acetate in extracts of lactating rat mammary gland. *Biochem. J.* **83** 41.

GEN Spencer A.F. and Lowenstein J.M. (1962) The supply of precursors for the synthesis of fatty acids. *J. biol. Chem.* **237**, 3640.

GEN Vagelos P.R. (1964) Lipid metabolism. *Ann. Rev. Biochem.* **33**, 139.

GEN Saver F., Pugh E.L., Wakil S.J., Delaney R. and Hill R.L. (1964) 2-Mercaptoethylamine and β-alanine as components of acyl carrier protein. *Proc. Nat. Acad. Sci. U.S.* **52**, 1360.

GEN Wakil S.J. (1964) Synthesis of fatty acids in animal tissues. *Metabolic and Physiological Significance of Lipids*. p. 3.

GEN Majerus P.W., Alberts A.W. and Vagelos P.R. (1965) Acyl carrier protein. IV. The identification of 4-phosphopantetheine as the prosthetic group of the acyl carrier protein. *Proc. Nat. Acad. Sci. U.S.* **53**, 410.

GEN Pugh E.L. and Wakil S.J. (1965) Studies on the mechanism of fatty acid synthesis. The prosthetic group of the acyl carrier protein and the mode of attachment to the protein. *J. biol. Chem.* **240**, 4727.

GEN Williamson I.P., Goldman J.K. and Wakil S.J. (1966) Studies on the fatty acid synthesising system from pigeon liver. *Fed. Proc.* **25**, 340.

GEN See Lynen F. (1967) 97 GEN.

GEN Simoni R.D., Criddle R.S. and Stumpf P.K. (1967) Fat metabolism in higher plants. Purification and properties of plant and bacterial acyl carrier proteins. *J. biol. Chem.* **242**, 573.

GEN Majerus P.W. (1967) Acyl carrier protein. Structural requirements for function in fatty acid biosynthesis. *J. biol. Chem.* **242**, 2325.

GEN Yang P.C., Butterworth P.H.W., Bock R.M. and Porter J.W. (1967)

Further studies on the properties of the pigeon liver fatty acid synthetase. *J. biol. Chem.* **242**, 3501.

GEN Burton D.N., Collins J.M. and Porter J.W. (1969) Biosynthesis of the fatty acid synthetase by isolated rat liver cells. *J. biol. Chem.* **244**, 1076.

GEN Shrago E., Spennetta T. and Gordon E. (1969) Fatty acid synthesis in human adipose tissue. *J. biol. Chem.* **244**, 2761.

GEN Stumpf P.K. (1969) Metabolism of fatty acids. *Ann. Rev. Biochem.* **38**, 159.

A–B Hatch M.D. and Stumpf P.K. (1961) Fat metabolism in higher plants. XVI. Acetyl coenzyme A carboxylase and acyl coenzyme A—malonyl coenzyme A transcarboxylase from wheat germ. *J. biol. Chem.* **236**, 2879.

A–B Williamson I.P. and Wakil S.J. (1966) Studies on the mechanism of fatty acid synthesis. XVII. Preparation and general properties of acetyl coenzyme A and malonyl coenzyme A—acyl carrier protein transacylases. *J. biol. Chem.* **241**, 2326.

A–B Gregolin C., Ryder E., Kleinschmidt A.K., Warner R.C. and Lane M.D. (1966) Molecular characteristics of liver acetyl CoA carboxylase. *Proc. Nat. Acad. Sci. U.S.* **56**, 148.

A–B Easter D.J. and Dils R. (1968) Fatty acid biosynthesis. Properties of acetyl CoA carboxylase in lactating rabbit mammary gland. *Biochim. biophys. Acta*, **152**, 653.

A–B Heinstein P.F. and Stumpf P.K. (1968) Properties of wheat germ acetyl-CoA carboxylase. *Fed. Proc.* **27**, 647.

A–B Alberts A.W. and Vagelos P.R. (1968) Acetyl-CoA carboxylase. I. Requirements for two protein fractions. *Proc. Nat. Acad. Sci. U.S.* **59**, 561.

A–B Miller A.L. and Levy H.R. (1969) Rat mammary acetyl CoA carboxylase. Isolation and characterisation. *J. biol. Chem.* **244**, 2334.

CD–E Goldman P., Alberts A.W. and Vagelos P.R. (1963) The condensation reaction of fatty acid synthesis. Identification of the protein bound product of the reaction and its conversion to long chain fatty acids. *J. biol. Chem.* **238**, 3579.

CD–E Alberts A.W., Majerus P.W. and Vagelos P.R. (1965) Acyl carrier protein. VI. Purification and properties of β-ketoacyl carrier protein synthetase. *Biochemistry*, **4**, 2265.

CD–E Toomey R.E. and Wakil S.J. (1966) Studies on the mechanism of fatty acid synthesis. XVI. Preparation and general properties of acyl-malonyl acyl carrier protein-condensing enzyme from *Escherichia coli*. *J. biol. Chem.* **241**, 1159.

E–F Wakil S.J. and Bressler R. (1962) Studies on the mechanism of fatty acid synthesis. Reduced triphosphopyridine nucleotide-acetoacetyl coenzyme A reductase. *J. biol. Chem.* **237**, 687.

E–F Alberts A.W., Majerus P.W., Talamo B. and Vagelos P.R. (1964) Acyl carrier protein. II. Intermediary reactions of fatty acid synthesis. *Biochemistry*, **3**, 1563.

E–F Toomey R.E. and Wakil S.J. (1966) Studies on the mechanism of fatty acid synthesis. XV. Preparation and general properties of β-ketoacyl acyl carrier protein reductase from *Escherichia coli*. *Biochim. biophys. Acta*, **116**, 189.

F–G Stern J.R., Del Campillo A. and Raw I. (1956) Enzymes of fatty acid metabolism. General introduction; crystalline crotonase. *J. biol. Chem.* **218**, 971.

F–G Stern J.R. and Del Campillo A. (1956) Enzymes of fatty acid metabolism. Properties of crystalline crotonase. *J. biol. Chem.* **218**, 985.

F–G Majerus P.W., Alberts A.W. and Vagelos P.R. (1965) Acyl carrier protein. III. An enoyl hydrase specific for acyl carrier protein thioesters. *J. biol. Chem.* **240**, 618.

G–H Hsu R.Y., Wasson G. and Porter J.W. (1965) The purification and properties of the fatty acid synthetase of pigeon liver. *J. biol. Chem.* **240**, 3736.

G–H Weeks G. and Wakil S.J. (1968) Studies on the mechanism of fatty acid synthesis. Preparation and general properties of enoyl acyl carrier protein reductases from *Escherichia coli*. *J. biol. Chem.* **243**, 1180.

175 GEN Harlan W.R. and Wakil S.J. (1963) Synthesis of fatty acids in animal tissues. Incorporation of C^{14}acetyl coenzyme A into a variety of long chain fatty acids by subcellular particles. *J. biol. Chem.* **238**, 3216.

GEN Ernin J. and Bloch K. (1964) Biosynthesis of unsaturated fatty acids in microsomes. *Science* **143**, 1006.

GEN See 174 A–H.

176 GEN Chang Y-Y. and Kennedy E.P. (1967) Pathways for the synthesis of glycerophosphatides in *Escherichia coli*. *J. biol. Chem.* **242**, 516.

GEN Pieringer R.A. and Bonner H., Jr. and Kunnes R.S. (1967) Biosynthesis of phosphatidic acid, lysophosphatidic acid, diglyceride and triglyceride by fatty acyl transferase pathways in *E. coli*. *J. biol. Chem.* **242**, 2719.

A–B Baranowski T. (1949) Crystalline glycerophosphate dehydrogenase from rabbit muscle. *J. biol. Chem.* **180**, 535.

A–B Miller O.N., Huggins C.G. and Arai K. (1953) Studies on the metabolism of 1,2 propanediol-1-phosphate. *J. biol. Chem.* **202**, 263.

A–B Fondy T.P., Levin L., Sollohub S.J. and Ross C.R. (1968) Structural studies on nicotinamide adenine dinucleotide-linked L-glycerol 3-phosphate dehydrogenase crystallised from rat skeletal muscle. *J. biol. Chem.* **243**, 3148.

A–B Kim S.J. and Anderson B.M. (1968) Properties of the nicotinamide adenine dinucleotide-binding sites of L-α-glycerophosphate dehydrogenase. *J. biol. Chem.* **243**, 3351.

A–B Telegdi M. (1968) The mechanism of action of rabbit muscle α-glycerophosphate dehydrogenase. *Biochim. biophys. Acta*, **159**, 227.

A–B Kim S.J. and Anderson B.M. (1969) Coenzyme binding to L-α-glycerophosphate dehydrogenase. *J. biol. Chem.* **244**, 1547.

B–C Kornberg A. and Pricer W.E. (1953) Enzymatic esterification of α-glycerophosphate by long chain fatty acids. *J. biol. Chem.* **204**, 345.

B–C Barden R.E. and Cleland W.W. (1969) L-Acylglycerol 3-phosphate acyl transferase from rat liver. *J. biol. Chem.* **244**, 3677.

D Hill E.E., Husbands D.R. and Lands W.E. (1968) The selective incorporation of ^{14}C-glycerol into different species of phosphatidic acid, phosphatidyl ethanolamine and phosphatidyl choline. *J. biol. Chem.* **243**, 4440.

B–D Lanos W.E.M. and Hart P. (1964) Metabolism of glycerolipids. V. Metabolism of phosphatidic acid. *J. Lipid. Res.* **5**, 81.

B–D Mårtensson E. and Kanfer J. (1968) The conversion of L-glycerol ^{14}C-3-phosphate into phosphatidic acid by a solubilised preparation from rat brain. *J. biol. Chem.* **243**, 497.

C–D Pieringer R.A. and Hokin L.E. (1962) Biosynthesis of phosphatidic acid from lysophosphatidic acid and palmityl coenzyme A. *J. biol. Chem.* **237**, 659.

D–E Smith S.W., Weiss S.B. and Kennedy E.P. (1957) The enzymatic dephosphorylation of phosphatidic acids. *J. biol. Chem.* **228**, 915.

D–E Coleman R. and Hubscher G. (1962) Metabolism of phospholipids. V. Studies of phosphatidic acid phosphatase. *Biochim. biophys. Acta*, **56**, 479.

E–F Weiss S.B., Kennedy E.P. and Kiyasu J.Y. (1960) The enzymatic synthesis of triglycerides. *J. biol. Chem.* **235**, 40.

E–F Goldman P. and Vagelos P.R. (1961) The specificity of triglyceride synthesis from diglycerides in chicken adipose tissue. *J. biol. Chem.* **236**, 2620.

E–F Kern F. and Borgstrom B. (1965) Quantitative study of the pathways of triglyceride synthesis by hamster intestinal muscle. *Biochim. biophys. Acta*, **98**, 520.

177 GEN Kornberg A. and Pricer W.E. (1953) Enzymatic synthesis of the coenzyme derivatives of long chain fatty acids. *J. biol. Chem.* **204**, 329.

GEN Mahler H.R., Wakil S.J. and Bock R.M. (1953) Studies on fatty acid oxidation. I. Enzymatic oxidation of fatty acids. *J. biol. Chem.* **204**, 453.

GEN Lynen F. and Ochoa S. (1953) Enzymes of fatty acid metabolism. *Biochim. biophys. Acta*, **12**, 299.

GEN Green D.E. (1963) Fatty acid oxidation. *Progress in Chemistry of Fats and Other Lipids*, vol. 6, chap. 3.

GEN Rossi C.R. and Gibson D.M. (1964) Activation of fatty acids by a guanosine triphosphate-specific thiokinase from liver mitochondria. *J. biol. Chem.* **239**, 1694.

GEN Galzigna L., Rossi C.R., Sartorelli L. and Gibson D.M. (1966) A GTP-dependent acyl-coenzyme A synthetase from rat liver mitochondria. *Fed. Proc.* **25**, 339.

GEN Rossi C.R., Galzigna L., Alexandre A. and Gibson D.M. (1967) Oxidation of long chain fatty acids by rat liver mitochondria. *J. biol. Chem.* **242**, 2102.

A–C & B–D Crane F.L., Mii S., Hauge J.G., Green D.E. and Beinert H. (1956) On the mechanism of dehydrogenation of fatty acyl derivatives of coenzyme A. I. The general fatty acyl coenzyme A dehydrogenase. *J. biol. Chem.* **218**, 701.

A–C & B–D Crane F.L. and Beinert H. (1956) On the mechanism of dehydrogenation of fatty acyl derivatives of coenzyme A. II. The electron transferring flavoprotein. *J. biol. Chem.* **218**, 717.

A–C & B–D Hauge J.G., Crane F.L. and Beinert H. (1956) On the mechanism of dehydrogenation of fatty acyl derivatives of coenzyme A. *J. biol. Chem.* **219**, 727.

B–D Green D.E., Mii S., Mahler H.R. and Bock R.M. (1954) Studies on the fatty acid oxidising system of animal tissues. III. Butyryl coenzyme A dehydrogenase. *J. biol. Chem.* **206**, 1.

C–E & D–F Stern J.R. and Del Campillo A. (1953) Enzymatic reaction of crotonyl coenzyme A. *J. Amer. chem. Soc.*, **75**, 2277.

C–E & D–F Wakil S.J. and Mahler H.R. (1954) Studies on fatty acid oxidising system in animal tissues. *J. biol. Chem.* **207**, 125.

H Stern J.R. (1956) Optical properties of acetoacetyl-S-coenzyme A and its metal chelates. *J. biol. Chem.* **221**, 33.

E–G & F–H Wakil S.J., Green D.E., Mii S. and Mahler H.R. (1954) Studies on the fatty acid oxidising system of animal tissues. β-Hydroxyacyl coenzyme A dehydrogenase. *J. biol. Chem.* **207**, 631.

F–H Green D.E., Denan J.G. and Leloir F. (1937) The β-hydroxybutyric dehydrogenase of animal tissues. *Biochem. J.* **31**, 934.

F–H Lehninger A.L. and Greville G.D. (1953) The enzymic oxidation of d- and l-β-hydroxybutyrate. *Biochim. biophys. Acta*, **12**, 188.

F–H Wakil S.J. (1955) D(—)β-Hydroxybutyryl CoA dehydrogenase. *Biochim. biophys. Acta*, **18**, 314.

F–H Stern J.R. (1957) Crystalline β-hydroxybutyryl dehydrogenase from pig heart. *Biochim. biophys. Acta*, **26**, 448.

F–H Marcus A., Vennesland B. and Stern J.R. (1958) The enzymatic transfer of hydrogen. VII. The reaction catalysed by β-hydroxybutyryl dehydrogenase. *J. biol. Chem.* **233**, 722.

G–J & H–K Goldman D.S. (1954) Studies on fatty acid oxidising system in animal tissues. The β-ketoacyl coenzyme A cleavage enzymes. *J. biol. Chem.* **208**, 345.

H–J Stern J.R., Coon M.J. and Del Campillo A. (1953) Acetoacetyl coenzyme A as intermediate in the enzymatic breakdown and synthesis of acetoacetate. *J. Amer. chem. Soc.* **75**, 1517.

178 GEN Kornberg A. and Pricer W.E. (1952) Studies on the enzymatic synthesis of phcspholipids. *Fed. Proc.* **11**, 242.

GEN Kennedy E.P. and Weiss S.B. (1956) The function of cytidine coenzymes in the biosynthesis of phospholipids. *J. biol. Chem.* **222**, **193**.

GEN Ansell G.B. and Hawthorne J.N. (1964) Phospholipids: chemistry, metabolism and function.

GEN Hutchinson H.T. and Cronan J.E., Jr. (1968) The synthesis of cytidine diphosphate diglyceride by cell free extracts of yeast. *Biochim. biophys. Acta*, **164**, 606.

A–B Wittenberg J. and Kornberg A. (1953) Choline phosphokinase. *J. biol. Chem.* **202**, 431.

D (& H) *See* Hill E.E. *et al.* 176 D.

A–D Sherr S.I. and Law J.H. (1965) Phosphatidylcholine synthesis in *agrobacterium tumefaciens*. II. Uptake and utilisation of choline. *J. biol. Chem.* **240**, 3760.

A–D Pennington R.J. and Worsfold M. (1969) Biosynthesis of lecithin by skeletal muscle. *Biochim. biophys. Acta*, **176**, 774.

A–D & E–H Kennedy E.P. (1957) Biosynthesis of phospholipids. *Fed. Proc.* **16**, 847.

A–D & E–H Kanoch H. (1969) Biosynthesis of molecular species of phosphatidyl choline and phosphatidyl ethanolamine from radioactive precursors in rat liver slices. *Biochim. biophys. Acta*, **176**, 756.

C Borkenhagen L.F. and Kennedy E.P. (1957) The enzymatic synthesis of cytidine diphosphate choline. *J. biol. Chem.* **227**, 951.

B–C Williams-Ashman H.G. and Banks J. (1956) Participation of cytidine coenzymes in the metabolism of choline by seminal vesicle. *J. biol. Chem.* **223**, 509.

C–D Weiss S.B., Smith S.W. and Kennedy E.P. (1958) The enzymatic formation of lecithin from cytidine diphosphate choline and D-1,2-diglyceride. *J. biol. Chem.* **231**, 53.

E–H Kanfer J. and Kennedy E.P. (1964) Metabolism and function of bacterial lipids. II. Biosynthesis of phospholipids in *Escherichia coli*. *J. biol. Chem.* **239**, 1720.

E–H Merke I. and Lands W.E.M. (1965) Metabolism of glycerolipids. Synthesis of phosphatidyl ethanolamine *J. biol. Chem.* **238**, 905.

179 GEN Cooksey K.E. and Greenberg D.M. (1961) Studies on the substrate specificity of the phosphatide methylating system of microsomes. *Biochem biophys. Res. Commun.* **6**, 256.

D Bremer J. and Greenberg D.M. (1957) Some factors influencing phosphatidyl-choline formation. *J. biol. Chem.* **227**, 107.

D Tattrie N.H. (1959) Positional distribution of saturated and unsaturated fatty acids on egg lecithin. *J. Lipid Res.* **1**, 60.

A–D Gibson K.D., Wilson J.D. and Udenfriend S. (1961) The enzymatic conversion of phospholipid ethanolamine to phospholipid choline in rat liver. *J. biol. Chem.* **236**, 673.

A–D Bremer J. and Greenberg D.M. (1961) Methyl transferring enzyme system of microsomes in the biosynthesis of lecithin (phosphatidylcholine). *Biochim. biophys. Acta*, **46**, 205.

E Wilson J.D., Gibson K.D. and Udenfriend S. (1960) Studies on the precursors of the methyl groups of choline in rat liver. *J. biol. Chem.* **235**, 3213.

E Bremer J., Figard P.H. and Greenberg D.M. (1960) The biosynthesis of choline and its relation to phospholipid metabolism. *Biochim. biophys. Acta*, **43**, 477.

D–E Hanahan D.J. and Chaikoff I.L. (1947) A new phospholipid-splitting enzyme specific for the ester linkage between the nitrogenous base and the phosphoric acid grouping. *J. biol. Chem.* **169**, 699.

D–E Tookey H.L. and Balls A.K. (1956) Plant phospholipase D. Studies on cotton seed and cabbage phospholipase D. *J. biol. Chem.* **218**, 213.

D–E Einset E. and Clark W.L. (1958) The enzymatically catalysed release of choline from lecithin. *J. biol. Chem.* **231**, 703.

180 C De Haas G.H. and Van Deenan L.L.M. (1965) Structural identification of isomeric lysolecithin. *Biochim. biophys. Acta*, **106**, 315.

B–C Hanahan D.J., Rodbell M. and Turner L.D. (1964) Enzymatic formation of monopalmitoleyl- and monopalmitoyllecithin (lysolecithins). *J. biol. Chem.* **206**, 431.

B–C Long C. and Penny I.F. (1957) The structure of the naturally occurring phosphoglycerides. 3. Action of moccasin-venom phospholipase A on ovolecithin and related substances. *Biochem. J.* **65**, 382.

B–C Hanahan D.J., Brockerhoff H. and Barron E.J. (1960) The site of attack of phospholipase (lecithinase) A on lecithin. *J. biol. Chem.* **235**, 1917.

B–C De Haas G.H., Daemen F.J.M. and Van Deenen L.L.M. (1962) The site of action of phosphatide acyl-hydrolase (phospholipase A) on mixed-acid phosphatides containing a poly-unsaturated fatty acid. *Biochem. biophys. Acta*, **65**, 260.

B–C Gallai-Hatchard J.J. and Thompson R.H.S. (1965) Phospholipase A activity of mammalian tissues. *Biochim. biophys. Acta*, **98**, 128.

C–D Shapiro B. (1953) Purification and properties of lysolecithinase from pancreas. *Biochem. J.* **53**, 663.

C–D Dawson R.M.C. (1958) The identification of two lipid components in liver which enable *Penicilium notatum* extracts to hydrolyse lecithin. *Biochem. J.* **68**, 352.

C–D Doery H.M. and Pearson J.E. (1964) Phospholipase B in snake venoms and bee venom. *Biochem. J.* **92**, 599.

E–F Macfarlane M.G. and Knight B.C.J. (1942) The lecithinase activity of bacterial toxins. 1. The lecithinase activity of *Clostridium welchii* toxins. *Biochem. J.* **35**, 884.

E–F Macfarlane M.G. (1950) The biochemistry of bacterial toxins. Variation in haemolytic activity of immunologically distinct lecithinase towards erythrocytes from different species. *Biochem. J.* **47**, 270.

E–F Long C. and Maguire M.F. (1954) The structure of naturally occurring phosphoglycerides. Evidence derived from a study of the action of phospholipase C. *Biochem. J.* **57**, 223.

E–F Hanahan D. and Vercamer R. (1954) The action of lecithinase D on lecithin. The enzymatic preparation of D-1,2-dipalmitolein and D-1,2-dipalmitin. *J. Amer. chem. Soc.* **76**, 1804.

E–F Bangham A.D. and Dawson R.M.C. (1962) Electrokinetic requirements for the reaction between *Cl. perfringens* α-toxin (phospholipase C) and phospholipid substrates. *Biochim. biophys. Acta*, **59**, 103.

E–G *See* Tookey H.L. *et al.* 179 D–E.

E–G Davidson F.M. and Long C. (1958) The structure of the naturally occurring phosphoglycerides. 4. Action of cabbage-leaf phospholipase D on ovolecithin and related substances. *Biochem. J.* **69**, 458.

181 A–B Kanfer J., Kennedy E.P. (1964) Metabolism and function of bacterial lipid. II. Biosynthesis of phospholipid in *E. coli. J. biol. Chem.* **239**, 1720.

A–B Sherr S.I. and Law J.H. (1965) Phosphatidylcholine synthesis in *Agrobacterium Tumefaciens. J. biol. Chem.* **240**, 3760.

C–F Borkenhagen L.F., Kennedy E.P. and Fielding L. (1961) Enzymatic formation and decarboxylation of phosphatidylserine. *J. biol. Chem.* **236**, PC 28.

C–F Hübscher G. (1962) Metabolism of phospholipids. VI. The effect of metal ions on the incorporation of L-serine into phosphatidylserine. *Biochim. biophys. Acta*, **57**, 555.

182 GEN Agranoff B.W., Bradley R.M. and Brady R.O. (1958) The enzymic synthesis of inositol phosphatide. *J. biol. Chem.* **233**, 1077.

GEN Paulus H. and Kennedy E.P. (1960) The enzymatic synthesis of inositol monophosphatide. *J. biol. Chem.* **235**, 1303.

GEN Thompson W., Strickland K.P. and Rossiter R.J. (1963) Biosynthesis of phosphatidylinositol in rat brain. *Biochem. J.* **87**, 136.

GEN Hawthorne J.N. and Kemp P. (1964) The brain phosphoinositols. *Advances in Lipid Res.* **2**, 127.

GEN Petzold G.L. and Agranoff B.W. (1965) Studies on the formation of CDP-diglyceride. *Fed. Proc.* **24**, 476.

183 GEN Zabin I. (1957) Biosynthesis of ceramide by rat brain homogenates. *J. Amer. chem. Soc.* **79**, 5834.

GEN Brady R.O. and Koval G.J. (1958) The enzymatic synthesis of sphingosine. *J. biol. Chem.* **233**, 26.

GEN Brady R.O., Formica J.W. and Koval G.J. (1958) Enzymic synthesis of sphingosine. Further studies on the mechanism of the reaction. *J. biol. Chem.* **233**, 1072.

GEN Sribney M. and Kennedy E.P. (1958) The enzymatic synthesis of sphingomyelin. *J. biol. Chem.* **233**, 1315.

GEN Fujino Y. and Negishi T. (1968) Investigation of the enzymatic synthesis of sphingomyelin. *Biochim. biophys. Acta*, **152**, 428.

184–185 GEN Kennedy E.P. (1957) Biosynthesis of phospholipids. *Fed. Proc.* **16**, 847.

GEN Hanahan D.J. and Thompson G.A. (1963) Complex lipids. *Ann. Rev. Biochem.* **32**, 215.

GEN Rossiter R.J. (1968) Metabolism of phosphatides in metabolic pathways, vol. 2. Ed. D.M. Greenberg. Academic Press, New York, p. 69.

186 GEN Popjáck G. (1958) Biosynthesis of cholesterol and related substances. *Ann. Rev. Biochem.* **27**, 533.

GEN Bloch K. (1958) Biogenesis and transformations of squalene. *Proc. 4th Intern. Congr. Biochem. 1958. Vienna.*

GEN Rudney H., Wolstenholme G.E.W. and O'Connor M. (1959) *Ciba Foundation Symposium on the Biosynthesis of Terpenes and Sterols.* Churchill, p. 75.

GEN Popjáck G. and Cornforth J.W. (1960) The biosynthesis of cholesterol. *Advanc. Enzymol.* **22**, 281.

GEN Wright L.D. (1961) Biosynthesis of isoprenoid compounds. *Ann. Rev. Biochem.* **30**, 525.

GEN Popjáck G. (1964) Recent advances in the study of sterol biosynthesis. *Metabolic and physiolog. Sig. of Lipids.* p. 47.

GEN Frantz I.D. and Schroepper G.J. (1967) Sterol biosynthesis. *Ann. Rev. Biochem.* 691.

A Higgins M.J.P. and Kekwick R.G.O. (1969) The role of malonyl coenzyme A in isoprenoid biosynthesis. *Biochem. J.* **113**, 36P.

A–B Rudney H. and Ferguson J., Jr. (1957) The biosynthesis of β-hydroxy-β-methyl glutaryl coenzyme A. *J. Amer. chem. Soc.* **79**, 5580.

A–B Rudney H. (1957) Biosynthesis of β-hydroxy-β-methyl glutaric acid. *J. biol. Chem.* **227**, 363.

A–B Ferguson J.J. and Rudney H. (1959) The biosynthesis of β-hydroxy-β-methyl glutaryl coenzyme A in yeast. *J. biol. Chem.* **234**, 1072.

A–B Eggerer H. (1965) Zum Mechanismus der biologischen Umwandlung von Citronensaüre. *Biochem. Z.* **343**, 111.

A–B Saver F. and Erfle J.D. (1966) Acetoacetate synthesis by guinea pig liver fractions. *J. biol. Chem.* **241**, 30.

D Siddiqi M.A. and Rodwell V. (1962) Bacterial metabolism of mevalonic acid. Conversion to acetoacetate. *Biochem. biophys. Res. Commun.* **8**, 110.

D Brodie J.D., Wasson G. and Porter J.W. (1963) The participation of malonyl coenzyme A in the biosynthesis of mevalonic acid. *J. biol. Chem.* **238**, 1294.

D Gosselin L. and Lynen F. (1964) Utilisation of 2-(^{14}C) mevalonic acid by a rat liver enzyme system in the absence of nicotinamide-adenine nucleotides. *Biochem. Z.* **340**, 186.

D Cornforth J.W. and Cornforth R.H. (1969) Chemistry of mevalonic acid. *Biochem. J.* **113**, 19P.

A–D Rudney H. (1959) The biosynthesis of β-hydroxy-β-methylglutaryl coenzyme A and its conversion to mevalonic acid. *Biosynthesis of Terpenes and Sterols.* p. 75.

B–C Brodie J.D. and Porter J.W. (1960) The synthesis of mevalonic acid by non-particulate avian and mammalian enzyme systems. *Biochim. biophys. Res. Commun.* **3**, 173.

B–C Bucher N.L.R., Overpath P. and Lynen F. (1960) β-Hydroxy-β-methyl glutaryl coenzyme A reductase; cleavage and condensing enzymes in relation to cholesterol formation in rat liver. *Biochim. biophys. Acta*, **40**, 491.

B–C Linn T.C. (1967) The demonstration and solubilisation of β-hydroxy-β-methyl glutaryl coenzyme A reductase from rat liver microsomes. *J. biol. Chem.* **242**, 984.

C–D Schlesinger M.J. and Coon M.J. (1961) Reduction of mevaldic acid to mevalonic acid by a partially purified enzyme from liver. *J. biol. Chem.* **236**, 2421.

D–E Tchen T.T. (1958) Mevalonic kinase: purification and properties. *J. biol. Chem.* **233**, 1100.

D–E Markley K. and Smallman E. (1961) Mevalonic kinase in rabbit liver. *Biochim. biophys. Acta*, **47**, 327.

D–E Hellig H. and Popjáck G. (1961) Studies on the biosynthesis of cholesterol. XIII. Phosphomevalonic kinase from liver. *J. Lipid. Res.* **2**, 235.

D–E Dorsey J.K. and Porter J.W. (1968) The inhibition of mevalonic kinase by geranyl and farnesyl pyrophosphates. *J. biol. Chem.* **243**, 4667.

E–F Henning U., Möslein E.M. and Lynen F. (1959) Biosynthesis of terpenes. Formation of 5-pyrophosphomevalonic acid by phosphomevalonic kinase. *Arch. biochem. Biophys.* **83**, 259.

E–F Bloch K., Chaykin S., Phillips A.H. and De Waard A. (1959) Mevalonic acid pyrophosphate and isopentenylpyrophosphate. *J. biol. Chem.* **234**, 2595.

E–F Levy H.R. and Popjáck G. (1960) Studies on the biosynthesis of cholesterol. Mevalonic kinase and phosphomevalonic kinase from liver. *Biochem J.* **75**, 417.

187 A–B *See* Bloch K., Chaykin S., Phillips A.H. and De Waard A. 186 E–F.

A–B & B–E Cornforth J.W., Cornforth R.H., Popjáck G. and Yengoyan L. (1966) Studies on the biosynthesis of cholesterol. XX. Steric course of decarboxylation of 5′-pyrophosphomevalonate and of the carbon to carbon bond formation in the biosynthesis of farnesyl pyrophosphate. *J. biol. Chem.* **241**, 3970.

A–E Rilling H.C. and Bloch K. (1959) On the mechanism of squalene biogenesis from mevalonic acid. *J. biol. Chem.* **234**, 1424.

C Goodman D.S., Dew S. and Popjáck G. (1961) Studies on the biosynthesis of cholesterol. Synthesis of allyl pyrophosphates, from mevalonate and then conversion into squalene with liver enzymes. *J. Lipid. Res.* **1**, 286.

B–C Agranoff B.W., Eggerer H., Henning U. and Lynen F. (1954) Isopentyl pyrophosphate isomerase. *J. Amer. chem. Soc.* **81**, 1254.

B–C Agranoff B.W., Eggerer H., Henning U. and Lynen F. (1960) Biosynthesis of terpenes. Isopentyl pyrophosphate isomerase. *J. biol. Chem.* **235**, 326.

B–C Holloway P.W. and Popjáck G. (1968) Isopentyl pyrophosphate isomerase from liver. *Biochem. J.* **106**, 835.

B–D Dorsey J.R., Dorsey J.A. and Porter J.W. (1966) The purification and properties of pig liver geranyl pyrophosphate synthetase. *J. biol. Chem.* **241**, 5353.

E Lynen F., Eggerer H., Henning U. and Kassel J. (1959) Farnesyl-pyrophosphat und 3-Methyl-Δ^3-butenyl-1-pyrophosphat, die biologischen Vorstufen des Squalens. *Angew. Chem.* **70**, 738.

E Donninger C. and Popjáck G. (1965) Studies on the biosynthesis of cholesterol. XVIII. The stereospecificity of mevaldate reductase and the biosynthesis of assymetrically labelled farnesyl pyrophosphate. *Proc. Roy. Soc.* B. **163**, 465.

188 GEN Coon M.J., Kupiecki F.P., Dekker E.E., Schlesinger M.J. and Del Campillo A. (1959) *Ciba Symposium on the Biosynthesis of Terpenes and Sterols.* Churchill, p. 62.

GEN Samuelsson B. and Goodman, De W.S. (1964) Stereochemistry at the centre of squalene during its biosynthesis from farnesyl pyrophosphate and subsequent conversion to cholesterol. *J. biol. Chem.* **239**, 98.

GEN Cornforth J.W., Cornforth R.H., Donninger C., Popjáck G., Ryback G. and Schroepfer G.J. (1965) Studies on the biosynthesis of cholesterol. XVII. The asymmetric synthesis of a symmetrical molecule. *Proc. Roy. Soc.* B. **163**, 436.

GEN Scallen T.J., Dean W.J. and Schuster M.W. (1968) Enzymatic conversion of squalene to cholesterol by an acetone powder of rat liver microsomes. *J. biol. Chem.* **243**, 5202.

A–B Krishna G., Feldbruegge D.H. and Porter J.W. (1962) The conversion of farnesyl pyrophosphate to squalene by soluble extracts of microsomes. *Biochem. biophys. Res. Commun.* **3**, 591.

A–B Cornforth J.W., Cornforth R.H., Donninger C., Popjáck G., Ryback G. and Schroepfer G.J., Jr., (1963) Stereospecific insertion of hydrogen atom into squalene from reduced nicotinamide-adenine dinucleotide. *Biochem. biophys. Res. Commun.* **11**, 129.

A–B Cornforth J.W., Cornforth R.H., Donninger C. and Popjáck G. (1965) Studies on the biosynthesis of cholesterol. XIX. Steric course of hydrogen eliminations and of C–C bond formations in squalene biosynthesis. *Proc. Roy. Soc.* B. **163**, 492.

A–B Rilling H.C. (1966) A new intermediate in the biosynthesis of squalene. *J. biol. Chem.* **241**, 3233.

A–B Popjáck G., Edmond J., Clifford K. and Williams V. (1969) Biosynthesis and structure of a new intermediate between farnesyl pyrophosphate and squalene. *J. biol. Chem.* **244**, 1897.

B–C Tchen T.T. and Bloch K. (1957) On the mechanism of enzymatic cyclization of squalene. *J. biol. Chem.* **226**, 931.

B–C Maudgal R.K., Tchen T.T. and Bloch K. (1958) 1-2-methyl shifts in the cyclisation of squalene to lanosterol. *J. Amer. chem. Soc.* **80**, 2589.

B–C Cornforth J.W., Cornforth R.H. Pelter A., Horning M.G. and Popjáck G. (1959) Studies on the biosynthesis of cholesterol. VII. Rearrangement of methyl groups during enzymic cyclisation of squalene. *Tetrahedron*, **5**, 311.

B–C Krishna G., Feldbruegge D.H. and Porter J.W. (1964) An enzyme-bound intermediate in the conversion of farnesyl pyrophosphate to squalene. *Biochem. biophys. Res. Commun.* **14**, 362.

B–C Corey E.J., Russey W.E. and de Montellano P.R.O. (1966) 2,3-Oxidosqualene, an intermediate in the biological synthesis of sterols from squalene. *J. Amer. chem. Soc.* **88**, 4750.

B–C Dean P.D.G., de Montellano P.R.O., Bloch K. and Corey E.J. (1967) A soluble 2,3-oxidosqualene sterol cyclase. *J. biol. Chem.* **242**, 3014.

B–C Yamamoto S. and Bloch K. (1969) On the enzymes catalysing the transformation of squalene to lanosterol. *Biochem. J.* **113**, 19P.

B–C Yamamoto S., Lin K. and Bloch K. (1969) Some properties of the microsomal 2,3-oxidosqualene sterol cyclase. *Proc. Nat. Acad. Sci. U.S.* **63**, 110.

189 GEN Olson J.A., Jr., Linoberg M. and Bloch K. (1957) On the demethylation of lanosterol to cholesterol. *J. biol. Chem.* **226**, 941.

GEN Danielsson H. (1963) Present state of research on catabolism and excretion of cholesterol. *Advanc. Lipid. Res.* **1**, 335.

GEN Goodman D.S., Avigan J. and Steinberg D. (1963) Studies on

cholesterol biosynthesis. Reduction of lanosterol to 24,25-dihydrolanosterol by rat liver homogenates. *J. biol. Chem.* **238**, 1283.
GEN Chesterton C.J. (1968) Distribution of cholesterol precursors and other lipids among rat liver intracellular structures. *J. biol. Chem.* **243**, 1147.
A–B Gautschi F. and Bloch K. (1958) Synthesis of isomeric 4,4-dimethyl cholesterols and identification of a lanosterol metabolite. *J. biol. Chem.* **233**, 1343.
C–D Lindberg M., Gautschi F. and Bloch K. (1963) Ketonic intermediates in the demethylation of lanosterol. *J. biol. Chem.* **238**, 1661.
D–E Frantz I.D., Jr., Sanghvi A.T. and Schroepfer G.J., Jr. (1964) Irreversibility of the biogenetic sequence from Δ^7-cholesten-3β-ol through $\Delta^{5,7}$-cholestadien-3β-ol to cholesterol. *J. biol. Chem.* **239**, 1007.
D–E Dempsey M.E., Seaton J.D., Schroepfer G.J. and Trockman E.W. (1964) The intermediary role of $\Delta^{5,7}$-cholestadien-3β-ol in cholesterol biosynthesis. *J. biol. Chem.* **239**, 1381.
E–F Avigan J., Goodman De W.S. and Steinberg D. (1963) Studies on cholesterol biosynthesis. IV. Reduction of lanosterol to 24,25-dihydrolanosterol by rat liver homogenates. *J. biol. Chem.* **238**, 1283.

190

A–B Olson J.A. and Anfinsen C.B. (1952) The crystallisation and characterisation of L-glutamic acid dehydrogenase. *J. biol. Chem.* **197**, 67.
A–B Barban S. (1954) Studies on the metabolism of the treponemata. I. Amino acid metabolism. *J. Bacteriol.* **68**, 493.
A–B Vallee B.L., Adelstein S.J. and Olson J.A. (1955) Glutamic dehydrogenase of beef liver, a zinc metalloenzyme. *J. Amer. chem. Soc.* **77**, 5196.
A–B Snoke J.E. (1956) Chicken liver glutamic dehydrogenase. *J. biol. Chem.* **223**, 271.
A–B Bulen W.A. (1956) The isolation and characterisation of glutamic dehydrogenase from corn leaves. *Arch. biochem. Biophys.* **62**, 173.
D–G Cohen P.P. (1951) Transaminases. *The Enzymes*, 1st ed., vol. 1. Academic Press, New York, p. 1040.
D–G Wilson D.G., King K.W. and Burris R.H. (1954) Transamination reactions in plants. *J. biol. Chem.* **208**, 863.
D–G Cohen P.P. (1955) Estimation of animal transaminases. *Methods in Enzymology*, vol. 2. p. 178.
D–G Jenkins W.T., Yphantis D.A. and Sizer I.W. (1959) Glutamic aspartic transaminase. I. Assay, purification and general properties. *J. biol. Chem.* **234**, 51.
D–G Ellis R.J. and Davies D.D. (1961) Glutamic-oxaloacetic transaminase of cauliflower. I. Purification and specificity. *Biochem. J.* **78**, 615.
K–L Elliot W.H. (1953) Isolation of glutamine synthetase and glutamotransferase from green peas. *J. biol. Chem.* **201**, 661.
K–L Lajtaa A., Mela P. and Waelsh H. (1953) Manganese dependent glutamotransferase. *J. biol. Chem.* **205**, 553.
K–L Levintow L. and Meister A. (1954) Specificity and mechanism of glutamine synthesis system. *Fed. Proc.* **13**, 251.
K–L *See* Huennekens F.M. and Osborn M.J. (1959) p. 106.
K–L Meister A. (1957) Glutamine synthesis. *Biochemistry of the Amino Acids*, p. 446.
K–L Boyer P.D., Mills R.C. and Fromm H.J. (1959) Hypothesis for and some kinetic studies with glutamine synthetase and acetate thiokinase. *Arch. biochem. Biophys.* **81**, 249.
K–L Pamiljans V., Krishnaswamy P.R. and Meister A. (1962) Studies on the mechanism of glutamine synthesis; isolation and properties of the enzyme from sheep brain. *Biochemistry*, **1**, 153.
K–L Krishnaswamy P.R., Pamiljans V. & Meister A. (1962) Studies on the mechanism of glutamine synthesis: evidence for the formation of enzyme bound activated glutamic acid. *J. biol. Chem.* **237**, 2932.

K–L *See* Friedkin M. (1963) p. 107.

191 B Bright H.S. and Ingraham L.L. (1960) The preparation of crystalline β-methyl aspartate. *Biochim. biophys. Acta*, **44**, 586.

A–B Barker H.A., Smyth R.D., Wawszkiewicz E.J., Lee M.N. and Wilson R.M. (1958) Enzymic preparation and characterisation of an α-L-β-methyl-aspartic acid. *Arch. biochem. Biophys.* **78**, 468.

A–B Weissbach H., Toohey J. and Barker H.A. (1959) Isolation and properties of B_{12} coenzymes containing benzimidazole or dimethyl benzimidazole. *Proc. Nat. Acad. Sci. U.S.* **45**, 521.

A–B Hogenkamp H.P.C. (1968) Enzymatic reactions involving corrinoids. *Ann. Rev. Biochem.* **37**, 225.

A–C Munch-Petersen A. and Barker H.A. (1958) The origin of the methyl group in mesaconate formed from glutamate by extracts of *Clostridium tetanomorphum*. *J. biol. Chem.* **230**, 649.

B–C Barker H.A., Smyth R.D., Wilson R.M. and Weissbach H. (1959) The purification and properties of β-methyl aspartase. *J. biol. Chem.* **234**, 320.

G–K Abramsky T. and Shemin D. (1965) The formation of isoleucine from β-methylaspartic acid in *Escherichia coli*. *J. biol. Chem.* **240**, 2971.

192 AB–FG Ellis R.J. and Davies D.D. (1961) Glutamic-oxalacetate transaminase of cauliflower. I. Purification and specificity. *Biochem. J.* **78**, 615.

AB–FG Goldstone A. and Adams E. (1962) Metabolism of γ-hydroxyglutamic acid. I. Conversion to α-hydroxy-γ-ketoglutarate by purified glutamic : aspartic transaminase of rat liver. *J. biol. Chem.* **237**, 3476.

AB–FG Cooper R.A. and Kornberg H.L. (1965) Some characteristics of the utilisation of L-glutamic acid by Ehrlich ascites tumour cells. *Biochim. biophys. Acta*, **104**, 618.

AB–FG Bhargava M.M. and Sreenivasan A. (1968) Two forms of aspartate aminotransferase in rat liver and kidney mitochondria. *Biochem. J.* **108**, 619.

J Dixon R.O.D. and Fowden L. (1961) γ-Aminobutyric acid metabolism in plants. 2. Metabolism in higher plants. *Ann. Botany (Lond.)* **25**, 313.

H–J Roberts E. and Frankel S. (1951) Further studies of glutamic acid decarboxylase in brain. *J. biol. Chem.* **190**, 505.

H–J Davison A.N. (1956) Amino acid decarboxylases in rat liver and brain. *Biochim. biophys. Acta*, **19**, 66.

H–J Shukuya R. and Schwert G.W. (1960) Glutamic acid decarboxylase. I. Isolation procedures and properties of the enzyme. *J. biol. Chem.* **235**, 1649.

J–K Scott E.M. and Jakoby W.B. (1959) Soluble γ-aminobutyric-glutamic transaminase from *Pseudomonas fluorescens*. *J. biol. Chem.* **234**, 932.

K–L Jakoby W.B. and Scott E.M. (1959) Aldehyde oxidation. III. Succinic semialdehyde dehydrogenase. *J. biol. Chem.* **234**, 937.

K–L Nirenberg M.W. and Jakoby W.B. (1960) Enzymatic utilisation of γ-hydroxy butyric acid. *J. biol. Chem.* **235**, 954.

K–L Albers R.W. and Koval G.J. (1961) Succinic semialdehyde dehydrogenase: purification and properties of the enzyme from monkey brain. *Biochim. biophys. Acta*, **52**, 29.

193 GEN Coon M.J. and Robinson W.G. (1958) Amino acid metabolism (proline and hydroxy-proline). *Ann. Rev. Biochem.* **27**, 565.

A–D Strecker H.J. (1960) The interconversion of glutamic acid and proline. Δ'-Pyrroline-5-carboxylic acid dehydrogenase. *J. biol. Chem.* **235**, 3218.

A–E Johnson A.B. and Strecker H.J. (1962) The interconversion of glutamic acid and proline. IV. The oxidation of proline by rat liver mitochondria. *J. biol. Chem.* **237**, 1876.

C Yura T. and Vogel H.J. (1957) On the glutamic family and lysine in *neurospora*. Enzymic formation of glutamic γ-semialdehyde and α-aminoadipic γ-semialdehyde from penta-and hexa-homoserine. *Biochim. biophys. Acta*, **24**, 648.

B–C Fincham J.R.S. (1953) Ornithine transaminase in *neurospora* and its relation to the biosynthesis of proline. *Biochem. J.* **53**, 313.

B–C Meister A. (1954) Enzymic transamination reactions involving arginine and ornithine. *J. biol. Chem.* **206**, 587.

B–E Taggart J.V. and Krakaur R.B. (1949) Studies on the cyclophorase system. The oxidation of proline and hydroxyproline. *J. biol. Chem.* **177**, 641.

C–D Vogel H.J. and Davis B.D. (1952) Glutamic γ-semialdehyde and Δ-pyrroline-5-carboxylic acid, intermediates in the biosynthesis of proline. *J. Amer. chem. Soc.* **74**, 109.

D–E Yura T. and Vogel H.J. (1955) On the biosynthesis of proline in *Neurospora crassa:* enzymic reduction of Δ'-pyrroline-5-carboxylase. *Biochim. biophys. Acta*, **17**, 582.

D–E Meister A., Radhakrishnan A.N. and Buckley S.D. (1957) Enzymatic synthesis of L-pipecolic acid and L-proline. *J. biol. Chem.* **229**, 789.

D–E Yura T. and Vogel H.J. (1959) Pyrroline-5-carboxylate reductase of *Neurospora crassa*. *J. biol. Chem.* **234**, 335.

D–E Adams E. and Goldstone A. (1960) Hydroxyproline metabolism. *J. biol. Chem.* **235**, 3499.

D–E Peisach J. and Strecker H.J. (1962) The interconversion of glutamic acid and proline. The reduction of Δ'-pyrroline-5-carboxylic acid to proline. *J. biol. Chem.* **237**, 2255.

E–F Wolf G. and Berger C.R.A. (1958) The metabolism of hydroxyproline in the intact rat. Incorporation of hydroxy proline into protein and urinary metabolites. *J. biol. Chem.* **230**, 231.

194 GEN Adams E. (1959) Hydroxyproline metabolism. *J. biol. Chem.* **234**, 2073.

GEN Singh R.M.M. and Adams E. (1964) Alpha-ketoglutaric semialdehyde: a metabolic intermediate. *Science*, **144**, 67.

A–C Radhakrishnan A.N. and Meister A. (1957) Conversion of hydroxyproline to pyrrole-2-carboxylic acid. *J. biol. Chem.* **226**, 559.

D–E Adams E. and Rosso G. (1967) α-Ketoglutaric semialdehyde dehydrogenase of *Pseudomonas*. Properties of the purified enzyme induced by hydroxyproline and of the glucarate-induced and constitutive enzymes. *J. biol. Chem.* **242**, 1802.

195 A–D Adams E., Friedman R. and Goldstone A. (1958) Animal metabolism of hydroxyproline: isolation and enzymic reactions of γ-hydroxyglutamic semialdehyde. *Biochim. biophys. Acta*, **30**, 212.

B Adams E. and Goldstone A. (1960) Hydroxy proline metabolism. Enzymic preparation and properties of Δ'-pyrroline-3-hydroxy-5-carboxylic acid. *J. biol. Chem.* **235**, 3504.

D Goldstone A. and Adams E. (1964) Further metabolic reactions of α-hydroxyglutamate: amidation to γ-hydroxyglutamine; possible reduction to hydroxyproline. *Biochem. biophys. Res. Commun.* **16**, 71.

D–E Goldstone A. and Adams E. (1962) Metabolism of γ-hydroxyglutaric acid. Conversion of α-hydroxy-γ-ketoglutarate by purified glutamic-aspartic transaminase of rat liver. *J. biol. Chem.* **237**, 3476.

D–GH Kuratomi K. and Fukunaga K. (1960) A new enzymic reaction concerned in the metabolism of γ-hydroxyglutamic acid. *Biochim. biophys. Acta*, **43**, 562.

F Wolf G. and Berger C.R.A. (1958) The metabolism of hydroxyproline in the intact rat. Incorporation of hydroxyproline into protein and urinary metabolism. *J. biol. Chem.* **230**, 231.

E–FG Dekker E.E. (1960) Enzymic formation of glyoxylic acid from γ-hydroxyglutamic acid. *Biochim. biophys. Acta*, **40**, 174.

E–FG Rosso R.G. and Adams E. (1967) 4-Hydroxy-2-ketoglutarate aldolase of rat liver. Purification, binding of substrates and kinetic properties. *J. biol. Chem.* **242**, 5524.

E–GH Maitra U. and Dekker E.E. (1961) Enzymic reactions in mammalian metabolism of γ-hydroxyglutamic acid. *Biochim. biophys. Acta*, **51**, 416.

E–GH Bouthillier L.P., Binette Y. and Pouliot G. (1961) Transformation de l'acide γ-hydroxyglutamique en alanine et en acide glyoxylique. *Can. J. Biochem. Physiol.* **39**, 1595.

E–GH Dekker E.E. and Maitra U. (1962) Conversion of γ-hydroglutamate to glyoxalate and alanine; purification and properties of the enzyme system. *J. biol. Chem.* **237**, 2218.

196 GEN Vogel H.J., Abelson P.H. and Bolton E.T. (1953) On ornithine and proline synthesis in *E. coli*. *Biochim. biophys. Acta*, **11**, 584.

B Vyas S. and Maas W.K. (1963) Feedback inhibition of acetylglutamate synthetase by arginine in *E. coli*. *Arch. biochem. Biophys.* **100**, 542.

B–D Baich A. and Vogel H.J. (1962) *N*-Acetyl-γ-glutamokinase and *N*-acetyl glutamic γ-semialdehyde dehydrogenase: repressible enzymes of arginine synthesis in *Escherichia coli*. *Biochem. biophys. Res. Commun.* **7**, 491.

E–F Vogel H.J. and Bonner D.M. (1956) Acetylornithinase of *Escherichia coli*: partial purification and some properties. *J. biol. Chem.* **218**, 97.

G–J Vogel H.J. (1955) *Amino Acid Metabolism*. (Eds. W.D. McElroy and B. Glass.) Johns Hopkins Press, Baltimore, U.S., p. 335.

H–J Fincham J.R.S. (1953) Ornithine transaminase in *Neurospora* and its relation to the biosynthesis of proline. *Biochem. J.* **53**, 313.

197 A–B Lowenstein J.M. and Cohen P.P. (1956) Studies on the biosynthesis of carbamylaspartic acid. *J. biol. Chem.* **220**, 57.

A–B Reichard P. (1957) Ornithine carbamyl transferase from rat liver. *Acta chem. Scand.* **11**, 523.

A–B Shepherdson M. and Pardee A.B. (1960) Production and crystallisation of aspartate transcarbamylase. *J. biol. Chem.* **235**, 3233.

A–B Bernlohr R.W. (1966) Ornithine transcarbamylase enzymes: occurrence in *Bacillus licheniformis*. *Science* **152**, 87.

A–B Stalon V., Ramos F., Piérard A. and Wiame J.M. (1967) The occurrence of a catabolic and an anabolic ornithine carbamoyltransferase in *Pseudomonas*. *Biochim. biophys. Acta*, **139**, 91.

B–C Ratner S. (1955) Enzymatic synthesis of arginine. *Methods in Enzymology*, vol. 2. p. 356.

B–C Schuegraf A., Warner R. and Ratner S. (1960) Free energy changes of the arginosuccinate synthetase reaction and of the hydrolysis of the inner pyrophosphate bond of adenosine triphosphate. *J. biol. Chem.* **235**, 3597.

B–C Rochovansky O. and Ratner S. (1961) Biosynthesis of urea. Further studies of arginosuccinate synthetase reaction. *J. biol. Chem.* **236**, 2254.

B–D Ratner S. (1962) Nitrogen transfer from aspartic acid in the formation of amide, amidine and guanido groups. *The Enzymes*, 2nd ed., vol. 6. Academic Press, New York, p. 495.

C–D Walker J.B. (1953) An enzymatic reaction between canavanine and fumarate. *J. biol. Chem.* **204**, 139.

C–D Ratner S., Anslow W.P. and Petrack B. (1953) Biosynthesis of urea. VI. Enzymatic cleavage of argininosuccinic acid to arginine and fumaric acid. *J. biol. Chem.* **204**, 115.

C–D Walker J.B. (1955) Biosynthesis of canavanosuccinic acid from canavanine and fumarate in kidney. *Arch. biochem. Biophys.* **59**, 233.

198 GEN Cantoni G.L. and Durell J. (1957) Enzymic mechanism of creatine synthesis. *J. biol. Chem.* **225**, 1033.

A–B Oginsky E.L. and Gehrig R.F. (1952) The arginine dihydrolase system of *Streptococcus faecalis*. *J. biol. Chem.* **198**, 799.

A–B Ratner S. (1954) Urea synthesis and metabolism of arginine and citrulline. *Advanc. Enzymol.* **15**, 319.

A–B Petrack B., Sullivan L. and Ratner S. (1957) Behaviour of purified arginine desiminase from *S. faecalis*. *Arch. biochim. Biophys.* **69**, 186.

A–C See 199 AC–G.

B–C Knivett V.A. (1954) Phosphorylation coupled with anaerobic breakdown of citrulline. *Biochem. J.* **56**, 602.

B–C Jones M.E., Spector L. and Lipmann F. (1955) Carbamyl phosphate, the carbamyl donor in enzymatic citrulline synthesis. *J. Amer. chem. Soc.* **77**, 819.

B–C Burnett G.H. and Cohen P.P. (1957) Study of carbamyl phosphate ornithine transcarbamylase. *J. biol. Chem.* **229**, 337.

B–C Reichard P. (1957) Ornithine carbamyl transferase from rat liver. *Acta chem. Scand.* **11**, 523.

B–C Ravel J.M., Grona M.L., Humphreys J.S. and Shive W. (1959) Properties and biotin content of purified preparations of the ornithine-citrulline enzyme of *Streptococcus lactis*. *J. biol. Chem.* **234**, 1452.

E–G Ratner S. and Rochovansky O. (1956) Biosynthesis of guanidinoacetic acid. Purification and properties of transamidinase. *Arch. biochem. Biophys.* **63**, 277.

E–G Walker J.B. (1957) Studies on the mechanism of action of kidney transaminase. *J. biol. Chem.* **224**, 57.

G–J Cantoni G.L. and Scarano E. (1954) The formation of S-adenosylhomocysteine in enzymatic transmethylation reactions. *J. Amer. chem. Soc.* **76**, 4744.

G–J Cantoni G.L. and Vignos P.J. (1954) Enzymatic mechanisms of creatine synthesis. *J. biol. Chem.* **209**, 647.

J–K Kuby S.A., Noda L. and Lardy H.A. (1954) Adenosine triphosphate : creatine transphosphorylase. Isolation of the crystalline enzyme from rabbit muscle. *J. biol. Chem.* **209**, 191.

J–K Corri O., Traverso-Corri A., Lagarrigue M. and Marcus F. (1958) Enzymic phosphorylation of creatine by 1:3-diphosphoglyceric acid. *Biochem. J.* **70**, 633.

K–L Borsook H. and Dubnoff J.W. (1947) The hydrolysis of phosphocreatine and the origin of urinary creatinine. *J. biol. Chem.* **168**, 493.

K–L Van Pilsum J.F. and Hiller B. (1959) On the proposed origin of creatinine from creatine phosphate. *Arch. biochem. Biophys.* **85**, 483.

199 GEN Cohen P.P. and Brown G.N., Jr. (1960) Ammonia metabolism and urea biosynthesis. *Comparative Biochemistry*, vol. 2. p. 161.

GEN Baker J.E. and Thompson J.F. (1962) Metabolism of urea and ornithine cycle intermediates by nitrogen-starved cells of *Chlorella vulgaris*. *Plant Physiol.* **37**, 618.

A–B Marshall M., Metzenberg R.L. and Cohen P.P. (1958) Purification of carbamyl phosphate synthetase from frog liver. *J. biol. Chem.* **233**, 102.

A–B Jones M.E. and Spector L. (1960) The pathway of carbonate in the biosynthesis of carbamyl phosphate. *J. biol. Chem.* **235**, 2897.

A–B Duffield P.H., Kalman S.M. and Brauman J.I. (1969) Fixation of carbon dioxide by carbamyl phosphate synthetase of *Escherichia coli*. Evidence for a reversibly formed intermediate. *Biochim. biophys. Acta*, **171**, 189.

A–B O'Neal T.D. and Naylor A.W. (1969) Partial purification and properties of carbamoyl phosphate synthetase of Alaska pea. *Biochem. J.* **113**, 271.

AC–G Greenberg D.M., Bagot A.E. and Roholt O.A. (1956) Liver arginase: properties of highly purified arginase. *Arch. biochem. Biophys.* **62**, 446.

AC–G Bach S.J. and Killip J.D. (1961) Studies on the purification and the kinetic properties of arginase from beef, sheep and horse liver. *Biochim. biophys. Acta*, **47**, 336.

AC–G Cabello J., Basilio C. and Prajoux V. (1961) Kinetic properties of erythrocyte and liver arginase. *Biochim. biophys. Acta*, **48**, 148.

J–HG See 197 C–D.

200 C–D Erkama J. and Virtanen A.I. (1951) Aspartase. *The Enzymes*, 1st ed., vol. 1. Academic Press, New York, p. 1244.

C–D Williams W.J., Litwin J. and Thorne C.B. (1955) Further studies on the biosynthesis of γ-glutamyl peptides by transfer reactions. *J. Biol. Chem.* **212**, 427.

C–D Williams V.R. and Mcintyre R.T. (1955) Preparation and partial purification of the aspartase of *Bacterium cadaveris*. *J. biol. Chem* **217**, 467.

E–F Ravel J.M., Norton S.J., Humphreys J.S. and Shive W. (1962) Asparagine biosynthesis in *Lactobacillus arabinosus* and its control by asparagine through enzyme inhibition and repression. *J. biol. Chem.* **237**, 2845.

E–F Cedar H. and Schwartz J.H. (1969) The asparagine synthetase of *E. coli*. I. Biosynthetic role of the enzyme. Purification and characterisation of reaction products. II. Studies on mechanism. *J. biol. Chem.* **244**, 4112, 4122.

G–H David W.E. and Lichstein H.E. (1950) Aspartic acid decarboxylase in bacteria. *Proc. Soc. Expl. Biol. N.Y.* **73**, 216.

G–H Gale E.F. (1953) Amino acid decarboxylases. *Brit. med. Bull.* **9**, 135.

201 A–B Black S. and Gray N.M. (1953) Enzymic phosphorylation of L-aspartate. *J. Amer. chem. Soc.* **75**, 2271.

A–B Black S and Wright N.G. (1955) β-Aspartokinase and β-aspartyl phosphate. *J. biol. Chem.* **213**, 27.

D Teas H.J., Horowitz N.H. and Fling M. (1961) Homoserine as a precursor of threonine and methionine in *Neurospora*. *J. biol. Chem.* **172**, 651.

A–D Black S. and Wright N.G. (1953) Enzymatic reduction of β-aspartyl phosphate to homoserine. *J. Amer. chem. Soc.*, **75**, 5766.

A–F Cohen G.N., Nisman B., Hirsch M.L. and Wiesendanger S.B. (1954) Physiologie cellulaire: précisions sur la synthèse de L-thréonine á partir d'acide L-aspartique par des extraits de *Escherichia coli*. *Compt. Rend. Acad. Sci.* **238**, 1746.

C–D Yura T. and Vogel H.J. (1959) Anhydroxy-α-amino acid dehydrogenase of *Neurospora crassa*: partial purification and some properties. *J. biol. Chem.* **234**, 339.

D–E Watanabe Y., Konishi S. and Shimuar K. (1957) Biosynthesis of threonine from homoserine. IV. Homoserine kinase. *J. Biochem.* (*Tokyo*), **44**, 299.

E–F Cohen G.N. and Hirsch M.L. (1954) Threonine synthetase. A system synthesising L-threonine from L-homoserine. *J. Bact.* **67**, 182.

E–F Flavin M. and Slaughter C. (1960) Purification and properties of threonine synthetase of *neurospora*. Threonine synthetase mechanism: studies with isotopic hydrogen. *J. biol. Chem.* **235**, 1103, 1112.

*E–F Flavin M. and Kono T. (1960) Threonine synthetase mechanism: studies with isotopic water. *J. biol. Chem.* **235**, 1109.

E–F Watanabe Y. and Shimura K. (1960) Mechanism of L-threonine synthesis from *O*-phosphohomoserine. *J. Biochem.* (*Tokyo*), **47**, 266.

GH–J Lin S.-C.C. and Greenberg D.M. (1954) Enzymatic breakdown of threonine by threonine aldolase. *J. gen. Physiol.* **38**, 181.

GH–J Karasek M.A. and Greenberg D.M. (1957) Studies on the properties of threonine aldolase. *J. biol. Chem.* **227**, 191.

202 A–B Saure F.W. and Greenberg D.M. (1956) Purification and properties of serine and threonine dehydrase. *J. biol. Chem.* **220**, 787.

A–B Nishimura J.S. and Greenberg D.M. (1961) Purification and properties of L-threonine dehydrase of sheep-liver. *J. biol. Chem.* **236**, 2684.

A–B Friedland R.A. and Avery E.H. (1964) Studies on threonine and serine dehydrase. *J. biol. Chem.* **239**, 3357.

A–B Leitzmann C. and Bernlohr R.W. (1968) Threonine dehydratase of *Bacillus licheniformis*. Purification and properties. *Biochim. biophys. Acta*, **151**, 449.

A–B Burns R.O. and Zarlengo M.H. (1968) Threonine deaminase from *Salmonella typhimurium*. I. Purification and properties. *J. biol. Chem.* **243**, 178.

A–B Rabinowitz K.W., Shada J.D. and Wood W.A. (1968) The mechanism of action of 5′-adenylic acid-activated threonine dehydrase. *J. biol. Chem.* **243**, 3214.

C–D Hartshorne D. and Greenberg D.M. (1964) Studies on liver threonine dehydrogenase. *Arch. biochem. Biophys.* **105**, 173.

C–E Green M.L. and Elliott W.H. (1964) The enzymic formation of aminoacetone from threonine and its further metabolism. *Biochem. J.* **92**, 537.

E Elliott W.H. (1958) A new threonine metabolite. *Biochim. biophys. Acta*, **29**, 446.

E Elliott W.H. (1959) Aminoacetone: its isolation and role in metabolism. *Nature*, **183**, 1051.

E Elliott W.H. (1960) Aminoacetone formation by *Staphylococcus aureus*. *Biochem. J.* **74**, 90.

*E Neuberger R. and Tait G.H. (1962) Production of amine acetone by *Rhodopseudomonas spheroides*. *Biochem. J.* **84**, 317.

E Urata G. and Granck S. (1963) Biosynthesis of α-aminoketones and the metabolism of aminoacetone. *J. biol. Chem.* **238**, 811.

C–E Elliott W.H. (1958) A new threonine metabolite. *Biochim. biophys. Acta*, **29**, 446.

C–E Neuberger A. and Tait G.H. (1960) The enzymic conversion of threonine to aminoacetone. *Biochim. biophys. Acta*, **41**, 164.

F–G Crook E.M. and Law K. (1952) Glyoxylase: the role of the components. *Biochem. J.* **52**, 492.

203 A–B Flavin M., Delavier-Klutchko C. and Slaughter C. (1964) Succinic ester and amide of homoserine: some spontaneous and enzymatic reactions. *Science*, **143**, 50.

A–B Rowbury R.J. and Woods D.D. (1964) *O*-Succinylhomoserine as an intermediate in the synthesis of cystathionine by *E. coli*. *J. gen. Microbiol.* **36**, 341.

B–C Kaplan M.M. and Flavin M. (1966) Cystathionine γ-synthetase of *Salmonella*. Catalytic properties of a new enzyme in bacterial methionine biosynthesis.
Structural properties of a new enzyme in bacterial methionine biosynthesis. *J. biol. Chem.* **241**, 4463, 5781.

C–D Wijesundera S. and Woods D.D. (1962) The catabolism of cystathionine by *E. coli*. *J. gen. Microbiol.* **29**, 353.

C–D Delavier-Klutchko C. and Flavin M. (1965) Role of a bacterial cystathionine β-cleavage enzyme in disulphide decomposition. *Biochim. biophys. Acta*, **99**, 375.

D–E Durell J., Anderson D.G. and Cantoni G.L. (1957) The synthesis of methionine by enzymic transmethylation. Purification and properties of thetin homocysteine methylpherase. *Biochim. biophys. Acta*, **26**, 270.

D–E Maw G.A. (1958) Thetin-homocysteine transmethylase. Some further characteristics of the enzyme from rat liver. *Biochem J.* **70**, 168.

D–E Klee W.A., Richards H.H. and Cantoni G.L. (1961) The synthesis of methionine by enzymic transmethylation. Existence of two separate homocysteine methylpherases of mammalian liver. *Biochim. biophys. Acta*, **54**, 157.

D–E Larabee A.R., Rosenthal S., Cathou R.E. and Buchanan J.M. (1963)

Enzymatic synthesis of the methyl group of methionine. *J. biol. Chem.* **238**, 1025.

D–E Woods D.D. (1963) Vitamin B_{12} in the synthesis of methionine. *Biochem. J.* **89**, 4P.

D–E Weissbach H., Peterhofsky A., Redfield B.G. and Dickerman H. (1963) Studies on the terminal reaction in the biosynthesis of methionine. *J. biol. Chem.* **238**, 3318.

D–E *See* Kisliuk R.L. (1963) 107 F–G.

D–E Buchanan J.M., Ecford H.L., Loughlin R.E., McDougal B.M. and Rosenthal S. (1964) The role of vitamin B_{12} in methyl transfer to homocysteine. *Ann. N.Y. Acad. Sci.* **112**, 756.

D–E Taylor R.T. and Weissbach H. (1967) N^5-Methyltetrahydrofolate-homocysteine transmethylase. *J. biol. Chem.* **242**, 1502.

D–E Hogenkamp H.P.C. (1968) Enzymatic reactions involving corrinoids. *Ann. Rev. Biochem.* **37**, 225.

204 B Cantoni G.L. (1953) *S*-adenosylmethionine; a new intermediate formed enzymically from L-methionine and ATP. *J. biol. Chem.* **204**, 403.

A–B Cantoni G.L. and Durell J. (1957) Activation of methionine for transmethylation. The methionine activating enzyme. *J. biol. Chem.* **225**, 1033.

A–B Mudd S.H. and Cantoni G.L. (1958) Activation of methionine for transmethylation. The methionine activating enzyme of bakers yeast. *J. biol. Chem.* **231**, 481.

C Salvatore F. and Zappia V. (1968) Quantitative analysis of *S*-adenosyl homocysteine in liver. *Biochim. biophys. Acta*, **158**, 461.

B–C Shapiro S.K. and Yphantis D. A. (1959) Assay of *S*-methylmethionine and *S*-adenosylmethionine homocysteine transmethylases. *Biochim. biophys. Acta*, **36**, 241.

C–D de la Haba G. and Cantoni G.L. (1959) The enzymatic synthesis of *S*-adenosyl-L-homocysteine from adenosine and homocysteine. *J. biol. Chem.* **234**, 603.

D–E Yanofsky C. and Reissig J.L. (1953) L-Serine dehydrase of *Neurospora*. *J. biol. Chem.* **202**, 567.

D–E Selim A.S.H. and Greenberg D.M. (1959) An enzyme that synthesises cystathione and deaminates L-serine. *J. biol. Chem.* **234**, 1474.

E–FG Matsuo Y. and Greenberg D.M. (1958) A crystalline enzyme that cleaves homoserine and cystathione. Isolation procedure and some physiochemical properties. *J. biol. Chem.* **230**, 545.

E–FG Matsuo Y. and Greenberg D.M. (1959) A crystalline enzyme that cleaves homoserine and cystathione. Mechanism of action, reversibility and substrate specificity. *J. biol. Chem.* **234**, 516.

205 GEN Du Vigneaud V., Chandler J.P., Chon M. and Brown G.B. (1940) The transfer of the methyl group from methionine to choline and creatine. *J. biol. Chem.* **134**, 787.

GEN Du Vigneaud V., Chandler J.P., Cohn M. and Brown G.B. (1941) The utilisation of the methyl group of methionine in the biological synthesis of choline and creatine. *J. biol. Chem.* **140**, 625.

A–B Taylor R.T. and Weissbach H. (1966) Role of *S*-adenosyl methionine in vitamin B_{12} dependant methionine synthesis. *J. biol. Chem.* **241**, 3641.

206 GEN Weinhouse S., McElroy W.D. and Glass B. (Eds.) (1955) The synthesis and degradation of glycine. *Amino Acid Metabolism*. p. 637.

GEN Smith R.A., Shuster C.W., Zimmermann S. and Gunsalus I.C. (1956) Serine synthesis in *Escherichia coli*. *Bact. Proc.* **107**.

A–B Sugimo E. and Pizer L.I. (1968) The mechanism of end product inhibition of serine biosynthesis. I. Purification and kinetics of phosphoglycerate dehydrogenase. *J. biol. Chem.* **243**, 2081.

A–F Umbarger H.E. and Umbarger M.A. (1962) The biosynthetic pathway of serine in *Salmonella typhimurium*. *Biochim. biophys. Acta*, **62**, 193.

C–A Black S. and Wright N.G. (1956) Enzymatic formation of glyceryl and phosphoglyceryl methylthiol esters. *J. biol. Chem.* **221**, 171.

C–A Ichihara A. and Greenberg D.M. (1957) Studies on the purification and properties of D-glyceric acid kinase of liver. *J. biol. Chem.* **225**, 949.

C–E Stafford H.A., Magaldi A. and Vennesland B. (1954) The enzymatic reduction of hydroxypyruvic acid to D-glyceric acid in higher plants. *J. biol. Chem.* **207**, 621.

F Mitoma C. and Greenberg D.M. (1952) Studies on the mechanism of biosynthesis of serine. *J. biol. Chem.* **196**, 599.

C–F Ichihari A. and Greenberg D.M. (1957) Further studies on the pathway of serine formation from carbohydrate. *J. biol. Chem.* **224**, 331.

D–F Neuhaus F.C. and Byrne W.L. (1959) Metabolism of phosphoserine. Purification and properties of O-phosphoserine phosphatase. *J. biol. Chem.* **234**, 113.

D–F Borkenhagen L.F. and Kennedy E.P. (1959) The enzymatic exchange of L-serine with O-phospho-L-serine catalysed by a specific phosphatase. *J. biol. Chem.* **234**, 849.

F–G Kisliuk R.L. and Sakami W. (1955) A study of the mechanism of serine biosynthesis. *J. biol. Chem.* **214**, 47.

F–G Huennekens F.M., Hatefi Y. and Kay L. (1957) Manometric assay and cofactor requirements for serine hydroxymethylase. *J. biol. Chem.* **224**, 435.

F–G *See* Rabinowitz J.C. (1960) p. 106.

F–G Blakeley R.L. (1960) A spectrophotometric study of the reaction catalysed by serine transhydroxymethylase. *Biochem. J.* **77**, 459.

F–G Schirch L. and Mason N. (1962) Serine transhydroxymethylase: spectral properties of the E-bound pyridoxal-5-phosphate. *J. biol. Chem.* **237**, 2578.

F–G Schirch L. and Mason N. (1963) Serine transhydroxymethylase. *J. biol. Chem.* **238**, 1032.

F–G Nakano Y., Fujioka M. and Wada H. (1968) Studies on serine hydroxymethylase isoenzymes from rat liver. *Biochim. biophys. Acta*, **159**, 19.

F–G Schirch L. and Gross T. (1968) Serine transhydroxymethylase. Identification as the threonine and allothreonine aldolases. *J. biol. Chem.* **243**, 5651.

207 A–B *See* 206 F–G.

B–C Metzler D.E., Longenecker J.B. and Snell E.E. (1954) Reversible catalytic cleavage of hydroxy-amino acids by pyridoxal and metal salts. *J. Amer. chem. Soc.* **75**, 2786.

B–C *See* Selim A.S.M. and Greenberg D.M. 204 D–E.

B–C Alföldi L., Raskó I. and Kerekes E. (1968) L-Serine deaminase of *Escherichia coli*. *J. Bact.* **96**, 1512.

D–G Kornberg H.L. and Morris J.G. (1963) β-Hydroxyaspartate pathway: a new route for biosynthesis from glyoxylate. *Nature*, **197**, 456.

D–G Morris J.G. (1963) The assimilation of 2-C compounds other than acetate. *J. gen. Micro.* **32**, 167.

208 AB–D Snoke J.E., Yanari S. and Bloch K. (1953) Synthesis of glutathione from γ-glutamylcysteine. *J. biol. Chem.* **201**, 573.

AB–D Webster G.C. and Varner J.E. (1954) Peptide bond formation in higher plants. II. Studies on the mechanism of synthesis of γ-glutamylcysteine. *Arch. biochem. Biophys.* **52**, 22.

AB–D Mandeles S. and Bloch K. (1955) Enzymatic synthesis of γ-glutamylcysteine. *J. biol. Chem.* **214**, 639.

D–E Snoke J.E. (1955) Isolation and properties of yeast glutathione synthetase. *J. biol. Chem.* **213**, 813.

D–E Ito E. and Strominger J.L. (1962) Enzymatic synthesis of the peptide in bacterial uridine nucleotides. *J. biol. Chem.* **237**, 2696.

D–E Neuhaus F.C. (1962) Kinetic studies on D-ala-D-ala synthetase. *Fed. Proc.* **21**, 229.

209 GEN Neuberger A. and Scott J.J. (1953) Aminolaevulinic acid and porphyrin biosynthesis. *Nature*, **172**, 1093.

GEN Rimington C. (1956) The biosynthesis of haemoglobin. *Brit. med. J.* **2**, 189.

D Neuberger A. (1961) Aspects of the metabolism of glycine and of porphyrins. *Biochem. J.* **78**, 1.

AB–D Gibson K.D., Laver W.G. and Neuberger A. (1958) Initial stages in the biosynthesis of porphyrins. *Biochem. J.* **70**, 71.

AB–D Kikuchi G., Kumar A., Talmage P. and Shemin D. (1958) The enzymatic synthesis of δ-aminolevulinic acid. *J. biol. Chem.* **233**, 1214.

E–F Gibson K.D., Neuberger A. and Scott J.J. (1955) The purification and properties of δ-aminolaevulic acid dehydrase. *Biochem. J.* **61**, 618.

E–G Granick S. (1958) Porphyrin biosynthesis in erythrocytes. 1. Formation of δ-aminolevulinic acid in erythrocytes. *J. biol. Chem.* **232**, 1101.

E–G Granick S. and Mauzerall D. (1958) Porphyrin biosynthesis in erythrocytes. Enzymes converting δ-aminolaevulic acid to coproporphyrinogen. *J. biol. Chem.* **232**, 1119.

E–G Bogorad L. (1962) Intermediates in the enzymatic synthesis of uroporphyrinogen. *Fed. Proc.* **21**, 400.

211 A–B Rothschild H.A., Cori O. and Barron E.S.G. (1954) The components of choline oxidase and aerobic phosphorylation coupled with choline oxidation. *J. biol. Chem.* **208**, 41.

A–B Ebisuzaki K. and Williams J.N. (1955) Preparation and partial purification of soluble choline dehydrogenase from liver mitochondria. *Biochem. J.* **60**, 644.

A–B Rendina G. and Singer T.P. (1959) Studies on choline dehydrogenase. Extraction in soluble form, assay and some properties of the enzyme. *J. biol. Chem.* **234**, 1605.

B–C Rothschild H.A. and Barron E.S.G. (1954) The oxidation of betaine aldehyde by betaine aldehyde dehydrogenase. *J. biol. Chem.* **209**, 511.

C–D Klee W.A., Richards H.H. and Cantoni G.L. (1961) The synthesis of methionine by enzymic transmethylation. VII. Existence of two separate homocysteine methylpherases in mammalian liver. *Biochim. biophys. Acta*, **54**, 157.

D–F Mackenzie C.G. and Frisell W.R. (1958) The metabolism of dimethylglycine by liver mitochondria. *J. biol. Chem.* **232**, 417.

D–F Frisell W.R. and Mackenzie C.G. (1962) Separation and purification of sarcosine dehydrogenase and dimethylglycine dehydrogenase. *J. biol. Chem.* **237**, 94.

E–F Frisell N.R. and Mackenzie C.G. (1953) Inhibition by methoxyacetate of sarcosine oxidation and its application to an analysis of dimethylglycine oxidation. *Fed. Proc.* **12**, 206.

212 A–B Bruggemann J. and Waldschmidt M. (1962) Die serinsulphydrase aus huhnerleber: ruckieaktion und vergleich mit der cysteindesulphydrase. *Biochem. Z.* **335**, 408.

A–B Bruggemann J., Schlossmann K., Merkenschlarger M. and Waldschmidt M. (1962) Zur frage des vorkommens der serinsulphydrase. *Biochem. Z.* **335**, 392.

A–B Kredich N.M. and Tomkins G.M. (1966) The enzymic synthesis of L-cysteine in *Escherichia coli* and *Salmonella typhimurium*. *J. biol. Chem.* **241**, 4955.

A–B De Issaly I.S.M. (1968) Cysteine biosynthesis in *Pasteurella multocida*. Cysteine synthetase, purification and properties. *Biochim. biophys. Acta*, **151**, 473.

A–B Kredich N.M., Becker M.A. nad Tomkins G.M. (1969) Purification and characterisation of cysteine synthetase, a bifunctional protein complex from *Salmonella typhimurium*. *J. biol. Chem.* **244**, 2428.

B–C Romano A.H. and Nickerson W. (1954) Cystine reductase of pea seeds and yeasts. *J. biol. Chem.* **208**, 409.

E Soda K., Novogrodsky A. and Meister A. (1964) Enzymatic desulphination of cysteine sulfinic acid. *Biochemistry*, **3**, 1450.

D–E Sorbo B. and Ewetz L. (1965) The enzymatic oxidation of cysteine to cysteine sulphinate in rat liver. *Biochem. biophys. Res. Commun.* **18**, 359.

D–E Wainer A. (1965) The preparation of cysteine sulphonic acid from cysteine *in vitro*. *Biochim. biophys. Acta*, **104**, 405.

D–E Lombardini J.B., Singer T.P. and Boyer P.D. (1969) Cysteine oxygenase. Studies on the mechanism of the reaction with oxygen. *J. biol. Chem.* **244**, 1172.

D–J Wheldrake J.F. and Pasternak C.A. (1968) The oxidation of cysteine by mast-cell tumour P815 in culture. *Biochem. J.* **106**, 437.

G–J Tager J.M. and Rautanen N. (1955) Sulphite oxidation by a plant mitochondrial system. I. Preliminary observations. *Biochim. biophys. Acta*, **18**, 111.

G–J Macleod R.M., Farkas W., Fridovitch I. and Handler P. (1961) Purification and properties of hepatic sulphite oxidase. *J. biol. Chem.* **236**, 1841.

K–L Metaxas M.A. and Delwiche E.A. (1955) The L-cysteine desulphydrase of *E. coli*. *J. Bacteriol.* **70**, 735.

213 GEN Fromageot C. (1955) The metabolism of sulphur and its relations to general metabolism. *Harvey Lectures, Series I*.

A–B Soda K., Novogrodsky A. and Meister A. (1964) Enzymatic desulfination of cysteine sulfinic acid. *Biochemistry*, **3**, 1450.

A–D Hope D.B. (1955) Pyridoxal phosphate as the coenzyme of the mammalian decarboxylase for L-cysteine sulphinic acid and L-cysteic acids. *Biochem. J.* **59**, 497.

A–D Davison A.N. (1956) Amino acid decarboxylases in rat brain and liver. *Biochim. biophys. Acta*, **19**, 66.

A–D Sörbo B. and Heyman T. (1957) On the purification of cysteine-sulphinic acid decarboxylase and its substrate specificity. *Biochim. biophys. Acta*, **23**, 624.

D Anapara J. (1953) 2-Aminoethanesulfinic acid: an intermediate in the oxidation of cysteine *in vivo*. *J. biol. Chem.* **203**, 183.

E Frendo J. and Koj A. (1959) Taurine in human blood platelets. *Nature*, **183**, 685.

E Read W.O. and Welty J.D. (1962) Synthesis of taurine and isethionic acid by dog heart slices. *J. biol. Chem.* **237**, 1521.

F–G Wilson D.G., King K.W. and Burris R.H. (1954) Transamination reactions in plants. *J. biol. Chem.* **208**, 863.

H–J Meister A., Sober H.A. and Tice S.V. (1951) Enzymatic decarboxylation of aspartic acid to α-alanine. *J. biol. Chem.* **189**, 577.

H–J Bheemeswar B. (1955) Studies on transaminase and decarboxylase catalysed by extracts of the silkworm *Bombyx mori L. Nature*, **176**, 533.

214 GEN Meister A. (1957) Valine, isoleucine and leucine. *Biochemistry of the Amino Acids*. Academic Press, New York, p. 729.

GEN Satyanarayana T. and Radhakrishnan A.N. (1962) Biosynthesis of valine and isoleucine in plants. *Biochim. biophys. Acta*, **56**, 197.

GEN Armstrong F.B., Gordon M.L. and Wagner R.P. (1963) Biosynthesis of valine and isoleucine. Enzyme repression in *Salmonella*. *Proc. Nat. Acad. Sci. U.S.* **49**, 322.

GEN Armstrong F.B. and Wagner R.P. (1963) Repression of the valine-isoleucine pathway in *Salmonella*. *Proc. Nat. Acad. Sci. U.S.* **49**, 628.

C–N Abramsky Y.T. and Shemin D. (1965) The formation of isoleucine from β-methyl aspartic acid in *Escherichia coli*. *J. biol. Chem.* **240**, 2971.

D–H & E–J Armstrong F.B. and Wagner R.P. (1961) Biosynthesis of valine and isoleucine. V. Electrophoretic studies on the reductoisomerase and reductase of *Salmonella*. *J. biol. Chem.* **236**, 3252.

H–K Arfin S.M. (1969) Evidence for an enol intermediate in the enzymatic conversion of α,β-dihydroxyisovalerate to α-ketoisovalerate. *J. biol. Chem.* **244**, 2250.

H–K Wixom R.L. Wikman J.H. and Howell G.B. (1961) Studies in valine biosynthesis. III. Biological distribution of a dihydroxy acid dehydrase. *J. biol. Chem.* **236**, 3257.

H–L & J–L Kanamori M. and Wixom R.L. (1963) Studies in valine biosynthesis. V. Characteristics of the purified dihydroxyacid dehydratase from spinach leaves. *J. biol. Chem.* **238**, 998.

215 GEN Fones W.S., Waalkes T.P. and White J. (1951) The conversion of L-valine to glucose and glycogen in the rat. *Arch. biochem. Biophys.* **32**, 89.

GEN Coon M.J., Robinson W.G. and Bachhawat B.K. (1955) Enzymatic studies on the biological degradation of the branched chain amino acids. *Amino Acid Metabolism*. McElroy and Glass, p. 341.

GEN Coon M.J. (1955) Enzymatic synthesis of branched chain acids from amino acids. *Fed. Proc.* **14**, 762.

GEN Krebs H.A. (1964) The metabolic fate of amino acids. *Mammalian Protein Metabolism*. Academic Press, New York, p. 125.

D–F Connelly J.L., Danner D.J. and Bowden J.A. (1968) Branched chain α-keto acid metabolism. Isolation and purification and partial characterisation of bovine liver α-keto-isocaproic: α-ketomethyl valeric acid dehydrogenase. *J. biol. Chem.* **243**, 1198.

D–F Bowden J.A. and Connelly J.L. (1968) Branched chain α-ketoacid metabolism. Evidence for the common identity of α-ketoisocaproic acid and α-keto β-methylvaleric acid dehydrogenases. *J. biol. Chem.* **243**, 3526.

E–G Crane F.L., Mii S., Hauge J.G., Green D.E. and Beinert H. (1956) On the mechanism of dehydrogenation of fatty acyl derivatives of coenzyme A. I. The general fatty acyl coenzyme A dehydrogenase. *J. biol. Chem.* **218**, 701.

E–G Robinson W.G., Nagle R., Bachhawat B.K., Kupiecki F.P. and Coon M.J. (1957) Coenzyme A thiol esters of isobutyric, methacrylic, and β-hydroxyisobutyric acids as intermediates in the enzymatic degradation of valine. *J. biol. Chem.* **224**, 1.

G–J Stern J.R. (1961) Crotonase. *The Enzymes*, 2nd ed., vol. 5. Academic Press, New York, p. 511.

G–J *See* 174 F–G.

J–L Rendina G. and Coon M.J. (1957) Enzymic hydrolysis of the coenzyme A thiol esters of β-hydroxy propionic acid and β-hydroxyisobutyric acid. *J. biol. Chem.* **225**, 523.

K–M Lehninger A.L. and Greville G.D. (1953) The enzymic oxidation of d- and l- β-hydroxybutyrate. *Biochim. biophys. Acta*, **12**, 188.

K–M Stern J.R. (1957) Crystalline β-hydroxybutyryl dehydrogenase from pig heart. *Biochim. biophys. Acta*, **26**, 448.

L–N Robinson W.G. and Coon M.J. (1957) The purification and properties of β-hydroxyisobutyric dehydrogenase. *J. biol. Chem.* **225**, 511.

N Sokatch J.R., Sanders L.E. and Marshall V.P. (1968) Oxidation of methylmalonate semialdehyde to propionyl coenzyme A in *Pseudomonas aeruginosa* grown on valine. *J. biol. Chem.* **243**, 2500.

O–P Flavin M. and Ochoa S. (1957) Metabolism of propionic acid in animal tissues. Enzymatic conversion of propionate to succinate. *J. biol. Chem.* **229**, 965.

O–P Lane M.D., Halenz D.R., Kosow D.P. and Hegre C.S. (1960) Further studies on mitochondrial propionyl carboxylase. *J. biol. Chem.* **235**, 3082.

O–P Kaziro Y., Ochoa S., Warner R.C. and Chen J. (1961) Metabolism of propionic acid in animal tissues. Crystalline propionyl carboxylase. *J. biol. Chem.* **236**, 1917.

O–R Kennedy E.P. (1957) Metabolism of fatty acids. Enzymic oxidation of fatty acids: conversion of propionate to succinate. *Ann. Rev. Biochem.* **26**, 126.

P–Q Mazumder R. and Sasakawa T. (1961) Further studies on the enzymatic isomerisation of methyl malonyl CoA to succinyl CoA. *Fed. Proc.* **20**, 272.

P–Q Mazumder R., Sasakawa T., Kaziro Y. and Ochoa S. (1962) Metabolism of propionic acid in animal tissues. IX. Methylmalonyl coenzyme A racemase. *J. biol. Chem.* **237**, 3065.

Q–R Beck W.S., Flavin M. and Ochoa S. (1957) Metabolism of propionic acid in animal tissues. III. Formation of succinate. *J. biol. Chem.* **229**, 997.

Q–R Beck W.S. and Ochoa S. (1958) Metabolism of propionic acid in animal tissues. Further studies on the enzymatic isomerisation of methylmalonyl coenzyme A. *J. biol. Chem.* **232**, 931.

216 GEN Strassman M., Locke L.A., Thomas A.J. and Weinhouse S. (1956) A study of leucine biosynthesis in *Torulopsis utilis*. *Science*, **121**, 303.

GEN See Meister A. 214 GEN.

GEN Rafelson M.E., Jr. (1957) The biosynthesis of leucine in *Aerobacter aerogenes*. *Arch. biochem. Biophys.* **72**, 376.

C Calvo J.M., Kalyanpur M.G. and Stevens C.M. (1962) 2-Isopropylmalate and 3-isopropylmalate as intermediates in leucine biosynthesis. *Biochemistry* **1**, 1157.

C Strassman M. and Celi L.N. (1963) Enzymatic formation of α-isopropylmalic acid, an intermediate in leucine biosynthesis. *J. biol. Chem.* **238**, 2445.

C Jungwirth C., Gross S.R., Margolin P. and Umbarger H.E. (1963) The biosynthesis of leucine. Accumulation of β-carboxy-β-hydroxyisocaproate by leucine auxotrophs of *Salmonella typhimurium* and *Neurospora crassa*. *Biochemistry*, **2**, 1.

AB–C Butler G.W. and Shen L. (1963) Leucine biosynthesis in higher plants. *Biochim. biophys. Acta*, **71**, 456.

E Martin W.R., Coleman W.H., Wideburg N.E., Cantrell R., Jackson M. and Denison F.W. (1962) β-Carboxy-β-hydroxyisocaproic acid formation in microorganisms. *Biochim. biophys. Acta*, **62**, 165.

C–E Gross S.R., Burns R.O. and Umbarger H.E. (1963) The biosynthesis of leucine. II. The enzymic isomerisation of β-carboxy-β-hydroxyisocaproate, and α-hydroxy-β-carboxyisocaproate. *Biochemistry*, **2**, 1046.

C–F Burns R.O., Umbarger H.E. and Gross S.R. (1963) The biosynthesis of leucine. III. The conversion of α-hydroxy-β-carboxyisocaproate to α-ketoisocaproate. *Biochemistry*, **2**, 1053.

F–G Tanenbaum S.W. and Shemin D. (1950) A study of the transamination reaction using isotopic nitrogen. *Fed. Proc.* **9**, 236.

F–G Rowsell E.V. (1956) Transamination with L-glutamate and α-oxoglutarate in fresh extracts of animal tissue. Transamination with pyruvate and other α-keto acids. *Biochem. J.* **64**, 235, 246.

217 GEN Coon M.J. and Gurin S. (1949) Studies on the conversion of radioactive leucine to acetoacetate. *J. biol. Chem.* **180**, 1159.

GEN See Coon M.J. *et al.* (1955) 215 GEN.

GEN Bachawat B.K., Robinson N.G. and Coon M.J. (1956) Enzymatic carboxylation of β-hydroxyisovalerate coenzyme A. *J. biol. Chem.* **219**, 539.

A–B Taylor R.T. and Jenkins W.T. (1966) Leucine aminotransferase. Purification and characterisation. *J. biol. Chem.* **241**, 4396.

C Coon M.J. (1950) The metabolic fate of the isopropyl group of leucine. *J. biol. Chem.* **187**, 71.

B–C See Connelly J.L. *et al.* 215 D–F.

B–C See Bowden J.A. *et al.* 215 D–F.

C–D See 177 A–C, B–D.

D–E del-Campillo-Campbell A., Dekker E.E. and Coon M.J. (1959)

Carboxylation of β-methylcrotonyl coenzyme A by a purified enzyme from chicken liver. *Biochem. biophys. Acta*, **31**, 290.

D–E Rilling H.C. and Coon M.J. (1960) The enzymatic isomerisation of β-methylvinylacetyl coenzyme A and the specificity of a bacterial β-methyl crotonyl coenzyme A carboxylase. *J. biol. Chem.* **235**, 3087.

D–E Lynen F., Knappe J., Lorch E., Jütting G., Ringelmann E. and Lachance J.-P. (1961) Zur biochemischen Funktion des Biotins, Reingung und Wirkungsweise der β-Methyl-crotonyl-carboxylase. *Biochem. Z.* **335**, 123.

E–F Hilz H., Knappe J., Ringelmann E. and Lynen F. (1958) Methyl glutaconase eine neue Hydratase, die am Stoffwechsel verzweigter Carbonsaüren Beteiligit ist. *Biochem. Z.* **329**, 476.

F–G Bucher N.L.R., Overath P. and Lynen F. (1960) β-Hydroxy-β-methylglutaryl coenzyne A. Reductase, cleavage and condensing enzymes in relation to cholesterol formation in rat liver. *Biochim. biophys. Acta*, **40**, 491.

218 GEN Ehrensvärd G., Reio L. and Saluste E. (1949) On the origin of the basic amino acids. *Acta. chem. scand.* **3**, 645.

GEN Strassman M. and Weinhouse S. (1953) The biosynthesis of lysine by *Torulopsis utilis*. *J. Amer. chem. Soc.* **75**, 1680.

GEN Vogel H.J. (1960) Two modes of lysine synthesis among lower fungi: evolutionary significance. *Biochim. biophys. Acta*, **34**, 282.

GEN Vogel H.J. (1965) Lysine biosynthesis and evolution in *Evolving Genes and Enzymes*. Ed. V. Bryston and H.J. Vogel. Academic Press, New York, p. 25.

GEN Betterton H., Fjellstedt T., Matsuda M., Ogur M. and Tate R. (1968) Localisation of the homocitrate pathway. *Biochim. biophys. Acta*, **170**, 459.

B Strassman M. and Ceci L.N. (1964) Enzymatic formation of homocitric acid, an intermediate in lysine metabolism. *Biochem. biophys. Res. Commun.* **14**, 262.

A–B Hogg R.W. and Broquist H.P. (1968) Homocitrate formation in *Neurospora crassa*. Relation to lysine biosynthesis. *J. biol. Chem.* **243**, 1839.

A–C Strassman M. and Ceci L.N. (1966) Enzymatic formation of *cis*-homoaconitic acid, an intermediate in lysine biosynthesis in yeast. *J. biol. Chem.* **241**, 5401.

B–C Strassman M., Ceci L.N. and Maragoudakis M.E. (1965) Homoaconitase in yeast. *Fed. Proc.* **24**, 228.

F Weber M.A., Hoagland A.N., Klein J. and Lewis K. (1964) Biosynthesis of α-keto adipic acid by extracts of bakers yeast. *Arch. biochem. Biophys.* **104**, 257.

B–F Strassman M., Ceci L.N. and Silverman B.E. (1964) Enzymatic conversion of homoisocitric acid to α-ketoadipic acid. *Biochem. biophys. Res. Commun.* **14**, 268.

219 B Windsor E. (1951) α-Aminoadipic acid as a precursor to lysine in *neurospora*. *J. biol. Chem.* **192**, 607.

A–B Matsuda M. and Ogur M. (1969) Separation and specificity of the yeast glutamate-α-ketoadipate transaminase. *J. biol. Chem.* **244**, 3352.

A–F See Cánovas J.L., Ornston L.N. and Stanier R.Y. (1967) 75–1.

C See Yura T. and Vogel H.J. (1957) 193 C.

B–C Sagisaka S. and Shimura K. (1962) Studies in lysine biosynthesis. IV. Mechanism of activation and reduction of α-aminoadipate. *J. Biochem. (Tokyo)*, **52**, 155.

B–C Sagisaka S. and Shimura K. (1962) Studies in lysine biosynthesis. III. Enzymatic reduction of α-aminoadipate: isolation and some properties of the enzyme. *J. Biochem. (Tokyo)*, **51**, 398.

B–C Larson R.L., Sandine W.D. and Broquist H.P. (1963) Enzymic reduction of α-aminoadipic acid: relation to lysine biosynthesis. *J. biol. Chem.* **238**, 275.

B–F Kuo M.H., Saunders P.P. and Broquist H.P. (1964) Lysine biosynthesis in yeast: a new metabolite of α-aminoadipic acid. *J. biol. Chem.* **239**, 508.

E Trupin J.S. and Broquist H.P. (1965) Saccharopine, an intermediate of the aminoadipic acid pathway of lysine biosynthesis. *J. biol. Chem.* **240**, 2524.

E Jones E.E. and Broquist H.P. (1965) Saccharopine, an intermediate of the aminoadipic acid pathway of lysine biosynthesis. *J. biol. Chem.* **240**, 2531.

C–E Jones E.E. and Broquist H.P. (1966) Saccharopine, an intermediate of the aminoadipic acid pathway of lysine biosynthesis. Aminoadipic semialdehyde reductase. *J. biol. Chem.* **241**, 3430.

E–F Vaughan S.T. and Broquist H.P. (1965) Saccharopine dehydrogenase, a marker of the aminoadipic acid pathway of lysine biosynthesis. *Fed. Proc.* **24**, 218.

E–F Saunders P.P. and Broquist H.P. (1966) Saccharopine dehydrogenase. *J. biol. Chem.* **241**, 3435.

220 B Gilvarg C. (1962) The branching point of diaminopimelic acid synthesis. *J. biol. Chem.* **237**, 482.

AB–C Yugari Y. and Gilvarg C. (1965) The condensation step in diaminopimelate synthesis. *J. biol. Chem.* **240**, 4710.

AB–C Truffa-Bachi P., Patte J.-C. and Cohen G.N. (1967) Sur la dihydropicolinate synthetase *d'Escherichia coli* K12. *Compt. Rend.* **265**, 928.

C–D Farkas W. and Gilvarg C. (1965) The reduction step in diaminopimelic acid biosynthesis. *J. biol. Chem.* **240**, 4717.

A–F Edelman J.C. and Gilvarg C. (1961) Isotope studies on diaminopimelic acid synthesis in *E. coli*. *J. biol. Chem.* **236**, 3295.

E Gilvarg C. (1961) *N*-Succinyl-α-amino-ϵ-ketopimelic acid. *J. biol. Chem.* **236**, 1429.

E–F Peterkofsky B. and Gilvarg C. (1961) *N*-succinyl-L-diaminopimelic-glutamic transaminase. *J. biol. Chem.* **236**, 1432.

F Gilvarg C. (1959) *N*-succinyl L-diaminopimelic acid. *J. biol. Chem.* **234**, 2955.

221 GEN Davis B.D. (1952) Biosynthetic interrelation of lysine, diaminopimelic acid and threonine in mutants of *E. coli*. *Nature*, **169**, 534.

B Work E. (1951) The isolation of diaminopimelic acid from *Corynebacterium diphtheriae* and *Mycobacterium tuberculosis*. *Biochem. J.* **49**, 17.

B Wright C.D. and Cresson E.L. (1953) Isolation and characterisation of diaminopimelic acid from culture filtrate of an *E. coli* mutant. *Proc. Soc. exptl. Biol. Med.* **82**, 354.

B Powell J.F. and Strange R.E. (1957) α,ϵ-Diaminopimelic acid metabolism and sporulation in *Bacillus sphearicus*. *Biochem. J.* **65**, 700.

B Meadow D. and Work E. (1959) Biosynthesis of diaminopimelic acid in *E. coli*. *Biochem. J.* **72**, 400.

A–B Kindler S.H. and Gilvarg C. (1960) *N*-succinyl-L-α,ϵ-diaminopimelic acid deacylase. *J. biol. Chem.* **235**, 3532.

B–C Hoare D.S. and Work E. (1955) The stereoisomers of α,ϵ-diaminopimelic acid: their distribution in nature and behaviour towards certain enzyme preparations. *Biochem. J.* **61**, 562.

B–C Antia M., Hoare D.S. and Work E. (1957) The stereoisomers of α-diaminopimelic acid. Properties and distribution of diaminopimelic acid racemase, an enzyme causing interconversion of the *LL* and *meso*isomers. *Biochem. J.* **65**, 448.

C–D Dewey D.L. and Work E. (1952) Diaminopimelic acid and lysine. Diaminopimelic acid decarboxylase. *Nature*, **169**, 533.

C–D Dewey D.L., Hoare D.S. and Work E. (1954) Diaminopimelic acid decarboxylase in cells and extracts of *E. coli* and *Klebsiella aerogenes*. *Biochem. J.* **58**, 523.

C–D Patte J.-C., Loviny T. and Cohen G.N. (1962) Répression de la décarboxylase de l'acide *méso-α-ε*-diaminopimélique par la L-lysine, chez *E. coli. Biochim. biophys. Acta*, **58**, 359.

C–D White P.J. and Kelly B. (1965) Purification and properties of diaminopimelate decarboxylase from *Escherichia coli. Biochem. J.* **96**, 75.

C–D Grandgenett D.P. and Stahly D.P. (1968) Diaminopimelate decarboxylase of sporulating bacteria. *J. Bacteriol.* **96**, 2099.

222 GEN Rothstein M. and Miller L.L. (1954) The metabolism of L-lysine-6-^{14}C. *J. biol. Chem.* **206**, 243.

GEN Krebs H.A. (1964) The metabolic fate of amino acids. *Mammalian Protein Metabolism*. Academic Press, New York, p. 125.

D Rao D.R. and Rodwell V.W. (1962) Metabolism of pipecolic acid in a *Pseudomonas* species. *J. biol. Chem.* **237**, 2232.

A–B Soda K., Misona H. and Yamamoto T. (1966) L-Lysine-α-keto-glutaric acid transamination reaction: identification of the product from L-lysine. *Agric. and Biol. Chem. (Japan)*, **30**, 944.

A–D Lowy P. (1953) Conversion of lysine to pipecolic acid by *Phaseolus vulgaris. Arch. biochem. Biophys.* **47**, 228.

A–D Grobbelaar N. and Steward F.C. (1953) Pipecolic acid in *Phaseolus vulgaris:* evidence for its derivation from lysine. *J. Amer. chem. Soc.* **75**, 4341.

A–G Borsook H., Deasey C.L., Haagen-Smit A.J., Keighley G. and Lowy P.H. (1948) The degradation of L-lysine in guinea pig liver homogenate; formation of α-aminoadipic acid. *J. biol. Chem.* **176**, 1383, 1395.

C–G Sagisaka S. and Shimura K. (1959) Enzymic reduction of α-amino-adipic acid by yeast enzyme. *Nature*, **184**, 1709.

D–E Baginsky M.L. and Rodwell V.W. (1967) Metabolism of pipecolic acid in a *Pseudomonas* species. *J. Bact.* **94**, 1034.

D–G Rothstein M. and Greenberg D.M. (1960) The metabolism of pipecolic acid-2-^{14}C. *J. biol. Chem.* **235**, 714.

D–G Rothstein M., Cooksey K.E. and Greenberg D.M. (1962) Metabolic conversion of pipecolic acid to α-aminoadipic acid. *J. biol. Chem.* **237**, 2828.

D–F Basso L.V., Rao D.R. and Rodwell V.W. (1962) Metabolism of pipecolic acid in *Pseudomonas* species. *Δ'*-Piperideine-6-carboxylic acid and α-aminoadipic acid-γ-semialdehyde. *J. biol. Chem.* **237**, 2239.

223 GEN Ames B.N. (1955) *Amino Acid Metabolism*. Johns Hopkins Press, p. 375.

GEN Rhuland L.E. and Hamilton R.B. (1961) The functional pathway of lysine biosynthesis in *E. coli. Biochim. biophys. Acta*, **51**, 525.

A–C Borsook H., Deasy C.L., Haagen-Smit A.J., Keighley G. and Lowy P.H. (1948) The degradation of α-aminoadipic acid in guinea pig liver homogenate. *J. biol. Chem.* **176**, 1395.

A–C See Rothstein M. and Miller L.L. (1954) 222 GEN.

C–E Tustanoff E.R. and Stern J.R. (1960) Enzymic carboxylation of crotonyl-CoA and the metabolism of glutaric acid. *Biochem. biophys. Res. Commun.* **3**, 81.

C–H Nishizuka Y., Kuno S. and Hayaishi O. (1960) Enzymatic formation of acetyl-CoA and CO_2 from glutaryl CoA. *Biochim. biophys. Acta*, **43**, 357.

G–H Stern J.R., Drummond G.I. Coon M.J. and Del Campillo A. (1960) Enzymes of ketone body metabolism. Purification of an acetoacetate-synthesising enzyme from ox liver. *J. biol. Chem.* **235**, 313.

224 GEN Ames B.N., Mitchell H.K. and Mitchell M.B. (1953) Some new naturally occurring imidazoles related to the biosynthesis of histidine. *J. Amer. chem. Soc.* **75**, 1015.

GEN Magasanik B. (1956) Guanine as a source of the nitrogen-1-carbon-2 portion of imidazole ring of histidine. *J. Amer. chem. Soc.* **78**, 5449.

GEN Moyed H.S. and Magasanik B. (1957) The role of purines in histidine biosynthesis. *J. Amer. chem. Soc.* **79**, 4812.

GEN Neidle A. and Waelsh H. (1959) The origin of the imidazole ring of histidine in *E. coli*. *J. biol. Chem.* **234**, 586.

GEN *See* Ames B.N. and Garry B. (1959) 73–2.

GEN Ames B.N. and Hartman P.E. (1963) The histidine operon. *Cold Spring Harbor Symp. New York*, **28**, 349.

GEN Whitfield H.J., Jr., Smith D.W.E. and Martin R.G. (1964) Sedimentation properties of the enzymes of the histidine operon. *J. biol. Chem.* **239**, 3288.

A–B Ames B.N., Martin R.G. and Garry B.J. (1961) The first step of histidine biosynthesis. *J. biol. Chem.* **236**, 2019.

A–B *See* Martin R.G. (1963) 66–1.

C Smith D.W.E. and Ames B.N. (1965) Phosphoribosyl adenosine monophosphate, an intermediate in histidine biosynthesis. *J. biol. Chem.* **240**, 3056.

A–C Voll M.J., Apella E. and Martin R.G. (1967) Purification and composition studies of phosphoribosyladenine triphosphate: pyrophosphate phosphoribosyltransferase, the first enzyme of histidine biosynthesis. *J. biol. Chem.* **242**, 1760.

A–F Smith D.W.E. and Ames B.N. (1964) Intermediates in the early steps of histidine biosynthesis. *J. biol. Chem.* **239**, 1848.

D–E Margolies M.N. and Goldberger R.F. (1966) Isolation of the fourth enzyme (isomerase) of histidine biosynthesis from *Salmonella typhimurium*. *J. biol. Chem.* **241**, 3262.

D–E Margolies M.N. and Goldberger R.F. (1967) Physical and chemical characterisation of the isomerase of histidine biosynthesis in *Salmonella typhimurium*. *J. biol. Chem.* **242**, 256.

225 A–B Shedloosky A.E. and Magasanik B. (1962) A defect in histidine biosynthesis causing an adenine deficiency. *J. biol. Chem.* **237**, 3725.

D–F Ames B.N. (1951) The biosynthesis of histidine D-erythro-imidazole-glycerol phosphate dehydrase. *J. biol. Chem.* **228**, 131.

DFG Ames B.N. and Mitchell H.K. (1955) The biosynthesis of histidine. Imidazoleglycerol phosphate, imidazoleacetol phosphate and histidinol phosphate. *J. biol. Chem.* **212**, 687.

FG Ames B.N. and Horecker B. (1956) The biosynthesis of histidine: imidazoleacetol phosphate transaminase. *J. biol. Chem.* **220**, 113.

H Vogel H.J., Davis B.D. and Mingioli E.S. (1951) L-Histidinol, a precursor of L-histidine in *Escherichia coli*. *J. Amer. chem. Soc.* **73**, 1897.

G–H Ames B.N. (1957) The biosynthesis of histidine; L-histidinol phosphate phosphatase. *J. biol. Chem.* **226**, 583.

J Adams E. (1955) L-Histidinal, a biosynthetic precursor of histidine. *J. biol. Chem.* **217**, 325.

H–J Adams E. (1955) Synthesis and properties of an α-amino aldehyde, histidinal. *J. biol. Chem.* **217**, 317.

H–K Loper J.C. and Adams E. (1965) Purification and properties of histidinol dehydrogenase from *Salmonella typhimurium*. *J. biol. Chem.* **240**, 788.

226 A–B Mehler A.H., Hayaishi T. and Tabor H. (1955) Urocanic acid. *Biochem. Prep.* **4**, 50.

A–B Peterkovsky A. (1962) The mechanism of action of histidase: Amino-enzyme formation and partial reactions. *J. biol. Chem.* **237**, 787.

A–B Spolter P.D. and Baldridge R.C. (1963) The metabolism of histidine. On the assay of enzymes in rat liver. *J. biol. Chem.* **238**, 2071.

A–B Cornell N.W. and Villee C.A. (1968) Purification and properties of rat liver histidinase. *Biochim. biophys. Acta*, **167**, 172.

A–K Tabor H. and Hayaishi O. (1952) The enzymatic conversion of histidine to glutamic acid. *J. biol. Chem.* **194**, 171.

A–K Tabor H. (1955) Intermediates in histidine breakdown. *Methods in enzymology*, **6**, 581.

B–E Miller A. and Waelsch H. (1957) The conversion of urocanic acid to formamidinoglutaric acid. *J. biol. Chem.* **228**, 365.

C Brown D.D. and Kies M.W. (1959). The mammalian metabolism of L-histidine. II. The enzyme formation, stabilisation, purification and properties of 4(5)-imidazolone-5(4)propionic acid, the product of urocanase activity. *J. biol. Chem.* **234**, 3188.

C–D Brown D.D. and Kies M.W. (1959) The mammalian metabolism of L-histidine. 1. The enzymatic formation of L-hydantoin-5-propionic acid. *J. biol. Chem.* **234**, 3182.

C–D Hassall H. and Greenberg D.M. (1963) The oxidation of 4(5)-imidazolone-5(4)propionic acid to hydantoin-5-propionic acid by xanthine oxidase. *Biochim. biophys. Acta*, **67**, 507.

C–E Rao D.R. and Greenberg D.M. (1961) Studies on the enzymic decomposition of urocanic acid. IV. Purification and properties of 4(5)imidazole-5(4)propionic acid hydrolase. *J. biol. Chem.* **236**, 1758.

C–E Snyder S.H., Silva O.L. and Kies M.W. (1961) The mammalian metabolism of L-histidine. IV. Purification and properties of imidazolone propionic acid hydrolase. *J. biol. Chem.* **236**, 2996.

C–G Revel H.R.B. and Magasanik B. (1958) The enzymatic degradation of urocanic acid. *J. biol. Chem.* **233**, 930.

E Borek B.A. and Waelsch H. (1953) Enzymatic degradation of histidine. *J. biol. Chem.* **205**, 459.

E–FK Miller A. and Waelsch H. (1957) Formimino transfer from formamidino glutaric acid to tetrahydrofolic acid. *J. biol. Chem.* **228**, 397.

E–GK Magasanik B. and Bowser H.R. (1955) The degradation of histidine by *Aerobacter aerogenes*. *J. biol. Chem.* **213**, 571.

E–GK Wachsman J.T. and Barker H.A. (1955) The accumulation of formamide during the fermentation of histidine by *Clostridium tetanomorphum*. *J. Bacteriol.* **69**, 83.

H Tabor H. and Mehler A.H. (1954) Isolation of *N*-formyl-1-glutamic acid as an intermediate in the enzymatic degradation of L-histidine. *J. biol. Chem.* **210**, 559.

227 A–B Weissbach H., Lovenberg W. and Udenfriend S. (1961) Characteristics of mammalian histidine decarboxylating enzymes. *Biochim. biophys. Acta*, **50**, 177.

B–C Mann P.J.G. (1961) Further purification and properties of the amine oxidase of pea seedlings. *Biochem. J.* **79**, 623.

D Bouthillier L.P. and Goloner M. (1953) The metabolism of histamine-β-^{14}C. *Arch. biochem. Biophys.* **44**, 251.

C–D Mackler B., Mahler H.R. and Green D.E. (1954) Studies on metalloflavo proteins. Xanthine oxidase, a molybdoflavoprotein. *J. biol. Chem.* **210**, 149.

C–D Doisy R.J., Richert D.A. and Westerfield W.W. (1955) Comparative studies on various inhibitors on xanthine oxidase and related enzymes. *J. biol. Chem.* **217**, 307.

C–D De Renzo E.C. (1956) Chemistry and biochemistry of xanthine oxidase. *Advanc. Enzymol.* **17**, 293.

C–D Dikstein S., Bergmann F. and Henis Y. (1957) Studies on uric acid and related compounds. The specificity of bacterial xanthine oxidases. *J. biol. Chem.* **224**, 67.

E–F Rothberg S. and Hayaishi O. (1957) Studies on oxygenases. Enzymatic oxidation of imidazoleacetic acid. *J. biol. Chem.* **229**, 897.

F–H Hayaishi O., Tabor H. and Hayaishi T. (1957) *N*-formimino L-aspartic acid as an intermediate in the enzymatic conversion of imidazole acetic acid to formyl aspartic acid. *J. biol. Chem.* **227**, 161.

H–J Ohmura E. and Hayaishi O. (1957) Enzymatic conversion of formyl aspartic acid to aspartic acid. *J. biol. Chem.* **227**, 181.

228 A–B Hassall H. and Greenberg D.M. (1963) The bacterial metabolism of L-hydantoin-5-propionate to carbamyl glutamic acid and glutamic acid. *J. biol. Chem.* **238**, 3325.

E Hassall H. and Greenberg D.M. (1963) Studies on the enzymic decomposition of urocanic acid. V. The formation of 4-oxoglutaramic acid, a non enzymic oxidation product of 4(5)-imidazolone-5(4)propionic acid. *J. biol. Chem.* **238**, 1423.

229 GEN Davis B.D. (1950) Studies on nutritionally deficient bacterial mutants, isolated by means of penicillin. *Experientia*, **6**, 41.

GEN Davis B.D. (1951) Aromatic biosynthesis. I. The role of shikimic acid. *J. biol. Chem.* **191**, 315.

GEN Thomas R.C., Cheldelin V.H., Christenson B.E. and Wang C.H. (1953) Conversion of acetate and pyruvate to tyrosine in yeast. *J. Amer. chem. Soc.* **75**, 5554.

GEN Tatum E.L., Gross S.R., Ehrensvärd G. and Garnjobst L. (1954) Synthesis of aromatic compounds by *Neurospora. Proc. Nat. Acad. Sci. U.S.* **40**, 271.

GEN Davis B.D. (1955) Biosynthesis of aromatic amino acids. *Amino Acid Metabolism.* (McElroy and Glass (Eds.)), p. 799.

GEN Rafelson M.E., Jr., Ehrensvärd G. and Reio L. (1955) Formation of aromatic amino acids in *Aerobacter aerogenes. Exptl. cell. Res. Suppl.* **3**, 281.

GEN Neish A.C. (1960) Biosynthetic pathways of aromatic compounds. *Ann. Rev. Plant. Physiol.* **11**, 55.

GEN Umbarger E. and Davis B.D. (1962) Pathways of amino acid biosynthesis. *The Bacteria*, vol. 3. Academic Press, New York, p. 168.

GEN Gibson F. and Pittard J. (1968) Pathways of biosynthesis of aromatic amino acids and vitamins and their control in microorganisms. *Bact. Rev.* 465.

GEN Morell H. and Sprinson D.B. (1968) Shikimate kinase isoenzymes in *Salmonella typhimurium. J. biol. Chem.* **243**, 676.

C Weissbach A. and Hurwitz J. (1959) Formation of 2-keto-3-deoxyheptonic acid in extracts of *E. coli* B. *J. biol. Chem.* **234**, 705.

C Hurwitz J. and Weissbach A. (1959) Formation of 2-keto-3-deoxyheptonic acid in extracts of *E. coli* B. II. Enzymic studies. *J. biol. Chem.* **234**, 710.

AB–C Srinivasan P.R., Katagiri M. and Sprinson D.B. (1955) The enzymatic synthesis of shikimic acid from D-erythrose-4-phosphate and phosphenol pyruvate. *J. Amer. chem. Soc.* **77**, 4943.

AB–C Srinivasan P.R. and Sprinson D.B. (1959) 2-Keto-3-deoxy-D-arabino-heptonic acid-7-phosphate synthetase. *J. biol. Chem.* **234**, 716.

AB–C Doy C.H. and Brown K.D. (1965) Control of aromatic biosynthesis: the multiplicity of 7-phospho-2-oxo-3-deoxy-D-arabinoheptonate D-erythrose-4 phosphate lyase (pyruvate phosphorylating) in *E. coli* W. *Biochim. biophys. Acta*, **104**, 377.

AB–C Jensen R.A. and Nester E.W. (1966) Regulatory enzymes of aromatic amino acid biosynthesis, in *Bacillus subtilis*. Purification and properties of 3-deoxy-D-arabino-heptulosonate-7-phosphate synthetase. *J. biol. Chem.* **241**, 3365.

AB–C Staub M. and Dénes G. (1969) Purification and properties of the 3-deoxy-D-arabino-heptulosonate-7-phosphate synthetase (phenyl alanine sensitive) of *Escherichia coli* K.12. *Biochim. biophys. Acta*, **178**, 588.

AB–D Srinivasan P.R., Katagiri M. and Sprinson D.B. (1959) Conversion of phosphenol pyruvate and D-erythrose 4-phosphate to 5-dehydroquinic acid. *J. biol. Chem.* **234**, 713.

E Salamon I.I. and Davis B.D. (1953) Aromatic biosynthesis. IX. Isolation of a precursor of shikimic acid. *J. Amer. chem. Soc.* **75**, 5567.

E Weiss U., Davis B.D. and Mingioli E.S. (1953) Aromatic biosynthesis. X. Identification of an early precursor of 5-dehydroquinic acid. *J. Amer. chem. Soc.* **75**, 5572.

A–E Kalan E.B., Davis B.D., Srinivasan P.R. and Sprinson D.B. (1956) Conversion of various carbohydrates to 5-dehydroshikimic acid by bacterial extracts. *J. biol. Chem.* **223**, 907.

C–D Srinivasan P.R., Rothschild J. and Sprinson D.B. (1963) Enzymic conversion of 3-deoxy-D-arabinoheptulosonic acid-7-phosphate to 5-dehydroquinic acid. *J. biol. Chem.* **238**, 3176.

C–D Adlersberg M. and Sprinson D.B. (1964) Synthesis of 3,7-dideoxy-D-*threo*-hepto-2,6-diulosonic acid: a study of 5-dehydroquinic acid formation. *Biochemistry*, **3**, 1855.

D–E Mitsuhashi S. and Davis B.D. (1954) Aromatic biosynthesis. XII. Conversion of 5-dehydroquinic acid to 5-dehydroshikimic acid by 5-dehydroquinase. *Biochim. biophys. Acta*, **15**, 54.

D–E Balinsky D. and Davies D.D. (1961) Aromatic biosynthesis in higher plants. Preparation and properties of dehydroquinase. *Biochem. J.* **80**, 300.

230 B Srinivasan P.R., Shigeura H.T., Sprecher M., Sprinson D.B. and Davis B.D. (1956) The biosynthesis of shikimic acid from D-glucose. *J. biol. Chem.* **220**, 477.

B Morgan P.N., Gibson M.I. and Gibson F. (1963) The conversion of shikimic acid into certain aromatic compounds by cell free extracts of *Aerobacter aerogenes* and *E. coli*. *Biochem. J.* **89**, 229.

B Bohm B.A. (1965) Shikimic acid (3,4,5-trihydroxyl-1-cyclohexene-1-carboxylic acid). *Chem. Rev.* **65**, 435.

A–B Yaniv H. and Gilvarg C. (1955) Aromatic biosynthesis. XIV. 5-Dehydroshikimic reductase. *J. biol. Chem.* **213**, 787.

A–B Whiting G.C. and Coggins R.A. (1967) The oxidation of D-quinate and related acids by *Acetomonas oxydans*. *Biochem. J.* **102**, 283.

B–E Davis B.D. and Mingioli E.S. (1953) Aromatic biosynthesis. VII. Accumulation of two derivatives of shikimic acid by bacterial mutants. *J. Bact.* **66**, 129.

C Weiss U. and Mingioli E.S. (1956) Aromatic biosynthesis. XV. The isolation and identification of shikimic acid-5-phosphate. *J. Amer. chem. Soc.* **78**, 2894.

D Levin J.G. and Sprinson D.B. (1964) Enzymatic formation and isolation of 3-enolpyruvylshikimate-5-phosphate. *J. biol. Chem.* **239**, 1142.

E Gibson M.I. and Gibson F. (1962) A new intermediate in aromatic biosynthesis. *Biochim. biophys. Acta*, **65**, 160.

E Gibson M.I., Gibson F., Doy C.H. and Morgan P. (1962) The branch point in the biosynthesis of the aromatic amino acids. *Nature*, **195**, 1173.

E Gibson M. I. and Gibson F. (1964) Chorismic acid. *Biochem. J.* **90**, 248.

E Edwards J.M. and Jackman L.M. (1965) Chorismic acid. A branch point intermediate in aromatic synthesis. *Australian J. Chem.* **18**, 1227.

F Davis B.D. (1953) Autocatalytic growth of a mutant due to accumulation of an unstable precursor. *Science*, **118**, 251.

F Metzenberg R.L. and Mitchell H.K. (1958) The biosynthesis of aromatic compounds by *Neurospora crassa*. *Biochem. J.* **68**, 168.

D–F Katagiri M. and Sato R. (1953) Accumulation of phenylalanine by a mutant of *E. coli*. *Science*, **118**, 250.

D–F Morell H., Clark M.J., Knowles P.F. and Sprinson D.B. (1967) The enzymic synthesis of chorismic acid and prephenic acid from 3-enolpyruvylshikimate-5-phosphate. *J. biol. Chem.* **242**, 82.

E–F Cotton R.G.H. and Gibson F. (1968) The biosynthesis of phenylalanine and tyrosine in the pea. Chorismate mutase. *Biochim. biophys. Acta*, **156**, 187.

231 B Cerutti P. and Guroff G. (1965) Enzymic formation of phenypyruvic acid in *Pseudomonas*, and its regulation. *J. biol. Chem.* **240**, 3034.

A–B Weiss U., Gilvarg C., Mingioli E.S. and Davis B.D. (1954) Aromatic biosynthesis. XI. The aromatization step in the synthesis of phenylalanine. *Science*, **119**, 774.

A–B Gilvarg C. (1955) Prephenic acid and the aromatization step in the synthesis of phenylpyruvate. *Amino Acid Metabolism*. (McElroy and Glass Eds.), p. 812.

C Uchida M., Suzuki S. and Ichihara K. (1954) Studies on the metabolism of *p*-hydroxyphenylpyruvic acid. *J. Biochem.* (*Tokyo*), **41**, 41.

A–BC Miller D.A. and Simmonds S. (1957) Phenylalanine and tyrosine metabolism in *E. coli* strain K-12. *Science*, **126**, 445.

A–C Schwinck I. and Adams E. (1959) Aromatic biosynthesis. XVI. Aromatization of prephenic acid to *p-OH*-phenypyruvic acid, a step in tyrosine biosynthesis in *E. coli*. *Biochim. biophys. Acta*, **36**, 102.

C–E Kenney F.T. (1959) Properties of partially purified tyrosine-α-ketoglutarate transaminase from rat liver. *J. biol. Chem.* **234**, 2707.

C–E Hayashi S.-I., Granner D.K. and Tomkins G.M. (1967) Tyrosine aminotransferase purification and characterisation. *J. biol. Chem.* **242**, 3998.

C–E Igo R.P., Mahoney C.P. and Limbeck G.A. (1968) Studies on tyrosine-α-ketoglutarate transaminase from bovine thyroid and liver tissue. *Biochim. biophys. Acta*, **151**, 88.

C–E Valeriote F.A., Auricchio F., Tomkins G.M. and Riley D. (1969) Purification and properties of rat liver tyrosine aminotransferase. *J. biol. Chem.* **244**, 3618.

D–E Moss A.R. and Schoenheimer R. (1940) Conversion of phenylalanine to tyrosine in normal rats. *J. biol. Chem.* **135**, 415.

D–E Udenfriend S. and Cooper J.R. (1952) The enzymic conversion of phenylalanine to tyrosine. *J. biol. Chem.* **194**, 503.

D–E Mitoma C. and Leeper L.C. (1954) Enzymatic conversion of phenylalanine to tyrosine. *Fed. Proc.* **13**, 266.

D–E Coon M.J. and Robinson W.G. (1958) Amino acid metabolism. *Ann. Rev. Biochem.* **27**, 561.

D–E Kaufman S. and Levenberg B. (1959) Further studies on the phenylalanine-hydroxylation cofactor. *J. biol. Chem.* **234**, 2683.

D–E Kaufman S. (1959) Studies on the mechanism of the enzymatic conversion of phenylalanine to tyrosine. *J. biol. Chem.* **234**, 2677.

D–E Kaufman S. (1962) On the structure of the phenylalanine hydroxylation cofactor. *J. biol. Chem.* **237**, PC 2712.

D–E Kaufman S., Bridgers W.F., Eisenberg F. and Friedman S. (1962) The source of oxygen in the phenylalanine hydroxylase- and the dopamine-β-hydroxylase-catalysed reactions. *Biochem. biophys. Res. Commun.* **9**, 497.

D–E Kaufman S. (1963) The structure of the phenylalanine-hydroxylation cofactor. *Proc. Nat. Acad. Sci. U.S.* **50**, 1085.

D–E Kaufman S. (1964) Studies on the structure of the primary oxidation product formed from tetrahydropteridines during phenylalanine hydroxylation. *J. biol. Chem.* **239**, 332.

232 GEN Weinhouse S. and Millington R.H. (1949) Ketone body formation from tyrosine. *J. biol. Chem.* **181**, 645.

GEN Ravdin R.G. and Crandall D.I. (1951) The enzymic conversion of homogentisic acid to 4-fumarylacetoacetic acid. *J. biol. Chem.* **189**, 137.

GEN La Du B.N., Jr. and Greenberg D.M. (1951) The tyrosine oxidation system of liver. *J. biol. Chem.* **190**, 245.

GEN Le May-Knox M. and Knox W.E. (1951) The oxidation in liver of

L-tyrosine to acetoacetate through *p*-hydroxyphenylpyruvate and homogentisic acid. *Biochem. J.* **49**, 686.

GEN Lerner A.B. (1953) Metabolism of phenylalanine and tyrosine. *Advanc. Enzymol.* **14**, 73.

GEN Dische R. and Rittenberg D. (1954) The metabolism of phenylalanine-4-^{14}C. *J. biol. Chem.* **211**, 199.

GEN Ribbons D.W. (1965) The microbial degradation of aromatic compounds. *Ann. Rep. Chem. Soc.* **62**, 445.

A–B Kaufman S. (1959) Studies on the mechanism of the enzymatic conversion of phenylalanine to tyrosine. *J. biol. Chem.* **234**, 2677.

A–B Kaufman S. and Levenberg B. (1959) Further studies on the phenylalanine-hydroxylation cofactor. *J. biol. Chem.* **234**, 2683.

B–C Schepartz B. (1951) Transamination as a step in tyrosine metabolism. *J. biol. Chem.* **193**, 293.

B–C Cannellakis Z.N. and Cohen P.P. (1956) Purification studies on tyrosine-α-ketoglutaric acid transaminase. *J. biol. Chem.* **222**, 53.

B–C Rowswell E.V. (1956) Transaminations with pyruvate and other α-keto acids. *Biochem. J.* **64**, 246.

B–C Kenney F.T. (1959) Properties of partially purified tyrosine-α-ketoglutarate transaminase from rat liver. *J. biol. Chem.* **234**, 2707.

B–C Sentheshanmuganathan S. (1960) The purification and properties of the tyrosine-2-oxoglutarate transaminase of *Saccharomyces cerevisiae*. *Biochem. J.* **77**, 619.

C–D La Du B.N. and Zannoni J.G. (1956) The tyrosine oxidation system of liver. Further studies on *p*-hydroxyphenyl pyruvic acid. *J. biol. Chem.* **219**, 273.

C–D Hager S.E., Gregerman R.I. and Knox W.E. (1957) *p*-Hydroxyphenyl pyruvic acid oxidase of liver. *J. biol. Chem.* **225**, 935.

D–E Crandall D.I. and Halikis D.N. (1954) Homogentisic acid oxidase. *J. biol. Chem.* **208**, 629.

D–E Knox W.E. and Edwards S.W. (1955) Homogentisate oxidase of liver. *J. biol. Chem.* **216**, 479.

E–FG Edwards S.W. and Knox W.E. (1956) Homogentisate metabolism: isomerisation of maleylacetoacetate by an enzyme which requires glutathione. *J. biol. Chem.* **220**, 79.

E–FG Lack L. (1961) Enzymic *cis-trans* isomerisation of maleylpyruvic acid. *J. biol. Chem.* **236**, 2835.

233 A–C Pitt B.M. (1962) Oxidation of phenylpyruvate to aromatic aldehydes and oxalate. *Nature*, **196**, 272.

C–D Gunsalus C.F., Stanier R.Y. and Gunsalus I.C. (1953) The enzymatic conversion of mandelic acid to benzoic acid. *J. Bact.* **66**, 548.

234 GEN Mason H.S. (1948) The chemistry of melanin. Mechanisms of the oxidation of dihydroxyphenylalanine by tyrosinase. *J. biol. Chem.* **172**, 83.

GEN Lerner A.B. and Fitzpatrick T.B. (1950) Biochemistry of melanin formation. *Physiol. Rev.* **30**, 91.

GEN Lerner A.B. (1953) The metabolism of phenylalanine and tyrosine. Advanc. *Enzymol.* **14**, 73.

GEN Kertesz D. (1957) State of copper in polyphenoloxidase (tyrosinase). *Nature*, **180**, 506.

GEN Cromartie R.I.T. and Harley-Mason J. (1957) Melanin and its precursors. *Biochem. J.* **66**, 713.

GEN Fitzpatrick T.B., Miyamoto M. and Ishikawa K. (1967) The evolution of concepts of melanin biology. *Arch. Dermatology*, **96**, 305.

A–B Ikeda M., Fahien L.A. and Udenfriend S. (1966) A kinetic study of bovine tyrosine hydroxylase. *J. biol. Chem.* **241**, 4452.

A–B Petrack B., Sheppy F. and Fetzer V. (1968) Studies on tyrosine hydroxylase from bovine adrenal medulla. *J. biol. Chem.* **243**, 743

235 GEN Udenfriend S. and Wyngaarden J.B. (1956) Precursors of adrenal epinephrine and norepinephrine *in vivo*. *Biochem. biophys. Acta*, **20**, 48.

A–B See 234 A–B.
B–C Hartman W.J., Pogrund R.S., Drell W. and Clark W.G. (1955) Studies on the biosynthesis of arterenol. *J. Amer. chem. Soc.* **77**, 816.
C–D Levin E.Y. and Kaufman S. (1961) Studies on the enzyme catalysing the conversion of 3,4-dihydroxyphenylethylamine to norepinephrine. *J. biol. Chem.* **236**, 2043.
D–E Axelrod J. (1962) Purification and properties of phenylethanolamine *N*-methyl transferase. *J. biol. Chem.* **237**, 1657.

236 GEN Morton M.E. and Chaikoff I.L. (1942) The *in vitro* formation of thyroxine and diodotyrosine by thyroid tissue. *J. biol. Chem.* **144**, 565.
GEN Taurog A. (1964) The biosynthesis of thyroxine. *Mayo. clin. Proc.* **39**, 569.
A–B Serif G.S. and Kirkwood S. (1958) Enzyme systems concerned with the synthesis of monoiodotyrosine. Further properties of the soluble and mitochondrial systems. *J. biol. Chem.* **233**, 109.
A–B Cunningham B.A. and Kirkwood S. (1961) Enzyme systems concerned with the synthesis of monoiodotyrosine. Ion requirements of the soluble system. *J. biol. Chem.* **236**, 485.

237 B Rivera A. and Srinivasan P.R. (1963) The role of 3-enolpyruvyl shikimate 5-phosphate in the biosynthesis of anthranilate. *Biochemistry*, **2**, 1063.
A–B Edwards J.M., Gibson F., Jackman L.M. and Shannon J.B. (1964) The source of the nitrogen atom for the biosynthesis of anthranilic acid. *Biochim. biophys. Acta*, **93**, 78.
A–B De Moss J.A. (1965) The conversion of shikimic acid to anthranilic acid by extracts of *Neurospora crassa*. *J. biol. Chem.* **240**, 1231.
A–B Ito J. and Yanofsky C. (1966) The nature of the anthranilic acid synthetase complex of *Escherichia coli*. *J. biol. Chem.* **241**, 4112.
A–E Gaertner F.H. and De Moss J.A. (1969) Purification and characterisation of a multienzyme complex in the tryptophan pathway of *Neurospora crassa*. *J. biol. Chem.* **244**, 2716.
D–E Creighton T.E. and Yanofsky C. (1966) Indole-3-glycerol phosphate synthetase of *Escherichia coli*, an enzyme of the tryptophan operon. *J. biol. Chem.* **241**, 4616.
E–F Meduski J.W. and Zamenhof S. (1969) Studies on tryptophan synthetases from various strains of *Bacillus subtilis*. *Biochem. J.* **112**, 285.

238 GEN Ijichi H., Ichiyama A. and Hayaishi O. (1966) Studies on the biosynthesis of nicotinamide adenine dinucleotide. *J. biol. Chem.* **241**, 3701.
GEN Deguchi T., Ichiyama A., Nishizuka Y. and Hayaishi O. (1968) Studies on the biosynthesis of nicotinamide-adenine dinucleotide in the brain. *Biochim. biophys. Acta*, **158**, 382.
A–B Knox W.E. and Mehler A.H. (1951) The adaptive increase of the tryptophan peroxidase-oxidase system of liver. *Science*, **113**, 237.
A–B Hayaishi O., Rothberg S., Mehler A.H. and Saito Y. (1957) Studies on oxygenases. Enzymatic formation of kynurenine from tryptophan. *J. biol. Chem.* **229**, 889.
A–B Tanaka T. and Knox W.E. (1959) The nature and mechanism of the tryptophan pyrrolase (peroxidase-oxidase) reaction of *Pseudomonas* and of rat liver. *J. biol. Chem.* **234**, 1162.
A–B Ishimura Y., Nozaki M., Hayaishi O., Tamura M. and Yamazaki I. (1967) Evidence for an oxygenated intermediate in the tryptophan pyrrolase reaction. *J. biol. Chem.* **242**, 2574.
A–B Piras M. and Knox W.E. (1967) Tryptophan pyrrolase of liver. II. The activating reactions in crude preparations from rat liver. *J. biol. Chem.* **242**, 2952.
A–B Yamamoto S. and Hayaishi O. (1967) Tryptophan pyrrolase of rabbit intestine. D- and L-Tryptophan-cleaving enzyme or enzymes. *J. biol. Chem.* **242**, 5260.

A–B Tokuyama K. (1968) Further studies on bacterial and liver tryptophan pyrrolase. *Biochim. biophys. Acta*, **151**, 76.

A–B Poillon W.N., Maeno H., Koike K. and Feigelson P. (1969) Tryptophan oxygenase. Purification, composition and subunit structure. *J. biol. Chem.* **244**, 3447.

E Mehler A.H. and May E.L. (1956) Studies with carboxyl-labelled 3-hydroxy-anthranilic acid and picolinic acid, *in vivo* and *in vitro*. *J. biol. Chem.* **223**, 449.

A–E Gholson R.K., Henderson L.M., Mourkides G.A. and Hill R.J. (1959) The metabolism of D,L-tryptophan-2-^{14}C by the rat. *J. biol. Chem.* **234**, 96.

B–C Hayaishi O. and Stanier R.Y. (1951) The bacterial oxidation of tryptophan. *J. Bact.* **62**, 691.

B–C Jakoby W.B. (1954) Kynurenine formamidase from *neurospora*. *J. biol. Chem.* **207**, 657.

C–D De Castro F.T., Price J.M. and Brown R.R. (1956) Reduced triphosphopyridinenucleotide requirement for the enzymatic formation of 3-hydroxykynurenine from L-kynurenine. *J. Amer. chem. Soc.* **78**, 2904.

C–D Saito Y., Hayaishi O. and Rothberg S. (1957) Studies on oxygenases. Enzymatic formation of 3-hydroxy-L-kynurenine from L-kynurenine. *J. biol. Chem.* **229**, 921.

E–F Decker R.H., Kang H.H., Leach F.R. and Henderson L.M. (1961) Purification and properties of 3-hydroxyanthranilic acid oxidase. *J. biol. Chem.* **236**, 3076.

G Andreoli A.J., Ikeda M.V., Nishizuka Y.Y. and Hayaishi O. (1963) Quinolinic acid. A precursor to nicotinamide adenine dinucleotide. *Biochem. biophys. Res. Commun.* **12**, 92.

239 GEN Imsande J. (1961) Pathway of diphosphopyridine nucleotide biosynthesis in *E. coli*. *J. biol. Chem.* **236**, 1494.

GEN Nishizuka Y. and Hayaishi O. (1963) Enzymic synthesis of niacin nucleotides from 3-hydroxyanthranilic acid in mammalian liver. *J. biol. Chem.* **238**, PC 483.

A–B Packman P.M. and Jakoby W.B. (1967) Crystalline quinolinate phosphoribosyl transferase. Properties of the enzyme. *J. biol. Chem.* **242**, 2075.

B–C & D–E Imsande J. and Handler P. (1961) Biosynthesis of diphosphopyridine nucleotide. Nicotinic acid mononucleotide pyrophosphorylase. *J. biol. Chem.* **236**, 525.

B–C Smith L.D. and Gholson R.K. (1969) Allosteric properties of bovine liver nicotinate phosphoribosyl transferase. *J. biol. Chem.* **244**, 68.

B–E Preiss J. and Handler P. (1958) Biosynthesis of diphosphopyridine nucleotide. II. Enzymatic aspects. *J. biol. Chem.* **233**, 493.

B–E Kahn V. and Blum J.J. (1968) Studies on nicotinic acid metabolism in *Astasia longa*. I. Incorporation of nicotinic acid into pyridine derivatives in exponentially growing and in synchronised cultures. *J. biol. Chem.* **243**, 1441, 1448.

E–F Kornberg A. (1950) Enzymatic synthesis of triphosphopyridine nucleotide. *J. biol. Chem.* **182**, 805.

E–F Wang T.P., Kaplan N.O. and Stolzenbach F.E. (1958) Enzymatic preparation of triphosphopyridine nucleotide from diphosphopyridine nucleotide. *J. biol. Chem.* **211**, 465.

240 GEN Renson J., Daly J., Weissbach H., Witkop B. and Udenfriend S. (1966) Enzymatic conversion of 5-tritiotryptophan to 4-tritio-5-hydroxytryptophan. *Biochem. biophys. Res. Commun.* **25**, 504.

GEN Grahame-Smith D.G. (1967) The biosynthesis of 5-hydroxytryptamine in brain. *Biochem. J.* **105**, 351.

A–B Friedland R.A., Wadzinski I.M. and Waisman H.A. (1961) The enzymatic hydroxylation of tryptophan. The effect of aromatic amino acids on the hydroxylation of tryptophan. *Biochem. biophys. Res. Commun.* **5**, 94; **6**, 227.

A–B Renson J., Weissbach H. and Udenfriend S. (1962) Hydroxylation of tryptophan by phenylalanine hydroxylase. *J. biol. Chem.* **237**, 2261.

B–C Udenfriend S., Clark C.T. and Titus E. (1953) 5-Hydroxytryptophan decarboxylase: a new route of metabolism of tryptophan. *J. Amer. chem. Soc.* **75**, 501.

B–C Buzard J.A. and Nytch P.D. (1957) Some characteristics of rat kidney 5-hydroxytryptophan decarboxylase. *J. biol. Chem.* **227**, 225.

C–D Bessman S.P. and Lipmann F. (1953) The enzymic transacetylation between aromatic amines. *Arch. biochim. Biophys.* **46**, 252.

C–D Tabor H., Mehler A.H. and Stadtman E.R. (1953) The enzymatic acetylation of amines. *J. biol. Chem.* **204**, 127.

C–D Weissbach H., Bogdanski D.F., Redfield B.G. and Udenfriend S. (1957) Studies on the effect of vitamin B_6 on 5-hydroxy-tryptamine (serotonin) formation. *J. biol. Chem.* **227**, 617.

C–D Weissbach H., Redfield B.G. and Axelrod J. (1961) The enzymic acetylation of serotonin and other naturally occurring amines. *Biochim. biophys. Acta*, **54**, 190.

D–E Axelrod J. and Weissbach H. (1961) Purification and properties of hydroxyindole-*O*-methyl transferase. *J. biol. Chem.* **236**, 211.

241 GEN Weissbach H., King W., Sjoerdsma A. and Udenfriend S. (1959) Formation of indole-3-acetic acid and tryptamine in animals. A method for estimation of indole-3-acetic acid in tissues. *J. biol. Chem.* **234**, 81.

A–E Kaper J.M. and Veldstra H. (1958) On the metabolism of tryptophan by *Agrobacterium tumefaciens*. *Biochim. biophys. Acta*, **30**, 401.

A–G Kosuge T., Heskett M.G. and Wilson E.E. (1966) Microbial synthesis of and degradation of indole-3-acetic acid. I. The conversion of L-tryptophan to indole-3-acetamide by an enzyme system from *Pseudomonas savastanoi*. *J. biol. Chem.* **241**, 3738.

B–D Singer T.P. and Pensky J. (1952) Isolation and properties of the α-carboxylase of wheat germ. *J. biol. Chem.* **196**, 375.

B–D Holzer H. and Beauchamp K. (1961) Nachweis und Charakterisierung von α-Lactyl-thiaminpyrophosphat ('actives Pyruvat') und α-Hydroxyäthyl-thiaminpyrophosphat ('activer Acetaldehyd'). *Biochim. biophys. Acta*, **46**, 225.

D–E Tanenbaum S.W. (1956) The metabolism of *Acetobacter peroxidans*. I. Oxidative enzymes. *Biochim. biophys. Acta*, **21**, 335.

D–E King T.E. and Cheldelin J.H. (1956) Oxidation of acetaldehyde by *Acetobacter suboxydans*. *J. biol. Chem.* **220**, 177.

D–E Jakoby W.B. (1958) Aldehyde oxidation. Dehydrogenase from *Pseudomonas fluorescens*. *J. biol. Chem.* **232**, 75.

242 A–B Miller I.L. and Tsuchida M. and Adelberg E.A. (1953) The transamination of kynurenine. *J. biol. Chem.* **203**, 205.

A–C Tashiro M., Tsukadi K., Kobayashi S. and Hayaishi O. (1961) A new pathway of D-tryptophan metabolism: enzymic formation of kynurenic acid via D-kynurenine. *Biochem. biophys. Res. Commun.* **6**, 155.

A–D Dagley S. and Johnson P.A. (1963) Microbial oxidation of kynurenic, xanthurenic and picolinic acids. *Biochim. biophys. Acta*, **78**, 577.

C–D Taniuchi H., Hayaishi O. (1963) Studies on the metabolism of kynurenic acid. III. Enzymatic formation of 7,8-dihydroxykynurenic acid from kynurenic acid. *J. biol. Chem.* **238**, 283.

244 GEN Gholson R.K., Henderson L.M., Mourkides G.A., Hill R.J. and Koeppe R.E. (1959) The metabolism of DL-tryptophan-α-C^{14} by the rat. *J. biol. Chem.* **234**, 96.

GEN Gholson R.K., Nishizuka Y., Ichiyama A., Kawai H., Nakamura S. and Hayaishi O. (1962) New intermediates in the catabolism of tryptophan in mammalian liver. *J. biol. Chem.* **237**, PC 2043.

GEN Sakami W. and Harrington H. (1963) Amino acid metabolism. *Ann. Rev. Biochem.* **32**, 355.

A–E Gholson R.K., Nishizuka Y., Ichiyama A., Kawai H., Nakamura A. and Hayaishi O. (1962) New intermediate in the catabolism of tryptophan in mammalian liver. *J. biol. Chem.* **237**, PC 2043.

A–E Nishizuka Y., Ichiyama A., Gholson R.H. and Hayaishi O. (1965) Enzymic formation of glutaric acid from 3-hydroxyanthranilic acid. *J. biol. Chem.* **240**, 733.

E–G Besrat A., Polan C.E. and Henderson L.M. (1969) Mammalian metabolism of glutaric acid. *J. biol. Chem.* **244**, 1461.

245 GEN Stanier R.Y. and Hayaishi O. (1951) The bacterial oxidation of tryptophan: a study in comparative biochemistry. *Science*, **114**, 326.

B–C Knox W.E. and Mehler A.H. (1951) The adaptive increase of the tryptophan peroxidase-oxidase system of liver. *Science*, **113**, 237.

B–C Knox W.E. (1953) The relation of liver kynureninase to tryptophan metabolism in pyridoxine deficiency. *Biochem. J.* **53**, 379.

C–D Ichihara A., Adachi K., Hosokawa K. and Takeda Y. (1962) The enzymatic hydroxylation of aromatic carboxylic acids: substrate specificities of anthranilate and benzoate oxidases. *J. biol. Chem.* **237**, 2296.

C–D Taniuchi H., Hatanaka M., Kuno S., Hayaishi O., Nakajima M. and Kurihara N. (1964) Enzymatic formation of catechol from anthranilic acid. *J. biol. Chem.* **239**, 2204.

J Dagley S. and Gibson D.T. (1965) The bacterial degradation of catechol. *Biochem. J.* **95**, 466.

J–K Kojima Y., Itada N. and Hayaishi O. (1961) Metapyrocatechase: a new catechol-cleaving enzyme. *J. biol. Chem.* **236**, 2223.

L Bayly R.C. and Dagley S. (1969) Oxoenoic acids as metabolites in the bacterial degradation of catechols. *Biochem J.* **111**, 303.

J–NO Dagley S., Chapman P.J., Gibson D.T. and Wood J.M. (1964) Degradation of the benzene nucleus by bacteria. *Nature*, **202**, 775.

246 GEN Hayaishi O., Katagiri M. and Rothberg S. (1957) Studies on oxygenases. Pyrocatechase. *J. biol. Chem.* **229**, 905.

GEN Ornston L.N. and Stanier R.Y. *See* 33–1.

GEN Ornston L.N. (1966) The conversion of catechol and protocatechuate to β-ketoadipate by *Pseudomonas putida*. *J. biol. Chem.* **241**, 3787.

GEN Ornston L.N. *See* 33–2.

GEN Ornston L.N. (1966) The conversion of catechol and protocatechuate to β-ketoadipate by *Pseudomonas putida*. *J. biol. Chem.* **241**, 3800.

GEN Hayaishi O. (1966) Crystalline oxygenases of *Pseudomonas*. *Bact. Rev.* **30**, 720.

D–E Sistrom W.R. and Stanier R.Y. (1954) The mechanism of formation of β-ketoadipic acid by bacteria. *J. biol. Chem.* **210**, 821.

H–J Katagiri M. and Hayaishi O. (1957) Enzymatic degradation of β-ketoadipic acid. *J. biol. Chem.* **226**, 439.

247 GEN Lack L. (1959) The enzymatic oxidation of gentisic acid (1959) *Biochim. biophys. Acta*, **34**, 117.

D–H Hopper D.J., Chapman P.J. and Dagley S. (1968) Enzymic formation of D-malate. *Biochem. J.* **110**, 798.

D–E Lack L. (1961) Enzymic *cis-trans* isomerization of maleylpyruvic acid. *J. biol. Chem.* **236**, 2835.

248 C–D Stanier R.Y. and Ingraham J.L. (1954) Protocatechuic acid oxidase. *J. biol. Chem.* **210**, 799.

C–D Gross S.R., Gafford R.D. and Tatum E.L. (1956) The metabolism of protocatechuic acid by *neurospora*. *J. biol. Chem.* **219**, 781.

C–D Fujisawa H. and Hayaishi O. (1968) Protocatechuate 3,4-dioxygenase. *J. biol. Chem.* **243**, 2673.

249 GEN Dagley S. and Patel M.D. (1957) Oxidation of *p*-cresol and related compounds by a *Pseudomonas*. *Biochem. J.* **66**, 227.

GEN Bayley R.C., Dagley S. and Gibson D.T. (1966) The metabolism of cresols by species of *Pseudomonas*. *Biochem. J.* **101**, 293.

GEN Dutton P.L. and Evans W.C. (1969). The metabolism of aromatic compounds by *Rhodopseudomonas palustris*. A new reductive method of aromatic ring metabolism. *Biochem. J.* **113**, 525.

Author Index

Subject Index